Civil Engineering and

Engineering Mechanics Series

N. M. Newmark and W. J. Hall, Editors

ELEMENTARY STRUCTURAL MECHANICS, by Tung Au
FOUNDATION DESIGN, by Wayne C. Teng
THEORY OF PRESTRESSED CONCRETE, by Michael Chi and Frank A. Biberstein

PRENTICE-HALL INTERNATIONAL, INC.
London Tokyo Sydney Paris
PRENTICE-HALL OF CANADA, LTD.
PRENTICE-HALL DE MEXICO, S.A.

FOUNDATION DESIGN

Civil Engineering and
Engineering Mechanics Series

N. M. Newmark and W. J. Hall, Editors

FOUNDATION DESIGN

Wayne C. Teng

Teng and Associates
Consulting Engineers, Chicago, Illinois

Prentice-Hall, Inc.

Englewood Cliffs, New Jersey

Current printing (last digit):
13

Library of Congress Catalog Card Number 62–19407

Printed in the United States of America
32980C

FOUNDATION DESIGN

Wayne C. Teng

Teng and Associates
Consulting Engineers, Chicago, Illinois

Prentice-Hall, Inc.

Englewood Cliffs, New Jersey

Current printing (last digit):
13

Library of Congress Catalog Card Number 62–19407

Printed in the United States of America
32980C

PREFACE

This book is written primarily for two groups of men: practicing engineers who frequently or occasionally design and supervise the construction of foundations and advanced students preparing for engineering practice.

A wealth of information in theory of soil mechanics and in practice of foundation engineering has been developed. Such information has become a basic tool for engineers. At the present time, a practicing engineer can no longer be excused for unwarranted wastefulness or inadequacy of foundation design. Instead, he must be capable of dealing with soil and foundation problems under normal conditions with confidence. At the same time, he should be able to recognize the unusual conditions which require specialists' guidance.

The purpose of the book is to provide essential data for foundation design under ordinary circumstances. The material is presented for convenient application. The background theories are generally presented in concise forms of formulae or charts. Limitations of these data are briefly pointed out to aid the student in recognizing the unusual conditions.

Since this book is a text on design and application, complicated and highly theoretical materials are excluded. For students desiring study of the theories, reference is made to texts of soil mechanics and the original articles.

The book is divided into three parts: Part 1, General Principles; Part 2, Foundations; Part 3, Retaining Structures. Part 1 contains the basic concepts and tools applicable to all foundations and retaining structures. For the convenience of practicing engineers, a brief review of some basic principles of soil mechanics is included. As a text book for the course of foundation design, some portion of Part 1 may be omitted.

Parts 2 and 3 deal with common types of foundations and retaining structures. All pertinent data pertaining to one type of foundations or retaining structures are presented in one chapter, and, generally, a complete design procedure is itemized near the beginning of the chapter. When considered helpful to the reader, numerical examples are given which are designated thus DE 8–2, indicating Design Example 2 in Chapter 8.

Although the material in this book generally covers the more commonly used foundation practice, exception has been taken in presenting the treatise on combined footings and mat foundations. In addition to the conventional methods of design, highly theoretical analyses are also included. The reason for this is that the conventional methods do not furnish all the necessary information regarding the internal stresses of the footings and mat foundations.

The author acknowledges the use of many materials from various technical publications. The source of information is acknowledged by a key thus: (Terzaghi, 1955), which denotes that the book or article can be found in the References under the author's name (Terzaghi) and the year of publication (1955).

Grateful appreciation is due to Dr. George E. Triandafilidis, who has reviewed the manuscript and offered constructive criticism. The author is deeply indebted to Dr. Ralph B. Peck whose teaching in foundation engineering has bestowed upon the author the basic philosophy in engineering practice. The manuscript was also reviewed by Dr. N. M. Newmark, Head of Civil Engineering Department, University of Illinois, and Dr. K. H. Chu, Professor of Civil Engineering, Illinois Institute of Technology.

The author also acknowledges the use of the illustrations for the following plates: Dr. D. U. Deere, University of Illinois, Plate 1; Mr. G. J. Higgins, Raymond Concrete Pile Co., Plates 2 and 8; Dr. J. Brinch Hansen, The Danish Geotechnical Institute, Plate 4; Moretrench Corp., Plate 5; Dr. G. E. Triandafilidis, University of Illinois, Plate 6; Commonwealth Edison Co., Plate 7; Calweld, Inc., Plate 9; *Engineering News-Record,* Plate 10; Messrs. E. E. White, Spencer, White, and Prentis, Plate 13.

The author is dedicating this book to his wife for her encouragement and assistance in preparing this work.

WAYNE C. TENG

CONTENTS

PART ONE GENERAL PRINCIPLES

1 SOILS, ROCKS, AND SOIL MOISTURE 3

1–1: Components of Soils, 3. 1–2: Relationships between Void Ratio, Water Content, and Unit Weight, 4. 1–3: Engineering Properties of Soils, 5. 1–4: Grain Size of Soils, 7. 1–5: Shear Strength, 8. 1–6: Permeability, 9. 1–7: Engineering Properties of Granular Soils, 10. 1–8: Engineering Properties of Cohesive Soils, 13. 1–9: Engineering Properties of Silt and Loess, 18. 1–10: Engineering Properties of Organic Soils, 19. 1–11: Engineering Properties of Rocks, 19. 1–12: Typical Engineering Properties of Igneous and Metamorphic Rocks, 21. 1–13: Engineering Properties of Sedimentary Rocks, 22. 1–14: Ground Water and Associated Phenomena, 23.

2 SOIL EXPLORATION 29

2–1: Purpose of Soil Exploration, 29. 2–2: Planning an Exploration Program, 30. 2–3: Available Information, 30. 2–4: Reconnaissance, 32. 2–5: Preliminary Design Data, 32. 2–6: Common Types of Boring, Sampling, and Testing, 33. 2–7: Standard Penetration Test, 37. 2–8: Thin-walled Tube (Shelby Tube) Sampling, 40. 2–9: Vane Test, 42. Number of Borings, 42. 2–11: Depths of Borings, 43. 2–12: Ground Water Measurement, 43. 2–13: Geologic Profile: Soil Profile, 45. 2–14: Common Soil Tests, 46. 2–15: Example of a Soil Exploration Program, 49.

3 *LOADS, BEARING CAPACITY, AND SETTLEMENT* 53

3–1: Types of Loads, 53. 3–2: Calculation of Loads, 56. 3–3: Bearing Capacity, 57. 3–4: Causes of Settlement, 60. 3–5: Consolidation, 60. 3–6: Differential Settlement, 62. 3–7: Calculation of Loads for Settlement Analysis, 65. 3–8: Settlement Cracks, 66. 3–9: Improving Bearing Capacity by Compaction, 67.

4 *LATERAL PRESSURES* 73

4–1: Basic Concepts, 73. 4–2: Earth Pressure Theories, 76. 4–3: Rankine Theory, 77. 4–4: Wedge Theory, 78. 4–5: Hansen Theory, 84. 4–6: Determination of Soil Properties for Earth Pressure Computation, 85. 4–7: Overcompaction, 86. 4–8: Seepage Pressure, 86. 4–9: Surcharge Load, 88. 4–10: Ice Thrust, 90. 4–11: Earth Pressure during Earthquakes, 92. 4–12: Wave Pressure, 93. 4–13: Other Lateral Forces, 95.

5 *FOUNDATION DRAINAGE AND WATERPROOFING* 97

5–1: Introduction, 97. 5–2: Methods of Dewatering, 97. 5–3: Stability of Bottom of Excavation, 103. 5–4: Foundation Drainage, 105. 5–5: Criteria for Selection of Filter Material, 106. 5–6: Waterproofing, 107. 5–7: Dampproofing, 109. 5–8: Waterstops, 109.

PART TWO FOUNDATIONS

6 SPREAD FOOTINGS 113

6–1: Use of Spread Footings, 113. 6–2: Common Types of Footings, 113. 6–3: Design of Footings, 115. 6–4: Depth of Footings, 115. 6–5: Bearing Capacity of Soils under Footings and Mat Foundations, 117. 6–6: Footing Size Proportions, 124. 6–7: Stress on Lower Strata, 125. 6–8: Settlement of Footings, 128. 6–9: Eccentric Loading, 130. 6–10: Inclined Load, 135. 6–11: Footings on Slopes, 137. 6–12: Uplift of Footings, 137. 6–3: Structural Design of Footings, 138. 6–14: Fixity of Column Base and Rotation of Footing, 141. 6–15: Construction, 142. 6–16: Design Example, 144.

7 STRAP FOOTINGS, COMBINED FOOTINGS, AND MAT FOUNDATIONS 151

7–1: Uses of Strap Footings, Combined Footings, and Mat Foundations, 151. 7–2: Common Types and Arrangement of Strap Footings, Combined Footings, and Mat Foundations, 153. 7–3: Design of Strap Footings, 154. 7–4: Design Methods for Combined Footings and Mat Foundations, 159. 7–5: Design of Combined Footings, 161. 7–6: Allowable Bearing Pressure for Mat Foundations, 174. 7–7: Design of Mat Foundations, 174. 7–8: Coefficient of Subgrade Reaction, 185. 7–9: General Considerations in Design of Mat Foundations, 189. 7–10: Construction of Mat Foundation, 190.

8 PILE FOUNDATIONS 193

8–1: Use of Piles, 193. 8–2: Types of Piles, 193. 8–3: Timber Piles, 196. 8–4: Precast Concrete Piles, 198. 8–5: Cast-in-place Concrete Piles, 200. 8–6: Composite Piles, 202. 8–7: Steel Piles, 203. 8–8: Design of Pile Foundations, 209. 8–9: Determination of Type and Length of Piles, 210. 8–10: Pile Capacity, 211. 8–11: Pile Spacing and Group Action, 216. 8–12: Stress on Lower Strata, 218. 8–13: Settlement Analysis, 220. 8–14: Design of Pile Caps, 223. 8–15: Uplift, 225. 8–16: Lateral Load, 226. 8–17: Batter Piles, 227. 8–18: Negative Skin Friction, 239. 8–19: Test Piles, 240. 8–20: Construction of Pile Foundations, 244. 8–21: Damage, Alignment, and Effect of Pile Driving, 250.

9 DRILLED CAISSONS 255

9–1: Use of Drilled Caissons, 255. 9–2: Types of Drilled Caissons, 256. 9–3: Design of Drilled Caissons, 258. 9–4: Bearing Capacity of Drilled Caissons, 258. 9–5: Skin Friction of Caisson Shaft, 261. 9–6: Stress on Lower Strata, 262. 9–7: Settlement of Drilled Caissons, 263. 9–8: Design of Elements of Drilled Caissons, 264. 9–9: Bending Moment and Eccentricity, 270. 9–10: Horizontal Force on Drilled Caissons, 272. 9–11: Construction of Drilled Caissons, 273. 9–12: Problems Concerning Construction of Drilled Caissons, 281. 9–13: Design Example, 281.

10 CAISSONS 289

10–1: Use of Caisons, 289. 10–2: Types of Caissons, 290. 10–3: Size and Shape of Caissons, 292. 10–4: Design of Caissons, 293. 10–5: Cutting Edge, 298. 10–6: Construction of Caissons, 299. 10–7: Sand Island Method, 301. 10–8: Pneumatic Caissons, 303. 10–9: Box Caissons (Floating Caissons), 307.

PART THREE RETAINING STRUCTURES

11 *RETAINING WALLS* 311

11–1: Common Uses of Retaining Walls, 311. 11–2: Principal Types of Retaining Walls, 313. 11–3: Design of Retaining Walls, 313. 11–4: Proportions of Retaining Walls, 314. 11–5: Earth Pressure Computation, 316. 11–6: Stability of Retaining Walls, 317. 11–7: Design of Structural Components, 325. 11–8: Backfill Drainage, 332. 11–9: Settlement and Tilting of Retaining Walls, 333. 11–10: Construction of Retaining Walls, 334. 11–11: Design Example, 337.

12 *SHEETPILING WALLS* 347

12–1: Use of Sheetpiling Walls, 347. 12–2: Common Types of Sheet Piles, 347. 12–3: Common Types of Sheetpiling Walls, 351. 12–4: Durability of Steel Sheetpiling, 352. 12–5: Design of Sheetpiling Walls, 354. 12–6: Lateral Pressure Acting on Sheetpiling Walls, 355. 12–7: Design of Cantilever Sheetpiling Walls, 358. 12–8: Design of Anchored Sheetpiling Wall, 364. 12–9: Stability of Sheetpiling in Cohesive Soils, 368. 12–10: Rowe's Theory of Moment Reduction, 369. 12–11: Wales and Tie Rods, 371. 12–12: Methods for Reducing Lateral Pressure, 373. 12–13: Types of Anchorage, 374. 12–14: Capacity of Deadmen, 375. 12–15: Location of Anchorage, 377. 12–16: Construction of Sheetpiling Walls, 378. 12–17: Design Examples, 379.

13 *BRACED COFFERDAMS* 389

13–1: Common Uses of Braced Cofferdams, 389. 13–2: Common Types of Braced Cofferdams, 390. 13–3: Design of Braced Cofferdam, 391. 13–4: Lateral Pressure on Braced Cofferdams, 392. 13–5: Components of Braced Cofferdams, 396. 13–6: Stability of Braced Cofferdams, 397. 13–7: Piping, 401. 13–8: Design Example, 403.

14 CELLULAR COFFERDAMS 409

14–1: Common Types of Cofferdams, 409. 14–2: Common Uses of Cellular Cofferdams, 409. 14–3: Common Types of Cellular Cofferdams, 412. 14–4: Components of Cellular Cofferdams, 413. 14–5: Design of Cellular Cofferdams, 415. 14–6: Material Survey and Tests, 415. 14–7: Average Width of Cofferdams, 416. 14–8: Stability Analysis—Cofferdams on Rock, 417. 14–9: Stability Analysis—Cofferdams in Deep Soil, 420. 14–10: Hansen's Theory of Stability of Cellular Cofferdams, 421. 14–11: Interlock Stress, 422. 14–12: Construction of Cellular Cofferdams, 423. 14–13: Design Example, 425.

APPENDIX 431

A–1: Soil Classification, 431. A–2: Unified Soil Classification System, 431. A–3: Highway Research Board (or AASHO) Soil Classification System, 432. A–4: Pedological Classification System, 432. A–5: Classification of Soils by Geological Origins, 435.

GEOPHYSICAL MAPS OF CONTINENTAL U.S.A. 443

Distribution of Soils, 444. Earthquake Zones, 445. Physical Divisions, 446.

REFERENCES 449

AUTHOR INDEX 457

INDEX 459

PART I
GENERAL PRINCIPLES

Plate One

A Profile of the Earth's Crust

The earth's crust is made of natural materials ranging from loose and incoherent *soils* to massive and hard *rocks*. Within such a wide range there are innumerous varieties of earth materials, each of which differs from the others in different degrees. The physical properties of earth materials are further complicated by the presence of water. For a given job the pertinent engineeering properties of all earth materials encountered should be determined. This chapter summarizes the more significant properties involved in the common foundation practice.

1 SOILS, ROCKS, AND SOIL MOISTURE

1-1 Components of Soils

Soil contains three components, namely, air, water, and solid matter:

1. The air content of a soil has little engineering significance; therefore it is not commonly determined.
2. The water content (or moisture content) of a soil is defined as the ratio between the weight of water and the weight of the solid matter. The latter is equal to the weight of oven-dried soil. The water content influences the engineering properties of a soil.
3. The solid matter of a soil is primarily composed of mineral aggregates (soil grains). In some cases the soil also contains organic material. The mineral aggregates are derived from rocks as a result of rock weathering.

The intergranular space which is occupied by air and water (or water alone when saturated) is known as a void. The amount of voids in a soil is expressed by its *void ratio* which is the ratio between the volume of voids and the volume of solid matter. The term *porosity* (percentage of voids) is also used. It represents the ratio between the volume of voids and the total volume including solid matter and voids. The void ratio or porosity is an important soil property. For instance, a soil having an excessive amount of voids is weak, compressible, and pervious.

The relative amount of water in the voids is defined by the *degree of saturation* which is the ratio between the volume of water and the volume of voids. A soil is fully saturated, or at 100 per cent saturation, if all the voids are filled with water.

1-2 Relationships between Void Ratio, Water Content, and Unit Weight

Equations relating the void ratio, water content, unit weight, and other terms are very useful because in practice it is often necessary to compute one if the others are known. These relationships can be readily derived by definition of the terms. Notations to be used in the equations are:

e = void ratio = volume of voids/volume of solid matter, expressed as a decimal;

G = specific gravity of solid matter = 2.65 (average for common soil minerals);

n = porosity = volume of voids/total volume, expressed as a decimal;

w = water content = weight of water/weight of solid matter, expressed as a percentage;

V = volume of soil sample;

V_v = volume of voids in the soil sample;

V_s = volume of solid matter in the soil sample;

S = degree of saturation = volume of water/volume of voids, expressed as a percentage;

γ_b = bulk unit weight = unit weight of soil plus the weight of water in the voids;

γ_s = saturated unit weight of soil if water fills up all the voids;

γ_d = dry unit weight = unit weight of oven-dried sample;

γ_w = unit weight of water = 1 g/cc = 62.5 pcf (65 pcf for sea water);

γ_G = unit weight of solid matter = $G\gamma_w$;

γ' = buoyant weight, also known as effective weight, $\gamma_e = \gamma_s - \gamma_w$.

By definition,

$$e = \frac{V_v}{V_s} = \frac{V_v}{V - V_v} = \frac{V_v/V}{1 - V_v/V}$$

$$n = \frac{V_v}{V}$$

Therefore,

$$e = \frac{n}{1 - n} \tag{1-1a}$$

$$n = \frac{e}{1 + e} \tag{1-1b}$$

Referring to Fig. 1-1, if the volume of solid matter is unity, the volume of air plus water is e by definition. The total volume of the sample is then $1 + e$.

The weight of the solid matter is equal to the volume of the solid matter $\times$ specific gravity $\times$ unit weight of water $= 1 \times G \times \gamma_w = G\gamma_w$. By the same reasoning, the weight of water in the sample equals $eS\gamma_w$.

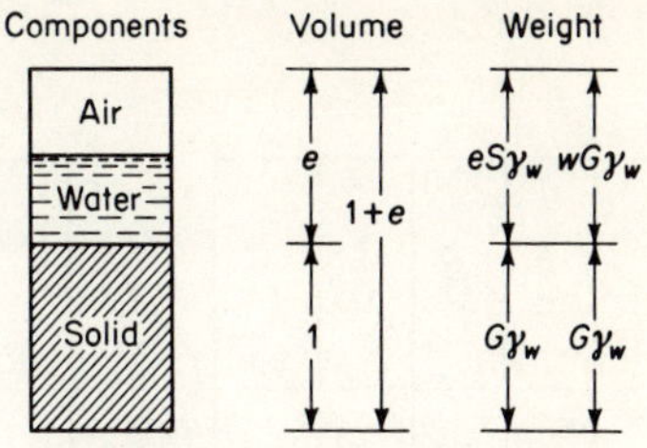

Fig. 1-1 Components of soil.

By definition again, water content is equal to the weight of water divided by the weight of solid matter. If the weight of solid matter is $G\gamma_w$, it follows that the weight of water is $wG\gamma_w$. From this, the following equations are derived.

$$eS = wG \tag{1-2}$$

Bulk unit weight
$$\gamma_b = \frac{eS + G}{1 + e}\gamma_w \tag{1-3}$$

$$= \frac{1 + w}{1 + e}G\gamma_w \tag{1-4}$$

Dry unit weight
$$\gamma_d = \frac{G\gamma_w}{1 + e} \tag{1-5}$$

Saturated unit weight
$$\gamma_s = \frac{G + e}{1 + e}\gamma_w \tag{1-6}$$

$$= \gamma_d(1 + w) \tag{1-7}$$

Buoyant unit weight
$$\gamma' = \gamma_s - \gamma_w$$

$$= \frac{G - 1}{1 + e}\gamma_w \tag{1-8}$$

1-3 Engineering Properties of Soils

The properties of soils are complex and variable. For a given engineering application, certain properties are more significant than others. The important engineering properties may be grouped into the following categories each of which is discussed in a separate section or sections.

Basic properties:	Unit weight, void ratio, and water content (Sec. 1-1 & 1-2); Grain size distribution (Sec. 1-4)
Strength:	Shear strength (Sec. 1-5)
Compressibility:	Consolidation (Sec. 3-5)
Seepage:	Permeability (Sec. 1-6)
Compaction characteristics:	(Sec. 3-9)

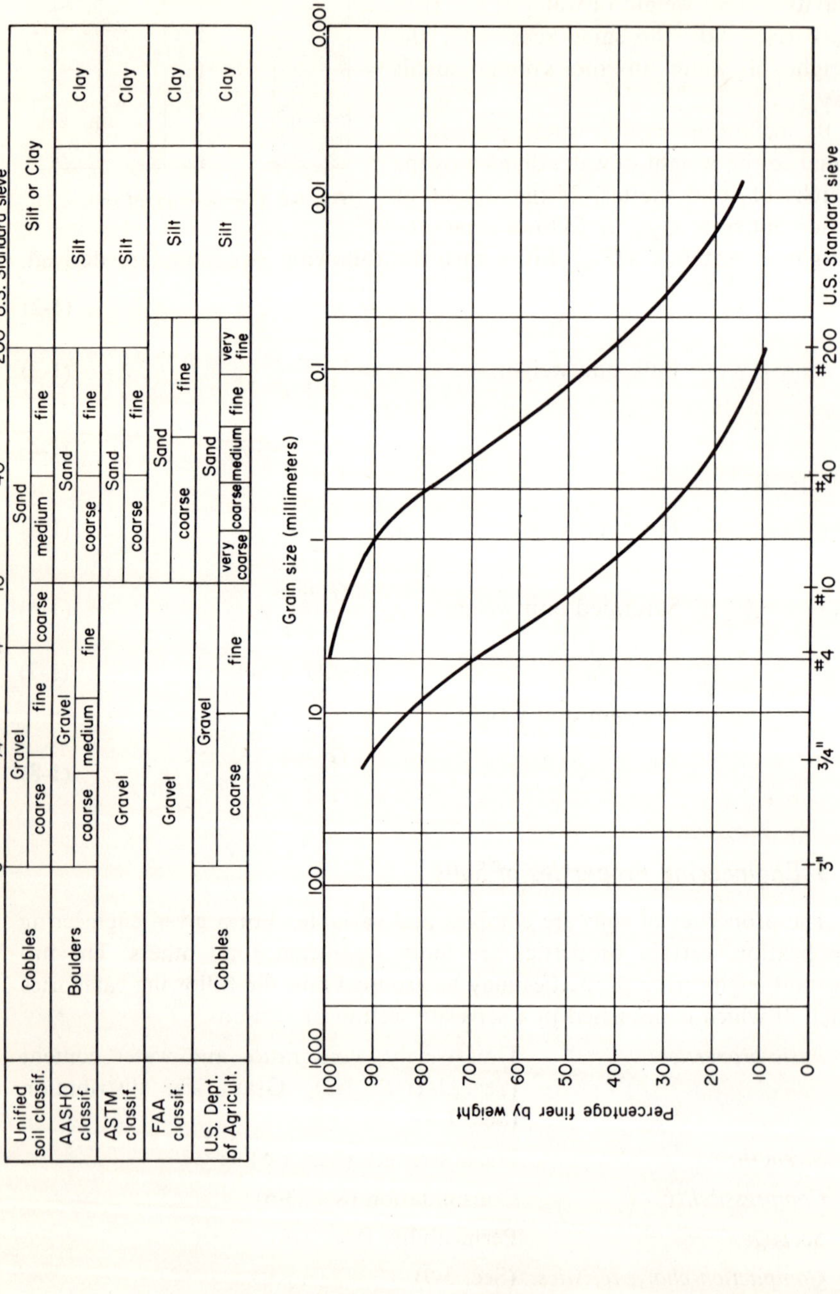

Fig. 1-2 Grain size distribution.

The engineering properties pertaining specifically to granular soils, cohesive soils, silt and loess, and organic soils are discussed in Sec. 1-7 to 1-10.

1-4 Grain Size of Soils

A soil may contain various sizes of grains ranging from large boulders, gravel, and sand to the size of silt, clay, and colloids. The dividing lines between these size limits are arbitrary and vary with different classification systems. In the upper part of Fig. 1-2, the grain size classifications used by different agencies are tabulated for ready comparison. The following ranges are typical:

Boulders, cobbles	Greater than 3 in.
Gravel	Retained on No. 4 or No. 10 sieve
Sand	Retained on No. 200 sieve*
Silt and clay	Passing No. 200 sieve*

The engineering properties of a soil depend largely on the proportion or distribution of the various grain sizes. Soils composed entirely of sand, or larger grains, possess markedly different characteristics than those containing silt and clay particles. For an accurate classification, the proportion of grain sizes should be determined by laboratory tests. Results of such tests are generally plotted on a semi-logarithmic graph as illustrated in Fig. 1-2. However, the approximate proportion of grain sizes can be estimated by the following methods:

Gravel vs. sand: Gravel is larger than the size of lead in the ordinary pencil. The exact differentiation of these two sizes is seldom important.

Sand vs. silt: Dry samples of fine sand and silt appear like dust. They may be differentiated by dispersing a spoonful of sample in a glass of water and measuring the time required for grains to settle. Sand grains will settle in a matter of one-half to one minute, whereas silt grains take one-quarter to one hour.

Silt vs. clay: These two soils may be differentiated by one of the following methods:

Dispersion test. Silt particles take one-quarter to one hour to settle, whereas clay particles remain in suspension from several hours to several days.

Shaking test. A spoonful of soil is mixed with water until it becomes a paste. Place it in the palm of the hand or a dish and shake it by horizontal abrupt motion. If silt particles are predominant, the sample will appear shiny.

---

* No. 200 sieve is about the smallest particle visible to the naked eye.

Rolling test. Only clayey soils can be rolled into thin threads (about $\frac{1}{8}$ in. diameter) if mixed with a proper amount of water.

Dry strength test. A small soil sample is thoroughly dried in an oven or in air and tested for breaking strength. If it contains no clay particles the sample breaks readily into powder.

1-5 Shear Strength

Engineering materials may be subjected to tension, compression, shear, or to combinations of these. Soils and rocks, however, are almost never required to resist tension. A compression failure of a mass of soil or rock, in reality, is a shear failure along a certain rupture plane or zone. Therefore, the structural strength of soil and rock is basically a problem of shear strength.

The shear strength of a given soil or rock may be expressed by Coulomb's equation:

$$s = c + \sigma' f = c + \sigma' \tan \varphi \tag{1-9}$$

where s = shear strength or shear resistance, psf;

c = cohesion, psf;

σ' = intergranular pressure acting perpendicular to the shear plane, psf;

= $(\sigma - u)$, in this case σ = total pressure, and u = pore water pressure (Sec. 1-14);

f = coefficient of friction;

φ = angle of internal friction of the soil, degrees.

In the preceding equation the first term c represents the portion of shear strength which is independent of the normal stress. The second term represents the frictional resistance between soil grains which is approximately

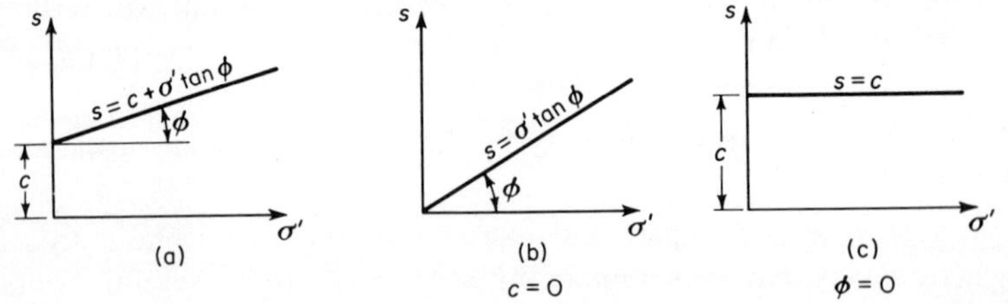

Fig. 1-3 Shear strength of soils.

proportional to the normal pressure, σ', on the surfaces. This equation is plotted by a graph shown in Fig. 1-3(a) in which the shear strength, s, is shown on the vertical axis and the normal stress, σ', on the horizontal axis.

Since a coarse-grained soil (sand and gravel) has no cohesion, its shear strength depends solely on the internal friction between grains. This type of soil is called *granular*, *cohesionless*, *noncohesive*, or *frictional* soil. Figure 1-3(b) shows the shear strength diagram for this type of soil. On the other hand, soils containing large amounts of fine grains (clay, silt, and colloid) are called *fine-grained* or *cohesive soils*.

The shear strength of cohesive soils, as determined by laboratory tests, must be used with caution; a variety of changes may take place during and after construction. The shear strength should be computed for the most critical conditions which usually exist immediately after construction, or immediately upon load application. At that time, the shear strength consists of only cohesion. Therefore design procedure is often made on the assumption of $\varphi = 0$. This procedure is known as *$\varphi = 0$ analysis*, Fig. 1-3(c). The validity of this assumption in practical problems has been proven by theory, experiments, and experience (Skempton, 1948). As time goes on, the pore water slowly escapes from the voids. This brings about an increase in intergranular stress and consequently an increase in shear strength from $s = c$ to $s = c + \sigma' \tan \varphi$. This increase in shear strength introduces an increase in the factor of safety.

However, there are cases in which a decrease in shear strength takes place with time. Some of the factors that cause such decrease are:

1. *Unloading*. When the ground is excavated and hence the load on the underlying clay is reduced, there is a tendency for the clay to swell and slowly become soft.
2. *Increase in pore pressure*. Changes in ground water condition or in seepage pressure may increase the water pressure in the voids (pore pressure) which tends to reduce the intergranular pressure, σ'.
3. *Softening of fissures*. Many stiff clays contain a network of fissures and cracks. When these clays are exposed, water enters the fissures; when these clays are subjected to lateral expansion (i.e., retaining wall moving away from backfill) due to the construction of retaining walls, the fissures open up. The water gradually softens the surfaces between fissures and causes a large reduction in shear strength. The softening process is often very slow and the reduction often large. In a period of several decades the average shear strength of some stiff clays has reduced to only several hundred pounds per square foot. However, a Norwegian stiff fissured clay was reduced to zero cohesion in only two months.

1-6 Permeability

To define the state of being permeable, or the property of allowing fluids and gases to pass through a soil or rock, the term coefficient of permeability is used. Figure 1-4(a) shows schematically a soil sample submerged in water.

Since both ends of the sample are subjected to an equal water head, there is no flow of water through the sample. However, if one end of the sample, point a in Fig. 1-4(b), is subjected to a higher head than the other end, point b, the water flows through the sample from a to b at a velocity v:

$$v = ki \tag{1-10}$$

where v = discharge velocity = quantity of water that percolates across a unit area of the sample in a unit time;

k = coefficient of permeability, depending on the characteristics of the soil;

i = hydraulic gradient = h/l; h being the hydraulic head; l, the length of sample.

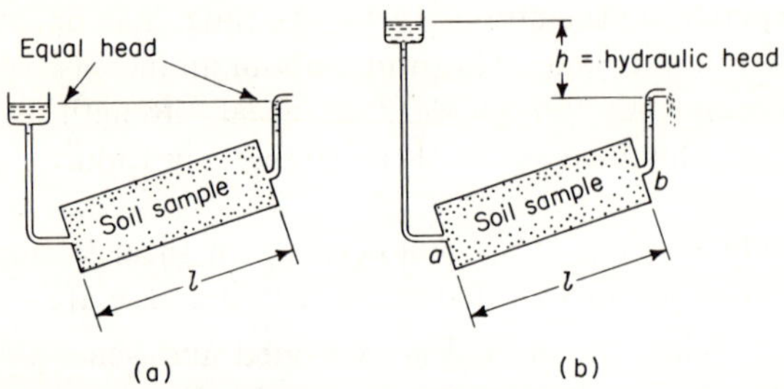

Fig. 1-4 Schematic diagram illustrating the seepage of water through soil.

The empirical equation above is known as Darcy's law. The value of k may be determined by laboratory methods or in the field by pumping tests. For ordinary foundation problems, the range of k values given by Peck, et al. (1953) may be useful.

Type of soil	*Permeability, k, cm/sec*	*Drainage quality*
Clean gravels	10^2 to 10^1	
Clean sands	1	Good
Clean sand and gravel mixtures	10^{-1} to 10^{-4}	
Very fine sands	10^{-5}	
Organic and inorganic silts, mixtures of sand, silt and clay, glacial till, stratified clay deposits	10^{-6}	Poor
Impervious soils, e.g., homogeneous clays below zone of weathering	10^{-7} to 10^{-9}	Very poor

1-7 *Engineering Properties of Granular Soils*

Granular soils are sand, gravel, cobbles, or mixtures of them. Fine sand is an exception, because its engineering properties are on the borderline between the granular and the fine-grained soils. A granular soil has the following significant engineering properties:

1. It is generally excellent foundation material for supporting structures and roads. Except for loose sand, the bearing capacity is large and the settlement is small. Settlement takes place shortly after application of load.
2. It is the best embankment material, because it has high shear strength, it is easy to compact, and it is not susceptible to frost action.
3. It is the best backfill material for retaining walls, basement walls, etc., because it exerts small lateral pressure, it is easy to compact, and it is easy to drain.
4. It cannot be used alone in earth embankment for dikes, reservoirs, etc., because of its high permeability. An excavation in such soils below ground level requires extensive dewatering.
5. It is prone to settlement under vibratory load.

The engineering properties of granular soils are largely influenced by the following factors: compactness, grain size and grain size distribution, and shapes of grains.

A. Compactness. The shear strength and compressibility of granular soils are most intimately related to the compactness of the grains, which is described as loose, medium (firm), or dense (compact). Quantitatively, the compactness is expressed in terms of *relative density:*

$$D_d = \frac{e_{max} - e}{e_{max} - e_{min}} \times 100$$

or

$$D_d = \frac{D_{max}(D - D_{min})}{D(D_{max} - D_{min})} \times 100$$

where D_d = relative density expressed in per cent,

e_{max} = void ratio in loosest state,

e_{min} = void ratio in most compact state,

e = in-place void ratio,

D_{max} = greatest dry density,

D_{min} = least dry density,

D = in-place dry density.

The compactness of a man-made fill is commonly expressed in terms of percentage of compaction. A representative soil sample is compacted in the laboratory to determine the maximum unit weight under a certain compaction procedure. If the actual in-place unit weight is equal to 95 per cent or 106 per cent of the maximum weight obtained in the laboratory, the fill is said to be compacted to 95 per cent or 106 per cent maximum density respectively. This procedure is discussed further in Sec. 3-9. It should be noted that a 100 per cent compaction does not correspond to 100 per cent relative density.

In practice, the compactness (relative density) of soil deposits is measured by a penetration test. Several types of static and dynamic penetration tests have been developed in different countries. The one most commonly used is the so-called *standard penetration test* (Terzaghi and Peck, 1948). This test consists of counting the number of hammer blows (140 lb hammer falling 30 in.) required to drive a standard sampler (called a split spoon) to a depth of 12 in. (see Sec. 2-7). For the purpose of foundation design, the relationships shown in Table 1-1 between the relative density D_d, penetration resistance N,

Table 1-1 RELATIVE DENSITY OF GRANULAR SOILS

Compactness		*Very loose*		*Loose*		*Medium*		*Dense*		*Very dense*	
Relative density D_d	0		15%		35%		65%		85%		100%
Standard penetration resistance, N=no. of blows per foot	0		4		10		30		50		
ϕ (degrees)*			28		30		36		41		
Unit weight, pcf											
moist		<100		95–125		110–130		110–140		>130	
submerged		< 60		55–65		60–70		65–85		> 75	
Identification in field		A reinforcing rod can be pushed into soil several feet						Difficult to drive 2″ × 4″ stake with a sledge hammer			

* Increase 5 degrees for soils containing less than 5% fine sand or silt.

and the angle of internal friction φ, may be used. The relationship between φ and D_d may be expressed approximately by the following equations (Meyerhof, 1956):

$$\varphi = 25 + 0.15\, D_d$$

for granular soil containing more than 5 per cent fine sand and silt.

$$\varphi = 30 + 0.15\, D_d$$

for granular soil containing less than 5 per cent fine sand and silt. In the equations above, D_d is expressed in per cent, not in decimal.

It should be pointed out here also that the φ-value of granular soils and standard penetration resistance N change only slightly upon saturation.

B. Grain size and grain size distribution. Other things being equal, the shear strength of granular soils increases with increasing size of grains. Under a shearing force the finer grains are easily rolled along, whereas the large cobbles and gravels wedge against each other.

Granular soils are said to be well-graded, poorly-graded (uniform), or gap-graded, depending upon the grain-size distribution:

Well-graded: Containing an assortment of grain sizes ranging from coarse to fine.

Poorly-graded (or uniform): Containing predominantly one or two sizes.

Gap-graded: Containing coarse grains and fine grains but lacking intermediate sizes.

In a well-graded soil the small grains tend to fill the voids between the large grains, therefore the soil is generally more compact and stable, and less permeable.

C. Shapes of grains. The shapes of sand and gravel grains can be examined by naked eye or with the aid of a magnifying glass. They are described as angular, subangular, subrounded, rounded, and very rounded. Figure 1-5 illustrates these general shapes. Soils containing angular grains have φ-values several degrees larger than ones containing rounded grains.

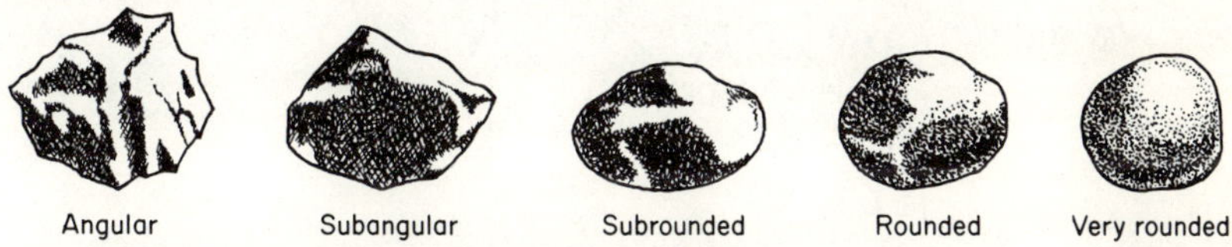

Fig. 1-5 Grain shapes of sand and gravel.

1-8 Engineering Properties of Cohesive Soils

Cohesive soils are those containing a large proportion of fine particles (clay size and colloidal size). Their shear strength is largely or entirely derived from cohesion. This type of soil includes clays, silty clays, and clays mixed with sand or gravel. It has the following significant engineering properties:

1. It often possesses low shear strength.
2. It is often plastic and compressible.
3. It loses part of its shear strength upon wetting.
4. It loses part of its shear strength upon disturbance.
5. It deforms plastically (creeps) under constant load. Creep is generally negligible when the shear stress is smaller than about 50 per cent of its shear strength and pronounced when the shear stress is greater than 75 per cent of its shear strength.

6. It shrinks upon drying and expands upon wetting. Seasonal changes are common.
7. It is very poor material for backfill because of large lateral pressure.
8. It is poor material for embankment because it has low shear strength and is more difficult to compact.
9. It is practically impervious.
10. Clay slopes are prone to landslide.

For a given job, where cohesive soils are encountered, the following properties should be determined:

Unit weight, void ratio, and water content
Shear strength
Plasticity
Compressibility
Sensitivity
Swelling properties

To adequately define the important engineering properties of a given cohesive soil, all the items above should be included. An example is given below.

Blue silty soft clay: $\gamma_s = 125$ psf
$w = 25.5$ per cent
$q_u = 0.45$ tsf
LL = 33
PL = 18
PI = 15
$C_c = 0.23$
sensitivity = 1.25

A. Shear strength. A cohesive soil is described as very soft, soft, medium, stiff, very stiff, or hard according to its shear strength. Quantitatively, each term corresponds to a *range* of shear strength. For example, a soft clay has a shear strength between 500 to 1000 psf.

For most practical cases the shear strength of cohesive soils is determined by unconfined compression tests. Only for large jobs and research work are the other types of shear tests justified. The procedure of an unconfined compression test is quite simple. A small soil sample (usually about 2 in. or 3 in. in diameter or square) is cut to a length of $1\frac{1}{2}$ to 2 times the diameter of the sample and is subjected to a uniform axial stress. The shear strength of a cohesive soil (under $\varphi = 0$ condition, Sec. 1-5) is equal to one-half the unconfined compressive strength,

$$s = c = \tfrac{1}{2}q_u \qquad \text{(1-11)}$$

The shear strength may also be approximated by the standard penetration test procedure, Table 1-2. However, the correlation between the penetration resistance and the shear strength of cohesive soils is very unreliable. Unless the project is very small, where the cost of taking tube samples and making unconfined compression tests is greater than the additional cost of construction using a more conservative design, the unconfined compression test should be used. The penetration test, however, should always be made as a guide and in comparison with the unconfined compression tests.

Table 1-2 SHEAR STRENGTH OF COHESIVE SOILS

Consistency		*Very soft*		*Soft*		*Medium*		*Stiff*		*Very stiff*		*Hard*
q_u = unconfined compression strength, tons per square ft	0		0.25		0.50		1.00		2.00		4.00	
Standard penetration resistance, N=no. of blows per ft	0		2		4		8		16		32	
Unit weight, pcf (saturated)			100–120			110–130			120–140			130+
Identification characteristics		Exudes from between fingers when squeezed in hand		Molded by light finger pressure		Molded by strong finger pressure		Indented by thumb		Indented by thumb nail		Difficult to indent by thumb nail

A stiff clay often possesses fissures, cracks and slickensides*, which affect the shear strength of the clay mass as a whole. In the first place, the fissures, etc., are planes of weakness. Furthermore, they are prone to softening by water. For discussion of reduction in shear strength, see Sec. 1-5.

B. Plasticity. To define the plasticity of cohesive soils, arbitrary indices have been chosen. These are the liquid limit, plastic limit, and plasticity index.

When a colloid of soil (very fine soil particles in suspension in water) is being dried out gradually, the material changes from a liquid state to a semiliquid (plastic) state. Further drying turns it to a solid state. The water content at which the soil changes from liquid state to plastic state is called

---

* Slickensides are polished and grooved surfaces as a result of relative movement of the soil or rock.

the *liquid limit* of the soil; that from plastic state to solid state is called the *plastic limit* of the soil. Liquid limit and plastic limit are known collectively as the Atterberg limits. The meaning of these limits are illustrated diagrammatically as follows:

Colloid state

Liquid state

—— Liquid limit (LL) ↑

Plastic state Plasticity index, PI = LL − PL

—— Plastic limit (PL) ↓

Solid state

Since the soil changes from one state to the other gradually, there is no sharp demarcation or abrupt change as the definitions may imply. Therefore, these limits are arbitrarily defined by certain test procedures, Sec. 2-15.

The difference between the liquid limit and the plastic limit is called the *plasticity index* (PI). The plasticity index represents the range of water content in which the soil remains plastic. A plastic soil has a large value of PI. In general, the plasticity index represents the relative amount of clay particles in the soil. A large PI indicates a large amount of clay size particles.

A highly plastic soil invariably exhibits some undesirable characteristics as a foundation material. It is often the cause of excessive foundation settlement, retaining wall movement, slope failure, etc.

C. Compressibility. When a cohesive soil is subjected to compression, some of the water and air is extruded from the voids of the soil. The voids in fine-grained soils are very small, therefore the water escapes very slowly. Consequently, the process of compression continues for a long period of time, often many years. This slow process of compression is called *consolidation.* As opposed to the compression of granular soils, cohesive soils require a long time to consolidate. The total compression due to consolidation is considerably larger than that caused by other factors and, hence, must be evaluated for any moderate to large size job.

Some natural deposits of cohesive soils have undergone heavy compression in the geologic history and therefore are relatively imcompressible. Such soils are known as *preconsolidated* or *overconsolidated.* Deposits which have not been subjected to previous compression are called *normally consolidated.* The amount of compression that a cohesive soil will experience depends upon the *compression index*, C_c and other factors:

$$S = \frac{C_c}{1 + e_0} H \log \frac{p_0 + \Delta p}{p_0}$$

where S = total compression of a layer with a thickness H under an additional pressure Δp; e_0 is the original void ratio of the soil and p_0 the existing soil pressure. For further discussion and a numerical example of this equation see Sec. 3-5.

The value of C_c of a given soil may be determined by a laboratory consolidation test. For the purpose of approximate calculation, the following empirical relations may be used for plastic normally consolidated clays.

$$C_c = 0.009\ (\mathrm{LL} - 10) \tag{1-12a}$$

where LL = liquid limit of the soil expressed in percentage not in decimal (Peck, et al., 1953) or

$$C_c = 0.30\ (e_0 - 0.27) \tag{1-12b}$$

where e_0 = natural void ratio of soil in place (Hough, 1957).

D. Sensitivity. Cohesive soil often loses a portion of its shear strength upon disturbance. The amount of strength loss due to thorough disturbance is expressed in terms of *sensitivity*. An undisturbed sample and a remolded sample of the soil are subjected to unconfined compression tests. The ratio between the undisturbed strength and the remolded strength is the sensitivity of the soil. A cohesive soil is described as insensitive, extra sensitive, etc., according to its sensitivity:

Insensitive clays	sensitivity = 1
Clays of low sensitivity	sensitivity = 1–2
Clays of medium sensitivity	sensitivity = 2–4
Sensitive clays	sensitivity = 4–8
Extra-sensitive clays	sensitivity = >8
Quick-clays	sensitivity = >16

Some clays have been reported to have sensitivity greater than 100. Sensitive soils may lose their strength due to construction operations. A well-known example is the drastic reduction in shear strength along the periphery of newly driven piles in clays. If the clay is not subjected to further disturbance, however, a large portion of the strength will be regained in a relatively short time.

E. Expansion and shrinkage. Some clays exhibit large volume changes: expansion (or swelling) upon wetting, and shrinkage upon drying. In some cases buildings have settled because of excessive shrinkage of the underlying clay and in other cases swelling is the problem. Seasonal swelling and shrinkage are caused by excess or deficiency of water.

Swelling (expansive) and shrinking clays are often characterized by high liquid limits and high plasticity indices as a result of high content of the more active clay minerals. Although such clays are often encountered at shallow depths, 60 ft and 200 ft thick layers have been reported.

This problem is often recognized by local experience. Where damages due to swelling and shrinking are known or suspected, soil samples should be tested to determine the *shrinkage limit*, *free swelling*, and *swelling pressure*. As a wet cohesive soil is dried out, the volume of the soil decreases with the decreasing water content. At a certain water content, however, the volume does not decrease upon further drying, and this water content is known as the shrinkage limit of the soil. The amount of expansion (or free swelling) is determined by admitting water to the soil sample and measuring the volume increase. Usually the measurement is made on a sample which is laterally confined and subjected to a normal pressure of 1.0 psi. If a normal pressure is applied to prevent expansion of the soil, this pressure is known as swelling pressure. Swelling pressures measured on some expansive clays exceed 10 tons per square foot.

Foundations on expansive clays often require unusual designs based on intelligent interpretation of skillfully conducted laboratory tests, sound engineering judgement, and local experience. Some local experiences have been reported in technical publications (e.g., Baracos and Bazozuk, 1957; Salas and Serratosa, 1957).

1-9 Engineering Properties of Silt and Loess

A. Silt is material with grain size passing No. 200 sieve but possesses no cohesion and plasticity. Its engineering properties are essentially those of fine sand. Because of the fine particle size, this soil has the following undesirable characteristics:

1. Low shear strength immediately after load application,
2. High capillarity and frost susceptibility,
3. Low permeability,
4. Low relative density–difficult to compact.

Furthermore, it is difficult to simulate test data on silts which would correspond to the field conditions. For medium to large jobs where foundations must be supported on silt deposits, the engineering properties should be determined by adequate laboratory and/or field tests conducted by specialists. For advanced study, reference is made to a report by K. Akai (1960).

B. Loess is the name given to wind-blown deposits having grain sizes of silt. Due to the peculiar process of deposition, loesses have completely different characteristics from the common silts which are deposited in water. The outstanding characteristics of a true loess are as follows (Clevenger, 1958):

1. It is commonly a loose deposit with dry unit weight as low as 65 pcf. Because of the presence of cementing material (clayey or calcareous), it exhibits moderate to high bearing capacity in dry or moist condition.

2. It subsides upon saturation due to loss of cementation. Structures supported on loess should be guarded against such danger.
3. It is capable of standing on nearly vertical bank.

If the cementation is destroyed by innundation or redeposition, the material is called modified loess which has all characteristics of silt.

1-10 Engineering Properties of Organic Soils

Any soil containing a sufficient amount of organic matter to influence its engineering properties is called an organic soil. The amount of organic matter is expressed in terms of *organic content* which is the ratio between the weight of organic matter and the ovendried weight of sample. The weight of organic matter can be determined by heating the sample to ignite the organic substances (McFarland, 1959).

Natural soil deposits may contain a very small percentage (up to 100 per cent) organic matter. Generally a relatively small percentage (as low as 2 per cent in some cases) will contribute sufficient undesirable characteristics. In some special applications (e.g. soil-cement), only a fraction of one per cent may render the soil undesirable.

Organic matters are derived principally from plant life and occasionally from animal organisms. They are found in the following forms:

Top soil (loam): the upper layer of ground, usually several inches deep.

Leached stratum: organic matter accumulated on an impervious layer from leaching through upper pervious soil.

Organic deposits: peat, swamp, lignite, coal, etc. In engineering literatures the term *muskeg* is used in Northern United States and Canada to denote a terrain consisting of swamp, bog, or other peat deposits.

Soils containing high organic matter will, evidently, have the following undesirable characteristics:

1. Low shear strength,
2. High compressibility,
3. Spongy structure which deteriorates rapidly; hence, results in subsidence without external load,
4. Acidity and other injurious characteristics to construction material.

Therefore, such materials should preferably not be used to support foundations.

1-11 Engineering Properties of Rocks

Engineers refer to the rock formation at some depth beneath a mantle of soil as bedrock, and the soil above the bedrock as overburden. In common foundation practice, the properties of bedrock fall into the problem of bearing capacity and permeability.

A. Factors affecting the engineering properties of rocks.

1. *Mineral constituents:* Rocks are made of one or more minerals, each possessing different strength and hardness from the others. The hardness of a mineral can be determined by scratching it with a fingernail, a copper coin, the blade of a knife, a sharp edge of glass, or a steel file. A mineral may be strong or weak in resisting crushing, shearing, and/or bending forces. Rocks consisting of soft and weak minerals have low bearing capacity.
2. *Texture and structure:* The texture of a rock is described as coarse-grained and fine-grained. The structure may be massive, dense, porous, or visecular (full of holes). The structure of an igneous rock may be columnal; limestone may be cavernous.
3. *Joints, bedding planes, and foliations:* Joints exist in every type of rock. They may be open and visible, or closed and indiscernible. Bedding planes are the boundaries between layers of sedimentary rocks. Foliations are characteristics of some metamorphic rocks which have banded or laminated structures. Joints, bedding planes, or foliations are planes of weakness in the rock formation.
4. *Weathering conditions:* The mineral constituents of rock may be altered by chemical weathering (decomposition) and/or physical weathering (disintegration). The weathered zone may be a few inches to several hundred feet deep. It may be near the present ground surface, but it might also have been covered by another rock formation. It may be detected by color stains and lack of luster, and it gives a dull sound when struck by a hammer. A weathered rock often is weak and becomes treacherous when exposed to moisture.
5. *Faults:* A fault is an inclined plane of break resulting from the relative movement that occurs when one side of the fault has moved up or down relative to the other side. A fault may be only a fine crack, or many feet wide. In a wide fault, the space is filled with crushed rock and soil. Faults may also be closely spaced and almost parallel. Faults only occur in areas known as zones of tectonic disturbance or volcanic activity. A fault is classified as active or dead, depending upon whether or not further movement is anticipated. This information can be obtained from the literature on the geology of the area.
6. *Cementation:* The mineral aggregates may be weakly or strongly cemented in any type of rock. Even a sound and massive igneous rock may have noncohesive or weakly cemented materials interbedded within it. Upon soaking in water, or exposure to the atmosphere, a weakly cemented rock may lose part or all of its cementation.

B. Bearing capacity of rocks. The bearing capacities of rocks are often determined by crushing a core sample on a testing machine. Samples for testing must be free from cracks and defects. In the rock formation, however, bedding planes, joints, and other planes of weakness do exist. The bearing capacity of a rock mass, including such weaknesses, is difficult to determine. Usually the bearing capacity to be used for design is restricted by local building code. If no such restrictions are stipulated, the design bearing capacity of a bedrock is commonly assumed to be $\frac{1}{5}$ to $\frac{1}{8}$ of the crushing strength (factor of safety = 5 to 8). Past experience demonstrates that these factors are satisfactory for normal conditions. However, for rocks containing extensive cracks and joints with wide open space, a higher factor should be used.

A great many bedrocks are stronger than the concrete used as foundation material. Therefore, the design values are often limited by that for concrete. In adverse conditions, such as placing concrete under water, the bearing value of concrete should be reduced. The bearing value is further complicated by the possibility of rock softening and accumulation of sediments or debris at the bottom of excavation.

C. Permeability of rock. If a piece of sound rock is used for determination of permeability, it will be found to be quite impermeable. However, joints, cracks, etc., in the natural rock formations permit seepage or free flow of water. The presence of such openings in rock formation nullifies the meaning of any laboratory tests on sound rock samples. For large jobs, the amount of seepage through rock formations can be determined by pumping tests. Otherwise, the seepage can only be approximately estimated by an experienced engineering geologist after a thorough examination of the nature and the extent of such openings.

Rocks are classified into three major groups, namely, igneous, metamorphic, and sedimentary. The most notable properties of each group are summarized in the following sections.

1-12 Typical Engineering Properties of Igneous and Metamorphic Rocks

Igneous rocks (granite, diorite, basalt, etc.) in general have the following characteristics:

1. Good structural characteristics–hard, dense, and durable–good construction materials;
2. High bearing capacity–good foundation material;
3. Joints in three dimensions–actual or potential joints are in three sets at approximately right angle to each other.

These joints divide the rock into prismatic blocks. In basalt, vertical joints create long columns adjoining each other. Pumice, tuff, volcanic ash, etc., are exceptions to the statement above. In the engineering sense, however, they are not considered as rocks.

The common metamorphic rocks (gneiss, schist, marble, slate, serpentine, etc.) have the following general characteristics:

1. Hard and strong if the rock is not weathered.
2. Jointed, folded, laminated or foliated–metamorphic rocks commonly have two or three sets of joints. The strength of the rock is greatly influenced by the joints and the folded, laminated or foliated structures.
3. Containing weak layers between very hard ones.

1-13 Engineering Properties of Sedimentary Rocks

By far the most common sedimentary rocks are limestones, sandstones, and shales:

A. Limestones.

1. The strength of limestones varies considerably, from soft calcareous limestones to hard limestones and dolomites. It may vary even within one limestone formation. The strength generally depends upon the texture of the rock. A limestone with porous or cavernous texture has very low compressive strength (as low as 3000 psi), and one with dense texture has very high strength (up to 40,000 psi).
2. Limestone grains are sometimes cemented together by clayey material, and the cementing strength may be reduced upon wetting.
3. Limestones usually contain fissures, cavities, and caverns which may be fully or partly filled with fine-grained soil, or may be empty. Three sets of joints are found in limestones. As in other sedimentary rocks, one set of the joints is parallel to the bedding planes. Cavities, fissures, and joints are planes of weakness.
4. Limestones may contain thin layers of sandstone. These layers are more permeable and, sometimes, weaker than the limestones.

B. Sandstones.

1. The strength of sandstones depends largely upon the degree of cementation and the type of cementing material:

Cementing material	*Usual color*	*Strength*
Iron oxide	Brown, red, orange	Variable, cement often in irregular bands
Clay	Dull, whitish grey	Low, treacherous when wet
Calcite $(CaCo)_3$	Grey, white, buff	Good
Silica $(SiO)_2$	White (often stained by iron oxide), buff, yellow, pink	Excellent

2. The durability is generally in proportion to the strength.
3. Three sets of joints exist in sandstones. Joints are generally spaced several feet apart.

C. Shales.

1. The strength of shale varies widely. Soft shales may be scratched by a fingernail or excavated by machine without the use of explosives. Hard shales, however, require blasting to excavate.
2. Shales have a closely laminated structure and a great tendency to split along the laminations. When wet, the shear strength along the laminations may be extremely low.
3. A shale often becomes soft or reduces to loose clay or silt after being soaked in water for several hours or days. Samples should be examined after alternate soaking and drying.
4. Core samples are difficult to obtain from soft shale formation. If accessible, samples may be obtained by hand-operated power tools. If reasonably good samples are possible to obtain, they should be subjected to unconfined compression tests to determine their structural strength.

1-14 *Ground Water and Associated Phenomena*

A. Ground water level. Natural soil deposits contain a certain amount of moisture in their voids. Near the ground surface, the voids are only partly filled with water, but at greater depths the soils and rocks are saturated. If a large perforated pipe is inserted in the ground, the free water level would correspond to that in the soil, and this level is called the *ground water level* or *ground water table*. It represents the level below which the soil and rock are submerged and above which water may rise by capillary action. The height of capillary rise depends on the size of the voids, which in turn depends on the grain size and gradation. In fine-grained soils, the capillary rise may exceed 20 ft. In gravel and coarse sand there is little capillary action.

Ground water is supplied by rain, snow, or other forms of precipitation. This source of water is known as meteoric water. Ground water is also derived from connate water (water which was entrapped in sediments during

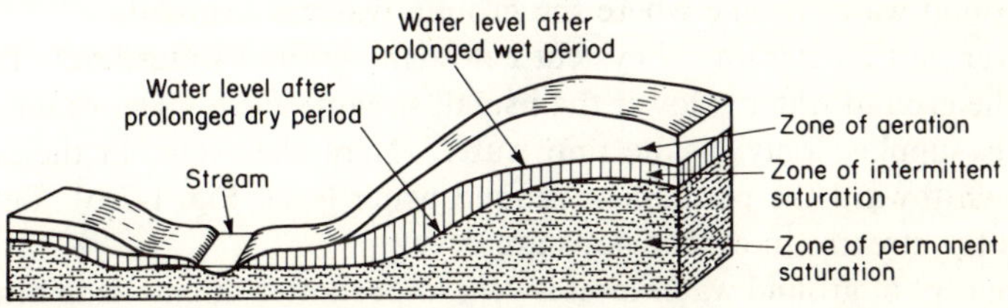

Fig. 1-6 Ground water profile.

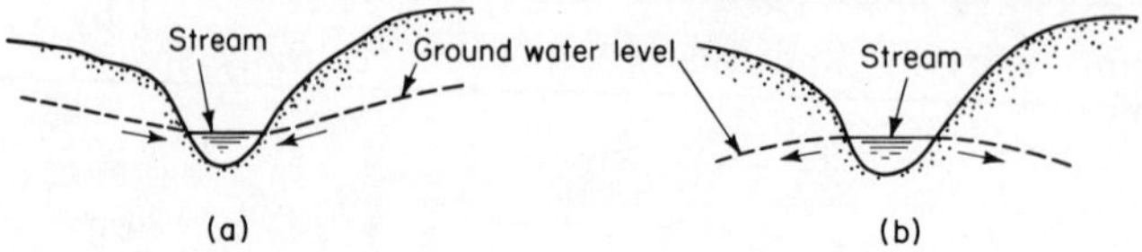

Fig. 1-7 Ground water and stream level: (a) ground water to stream (humid region); (b) stream to ground water (arid region).

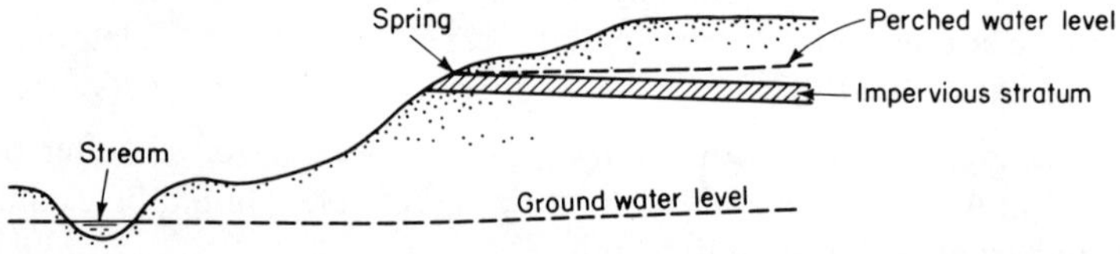

Fig. 1-8 Perched water.

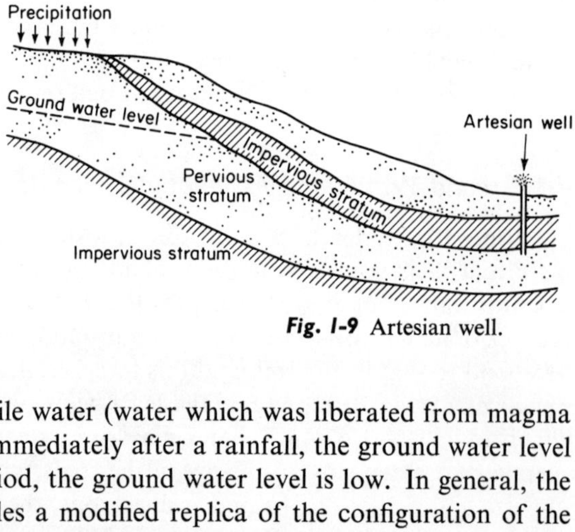

Fig. 1-9 Artesian well.

their deposition) and juvenile water (water which was liberated from magma or other igneous origin). Immediately after a rainfall, the ground water level is high, and after a dry period, the ground water level is low. In general, the ground water level resembles a modified replica of the configuration of the ground surface. The ground water is high where the ground surface is high; it dips down toward the stream valley and it may emerge as an open body of water, Fig. 1-6. Most streams and lakes are maintained by ground water flowing into them. Therefore, lakes, streams, etc., are nothing but low spots in the ground water profile where the ground water is exposed.

However, not all stream valleys cut below the ground water level. In arid regions the ground water is lower than small stream valleys. The water in the streams is supplied only by the rain water. Here the water in the stream seeps down toward the permanent ground water level, Fig. 1-7(b). Between rain-falls the streams in arid regions are dry.

The concept of ground water level is very important in engineering practice. For any foundation work, the ground water level (the maximum water level and sometimes the maximum and minimum water levels) must be determined with a reasonable accuracy. The river and lake levels are useful references for determining the ground water level in the area. However, the water level in a pond, etc., in some cases is not a true indication of the actual ground water level. A large body of free water may be retained on top of an impervious layer. Such water body is called *perched water*, Fig. 1-8. Below this impervious layer the soil may be dry.

Another phenomenon that may be encountered in foundation engineering is the *artesian well.* The geologic and hydraulic conditions effecting an artesian well are illustrated in Fig. 1-9. The water head which pushes the water up the well is a result of the configuration of the soil strata. If a foundation is constructed in an artesian area, it may be subjected to uplift pressure.

B. Effective pressure and pore pressure. At a level above the ground water, Fig. 1-10(a), the vertical pressure is equal to the unit weight of the moist soil times the depth Z. If the soil is submerged in water, Fig. 1-10(b), the total vertical pressure q at depth Z is

$$q = H\gamma_w + Z\gamma_s$$

where γ_w = unit weight of water and γ_s = the saturated unit weight of the soil. The latter consists of two parts: the unit weight of water plus the buoyant weight of the soil particles:

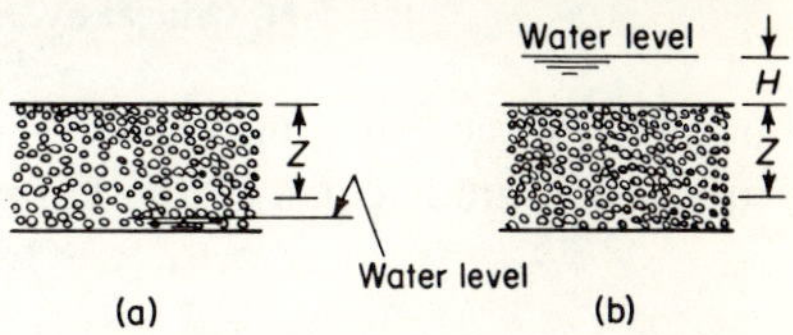

Fig. 1-10 Vertical pressure in soil.

$$\gamma_s = \gamma_w + \gamma'$$

As any object submerged in water, the buoyant weight, γ', is equal to the dry weight minus the weight of water displaced by the soil particles. Substituting the latter equation into the previous one,

$$q = H\gamma_w + Z\gamma_w + Z\gamma' = (H + Z)\gamma_w + Z\gamma'$$

This equation indicates that the total vertical pressure q is made of two components: (1) pressure called neutral pressure, pore pressure, or pore-water pressure which is due to the head of water $H + Z$, and (2) pressure called intergranular pressure or effective pressure which is due to the buoyant weight γ' of the soil grains. The neutral pressure is the water pressure in the voids. The effective pressure is transmitted from grain to grain.

The discussion above deals with hydrostatic conditions where there is no flow of water in the soil. When water percolates through a pervious soil* the force or head due to seepage water must be included. In such cases, the flow net method is useful, and reference should be made to standard text books of

* A pervious soil that transmits large quantity of water is known as an aquifer or waterbearing stratum.

soil mechanics (Terzaghi, 1942; Terzaghi and Peck, 1948; Tschebotariff, 1952; etc.). The following deals with one of the most common problems in seepage pressure.

C. Critical hydraulic gradient and quicksand. A quicksand is not due to an intrinsic property of the sand, but a condition of excessive seepage pressure. The basic principle of this condition is illustrated in Fig. 1-11(a). A container with a depth *d* is subjected to a hydraulic head *h*. The hydraulic uplift at the

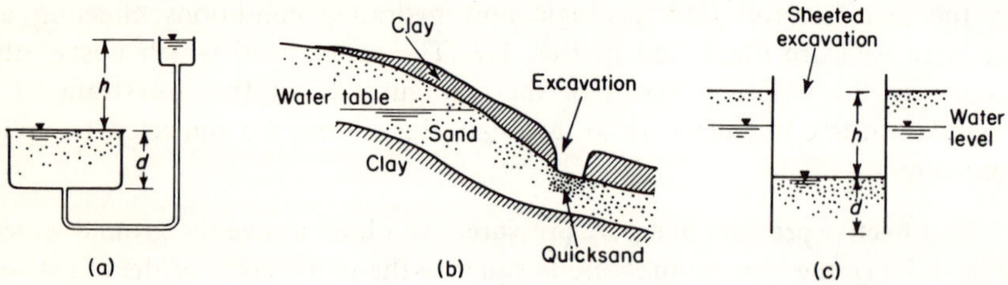

Fig. 1-11 Critical hydraulic gradient and quicksand.

bottom of the container is $h\gamma_w$ (γ_w being the unit weight of water). If the uplift pressure is equal to the buoyant weight of the soil above, that is

$$h\gamma_w = d\gamma' \tag{1-13}$$

the effective pressure between soil grains is zero; and hence the shear strength becomes zero. This condition is called quicksand. Any object placed on quicksand will sink because it has no bearing power as a result of zero shear strength. If the hydraulic head is greater than the buoyant weight of the soil, the grains will be uplifted, and the sand will appear like boiling water. This phenomenon is known as a boil.

Equation (1-13) may be rearranged by substituting Eq. (1-8) for γ'.

$$\frac{h}{d} = \frac{G - 1}{1 + e} \tag{1-14}$$

where G = specific gravity of soil grains and e = void ratio of the soil. The value of $(G - 1)/(1 + e)$ for a given soil is called the *critical hydraulic gradient*. Since the specific gravity G of sand is about 2.65 and the void ratio e ranges from 1.0 for loose sand to 0.25 for dense sand, the value of critical hydraulic gradient ranges from 0.8 to 1.3. For average conditions it may be taken as 1.0.

Quicksand may be caused by a large head in the natural soil strata, Fig. 1-11(b), or by artificial dewatering, Fig. 1-11(c).

Plate Two

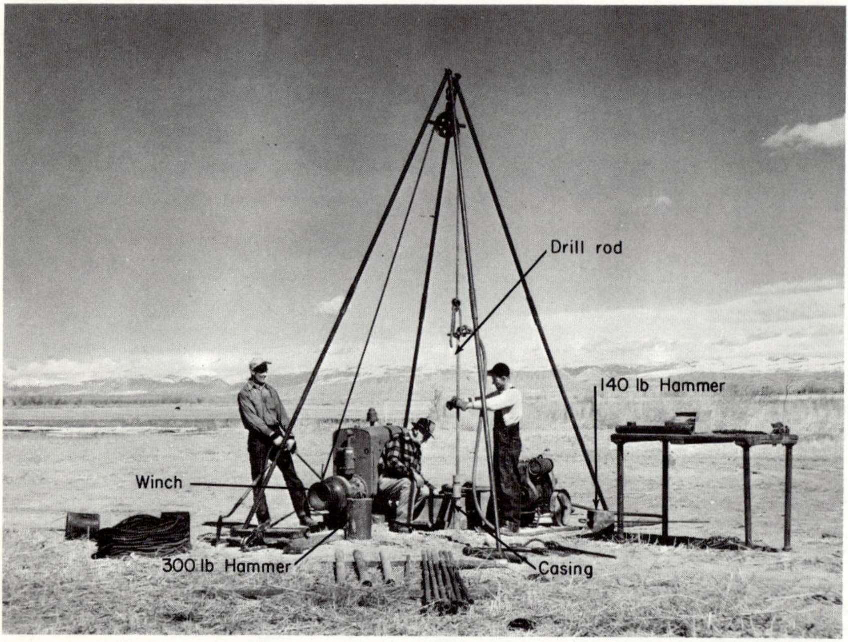

Making Borings from an A-frame Rig

Subsurface conditions at any given site must be adequately explored to obtain information required in design and construction. The arrangement of various soil strata and rock formations are explored by borings and the pertinent engineering properties of each type of the materials encountered are evaluated by tests.

Various boring techniques, sampling devices, and testing procedures are available for different purposes. For each job the engineer must work out an adequate exploration program. This chapter summarizes the information and criteria to aid the planning of such programs.

2 SOIL EXPLORATION

2-1 *Purpose of Soil Exploration*

The purposes for soil exploration are to obtain information as bases for:

New structures:

1. The selection of type and depth of foundation.
2. The determination of bearing capacity of the selected foundation.
3. The prediction of settlement of the selected foundation.
4. The establishing of the ground water level.
5. The evaluation of the earth pressure against walls, abutments.
6. The provisions against constructional difficulties.
7. The suitability of soil and the degree of compaction of fill (under slabs and pavements and against retaining walls).

Existing structures:

1. The investigation of the safety of the structure.
2. The prediction of settlement.
3. The determination of remedial measures if the structure is unsafe or will suffer detrimental settlement.

Highways and airfields:

1. The location of the road (and runways) both vertically and horizontally.
2. The location and selection of borrow material for fills and subgrade treatment.

3. The design and location of ditches, culverts, and drains.
4. The design of roadway sections.
5. The need and type of subgrade treatment.
6. The location of local source of construction materials for bases and wearing surfaces.

2-2 *Planning an Exploration Program*

An engineer planning a soil exploration program for a specific job must (1) have a clear idea of what he is trying to accomplish by the exploration (Sec. 2-1), (2) be well acquainted with current methods and procedures for soil boring, sampling, and testing, and (3) keep in mind the relative costs of soil exploration versus the cost of the foundation construction.

The planning of a soil exploration should always start by obtaining preliminary information. The procedures for obtaining such information for highway projects are discussed in the AASHO Manual. For buildings and similar projects, the following information should be obtained first.

Available information
Reconnaissance
Building code requirements
Preliminary design data

After this preliminary information is obtained and digested, a tentative exploration program is worked out. The first two or three borings should be scattered around the entire site to disclose the general characteristics of the subsoils. As the boring operation progresses, the balance of the boring program should be constantly revised so that the number and type or types of borings will furnish sufficient data concerning the arrangement of the successive soil strata, and that sufficient number of soil samples are taken for laboratory tests.

There is no hard and fast procedure for planning a boring program. Each condition must be weighed with common sense, good judgement, and relative economy. For example, if the job is small, it may be more economical to make the foundation design on conservative values rather than making elaborate borings and tests. An example of planning a soil exploration program is given in Sec. 2-15.

2-3 *Available Information*

For large and important projects, the engineer should get the published geological and topographical information before starting the soil exploration. In the United States, the following sources are available:

1. *U.S. Geological Survey*, Washington 25, D.C.

   Geologic Map of U.S. (Scale 1 in. = 40 mile). Rock units are distinguished by patterns printed in 23 colors.

   Status Index Maps: A series of maps showing the status of various phases of mapping in the United States. Each map is accompanied by a text which gives a detailed explanation.

   (a) Aerial Mosaics of U.S. show areas in the U.S. for which photomaps have been prepared from aerial photographs and agencies from which copies may be obtained.
   (b) Geologic Mapping in U.S. shows by color patterns the areas covered by published geologic maps.
   (c) Topographic Mapping in U.S. provides an index to topographic mapping in each state. On a base map the available quadrangles of topomaps are shown.
   (d) State Geological Index Maps are available for almost all of the states. Each published geologic map is outlined on a state base map, an explanatory key gives the source of publication.

2. *State Geologic Survey*. Most of the states have a Geological Survey or similar agencies that can supply information on availability of geologic maps and other references.
3. *Soil Survey Section of the Bureau of Plant Industry*, U.S. Department of Agriculture. The Agriculture Year Book of 1938, *Soil and Men* has an abundance of useful data. Areas which are not covered by these maps have often been mapped by individual farm maps. These maps indicate the soil type and series which can be invaluable aid for furnishing ground information. The regional soil scientists usually can furnish with soil profile descriptions, soil keys, nomenclature, and the type of parent material associated with the various soil series mapped in his region. The Highway Research Board has published several bulletins concerning the available information in United States.
4. *Hydrological Data.* U.S. Army Engineers map of areas and waterways; information regarding river and tidal levels; stream flow data and maximum flood levels.
5. *Soils Manual.* Several state highway departments have published such manuals (Michigan, Washington, Missouri, etc.).
6. *The Origin, Distribution and Airphoto Identification of U.S. Soils* (Belcher, et al.), U.S. Department of Commerce, 1946. See Plate A-1.
7. *Text Books on Geomorphology of U.S.* See Plate A-2 for physical divisions of U.S.

2-4 *Reconnaissance*

The engineer should always inspect the site to obtain the following data before actual exploration starts:

1. The general topographical characteristics–site on top of a hill, on a bluff, in a valley, on an abandoned lot with debris, etc.
2. The type of construction and conditions of the existing structures in the adjoining properties. Look for settlement cracks on exterior walls.
3. The soil profiles in highway or railroad cuts and quarries.
4. The high water marks on old buildings, bridge abutments, etc.
5. The soil conditions, ground water level, and depth of rocks. General information of this nature can often be obtained from the local people.
6. The depth of scour and history of flood levels (bridge foundations) from the local people.
7. Photographs of the site and adjacent structures.

2-5 *Preliminary Design Data*

The soil exploration and the preliminary design of the structure are so intimately associated that they should be started about the same time. Exploration made ahead of the preliminary design often results in inadequate information or unnecessary waste. The preliminary design data should include:

Buildings:

1. The size and height of building and the depth of basement.
2. The approximate arrangement of columns and bearing walls.
3. The approximate range of column and wall loads.
4. The type of framing–simple span structures, continuous or rigid frame structures, arches, shell structures, foundations for precision machinery, etc.
5. The type of exterior walls–brick and glass are sensitive to settlement whereas metal panels and sidings are more flexible.

Bridges:

1. The type and length of bridge span.
2. The approximate vertical and horizontal loads on the piers and abutments. For shorter span bridges the dead load can be estimated from typical design drawings (e.g., *Standard Plans for Highway Bridge Superstructures* by Bureau of Public Roads and publications by some

state highway departments). The live load can be readily obtained from chart in AASHO or AREA Codes.

2-6 *Common Types of Boring, Sampling, and Testing*

A soil exploration operation consists of three steps, namely boring (advancing a test hole in the ground), sampling (taking soil or rock sample from the test hole), and testing. These steps may be separated operations or combined in one. Testing may be done in the field or in a laboratory.

At least one soil sample should be taken in every 5 ft of depth of the test hole. First a soil sampler (split-spoon, Shelby tube, . . .) is driven or pushed into the ground surface to take a soil sample. The sample is visually examined and preserved for laboratory test. After the sample is taken, the test hole is advanced for about 4 ft. During the advancing of the hole, shavings and cuttings of soil brought up by the boring tools are examined. If soil shavings indicate change in soil characteristics, the depth where the change occurs should be recorded, and additional soil sample should be taken. The sampler is again advanced to take soil sample. In such alternative sequence, the test hole is advanced and soil samples are taken. In a certain critical layer or layers of soil, continuous sampling may be desired.

When test holes are carried into bedrock, generally continuous rock cores are taken by means of rotary drills. Core samples are brought up by the drill and are visually examined. These samples (core) may be delivered to the laboratory for unconfined compression tests. The general characteristics, particularly the percentage of sample recovered* from the test hole, are important information for foundation design and construction.

Water level in the test holes should be observed. Lack of information or misleading data concerning the ground water level often result in design fallacy and construction difficulties. The procedure for measuring ground water level is discussed in Sec. 2-12.

In Table 2-1, the more commonly used methods for boring, sampling, and testing are outlined. Further comments on the standard penetration test, thin-walled tube sampling, vane test, etc., are presented in the subsequent sections in this chapter. It may be noteworthy to emphasize that any portion of soil exploration (boring, sampling, or testing) on any project, no matter how small it may be, must be done by qualified personnel using appropriate equipment specially made for such purposes.

The sizes of casings and drill rods used in soil and core borings are standardized (Hvorslev, 1949). They are so designed that the loss in hole diameter is at a minimum when a given size of casing cannot be advanced

* Percentage of recovery = total length of rock sample recovered × 100 divided by total depth of hole advanced in the bedrock.

Table 2-1 COMMON TYPES OF BORING, SAMPLING, AND TESTING

1. Boring: Procedure for Advancing a Hole in Ground

Type	*Procedure*	*General uses*	*Remarks*
Test pit	Pit may be excavated by hand tools, backhoe, bulldozer, or Caisson drilling equipment	1. Preliminary exploration 2. For taking hand-cut soil samples	Unsuitable for exploration in granular soils below water level
Auger	Operated by hand or by power; 2 to 6 in. diameter, helical or post hole type	Auger boring and wash boring are the most common methods for advancing test holes	Limited depth if operated by hand
Wash boring	Hole advanced by water jets and a chopping bit		
Percussion drill	Jack hammer or other types of drilling equipment which bore a hole in soil or rock by a chopping bit	For drilling a hole in rock or hard material	
Rotary drill	Continuous rock core by means of rotary diamond drill bit attached to a core barrel and drill rod	Common type for bed rock	

2. Sampling: Method for Taking Soil and Rock Samples from Test Holes

Type	*Procedure*	*General uses*	*Remarks*
Highly disturbed samples	Auger boring, wash boring, percussion drilling, see Part 1 of this table	Shavings of soil brought up by auger, soil particles carried by wash water, or dust and chips from percussion drill are indications of types of soil encountered	Unsuitable for foundation exploration
Split spoon	A standard split spoon sample is a 2 in. O.D. $1\frac{3}{8}$ in. I.D. tube, 18 to 24 in. long; the tube is split longitudinally in the middle	1. Taking disturbed samples 2. Taking samples from hard soils 3. Taking samples from soft rocks	Sampler is also used for making standard penetration test (Sec. 2-7)
Thin-walled tube	16-guage seamless steel tube commonly 2 and 3 in. diameter; preferably pushed by static force instead of driven by hammer	Taking undisturbed samples from cohesive soils	Unsuitable for granular soils and hard materials
Core boring	Rotary drill, see Part 1 of this table	Taking continuous rock samples	
Hand-cut samples	Cut by hand from side of test pit	Samples are least disturbed; not commonly used because of large expenses involved	

Table 2-1 COMMON TYPES OF BORING, SAMPLING, AND TESTING *cont.*

3. Laboratory Testing of Soils

Properties of soil	*Type of test*	*ASTM* Designation of standard methods	*ASTM* Suggested methods	*AASHO Designation*
Grain size distribution	Mechanical analysis	D421, D422 D1140		T88
Consistency	Liquid limit (LL)	D423		T89
	Plastic limit (PL)	D424		T90
	Plasticity index (PI)	D424		T91
Unit weight	Specific gravity	D854		T100
Cohesive soils*				
Noncohesive soils*				
Moisture	Natural water content			
	Field moisture equivalent	D426		T93
	Centrifuge moisture equivalent	D425		T94
Shear strength				
Cohesive soils	Unconfined compression		Yes	
Noncohesive soils	Direct shear		Yes	
General	Triaxial		Yes	
Volume change	Shrinkage factors	D427		T92
	Volume change		Yes	T116
	Expansion pressure		Yes	
Compressibility	Consolidation		Yes	
Permeability	Permeability		Yes	
Compaction characteristics	Standard proctor	D698		T99
	Modified proctor		Proposed 1958	T180
California bearing ratio (CBR)			Yes	

* See T. W. Lambe, *Soil Testing for Engineers* (New York: John Wiley & Sons, Inc., 1951).

Table 2-1 COMMON TYPES OF BORING, SAMPLING, AND TESTING *cont.*

4. Field Testing of Soils

Purpose of test	*Type of test*	*ASTM* Designation of standard methods	*ASTM* Suggested methods	*AASHO Designation*
Compaction control	Moisture-density relations	D698	Proposed	T99, T180
	In-place density	D1556	Yes	T147 T181
	Penetrometer needle		Proposed 1958	
Shear strength (soft clay)	Vane test			
Relative density (granular soil)	Penetration test			
Permeability	Pumping test			
Bearing capacity				
Pavements	CBR		Yes	
	Plate bearing	D1195 D1196		
Footings	Plate bearing	D1194		
Piles (vertical load)	Load test	D1143		
Batter piles	Latteral load test		Yes	

further and it becomes necessary to use a smaller casing during continued advance of the bore hole. The four standard sizes used in soil exploration are designated as EX, AX, BX, and NX casings and E, A, B, and N drill rods. NX core bits, operated with N drill rods, will pass through NX casing and drill a hole large enough to admit BX casing, which in turn will admit BX core bits, drilling a hole large enough for AX casing, etc.

Casing and core barrel	*Drill rod*	*Core barrel bit (O.D.), in.*	*Diameter of bore hole, in.*	*Diameter of core sample, in.*
EX	E	$1\frac{7}{16}$	$1\frac{1}{2}$	$\frac{7}{8}$
AX	A	$1\frac{27}{32}$	$1\frac{7}{8}$	$1\frac{1}{8}$
BX	B	$2\frac{5}{16}$	$2\frac{3}{8}$	$1\frac{5}{8}$
NX	N	$2\frac{15}{16}$	3	$2\frac{1}{8}$

2-7 *Standard Penetration Test*

Because of the extreme difficulty in obtaining undisturbed samples from granular soils and in securing an undisturbed specimen from the sample, the engineering properties of such soils are determined by taking disturbed samples and by measuring the relative density by penetration test. The disturbed samples are useful for grain size analysis. The penetration tests are made at frequent intervals (for example, every 5 feet and at least one test in each of the different soil strata) along the depth of the boring. Dynamic and static penetration methods have been developed, and different tools and different procedures have been employed. The widely used method is the *standard penetration test.*

In the standard penetration test a soil sampler known as a *split spoon* is used, Fig. 2-1(d). It is an open-ended steel cylinder which splits longitudinally into two halves. These two halves are held together by a cutting shoe at the lower end and a coupling which connects the sampler to the drill rod. The split spoon is driven 18 in. into the ground by means of a 140 lb weight (hammer) falling a free height of 30 in. The number of hammer blows for each 6 in. penetration is recorded. The total number of blows required to drive the second and third 6 in. of penetration is called the *standard penetration resistance N* which represents number of blows per foot (Terzaghi and Peck, 1948). After the blow counts are recorded, the spoon is withdrawn from the bore hole and a representative sample is secured. These samples are kept in airtight jars with proper identification for visual examination and/or laboratory tests.

The correlation between the relative density of granular soil and the standard penetration resistance are shown in Table 1-1. In conjunction with this table, the following points must be considered.

1. The correlation does not give very reliable values in gravel and soils containing large gravels. In loose coarse gravel deposits the split spoon tends to slide into the large voids and gives low penetration resistance. Low resistance may be further facilitated when the split spoon rotates the round pebbles as it penetrates into the voids. On the other hand, excessively large resistance may be expected when the spoon is blocked by a large piece of gravel, or when a piece of gravel is wedged inside the spoon. Therefore, the correlation shown in Table 1-1 can be best considered only as a rough estimate in gravels and gravelly soils.

2. For standard penetration tests made at shallow depth, the number of blows are usually too low. At a greater depth, the same soil with same relative density would give higher penetration resistance. The influence of the weight of soil above (which is called overburden pressure) on the standard penetration resistance is shown in Fig. 2-2 which may be approximated by the following equation

(a)

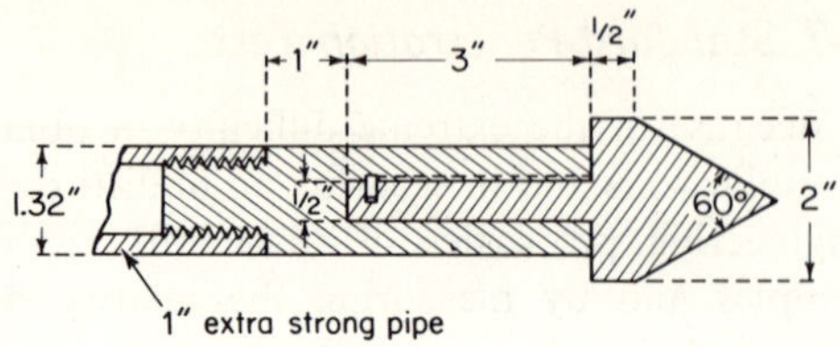

(b)

Fig. 2-1 (a) Open pit; (b) auger boring; (c) cone penetrometer used for sounding test; (d) split spoon sampler (Courtesy of Soil Testing Service, Inc.); (e) shelby tube; (f) core bit for rock samples.

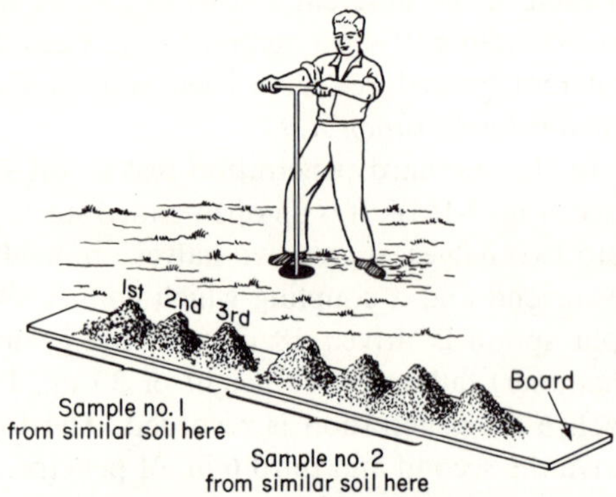

(c)

(d)

(e)

(f)

$$N = N'\left(\frac{50}{p + 10}\right) \tag{2-1}$$

where N = adjusted value of standard penetration resistance;
N' = standard penetration resistance as actually recorded;
p = effective overburden pressure, psi, not exceeding 40;
= weight of soil above the level at which the standard penetration test is made. Use buoyant weight for soil below water level.

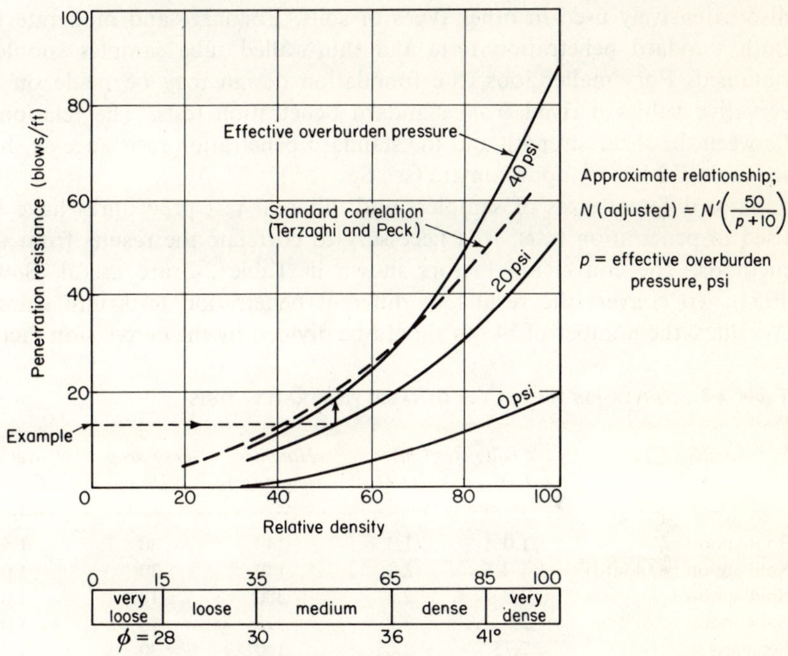

Fig. 2-2 Relationship between standard penetration resistance and relative density of sand. After Gibbs and Holtz, 1957.

The following example illustrates the use of the figure and the equation:

Given: $N' = 12$, as determined by standard penetration test at 30 ft below ground surface.
Water level at 20 ft below ground surface.

Assume $\gamma = 110$ pcf and $\gamma' = 70$ pcf

Then,

$$p = \frac{20 \times 110 + 70 \times 10}{144} = 20 \text{ psi}$$

Enter 12 on the vertical axis; proceed horizontally to the curve for $p = 20$ psi. The soil is found to be of medium density with an adjusted $N = 20$. This may be computed by the approximate equation (2-1) also,

$$N = 12\left(\frac{50}{20 + 10}\right) = 20$$

The value $N = 20$ should be used in foundation design. Due to the absence of substantial field data, the equation above should be used conservatively. It is recommended at the present time that, if the adjusted N value exceeds two times the actual test value (N'); the adjusted value N to be used in design should be divided by a factor of safety, say 2.

Standard penetration tests are not only useful in granular soils, they are also extensively used in other types of soils. For large and moderate jobs, both standard penetration tests and thin-walled tube samples should be obtained. For smaller jobs, the foundation design may be made on conservative values derived from standard penetration tests. The relationship between the shear strength and the standard penetration resistance as shown in Table 1-2 is very approximate (see Sec. 1-8).

Since different sizes of samplers and different test procedures have been used in penetration tests, it is necessary to correlate the results from these methods. The conversion factors shown in Table 2-2 are useful (Sowers, 1954). To convert the results of different penetration tests into standard N-values, the number of blows should be divided by the conversion factor.

Table 2-2 CONVERSION FACTOR FOR DYNAMIC PENETRATION TESTS

Sampler type	*Diameter (in.)*		*Hammer*	*Free drop*	*Conversion*
	I.D.	*O.D.*	*weight (lb)*	*(in.)*	*factor*
Split spoon	1.0	1.3	140	30	1.5
Split spoon (standard)	1.4–1.5	2.0	140	30	1.0
Split spoon	2.0	2.5	300	18	1.0
Split spoon	2.5	3.0	375	18	1.0
Seamless	2.75	2.65	140	30	1.0

2-8 *Thin-walled Tube (Shelby Tube) Sampling*

For moderate to large jobs the shear strength of the cohesive soils should be determined from relatively undisturbed samples. This is usually done by taking samples from the bore hole by means of a seamless thin-walled steel tube commonly known as a Shelby tube, Fig. 2-1(e). The tube is 2 in. or 3 in. in diameter and has a bevelled butting edge at the lower end. It is connected to the drill rod and pushed by static force into the bottom of the hole. When the tube is almost full (avoid over-penetration), it is withdrawn from the hole, removed from the drill rod, sealed at both ends with parafin, and shipped to soils laboratory for tests. When ready for tests, the tube samples are sawed into short lengths, not longer than 6 in. The samples are ejected from the tubes and subjected to unconfined compression tests, Atterberg tests, natural water content, etc.

Generally Shelby tube samples are taken from the cohesive soil strata in only a portion of the total test holes. Standard penetration tests are made in all other depths and test holes at frequent intervals for the full depth. The penetration resistance, in this case, is used hand in hand with the unconfined compression data. However, the latter is considered more reliable.

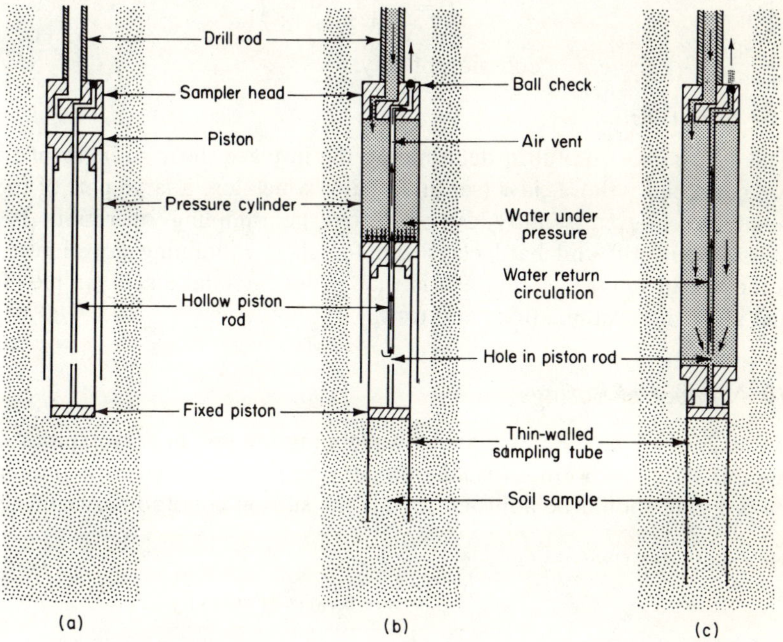

Fig. 2-3 Diagrammatic sketch of Osterberg piston: (a) sampler is set in drilled hole; (b) sampling tube is propelled hydraulically into soil; (c) pressure is released through hole in piston rod. After *Engineering News-Record*, April 24, 1952.

If consolidation tests are necessary, samples of 3 in. or larger diameters are taken. Some soils tend to drop out from the sampler while being withdrawn from the bore hole. In such cases, piston samplers may be used. Several types of such samplers are available, but the most advantageous is the one developed by Osterberg (1952). The principle of this type is illustrated in Fig. 2-3. The major advantage of the piston sampler are (1) it is capable of securing samples whereas the open sampler fails to do so, and (2) the sample is less disturbed.

2-9 *Vane Test*

A vane tester consists of a pair of thin steel blades connected to a vertical shaft, Fig. 2-4. The tester is pushed into the ground or into the bottom of a bore hole and a torque is applied on the shaft. If the shaft is kept free from the surrounding soil by means of a casing, the torque T required to shear the soil along the cylindrical surface with diameter d and height h is (Skempton,1950)

$$T = c\pi\left(\frac{d^2h}{2} + \frac{d^3}{6}\right)$$

where c = cohesion of clay.

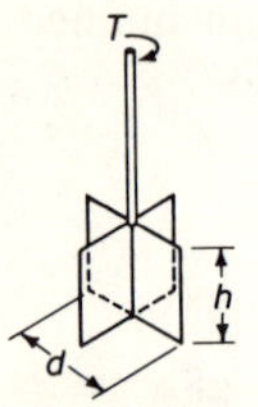

Fig. 2-4 Vane tester.

This test is most useful in determining the in-place shear strength of soft clays, particularly those clays (sensitive clays) which lose a large part of their strength when even slightly disturbed by the sampling operation. It is unsuitable for stiff and hard clays and for clays containing gravel pieces. Another disadvantage of this test is that it does not take soil samples for visual and other examinations and tests.

2-10 *Number of Borings*

Table 2-3 may be used as a guide in planning the preliminary exploration program. As a few borings (usually deeper ones) are completed, the preliminary program should be adjusted to suit the subsoil condition. A sufficient

Table 2-3 SUGGESTED NUMBER OF BORINGS

Project	*Distance between borings (ft)* — *Horizontal stratification of soil* — *Uniform*	*Average*	*Erratic*	*Minimum no. of borings each structure*
Multi-story buildings	150	100	50	4
1 or 2 story buildings	200	100	50	3
Bridge piers, abutments, television towers, etc.		100	25	1–2 for each foundation unit
Highways	1000	500	100	
Barrow pits (for compacted fill)	1000–500	500–200	100–50	

number of soil borings should be taken so that a soil profile can be drawn with reasonable accuracy to serve as the basis for the foundation design. If soil is extremely erratic, there is no need to take any more borings than necessary to ascertain this fact.

2-11 *Depths of Borings*

Highways and airfields. Minimum depth of borings is 5 ft but should extend below organic soil, muck, artificial fill, or compressible layers such as soft clays and silts.

Retaining walls and quay walls.

1. Below organic soil, muck, artificial fill, or any compressible layer;
2. Deeper than possible surface of sliding; and
3. Deeper than width of the base of wall.

Embankments and cuts.

1. Below organic soil, muck, artificial fill, or any compressible layer;
2. Deeper than possible surface of sliding; and
3. Equal to the width at bottom of cuts.

Structural foundations. The depth of borings depends upon the soil profile and the type of feasible foundation. Proceed as follows:

1. If no preliminary soil information is available, start with one or two deep borings to bedrock or to a depth equal to the width of the structure.
2. Analyze the boring results and determine the number and depths of additional borings. Borings should be carried to:
   (a) Below any organic soil, muck, artificial fill, or compressible layer;
   (b) Sufficiently deep for establishing the bottom elevation of foundation (footings, piles, or caissons); and
   (c) Sufficiently deep for checking the possibility of a weaker soil, at a greater depth, which may settle under the sustained load.

2-12 *Ground Water Measurement*

Ground water affects many important phases of foundation design and construction and must be determined in each job with reasonable accuracy. Since ground water is always effected by the weather conditions, the season, and the drainage conditions of the adjoining areas, the water level determined from the soil borings should always be adjusted with the climatological and hydrological data of the locale, in order to establish the highest and lowest levels during the life of the project.

The method of determining the ground water level in a bore hole varies with the permeability of the soil:

1. In permeable sand or gravel, the water level in the bore hole will seek its final level in a matter of minutes. The hole should be cleaned by horizontal jets and a steel tape coated with chalk can be lowered to the hole for measuring the water level.

2. In silt or silty sand, the time required for the ground water to reach its final level may be more than several days. In such cases approximate water level may be estimated by either one of the two methods.

(a) Bail out the water to the estimated water level and measure the levels at two or more equal consecutive time intervals. The final water level may be estimated by the method shown in Fig. 2-5.

(1) Determine rise or fall of water levels in two or more equal time intervals

$$T = T_1 - T_0 = T_2 - T_1 = T_3 - T_2 = \cdots$$

(2) The final water level:

$$H_0 = \frac{h_1^2}{h_1 - h_2}$$

or $$H_2 = \frac{h_2^2}{h_1 - h_2},$$

or $$H_3 = \frac{h_3^2}{h_2 - h_3}$$

(3) Fill or empty the casing to the computed level for final verification or correction

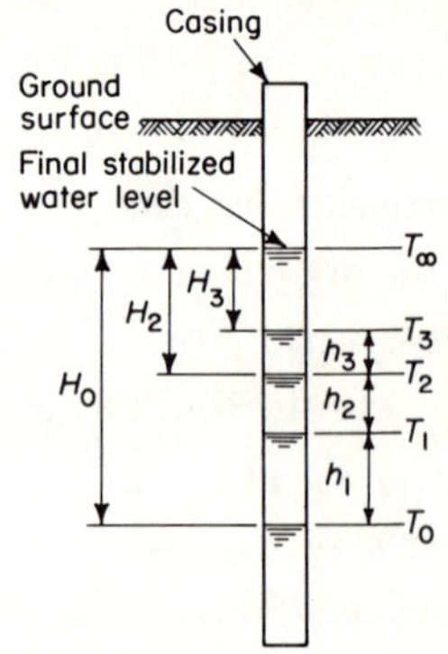

Fig. 2-5 Method for estimating final ground water level. After Hvorslev, 1949.

(b) Fill the hole with water and bail it out successively. After each bailing, measure the water level and determine whether it is rising or lowering. The true water level is one that lies between a rising and a lowering level.

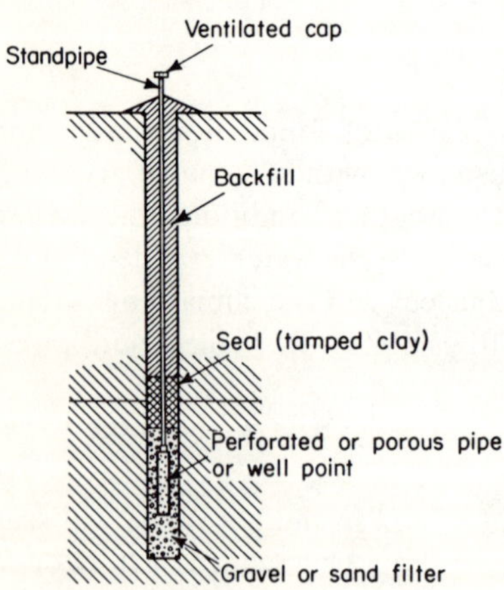

Fig. 2-6 Piezometer for observing ground water level. After Hvorslev, 1949.

3. In more impermeable soils, an observation pipe (piezometer) with ventilated cap or with a special pressure measuring device should be used. A simple piezometer is shown in Fig. 2-6. The time required for water in such an observation pipe to seek to a practically final level of equilibrium may exceed 10 weeks. In normal boring programs, the ground water level is measured at 24 hours after completion of the boring. This water level is often misleading in case of impervious soils. In this case, the water level in adjacent lakes, etc., may serve as basis for establishing permanent ground water table (Sec. 1-14). If the construction is below water table, an estimate of quantity of flow may be desired. This can be done by pumping or bailing water out of the bore hole, determining the rate and the amount of water being pumped or bailed out, and measuring the head of water lowered by pumping or bailing.

2-13 Geologic Profile: Soil Profile

The arrangement of various soil layers can be best shown in the form of a geologic profile or a soil profile. A geologic profile is a graphical representation of underground conditions along a given line on the ground surface. In order to clearly show the various soil layers, the vertical scale is usually made larger than the horizontal scale.

A soil profile is simple to construct. First, all borings along the profile are represented by vertical lines, with the spacing of the borings drawn to a convenient horizontal scale. Along each boring, the separate soil layers are shown at the correct elevations and are clearly identified. The boundaries between identical soil layers are connected to indicate the most likely stratification. An example of a geologic profile is shown in Fig. 2-7.

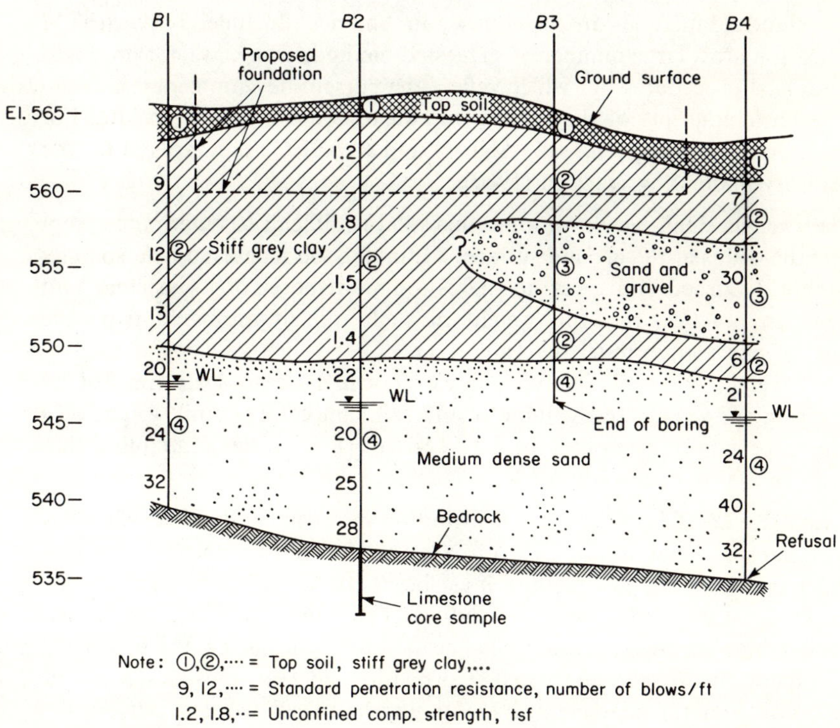

Fig. 2-7 Example of a geologic profile.

The reliability of a geologic profile as compared to the actual soil condition depends upon the nature of the ground and the spacing of the borings. If the soil conditions are erratic, the arrangement of various layers between the borings may differ considerably from the interpolation. Many natural soil deposits, notably glacial drifts, are extremely erratic. In such cases, the soil profile should be used with that point in mind.

On the soil profile, the ground water level, existing constructions, and the proposed structure should also be indicated. It is also helpful if the essential engineering data, such as the standard penetration resistance, unconfined compression strength, etc., are indicated on the profile, as shown in Fig. 2-7.

2-14 *Common Soil Tests*

Standard methods for testing soils for engineering purposes have been established by the American Society for Testing and Materials and the American Association of State Highway Officials. They are included in the following publications.

Procedures for Testing Soils (ASTM, 1958).

Highway Materials (AASHO, Parts I and II, 1955, Part III, 1958).

These standard methods are widely used in America. Included in the ASTM publication are a large number of suggested methods which are not proposed as standard procedures but which reflect the current development.

The basic concepts of the more common tests are presented below. For more detailed discussion, see ASTM and AASHO standards and the laboratory manual by Dawson (1960).

Unit weight. Unit weight of a granular soil is difficult to determine except where the soil is at the ground surface. Granular soil recovered by a sampler is highly disturbed and gives no indication whatsoever of its original unit weight. In practice, the unit weight of such soils is estimated from the results of penetration tests (see Table 1-1).

Unit weight of a cohesive soil, however, can be readily determined by measuring the weight and volume of the soil sample. The unit weight of a plastic clay may be computed by Eqs. (1-3) to (1-8) on the assumption that the clay is 100 per cent saturated.

Grain size analysis. Grain size distribution of a soil can be determined by sieve analysis down to the size of No. 200 sieve. For determination of smaller fractions, the wet method must be used: A soil sample is dispersed thoroughly in distilled water. The soil-water mixture is well shaken so that all soil grains are in suspension. By means of a hydrometer, the density of the suspension can be determined. Correlation between the density of the suspension and the diameter of the grains has been worked out on the assumption that all grains are spherical.

Water content. The natural water content of a soil sample is determined by weighing the sample before and after it is dried in the oven under controlled temperature.

Liquid limit. The liquid limit of a soil is the water content at the boundary between the liquid and plastic states. The standard equipment for liquid limit test is shown in Fig. 2-8(a). A soil sample (with grains passing No. 40 sieve) is thoroughly mixed with water and is placed in the dish to a thickness of 1 in. at the bottom of the dish. A groove of $\frac{1}{2}$ in. width is cut in the middle of the sample. The dish is lifted and dropped by turning the crank. The number of drops required to close this $\frac{1}{2}$ in. groove is recorded. The liquid limit is the water content at which 25 drops of the dish will close the $\frac{1}{2}$ in. groove.

Plastic limit. The plastic limit of a soil is the water content at the boundary between the plastic and semisolid states. The water content at the boundary is arbitrarily defined as the lowest water content at which the soil can be rolled into threads $\frac{1}{8}$ in. in diameter without the threads breaking into pieces. Figure 2-8(b) shows a soil sample being rolled into threads.

Unconfined compression test. A relatively undisturbed soil sample, usually secured by means of a thin-walled tube, is subjected to an axial compression in a manner similar to the test of a concrete cylinder. For plastic clays, the unconfined compression strength is taken at 20 per cent strain of the sample. The sample of a stiff soil, however, will break before reaching the 20 per cent strain. For most practical cases, the shear strength of a cohesive soil may be taken as one-half of its unconfined compression strength (Sec. 1-8). Figure 2-8(c) shows one type of unconfined compression machine.

Direct shear test. The test is conducted by means of a shear box or other variations of this aparatus. A shear box is a sample container which is split in the mid-height. When a normal force N is applied, the force required to start the movement of the upper half of the sample with respect to the lower half is measured. This test is very useful in measuring the relationship between the shear strength S and the angle of internal friction φ of granular soils by the following equation:

$$S = N \tan \varphi$$

Consolidation test. The consolidation test is intended to provide basic information for making settlement calculations including the time rate of the settlement. An undisturbed sample is carefully trimmed and fitted into a rigid ring. Porous stones are placed on the top and bottom of the sample. Vertical load (consolidation pressure) is applied in increments and the soil moisture is allowed to escape through the porous stones. The amount of compression of the sample at various time intervals is measured by means of

(a)

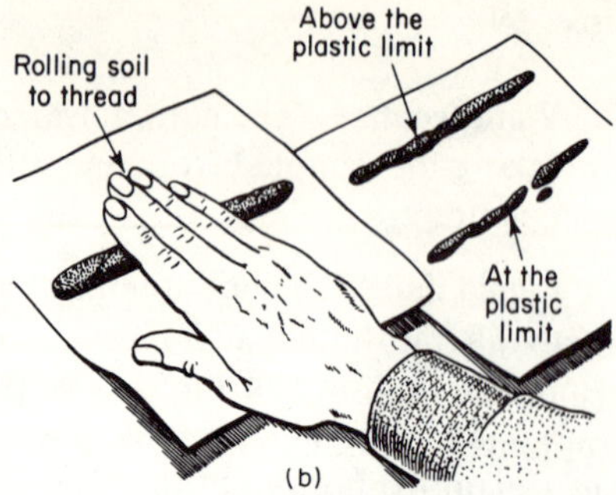

(b)

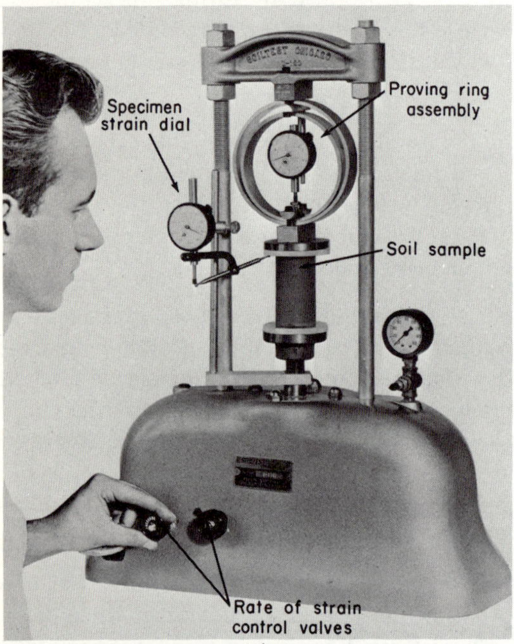

(c)

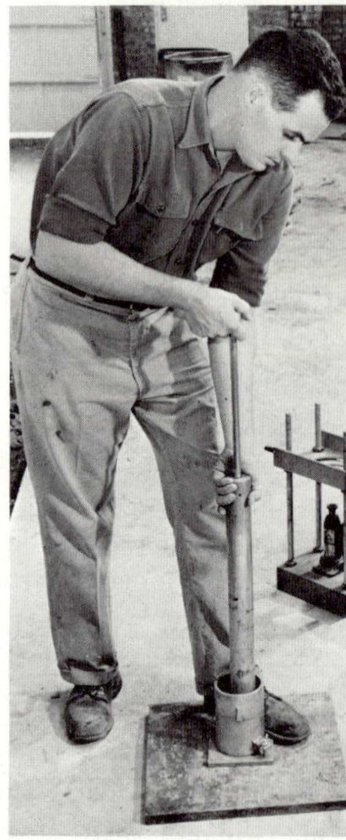

(e)

(d)

Fig. 2-8 (a) Dish used for determination of the liquid limit of soil (Test Lab Corporation); (b) testing plasticity of sample; (c) one type of unconfined compression machine (Soil Test, Inc.); (d) consolidation test (Soil Testing Services, Inc.); (e) compaction test (Soil Testing Services, Inc.)

a dial micrometer, Fig. 2-8(d). The results of this test are plotted in the form of *e-p* or *e*-log *p* graphs, *p* being the consolidation pressure and *e* being the corresponding void ratio of the soil (Sec. 3-5).

Compaction test. The purpose of a laboratory compaction test is to determine the moisture density relationship of a representative soil sample when compacted in a mold of a given size with a hammer of a given weight dropping a given height, Fig. 2-8(e). Several alternative procedures are available, each corresponding to a specific compactive effort.

Several samples of a soil are prepared at different water contents. Each sample is compacted according to the specified procedures. The unit dry weight and the water content of each compacted sample are determined, and the results are plotted in a graph similar to that shown in Fig. 2-9. This relationship is very useful in controlling artificially compacted fills (Sec. 3-9).

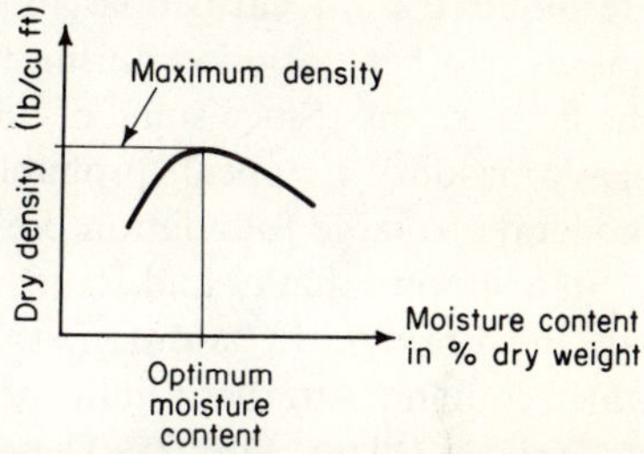

Fig. 2-9 Moisture-density relationship.

2-15 *Example of a Soil Exploration Program*

Soil exploration programs vary with the size and nature of the projects, the geologic conditions of the sites, and the type of foundations to be selected. Since the geologic (subsoil) conditions of the site are generally little known or completely unknown at the time of planning the soil exploration, there is no hard and fast rule for planning these programs. Any program in fact should be adjusted as soon as part of the boring and test is done. The following discussion and Design Example 2 illustrate a common case of soil exploration for buildings.

A scientific research laboratory was proposed in a large city in the northern U.S. It was a three-story reinforced concrete structure with an intention of having a partial basements. All floors were subjected to heavy equipment loading. With typical column spacings of 20 ft × 24 ft, the interior column footing is subjected to a total load of 215,000 lb. The live load was calculated with a permissible reduction in accordance with the City building codes. This structure is considered an ordinary institutional type which can withstand the standard differential settlement (Sec. 3-6).

The first step in planning consists of a library research. Published information concerning the general area of the site was studied (Sec. 2-3). It was found that the subsoil in this area was of glacial origin and the bedrock was limestone probably 35 to 60 ft below the ground surface. There were published articles concerning foundation conditions of several large buildings in this city. All of these buildings were supported on deep foundations carried to bedrock or hardpan.

The engineer made a trip to the site and found no heavy buildings in the close vicinity. All houses adjacent to the site were in good condition and showed no settlement cracks. He also noticed that the invert elevation of the sewer lines in the vicinity was higher than the level of the proposed basement. He inquired the local people and discovered that basements of the existing houses had been flooded during heavy rains due to the water back-up from the floor drains. Since some of the houses in the vicinity were constructed of heavy masonry, it appears probable that the subsoil was capable of supporting moderate to large foundation pressure.

Split spoon samples and standard penetration tests were required throughout the borings. In addition, thin-walled tube samples were required for water content, Atterberg limits and unconfined compression tests for soft to stiff clays (Plate DE 2). These informations were considered sufficient for determination of the feasibility of spread footings and the allowable bearing pressure. If the first portion of the exploration indicates the necessity of taking sample for consolidation tests, 3-in., thin-walled tube samples should be added.

To preclude the possibility of inadequate design due to a compressible layer at greater depth and to provide information for comparison between shallow spread footings versus deep foundations, two deep borings were scheduled to take 5 ft core samples from the bedrock. It was generally known in this area that the bedrock was sound and competent for supporting heavy foundation load. Should these two core borings indicate large difference in bedrocks elevation, additional core samples may be desirable.

It was thought the proposed basement would not require excessive construction cost if overhead plumbing could be used, provided that (1) the ground water was below the basement level, and/or (2) the soil adjacent to and immediately below the basement is relatively impervious. The procedures for measuring ground water level were outlined in the specifications.

DE 2
Soil Exploration
Sh. 1 of 1

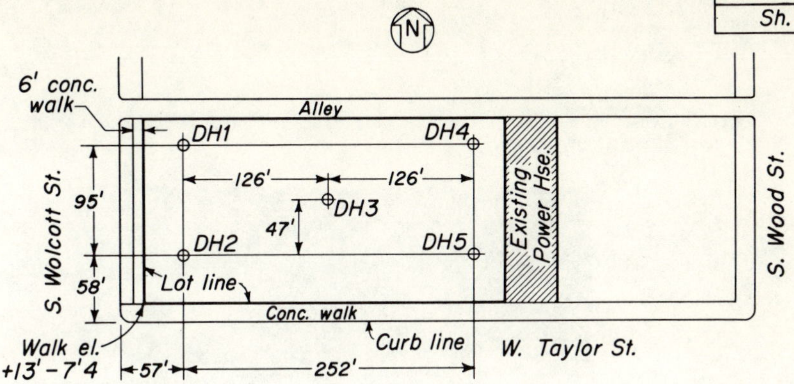

Plan of Soil Borings

All boring operations shall be done under the supervision of engineer

Depth of borings DH2, DH3, DH4, = 25' each.
DH1 and DH5 = 5' into bedrock

Type of sampling. One Shelby tube sample from each stratum of soft, medium and stiff clays in borings DH1 and DH5. Standard penetration method for balance of borings

Standard penetration method 2" O.D. 1 3/8" I.D. split spoon driven by 140 lb falling 30". One sample from each soil stratum but not farther apart than 5' intervals. Preserve one jar sample for each stratum, seal jars with parafin and submit them to engineer

Shelby tube samples. 2" diameter seamless 16 gage steel tubing with beveled cutting edge to be pushed into clay. Seal both ends with parafin and deliver them to a designated soils lab. for the following tests:
Natural moisture content, liquid limit, plastic limit and unconfined compression test

Core borings. Standard diamond drill method to obtain continuous core samples in soft rocks, use split spoon to obtain sample as much as practicable

Ground water levels. Use casing if hole caves in.

(1) In all cases: Measure water level 24 hr after completion of each boring

(2) In silty sand or silt: Fill the bore hole and dewater it successively. After each stage of dewatering, measure the water level and determine whether the level is rising or falling. The true water level is in between a rising and a falling one

(3) In clay: Install ground water observation pipe as required and directed by engineer

Plate Three

Building Cracks Caused by Excessive Foundation Settlement

Any material when stressed will experience a corresponding strain. The total vertical strain of foundation soil is called settlement. Settlement becomes objectionable when it impairs the strength or the appearance of the structure.

A foundation, like any other parts of a structure, must be designed against the danger of failure and the possibility of excessive settlement. Hence, the prerequisites for a satisfactory foundation design are (1) reasonably accurate calculation of loads, (2) adequate margin against collapse, and (3) free from settlement damages. These factors are discussed in this chapter.

3 LOADS, BEARING CAPACITY, AND SETTLEMENT

3-1 Types of Loads

A structure may be subjected to a combination of some or all of the following loads and forces.

Dead load includes the weight of the structure and all material permanently attached to it, such as the floor finish, exterior walls, and fireproofing. Permanent and fixed service equipment such as plumbing stacks and risers, electric feeders, heating, ventilating, and air conditioning systems, are usually considered as part of the dead load. If the weight of earth is directly supported by elements of the structure, it should be considered as dead load.

Live load includes all vertical loads that are not a permanent part of the structure but are expected to superimpose on the structure during a part or all of its useful life. Vertical loads due to wind or snow are not considered as live load. Human occupancy, partition walls, furniture, warehouse goods, and mechanical equipment are major live loads. The magnitude of live load to be used in the design of various buildings is usually stipulated in local building codes.*

Railroad and highway bridges as well as other structures subjected to traffic loadings are designed for the Cooper E loading and the highway H loading. Industrial floors subjected to a special type of industrial truck

* The values listed in the *American Standard Building Code Requirements For Minimum Design Loads in Buildings and Other Structures* (New York: American Standards Association) represent the most current data and should be consulted in establishing realistic live loads.

traffic must be designed to suit each specific truck loading. Reaction from industrial cranes sometimes constitutes a large portion of the live load.

The live loads due to human occupancy including furniture and appliances are often reduced for the design of long girders, columns and foundations. The amount of reduction varies with the floor area and number of floors. At any given time a large area is unlikely to be subjected to the full load stipulated in the codes. It is indeed highly improbable that every floor in a multistory building will have a full load at the same time. In each case, the local building codes should be consulted regarding the permissible reduction in live load.

It is generally considered unnecessary to add the impact effect to the foundation loads unless they are transmitted directly to the foundation.

Wind load acts on all exposed surfaces of structure. Overhanging parts are subjected to uplift pressure. The magnitude of design pressure is usually stipulated in local building codes.*

Snow load acts on ordinary roofs, either flat, pitched, or curved. The design load per square foot of horizontal projection of the roof is given in local building codes.*

Earth pressure is a lateral force acting permanently against the portion of substructure below ground surface. It should be treated as a basic load, similar to dead load. The calculation of earth pressure against basement walls is discussed in Chapter 4.

In the cases where the ground surface on one side of the structure is considerably higher than the other side, the stability of the structure due to the unbalanced earth pressure should be analyzed. The principle of analysis is very similar to that of retaining walls.

Water pressure may act laterally against basement walls and vertically against base slabs. Considering the substructure as a whole, the lateral hydrostatic pressure is always balanced, but the hydrostatic uplift or buoyancy force must be counteracted by the dead load of the structure. If the dead load is insufficient, some provision must be made to anchor the structure. The uplift force is often a problem during construction before all the dead load is available. In such cases, the basement may be flooded during the high water stage, or the site may be dewatered to a level such that the total hydrostatic uplift does not exceed the dead load.

Structures constructed in a swift river, lakefront, etc., are also subjected to forces due to current flow, ice floe, and wave forces. See Chapter 4 for discussion.

---

* In the absence of such information, consult the American Standards Association, work cited.

Earthquake motion may result in lateral force. In some areas of the world the earth's crust is unstable and undergoes readjustments from time to time. The readjustments occur in sudden movements known as earthquakes. Every structure constructed in these areas must be designed to resist the lateral force (inertia force) due to this motion. This lateral force may act on the structure in any horizontal direction. The effect of this lateral force may be evaluated in accordance with the *Uniform Building Code* (International Conference of Building Officials, 1961):

$$V = ZKCW$$

$$M = J \Sigma F_x h_x$$

where V = base shear or the total lateral force at the base of the structure.

M = base moment or the overturning moment at the base of the structure.

Z = coefficient depending on the severity of earthquake;*
= 0 for Zone No. 0,
= $\frac{1}{4}$ for Zone No. 1,
= $\frac{1}{2}$ for Zone No. 2,
= 1.0 for Zone No. 3.

K = numerical coefficient.*

$C = 0.05/\sqrt[3]{T}$, T being the fundamental period of vibration of the structure in seconds in the direction considered. The value of T should be properly substantiated by technical data. In the absence of such data, the value of T may be determined by: $T = 0.05H/\sqrt{D}$ (H = height of the main portion of the building in feet, D = dimension of the building in feet, in a direction parallel to the applied force). Use $T = 0.1 \times$ the total number of stories above exterior grade in all buildings in which the lateral resisting system consists of a moment-resisting space frame, provided that (1) the space frame resists 100 per cent of the required lateral force, and (2) this space frame is not enclosed by, or, adjoined by, more rigid elements which tend to prevent the frame from resisting lateral forces. (For the purpose of computing C, the value of T need not be less than 0.1 seconds.)

W = total dead load (including 25 per cent of the floor live load in storage and warehouse occupancies).

---

* See Earthquake Zones Map and accompanying table in Soil Maps of Continental U.S.A. at the end of this book.

$J = 0.5/\sqrt[3]{T^2}$. The required value of J should be not less than 0.33 nor more than 1.00.

F_x = lateral forces applied to a level designated as x, as computed by the following formula:

$$F_x = V\frac{W_x h_x}{\Sigma\, Wh}$$

where W_x = the portion of dead load which is located at the level designated as x.

h_x = height in feet above the base to the level designated as x.

The total base shear is assumed to distribute to the various resisting elements in proportion to their rigidities considering the rigidity of the horizontal bracing system as well as the rigidities of the vertical resisting elements.

Horizontal torsional moments due to an eccentricity between the center of mass and the center of rigidity should be taken into consideration.

Individual pile or caisson footings of every structure should be interconnected by ties. Each tie should be designed to carry a horizontal tension or compression equal to 10 per cent of the larger pile cap (or caisson) loading.

3-2 *Calculation of Loads*

Total loads acting on the foundation are calculated in three categories:

1. Normal load = dead load + snow load + live load (after reduction according to building codes) + vertical reaction due to lateral earth pressure* − buoyancy (if below permanent ground water and if water level is not lowered by subdrains). Traffic load is considered in the live load group.

The dead and live loads on columns are usually computed by tributary area method. This method assumes that a column carries all the load in a floor area which is enclosed by points at equal distance between adjacent columns. This assumption is considered sufficiently accurate for ordinary building frames. If long cantilevers, exceptionally unequal column spacings, or other unusual framings are used, a more accurate method may be necessary.

2. Maximum and minimum loads = dead load + live load (assume live load equal to zero when calculating minimum load) + vertical load due to unbalanced earth pressure, wind pressure, crane loading, etc. − buoyancy (below high water level or low water level). In earthquake zones the force arising from earthquake motion should replace that due to wind. This

* For example, vertical reaction against the bottom of a retaining wall due to lateral earth pressure.

assumes that the maximum wind pressure and the maximum earthquake motion do not occur simultaneously.

3. Horizontal load = wind pressure + horizontal force from traveling crane, etc. + unbalanced earth pressure. In earthquake zones, forces arising from earthquake motion is an important factor. The relative amount of horizontal load carried by each column depends on the method of framing.

3-3 Bearing Capacity

The conventional method of foundation design is based on the concept of bearing capacity, or allowable bearing pressure, of the soil. The bearing capacity is defined as the load or pressure developed under the foundation without introducing damaging movements in the foundation and in the superstructure supported on the foundation. Since damaging movements may result from foundation failure (collapse) as well as from excessive settlement, the following criteria must always be used in evaluating the bearing capacity.

1. Adequate factor of safety against failure (collapse).
2. Adequate margin against excessive settlements.

Although failures or collapses of foundation have been reported from time to time, by far the most common difficulty of foundations arises from excessive settlements. Therefore, this criteria warrants skillful and careful attention of the practicing engineer, Sec. 3-4 through 3-7.

In order to be able to provide an adequate factor of safety against foundation collapse, the so-called ultimate bearing capacity must be known. Usually a factor of safety of 3 is used for maximum load normally expected to act upon the foundation, as calculated by the first method in Sec. 3-2. A factor of safety of not less than 2 should be used for the maximum loads ever to be expected.

Theories of bearing capacity are developed on the principle of ultimate design in which a certain surface of failure (surface of rupture, or slip surface) is assumed. For spread footings, two common types of slip surface have been used: logarithmic spiral surface and circular surface. The spiral surface was used by Terzaghi in his classical work on shallow foundations and was later extended by Meyerhof to deep foundations, foundations on slope, and foundations subjected to inclined and eccentric loads. The validity of spiral surface has been proved by Jumikis (1956). The spiral surface reduces to a circular arc if the soil exhibits no frictional resistance ($\varphi = 0$).

Figure 3-1(a) illustrates the basic principle of the Terzaghi bearing capacity theory. When a load Q is applied centrally on the footing, the footing undergoes a certain amount of elastic and plastic settlement. An increase in the load Q tends to push the footing down, and a mass of soil within the triangle *abc* also moves down with the footing. This downward movement is resisted by the weight of the soil in the sliding wedges *acde* and *bcde* and by the shear resistance of the soil along the slip surfaces *cde*. For each set of assumed slip surfaces we can compute the corresponding load Q

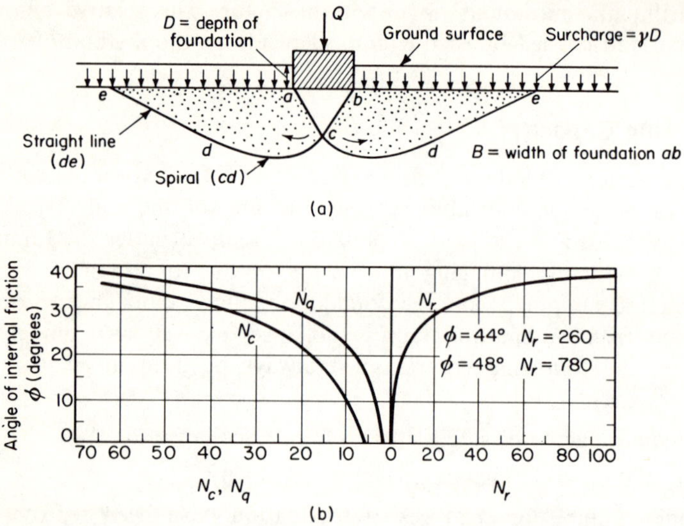

Fig. 3-1 Terzaghi bearing capacity theory: (a) slip surface; (b) bearing capacity factors in Eqs. (3-1a) and (3-1b).

that is necessary to cause the failure. The set of true slip surfaces is one that has the least resistance or requires smallest load Q. Terzaghi has expressed the bearing capacity values in the following general approximate equations (Terzaghi, 1942; Meyerhof, 1951).

Continuous footings:

$$q_{\text{ult}} = \frac{Q}{A} = cN_c + \gamma DN_q + 0.5\gamma BN_\gamma \tag{3-1a}$$

Square and circular footings:

$$q_{\text{ult}} = \frac{Q}{A} = 1.3cN_c + \gamma DN_q + 0.6\gamma BN_\gamma \tag{3-1b}$$

where q_{ult} = ultimate bearing capacity, psf;

Q = ultimate bearing capacity, lb;

A = area of footing, sq ft;

c = cohesion of soil, psf;

γ = moist unit weight of soil if above the water level;

= buoyant weight (submerged unit weight) if below the water level;

D = depth of foundation measuring from low side of the ground surface to bottom of footing;

B = width of footing (for rectangular footing B = the smaller side, for circular footing B = diameter);

N_c, N_q, N_γ, = Terzaghi's bearing capacity factors, see Fig. 3-1(b).

Research has extended Terzaghi bearing pressure formulae to include the effect of the shape and the depth of footing as well as the inclination of loads (Hansen, J. Brinch, 1957).

For granular soils: ($c = 0$):

$$q_{\text{ult}} = \frac{1}{2}\gamma N_\gamma B\left(1 - 0.3\frac{B}{L}\right)\left(1 - 1.5\frac{H}{V}\right)^2 + \gamma D N_q\left(1 + 0.2\frac{B}{L}\right)\left(1 + 0.1\frac{D}{B}\right)\left(1 - 1.5\frac{H}{V}\right) \qquad (3\text{-}2)$$

limitations: $B \leqq L$, $D \leqq 15B$, and $H \leqq V \tan\varphi$

For cohesive soils: ($\phi = 0$):

$$q_{\text{ult}} = 5c\left(1 + 0.2\frac{B}{L}\right)\left(1 + 0.2\frac{D}{B}\right)\left(1 - 1.3\frac{H}{V}\right) + \gamma D \qquad (3\text{-}3)$$

limitations: $B \leqq L$, $D \leqq 2.5B$ and $H \leqq 0.4\overline{V}$

where q_{ult} = ultimate bearing capacity = V/BL, psf;

γ = unit weight of soil, pcf;

BL = width × length of footing, ft;

H, V = horizontal and vertical components of the load acting on the footing, lb;

D = depth of footing measuring from ground surface to bottom of footing, ft;

N_γ, N_q = bearing capacity factors of soil (see Fig. 3-1);

φ = angle of internal friction of soil;

c = cohesion of soil = $\frac{1}{2}$ × unconfined compression strength, psf.

The calculation is greatly simplified if the slip surface is assumed to be a circular arc instead of a logarithmic spiral. For this reason, slip circle method is commonly used for analyzing the bearing capacity of multi-layer soil systems. It also gives results practically identical to Terzaghi's equation in the case of saturated clays loaded quickly. However, this method is extremely inaccurate for pure granular soils.

3-4 Causes of Settlement

Foundation settlement may be caused by some or a combination of the following reasons:

1. *Elastic compression* of the foundation and the underlying soil.

2. *Inelastic* (*plastic*) *compression* of the underlying soil.

In practice the elastic and inelastic compression of granular soils is determined by empirical rules, Chapter 6. The stress-strain relations of fine-grained soils may be determined by laboratory tests. In general, the inelastic compression is much larger than the elastic compression. The inelastic compression of fine-grained soils can be reasonably predicated by the Theory of Consolidation as discussed in Sec. 3-5.

3. *Ground water lowering;* especially repeated lowering and raising of water level in loose granular soils tends to compact the soil and cause settlement of the ground surface. Prolonged lowering of water level in fine-grained soils may introduce settlement because of the extrusion of water from the voids. Pumping water or draining water by tiles or pipes from granular soils without adequate filter material as protection may in a period of time carry a sufficient amount of fine particles away from the soil and cause settlement.

4. *Vibration* due to pile driving, blasting, and oscillating machineries may cause settlement in deposits of granular soils.

5. *Seasonal swelling* and shrinking of expansive clays.

6. *Ground movement on earth slopes*, e.g., surface erosion, slow creep or landslide.

7. *Other causes* such as adjacent excavation, mining subsidence, underground erosion, etc.

A certain amount of elastic and inelastic settlement of foundations is unavoidable and it should be taken into account in the design.

3-5 Consolidation

The term consolidation denotes a slow process of compression due to extrusion of water from the voids in the fine-grained soils. The consolidation characteristics of a soil are determined by laboratory tests, Sec. 2-14.

The results of consolidation tests are plotted in an *e*-*p* curve, or in an *e*-log *p* curve, Fig. 3-2, where *p* is the applied axial pressure and *e* is the corresponding void ratio of the sample. In this figure three curves are shown for the same clay. The solid lines show the relationship between the pressure and the void ratio for a so-called undisturbed sample and a completely remolded sample. These two curves meet at a point *b*. Because of the large influence of disturbance, even the most carefully conducted laboratory test does not represent the *e-p* relationship of the virgin soil in place. For a normally loaded clay of low sensitivity, the straight line *ab* is considered the true relationship between *e* and *p*. The point *a* is located by e_0 and p_0, e_0 being the void ratio of the soil in place prior to consolidation and p_0 being the overburden pressure on the soil in place.

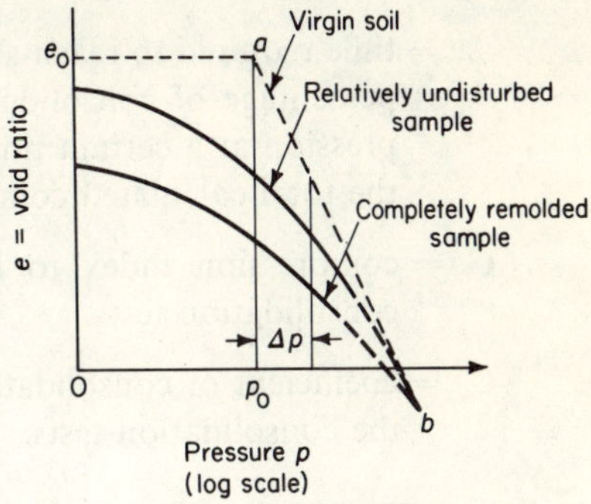

Fig. 3-2 *e*-log *p* curve.

For clays of high sensitivity and for preconsolidated clays, the procedure for constructing the virgin *e*-log *p* curves becomes more complex, and the reader is referred to standard text books on soil mechanics, e.g., Peck, et al. (1953) and a paper by Schmertman (1955).

The slope of the *e*-log *p* curve for thé virgin soil is called the *compression index*, C_c, for the soil.

$$C_c = \frac{e_0 - e}{\log (p/p_0)}$$

where e_0 and p_0 are as defined above, and *e* is the void ratio corresponding to any pressure *p* greater than p_0. Approximate value of C_c may be obtained from Eq. (1-12a) or (1-12b).

The magnitude and rate of settlement due to consolidation of normally consolidated fine-grained soils can be calculated by the following equations (Terzaghi, 1943):

$$S = \frac{e_0 - e}{1 + e_0} H$$

$$= C_c \frac{H}{1 + e_0} \log \frac{p_0 + \Delta p}{p_0} \tag{3-4}$$

$$t = \frac{T_v}{C_v} H^2 \tag{3-5}$$

where S = settlement due to consolidation.

H = thickness of the layer. If the soil is drained on top and bottom as in the consolidation test, half-thickness should be used.

e_0 = natural void ratio of the soil in place.

p_0 = weight of soil above mid-height of the consolidating layer.

Δ_p = consolidation pressure = net additional pressure.

T_v = time factor, a coefficient depending upon the percentage of consolidation (see Fig. 3-3).

t = time required to reach a certain percentage of consolidation. The percentage of consolidation is the ratio of the amount of compression at a certain time during the process of consolidation to the total calculated compression S.

C_c = compression index to be determined from the results of the consolidation tests.

C_v = coefficient of consolidation to be determined from the results of the consolidation tests.

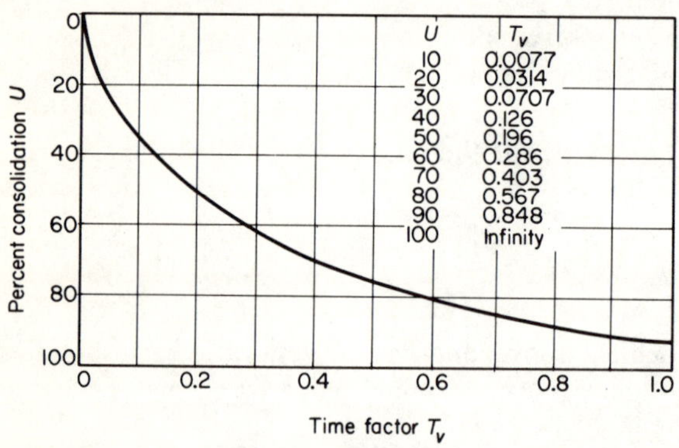

Fig. 3-3 Time factor as a function of percentage of consolidation.

After the soil has reached the calculated 100 per cent consolidation, the actual consolidation still continues. This phenomenon is called the *secondary consolidation.* There is no accurate method available for determining the amount and rate of the secondary consolidation. Experience has shown that for the majority of natural soil deposits the amount of secondary consolidation is generally relatively small as compared to the primary consolidation. However, the magnitude of secondary consolidation is often large for organic soils and most plastic clays.

Fortunately, more and more experience and settlement records are accumulated in large cities where construction is most active. This experience and records can be of great value to practicing engineers for their works in these cities.

3-6 *Differential Settlement*

Theoretically speaking, no damage will be done to a structure if it settles uniformly as a whole regardless of how large the settlement may be. The only damage would be to the connections of the underground utility lines. However, when the settlement is nonuniform, as is always the case, the difference of settlement between two adjacent columns, commonly referred to as differential settlement, may cause damage to the structure.

The amount of differential settlement that can be tolerated by a structure depends on a number of factors, including the type of construction and the column spacing. Simple-span frames can take considerably greater distortion than rigid frames. A fixed-end arch would suffer greatly if the abutments should settle or rotate. Different types of construction materials can withstand different degrees of distortion. For example, sheet metal wall panels and prefabricated curtain walls do not show distress as readily as brick masonry.

Differential settlement between foundations are a result of a number of causes. Soil characteristics are never uniform even in an apparently uniform soil deposit. The actual loads acting on foundations influence the magnitude of settlement. Therefore, it is impracticable to derive theoretical analyses for calculating accurately the differential settlement of foundations. Consequently, empirical rules have been established and found satisfactory in American practice. It is based upon the simple logic that if the maximum total settlement is kept within a reasonable limit, the differential settlement will be only a fraction depending upon the type of structure and pattern of loading (generally about three-quarters of this limit).

Types of structure	*Allowable maximum settlement (in.)*
Commercial and institutional buildings	1*
Industrial buildings	$1\frac{1}{2}$
Warehouses	2
Special machinery foundations	As required by manufacturer (often less than 0.02 in.)

Based on the same reasoning, the 1955 U.S.S.R. Building Code permits average settlements listed in Table 3-1. These values are considerably greater than those customarily allowed in the U.S.A.

The theory of elasticity indicates that the value of average settlement of a perfectly flexible foundation does not differ more than 7 per cent from that of a perfectly rigid foundation.† The difference is even smaller between the average settlement of a perfectly rigid foundation and that of a foundation of intermediate rigidity (Polshin and Tokar, 1957). Therefore, it is possible to

* Terzaghi and Peck (1948).

† Settlement of perfectly rigid foundation $= \pi/4 \times$ maximum settlement of perfectly flexible foundation.

Table 3-1 PERMISSIBLE SETTLEMENT–1955 U.S.S.R. BUILDING CODE*

Item No.	*Kind of building and type of foundation*	*Average settlement (cm)*
1.	Buildings with plain brick walls on continuous and separate foundations with the wall length L to the wall height H (H counted from the foundation footing)	
	$\frac{L}{H} \geqq 2.5$	8
	$\frac{L}{H} \leqq 1.5$	10
2.	Buildings with brick walls, reinforced with reinforced concrete or reinforced brick belts (not depending on the ratio of L/H)	15
3.	Framed buildings	10
4.	Solid reinforced concrete foundations of blast furnaces, smoke stacks, silos, water towers, etc.	30

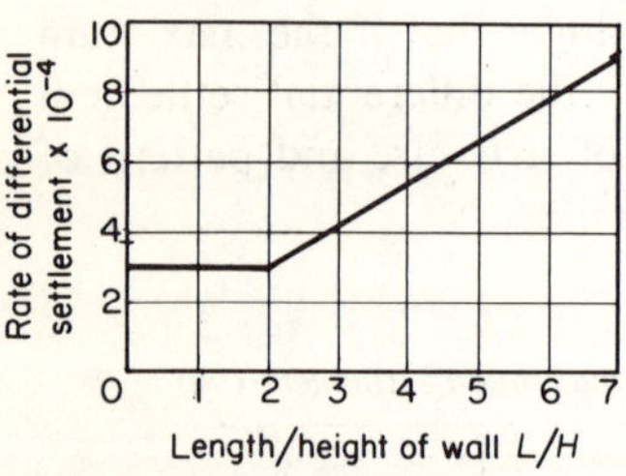

Fig. 3-4 Permissible differential settlement of brickwalls. After Polshin and Tokar, 1957.

estimate the actual average settlement if the settlement is calculated by assuming either a perfectly rigid or a perfectly flexible foundation.

Although it is often sufficient to design the foundation for the permissible total settlement, an understanding of the nature of differential settlement is helpful, and occasionally necessary. The following information is taken mostly from Polshin and Tokar (1957). Brick masonry will crack when the unit elongation amounts to 0.0005. Based on this criterion, the permissible differential settlement of brick walls is shown in Fig. 3-4, and as follows:

For $L/H \leqq 2$: Rate of differential settlement = 0.0003 in./in.

For $L/H = 8$: Rate of differential settlement = 0.0010 in./in.

Use lineal interpretation for intermediate values of L/H. The rate of differential settlement is defined as the slope, or the relative settlement, between two points divided by the horizontal distance.

The maximum differential settlements permitted by the 1955 U.S.S.R. Building Code are shown in Table 3-2.

* From Polshin and Tokar.

3-7 Calculation of Loads for Settlement Analysis

In order to keep the foundation settlement within the permissible limits, realistic loads actually acting upon the foundation must be computed. For this purpose, the loads are classified as permanent and transient types. The engineer must use his judgement in each practical case to determine what loads are permanent and what are transient. Dead load and all fixed equipment are permanent. Sometimes, one-half of the design live load is taken as being permanent. In some structures where occupancy is relatively short, such as sports stadiums, the total live load is of the transient type.

Table 3-2 MAXIMUM DIFFERENTIAL SETTLEMENT PERMITTED BY U.S.S.R. BUILDING CODE*

Item No.	*Description of standard value*	*Subsoil*	
		Sand and hard clay	*Plastic clay*
1.	Slope of crane way as well as tracks for bridge crane truck	0.003	0.003
2.	Difference in settlement of civil and industrial building column foundations:		
	(a) for steel and reinforced concrete structures,	$0.002L$	$0.002L$
	(b) for end rows of columns with brick cladding,	$0.007L$	$0.001L$
	(c) for structures where auxiliary strain does not arise during nonuniform settlement of foundations (L = distance between column centers)	$0.005L$	$0.005L$
3.	Relative deflection of plain brick walls:		
	(a) for multi-story dwellings and civil buildings at $L/H \leqq 3$	0.0003	0.0004
	at $L/H \geqq 5$	0.0005	0.0007
	(L = length of deflected part of wall; H = height of wall from foundation footing)		
	(b) for one-story mills	0.0010	0.0010
4.	Pitch of solid or ring-shaped foundations of high rigid structures (smoke stacks, water towers, silos, etc.) at the most unfavorable combination of loads	0.004	0.004

For structures supported on fine-grained soils, settlement takes place only under long time loading. Transient loading, if applied in very short duration or durations, brings little additional settlement. In such cases, the settlement should be calculated only under those loads which remain on the structure for a period, or a number of periods, long enough to cause consolidation settlement of the soil. This load may be referred to as the *service load.* On

* From Polshin and Tokar.

the other hand, structures supported on granular soils will settle almost immediately upon the application of load. In this case the service load should include the maximum load which is expected to occur at any time during the life of the structure.

Therefore, it appears reasonable to reduce the differential settlement due to live load variations by maintaining equal bearing pressure for all foundations under the service load. This may be done by the following procedure.

1. Determine the required bearing area, or the number of piles, for the column having the largest live load/dead load ratio. In the conventional method of design, the area A, or the number of piles N, is

$$A = \frac{\text{Dead load} + \text{Maximum live load}}{\text{Allowable bearing pressure}}$$

$$N = \frac{\text{Dead load} + \text{Maximum live load}}{\text{Allowable bearing capacity of pile}}$$

2. Compute for this same column the design bearing value:

$$q_d = \frac{\text{Service load}}{A} \quad \text{or} \quad \frac{\text{Service load}}{N}$$

3. Determine the area, or the number of piles, for all other columns by the use of q_d, that is,

$$\text{Bearing area, or number of piles} = \frac{\text{Service load}}{q_d}$$

3-8 Settlement Cracks

It is always helpful to differentiate settlement cracks from cracks caused by other factors such as shrinkage or structural deformation. For example, a reconnaissance site survey should always include an examination of existing buildings in the area to find out if settlement cracks have formed on the exterior walls. The engineer should also investigate the cracks on the building before underpinning or other methods of strengthening the structure or foundation is done.

Buildings often develop some cracks due to factors other than settlement. Heavy or large concrete members may have shrinkage cracks if not properly designed or constructed. Minor cracks are not uncommon in concrete structures due to improper details and faulty placement of concrete. Masonry blocks not properly made may also crack after construction and when the building is heated. Some of such cracks may be explained but others may challenge the judgement of best informed engineers. In cases where the

(a) (b) (c)

Fig. 3-5 Diagrammatic sketches showing the relationship between the nature of differential settlement and cracks.

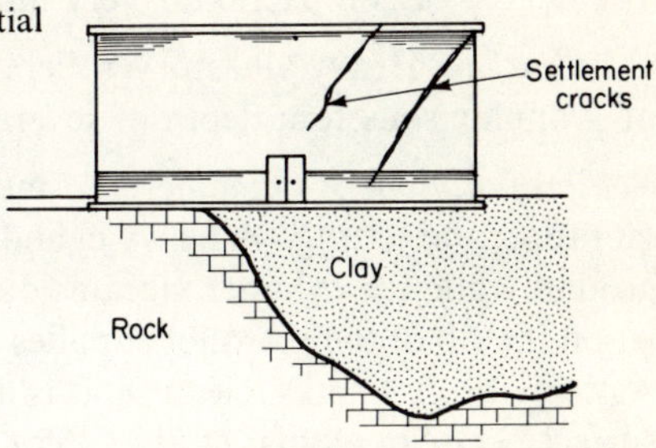

Fig. 3-6 Example of differential settlement.

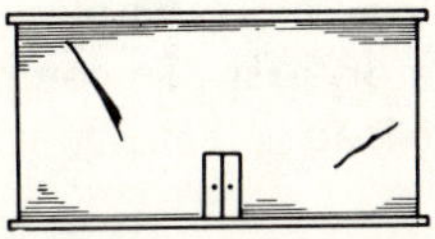

Fig. 3-7 Diagrammatic sketch showing wall cracks not caused by foundation settlement.

nature of the cracks does not lend to evident explanation, soil borings should be taken to determine the possibility and the source of settlement.

Settlement cracks are usually in diagonal position, although vertical cracks are also possible. They often start from the top, the bottom, or the end of the wall. The cracks generally start from the top if the end of the wall settles more than the rest, Figs. 3-5(a) and 3-5(b). If the middle portion of a wall settles more, the cracks open up near the bottom of the wall, Fig. 3-5(c).

When a building is supported on two different materials having large variations in compressibility, the possibility of detrimental differential settlement should be analysed, Fig. 3-6. In extreme cases an expansion joint may be necessary at the boundary of the different foundation soils.

While settlement cracks can often be traced to some type of relative movement and hence are extended to the edge of the wall, shrinkage or other types of cracks may be irregular or may terminate before reaching the edge of the wall, Fig. 3-7.

3-9 *Improving Bearing Capacity by Compaction*

Compaction may be utilized to improve the bearing capacity of natural soil deposits or man-made fills.

1. When soft or loose soil deposits are encountered, the deposits may be removed and replaced by a new compacted fill or may be compacted by various in-place compaction methods. Some of the more common methods are:

Flooding. Only very loose sand can be affected by flooding with water. The degree of compaction is very limited.

Vibration. Heavy vibratory rollers and compactors may compact a layer of granular soils to a depth of several feet.

Vibroflotation. A commercial method which combines the effect of vibration and jetting. A heavy cylinder, known as *vibroflot*, is inserted in the ground while the cylinder vibrates due to a rotary eccentric weight. A water jet on the tip of the vibroflot supplies a large amount of water under pressure. As the vibroflot sinks, clean sand is added into a crater that develops on the surface. (D'Appolonia, et al., 1955.)

Compaction by preloading. When a saturated fine-grained soil (clay or silt) is subjected to a long-time compression, this compressive force squeezes water out (consolidation) from the voids at a very slow rate. The rate and amount of consolidation can be determined by laboratory test. To compact by preloading the ground is loaded with earth fill. The load is removed when the desired percentage of consolidation is reached.

Sand drains. To accelerate the consolidation process, vertical sand drains may be installed at uniform spacing (Christie, 1959). Holes of 12 in. or greater diameter are bored and are filled with clean sand. The top of these drains are interconnected by sand trenches or blanket, Fig. 3-8.

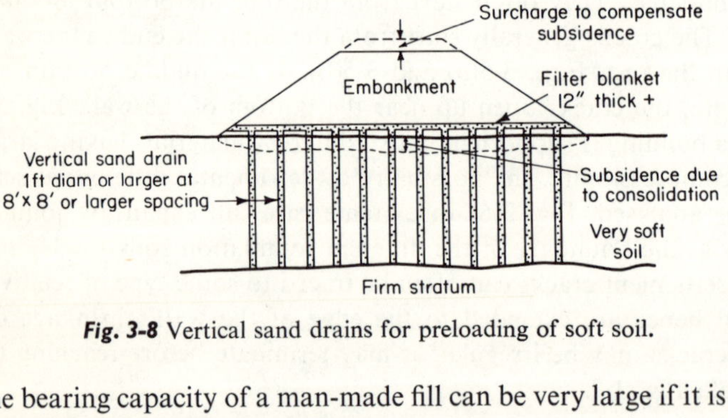

Fig. 3-8 Vertical sand drains for preloading of soft soil.

2. The bearing capacity of a man-made fill can be very large if it is properly designed and compacted. Qualitatively speaking, compaction increases the shear strength and decreases the compressibility of the soil.

The compactness of a given compacted soil is expressed in terms of percentage of compaction. Representative samples of the soil are tested in the laboratory to determine its maximum dry density under a specific compaction procedure (Sec. 2-14). The maximum dry density, say 110 pcf, is used as a basis for comparison. If the same soil is compacted in the field to a dry density of say, 100 pcf, then the percentage of compaction is said to be 100/110 = 91 per cent, or the soil is compacted to 91 per cent of maximum density. It should be noted that there are several types of laboratory compaction procedures each of which gives a different maximum density and optimum water content for a given soil. Therefore, when specifying the percentage of compaction, it is necessary to state the laboratory procedure to be used. Generally 95 per cent to 100 per cent compaction is specified for fills supporting foundations and floor slabs. In areas where settlement is of less importance, such as landscaped areas, 90 per cent compaction may suffice.

The percentage of compaction is a convenient, relative measure of the compactness, not a quantitative measure of the shear strength of any soil, although for a given soil the shear strength generally increases with the percentage of compaction. Obviously the properties are practically identical for a given type of soil if the material is compacted to the same percentage of compaction at the same water content.

Compaction of earth fill is accomplished by spreading fill materials at a moisture near the optimum water content, Fig. 2-9. If the soil taken from the borrow pit is too dry, additional water can be provided by sprinkling and mixing each layer. The thickness of each loose layer is usually 6 to 12 in. for clayey soils and up to 18 in. for granular soils. If the soil is too wet and slow drying, it is often a better economy to use material from another source.

The common types of compaction equipment are:

1. Rollers compact the soil by the large unit pressure on the contact area between the rollers and the soil.

(a) *Smooth-faced steel roller* is mainly a steel drum filled with water or sand and it may be towed by a tractor or self-propelled. This type is generally used to compact slag, rock, or coarse gravel.

(b) *Pneumatic tired roller* consists of four or more parallel wheels loaded by a large box of sand and towed by a tractor or self-propelled. The degree of compaction increases with the total wheel loads and, in particular, with the inflation pressure of the tires, ranging from 80 to more than 100 psi. This type is suitable for compacting granular soils. The pneumatic tired rollers may be mounted on wobbling axles.

(c) *Sheepsfoot roller* has series of tamping feet which break the soil lumps and compact them by the high foot pressure.

(d) *Grid roller* is an open roll covered with a grid or lacing of steel. The grid can break up hard chunks of soil and compact them.

(e) *Segmented wheel roller* is self-propelled and compact clays with the tamping action of the feet.

2. Mechanical tamper is used to compact soils in areas inaccessible to the large rollers, or where the heavy rollers are prohibited to avoid damage to the adjacent or underlying structures and utilities. Tampers are powered by compressed air, gasoline, or electricity, and vary in weight from about 30 to several thousand pounds.

3. Vibratory compactor is a cylindrical roller whose compactive effort is amplified by vibrating weight within the roller drum. The weight revolves at a constant speed powered by a gasoline engine mounted on the unit. This type of roller is very effective in compacting granular soils.

4. Earth moving equipment may be used for compaction purposes but is not considered very satisfactory.

Plate Four

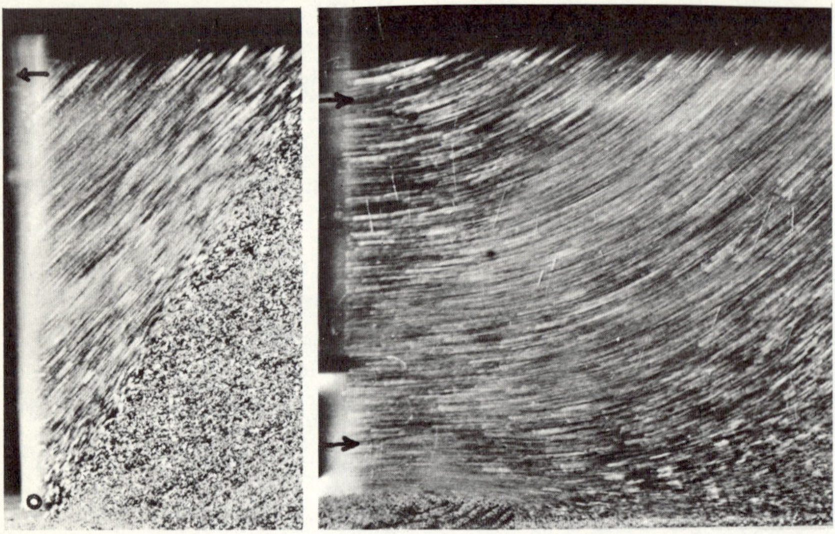

Motion of Sand Grains at the State of Failure:

(left) retaining wall is permitted to yield away from the soil (active pressure); (right) retaining wall is forced to move against the soil (passive pressure)

Substructures and foundations, such as retaining walls and basement walls, are subjected to lateral pressure where the ground level on one side differs from the ground level on the other side. Lateral pressure can also be caused by vertical loading (surcharge) applied adjacent to the substructure. The magnitude of such lateral pressure depends largely upon the characteristics of the soil and the type of retaining structure. This pressure must be calculated with sufficient accuracy in order to achieve the most economical design with an appropriate margin of safety.

In addition, substructures and foundations may be subjected to lateral pressures resulting from unbalanced water head, ice thrust, earthquake motion, and other external forces such as impact and mooring pull from vessels on water front structures.

4
LATERAL PRESSURES

4-1 Basic Concepts

A. Lateral strain. The magnitude and distribution of the earth pressure acting on a retaining structure or foundation depends largely upon the relative lateral strain of the soil behind the structure. A clear understanding of this relationship is necessary in order to compute the earth pressure correctly. In some cases an earth pressure computed on the basis of an incompatible condition of strain may err in excess of 100 per cent.

When the soil is prevented from lateral strain (expansion or contraction) by an unyielding retaining structure of great rigidity, the pressure is known as *earth pressure at rest.* Lateral pressure against basement walls is generally in this category.

If the retaining structure is permitted to move away from the soil allowing a lateral expansion of the soil, the earth pressure decreases with the increasing expansion. Further expansion will cause a shear failure of the soil in which a sliding wedge tends to move forward and downward, Fig. 4-1(a). At this state of failure the earth pressure is at the minimum value; additional deformation does not reduce the earth pressure any further. This minimum pressure is known as *active earth pressure.*

On the other hand if the retaining structure is forced to move backward toward the soil causing a lateral contraction of the soil, the force required to start the movement is greater than the earth pressure against a rigid and unyielding wall. A larger force is required to move a greater distance until a state of failure is reached where a sliding wedge is formed, Fig. 4-1(b). This wedge of soil moves backwards and upwards with respect to its original position. At this state of failure the earth pressure is at a maximum value

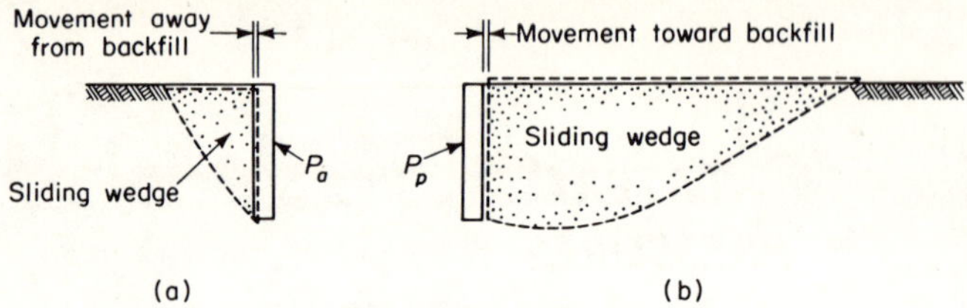

Fig. 4-1 (a) Active pressure; (b) passive pressure.

known as *passive earth pressure* or *passive resistance.* After this stage, no greater force is required to introduce further movement of the wedge.

The amount of movement of the retaining structure is sometimes called the yield. The relationship between the yield and the magnitude of earth pressure of sand is shown in Fig. 4-2. In this figure, the unit earth pressure at depth Z is equal to Kp, where K is, as discussed later, the ratio $p/\gamma Z$, γ = the unit weight of soil, Z = the depth, and p = the lateral earth pressure at depth Z. It is seen that the yield required to permit the lateral pressure to

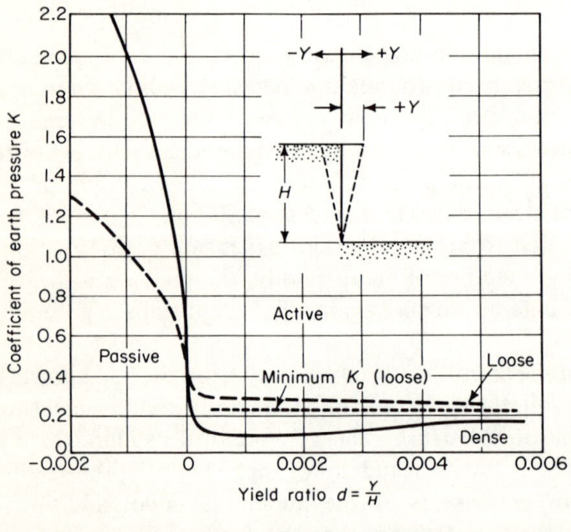

Fig. 4-2 Influence of wall movement on intensity of earth pressure. From Terzaghi.

reduce to the active earth pressure is rather small, being less than one-half of one per cent of the height (or the depth below surface of backfill). Retaining walls can readily tilt a fraction of an inch or more due to compression of soil under large toe pressure, unless the wall is supported directly on hard rock.

Therefore it is imperative in practical problems to determine the type of movement at the beginning of design. For structures that do not yield or deflect at all, the earth pressure at rest should be used. Active earth pressure is the minimum value to which a given structure will be subjected. It can be

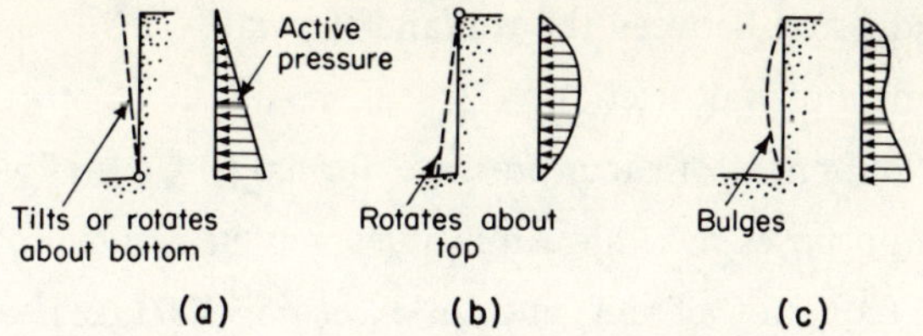

Fig. 4-3 Distribution of earth pressure.

used for design of ordinary retaining walls which would slide or tilt an amount equal to or greater than the minimum yield. For structures that deflect or move in a different fashion, the magnitude and distribution are also different. Figure 4-3 illustrates some notable examples.

If the retaining structure is prevented from outward movement near the ground surface but capable of large yield at the lower depth, the lateral pressure approaches the earth pressure at rest at the top and reduces to a very small or zero pressure at the bottom as shown in Fig. 4-3(b). In practice, this is often the case of braced or strutted open cut, Chapter 13. The top tier of struts (or bracing) prevents the lateral movement of soil from the beginning of excavation. As excavation progresses, the lower face moves inward to the excavation before additional struts are placed.

Another type of yield is shown in Fig. 4-3(c), representing a flexible sheet pile wall which is anchored at the top by tie rods and at the bottom by being embedded in the soil. The lateral pressure is small near the middle of wall due to lateral deflection of the flexible sheet piling.

B. Wall friction and adhesion. At the state of failure, the sliding wedge moves forward as well as downward as shown in Fig. 4-1(a). If the retaining structure is permitted outward movement but is held rigidly to prevent downward movement, there is a relative vertical movement between the back of the wall and the soil. Furthermore, if the back of the wall is rough and the shearing strain is large, the shear stress along the back of the wall may be equal to the total shear strength of the soil: $s = c + \sigma \tan \varphi$. In such a case, the back of the wall is said to be perfectly rough. On the other hand, if the back of the wall is smooth or if there is little relative shearing strain between the wall and the soil, there is no shearing resistance along the back of the wall, and the earth pressure acts perpendicular to the back of the wall. In this case, the wall is said to be perfectly smooth. In design work, the magnitude of wall friction must be established according to the anticipated condition. The shearing resistance s_w along the back of the wall is usually assumed to consist of two parts, namely:

$$s_w = c_a + \sigma \tan \delta$$

where s_w = shearing resistance along the back of wall,

c_a = adhesion between the soil and the wall,

δ = angle of wall friction,

σ = component of earth pressure normal to back of wall.

For design purposes the following values may be assumed:

$c_a = c$ (cohesion of soil) but not exceeding 1000 psf;

$\delta = 30°$ steel pile coated with tar or bitumen,

$= 20°$ concrete or brick walls,

$= 15°$ uncoated steel sheet pile,

$= 0°$ if the wall tends to move downward together with the soil,

$= 0°$ sheetpiling with small penetration or penetrated in soft or loose soil,

$= 0°$ if backfill is subjected to vibration.

C. Hydrostatic pressure. When a part or the entire depth of soil behind the retaining structure is submerged, the lateral pressure is considered to comprise of two components: one due to the hydrostatic pressure and the other due to the buoyant weight of soil. The buoyant weight of soil below water level is equal to the weight of soil particles in the atmosphere minus the weight of water displaced by these particles. The combined pressure is shown in Fig. 4-4.

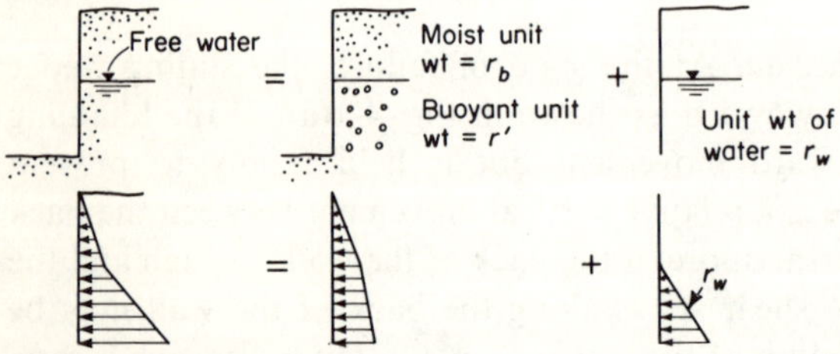

Fig. 4-4 Lateral pressure below water level.

If the water level is equal on both sides of the retaining structure, the net hydrostatic pressure is zero, of course.

4-2 *Earth Pressure Theories*

Earth pressure theories may be classified into four categories:

1. *The theory of elasticity* which is commonly used to calculate the vertical and lateral pressures within a mass of soil due to surcharge loads;
2. *The theory of plasticity* which was utilized in the Rankine theory;
3. *The wedge theory* which was first developed by Coulomb and later extended to more general conditions;

4. *Empirical rules* which have been derived for the design of highly indeterminate substructures such as anchored sheet piles and open cut bracings.

Both the Rankine theory and the wedge theory deal with a soil mass at a state of failure. Hence, these theoretical values exist only when the soil mass fails by internal shear. When the retaining structure is incapable of yielding sufficiently to permit such shear failure, the Rankine and wedge theories will give erroneous results, as already discussed.

4-3 Rankine Theory

Rankine theory deals with earth pressure within a soil mass under the following conditions:

1. The ground surface is a straight line (horizontal or sloping surface).
2. The soil mass is in the so-called Rankine state.

When a soil mass is allowed to expand (active earth pressures) or contract (passive earth pressure), rupture surfaces will form within the mass. If not interrupted by the back of retaining wall or other structure, these rupture surfaces will be a series of straight lines making an angle i with the horizontal:

$$\left.\begin{aligned} &\textit{Active earth pressure:} && i = 45 + \frac{\varphi}{2} \\ &\textit{Passive earth pressure:} && i = 45 - \frac{\varphi}{2} \end{aligned}\right\} \tag{4-1}$$

When the state above exists, the soil is said to be in the Rankine state, and the Rankine theory is applicable:

$$p_a = qK_a - 2c\sqrt{K_a} \tag{4-2}$$

$$p_p = qK_p + 2c\sqrt{K_p} \tag{4-3}$$

where p_a and p_p = unit active and passive earth pressure, respectively, at a depth Z;

q = vertical pressure or load due to the weight of soil above Z, using submerged weight for the portion below ground water level;

c = cohesive strength of the soil;

K_a and K_p = coefficient of active and passive earth pressure, respectively.

The values of K_a and K_p are shown in Fig. 4-5. It is seen that

$$K_a = \frac{1}{K_p} \tag{4-4}$$

An example of computation of earth pressure by Rankine theory is given in Fig. 4-6.

The value of K_0 (earth pressure at rest) is difficult to determine. Test results indicate:

$$\begin{aligned} K_0 &= 0.35 - 0.60 \quad &&\text{for sand and gravel} \\ &= 0.45 - 0.75 &&\text{for clay and silt} \\ &= 1.0 &&\text{for overconsolidated clays} \end{aligned}$$

According to Rankine theory, earth pressure increases lineally with the depth in the same manner as the lateral pressure exerted by a fluid. For this reason engineers often refer the design earth pressure as *equivalent fluid pressure*.

4-4 *Wedge Theory*

The principle of the wedge theory is illustrated in Fig. 4-7. It deals with a sliding wedge bounded by the ground surface on top, the back of the retaining structure on one side and the surface of rupture on the other side. The sliding surface may be assumed to be a plane as represented by the straight line *ef*, Fig. 4-7(a). If the soil possesses cohesion, vertical tension cracks are formed from the ground surface to a depth

$$H_0 = \frac{2c}{\gamma} \tan\left(45 + \frac{\varphi}{2}\right)$$

γ being the unit weight of soil. The sliding wedge *abfde* is analysed as a free body.

Since *ab* and *de* are cracks there would be no external force acting on these surfaces. As the wedge is moving downward along the sliding surface *fe*, there is a cohesive resistance C equal to the unit cohesion c times the length of the surface *fe*, and a frictional resistance R acting at an angle φ with the perpendicular line and being opposite to the movement of the wedge. The value of c and φ may be determined by soil tests but the magnitude of R is unknown. The forces acting against surface *bf* are the adhesion C_a and the frictional resistance P_a acting at an angle δ with the normal to *bf*. The value of C_a and δ must be estimated before the magnitude of P_a can be determined. The commonly used values of C_a and δ are discussed in Sec. 4-1(b).

The weight W of the sliding wedge *afde* is resisted by the cohesion C, the frictional resistance R, the adhesion C_a, and the active earth pressure P_a. All

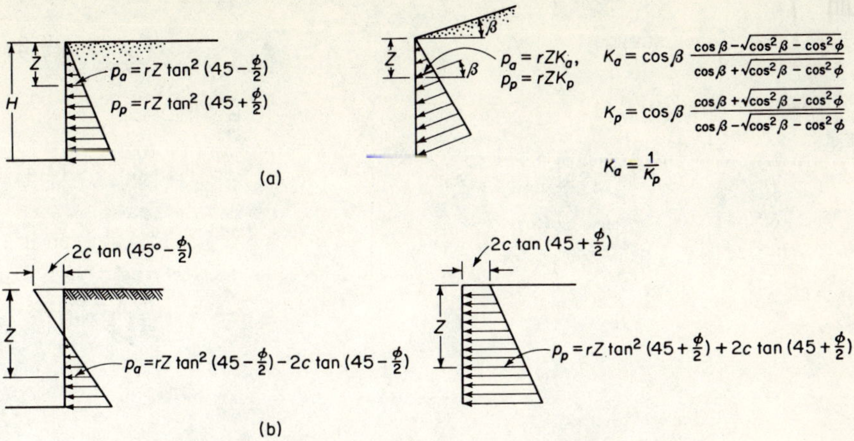

Fig. 4-5 Summary of commonly used Rankine formulae: (a) granular soil; (b) cohesive soil.

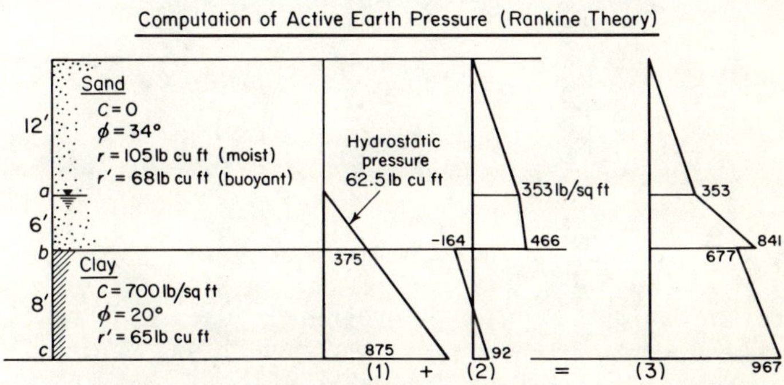

At a (12′ below ground surface):

$q = r \times 12 = 105 \times 12 = 1260$ lb/sq ft

$K_a = \tan^2 (45 - \phi/2) = \tan^2 (45 - 34/2) = 0.28$

$p_a = K_a q = 0.28 \times 1260 = 353$ lb/sq ft

At b (18′ below ground surface, in sand):

$q = 1260 + r' \times 6 = 1260 + 68 \times 6 = 1668$ lb/sq ft

$K_a = 0.28$

$p_a = 0.28q = 466$ lb/sq ft

At b (18′ below ground surface, in clay):

$q = 1668$ lb/sq ft

$\tan (45 - \phi/2) = 0.70$

$p_a = q \tan^2 (45 - \phi/2) - 2c \tan (45 - \phi/2)$

$= 1668 \times 0.49 - 2 \times 700 \tan (45 - 20/2) = -164$ lb/sq ft

At c (26′ below ground surface):

$q = 1668 + 65 \times 8 = 2188$ lb/sq ft

$p_a = 2188 \times 0.49 - 2 \times 700 \tan (45 - \phi/2) = 92$

Fig. 4-6 Numerical example (Rankine theory).

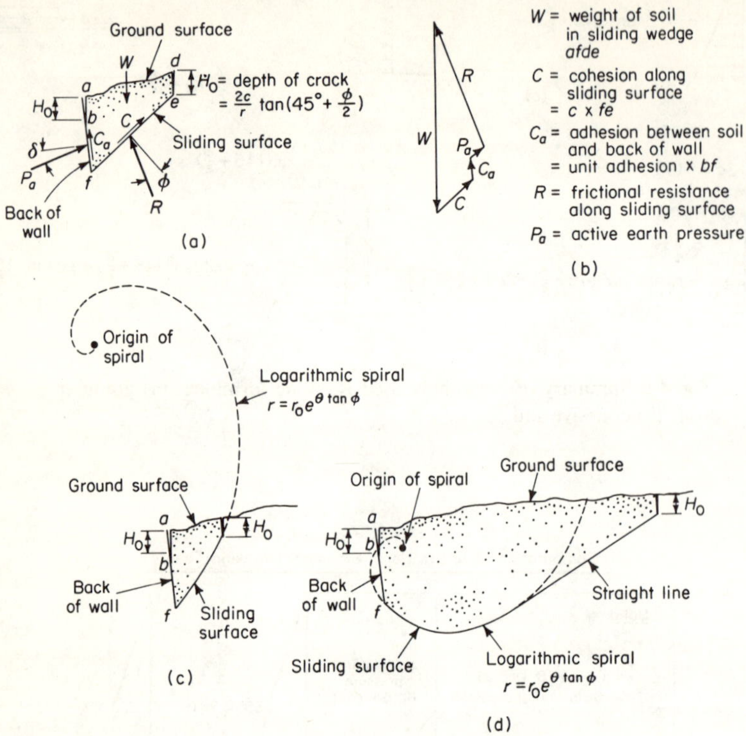

Fig. 4-7 Wedge theory.

the forces are known except for the magnitude of R and P_a which can be readily determined by graphical method as shown in Fig. 4-7(b).

For every assumed position of sliding surface there is a corresponding value of P_a. After several trial wedges, it is possible to determine the most critical surface of sliding *ef* which requires the largest value of P_a. This value is the active earth pressure. A number of short-cut methods have been developed and discussed in textbooks on soil mechanics.

Although the illustration above is made for active earth pressure, it can be applied to the case of passive earth pressure. The only difference is that the direction of all forces on *bf* and *fe* must be reversed because the passive wedge moves in the opposite direction with the active wedge.

A large number of experiments have been conducted to prove the validity of the wedge theory and it is found that the sliding surface is not plane or straight line but a logarithimic spiral or a combination of spiral and straight line. The equation for the spiral is

$$r = r_0 e^{\theta \tan \varphi}$$

Table 4-1 EARTH PRESSURE COEFFICIENTS,* K_a

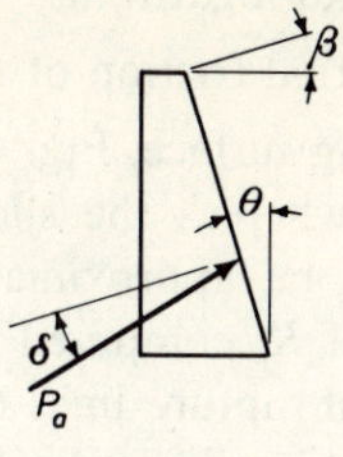

ϕ †	δ	β	θ	K_a Horizontal component	K_a Vertical component
32°	16°	0°	0°	0.134	0.038
			4° 46′	.146	.055
			18° 26′	.180	.123
		18° 26′	0°	.174	.050
			4° 46′	.193	.073
			18° 26′	.246	.169
32°	28°	0°	0°	.121	.065
			4° 46′	.131	.085
			18° 26′	.155	.163
		18° 26′	0°	.161	.086
			4° 46′	.177	.114
			18° 26′	.221	.232
38°	19°	0°	0°	.103	.035
			4° 46′	.115	.051
			18° 26′	.148	.113
		18° 26′	0°	.128	.044
			4° 46′	.144	.064
			18° 26′	.194	.148
38°	28°	0°	0°	.096	.051
			4° 46′	.107	.069
			18° 26′	.132	.139
		18° 26′	0°	.121	.064
			4° 46′	.136	.088
			18° 26′	.178	.187

* From *Civil and Structural Design* (TVA Projects).

† ϕ = angle of internal friction.

where r and θ = variables in the polar coordinate system,

r_0 = arbitrarily selected length,

e = base of natural logarithms,

φ = angle of internal friction of the soil.

It can be seen that the sliding surface, Fig. 4-7(c), for active earth pressure is practically a straight line whereas the sliding surface for passive earth pressure, Fig. 4-7(d), cannot be approximated by a straight line alone. Figure 4-8 shows the values of K_p computed by the wedge theory assuming straight rupture lines and spiral rupture lines (Terzaghi, 1953). It is obvious from this figure that for angle of wall friction δ less than about 20 degrees the value of K_p does not differ significantly. For large values of δ, although seldom justified in design, the values of K_p should be determined by the general wedge theory assuming spiral surface of rupture.

The general wedge theory is most useful in engineering practice because it is applicable to any configuration of ground surface and any slope of the back of wall. In cantilever walls where surface of rupture *ab* according to Rankine theory is interrupted, Fig. 4-9, the earth pressure acting on the line *ac* connecting the heel and the top of the wall is determined instead. In all cases, the direction of the earth pressure must be assumed before the wedge theory can be started.

The wedge theory gives the total lateral earth pressure but does not directly furnish the information concerning the distribution of the pressure. In practice this may be overcome by dividing the total height of the wall H into, say, four equal divisions, each equal to $H/4$. The total lateral pressure against the upper $H/4$, $H/2$, $3H/4$ and the full height of wall H is determined by the trial wedge method. Let these total pressures be P_{a1}, P_{a2}, P_{a3}, and P_{a4}, respectively. From these values, the lateral pressure may be distributed in a stepped fashion as shown in Fig. 4-10.

If the ground surface is plane and the soil is granular ($c = 0$), the following equations may be derived from the wedge theory assuming plane surface of sliding:

$$\left.\begin{matrix} p_a \\ p_p \end{matrix}\right\} = \gamma Z \left\{\begin{matrix} K_a \\ K_p \end{matrix}\right.$$

$$= \gamma Z \frac{\cos^2(\varphi \mp \theta)}{\cos^2\theta\cos(\theta \pm \delta)\left[1 \pm \sqrt{\dfrac{\sin(\delta + \varphi)\sin(\varphi \mp \beta)}{\cos(\theta \pm \delta)\cos(\theta - \beta)}}\right]^2} \tag{4-5}$$

where p_a and p_p = active and passive earth pressure at depth Z,

γ = unit weight of soil,

φ = angle of internal friction.

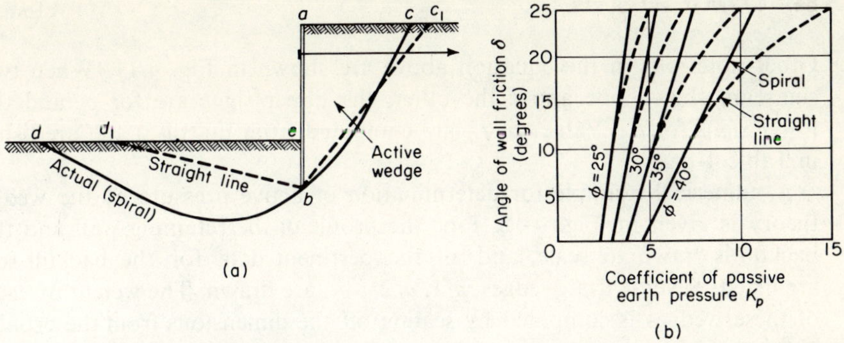

Fig. 4-8 Passive earth pressure; straight line versus spiral surface of sliding. Terzaghi, 1954.

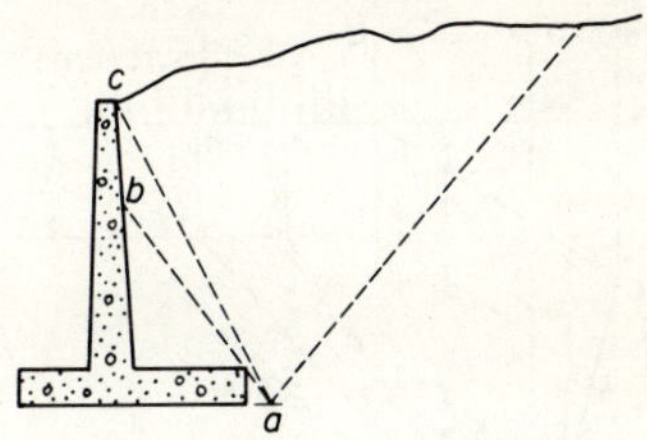

Fig. 4-9 Selection of sliding surfaces.

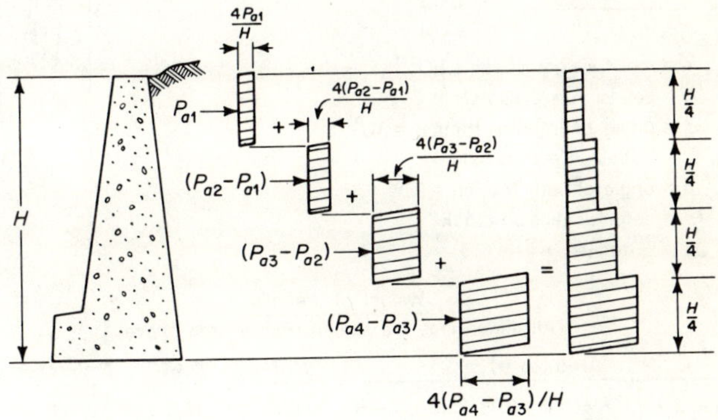

Fig. 4-10 Approximate pressure distribution (wedge theory).

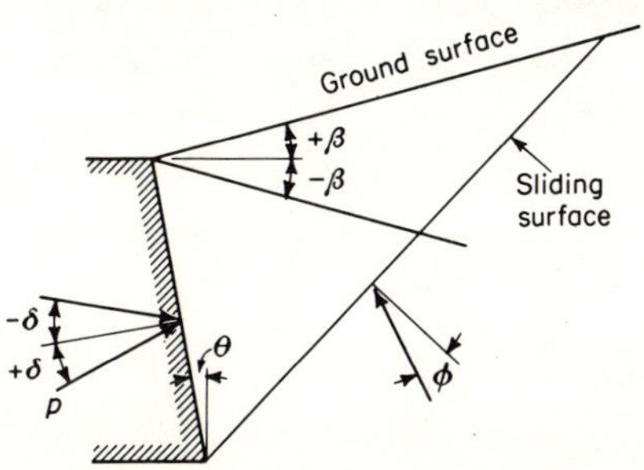

Fig. 4-11 Notations used in Eq. (4-5).

Other notations in the equation above are shown in Fig. 4-11. When two signs are shown one above the other, the upper signs are for p_a and the lower signs for p_p. Values of K_a are computed from Eq. (4-5) and are listed in Table 4-1.

A numerical example for determination of active pressure by the wedge theory is given in Fig. 4-12. First the profile of the retaining wall and the backfill is drawn to scale, and all the pertinent data for the backfill soil are listed. Several trial wedges, *af*1, *af*2, . . ., are drawn. The weight of each of these wedges is computed by scaling off the dimensions from the profile.

Graphical Determination of Active Earth Pressure (Wedge Theory)

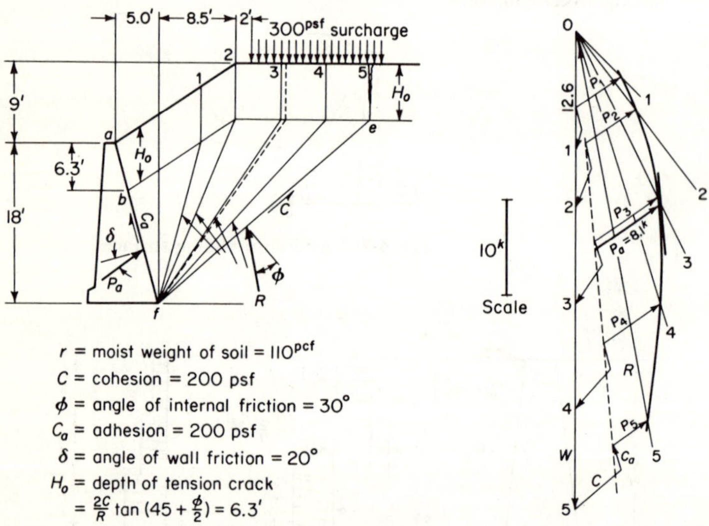

Weight of Wedges

(All dimensions are scaled from sketch above)

Wedge	Area (sq ft)	Wt (k)	Σ wt	Σ Surcharge	Total wt (k)
1	$\frac{1}{2} \times 13' \times 9.2 + \frac{1}{2}(11.5 + 9.7) \times 5.2 = 114.8$	12.6	12.6	0	12.6
2	$\frac{1}{2} \times 22.5 \times 2.5 + 6.3 \times 4.0 = 53.3$	5.9	18.5	0	18.5
3	$\frac{1}{2} \times 20.7 \times 5 + 6.3 \times 5.0 = 83.2$	9.2	27.7	0.9	28.6
4	" " = 83.2	9.2	36.9	2.4	39.3
5	" " = 83.2	9.2	46.1	3.9	50.0

Cohesion and Adhesion

Wedge	Length (ft)	C (k)	Length (ft)	C_a (k)
1	18.7	2.7	13.0	2.6
2	22.4	4.5		
3	24.8	5.0		
4	27.7	5.5		
5	31.3	6.3		

Fig. 4-12 Numerical example (wedge theory).

Often some of the trial wedges can be made in equal dimensions to reduce the computation work. The student may follow the example readily by referring to the discussion at the beginning of Sec. 4-4.

After the lateral pressure P_1, P_2, . . ., are determined for five trial wedges, the maximum value is obtained by drawing a line parallel to the line connecting the points of intersection between C_a and P_5, P_4, etc. This value, 8.1, is the active earth pressure whose magnitude is scaled off from the force polygon.

4-5 *Hansen Theory*

Realizing the fact that earth pressure depends largely on the type of lateral movement of the soil, J. Brinch Hansen has developed a general method to account for the various types of possible movement of retaining structures. He has proved by theory and experiments that the correct earth pressure can be determined if a compatible sliding wedge or rupture surface is used in the computation. The rupture surface may be a straight line, an arc, or a more complicated composite curve. In his extensive study entitled *Earth Pressure Calculation* (J. Brinch Hansen, 1953) tables and graphs are included to facilitate the use of the theory. Reference is made to the original publication for advanced students.

4-6 *Determination of Soil Properties for Earth Pressure Computation*

The values for the unit weight γ, the unit cohesion c, and the angle of internal friction φ, to be used in earth pressure computation should be determined by tests carried out on representative samples of the material corresponding to the conditions which will exist after construction. It is desirable to determine these values prior to the design. If not determined before, the values must be selected on the basis of available material and the construction should be carried out to satisfy the minimum assumption used in the design. For retaining walls of small to moderate height, satisfactory design pressure may be obtained from empirical rules without significant sacrifice in construction cost.

The values of c and φ may be determined by shear tests or triaxial tests for predominantly granular soils and may be used directly in the earth pressure computation. The actual earth pressure will not differ materially from the computed values. However, the problem becomes complex when dealing with cohesive soils. In this type of soil, conditions may change after completion of construction, as discussed in Sec. 1-8. The earth pressure must be computed for the least shear strength anticipated to exist at some time of the life of the retaining structure.

4-7 *Overcompaction*

Since loosely placed backfill material will eventually settle under its own weight and will not be capable of supporting structures, floors, and roads, it is common practice to compact the backfill in layers. Unfortunately the compaction may bring about ill effects, namely excessive lateral pressure due to the heavy weight of compaction equipment. To avoid excessive lateral pressure, the compaction work adjacent to a retaining structure should be done with light weight tampers. If heavy equipment is allowed to operate close by, the stresses and stability of the retaining structure should be investigated. Tests have indicated that against a rigid wall the lateral pressure due to an overcompacted backfill may be several times that due to a loose backfill (Sowers et al., 1957). For this reason, overcompaction by heavy equipment should be avoided within the area approximately equal to the sliding wedge.

4-8 *Seepage Pressure*

During a rain storm the soil behind the retaining structures contains a large amount of water. If no drainage is provided in the backfill, the water percolates through the backfill in a downward direction, continues under the base of the retaining structure and rises through the soil in front as shown by

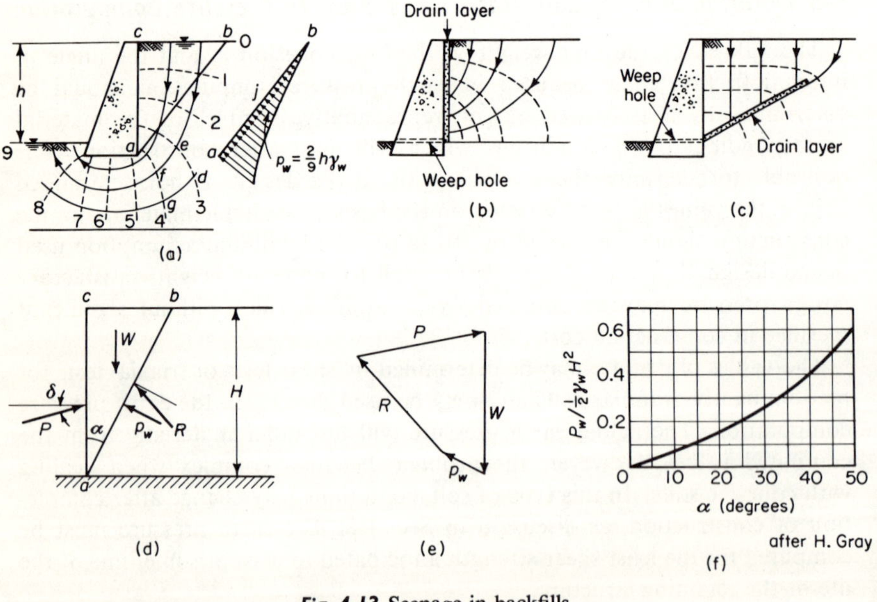

Fig. 4-13 Seepage in backfills.

solid lines and arrows in Fig. 4-13(a). The seepage water affects a retaining structure in several ways:

1. *Increase in weight of soil due to saturation or partial saturation.* Earth pressure is increased due to increase in weight of the sliding wedge. In silty and clayey soils the design may have to be based on the assumption that the entire backfill is fully saturated because of the tendency of these types of soil to retain water for a long period of time.
2. *Uplift on the surface of sliding.*
3. *Uplift against the base of retaining wall.*
4. *Reduction in passive resistance.* The seepage pressure exerted by the rising water in front of the retaining structure tends to uplift the soil grains and hence reduce the passive resistance. In extreme cases where the hydraulic head of this seepage flow is so high that it cancels out the weight of the soil grains (quicksand condition, Sec. 1-14), the passive resistance becomes zero.

In general, it is more economical to make provisions for reducing the seepage pressure than designing the retaining structure for the full amount. This can be done by using granular backfill and providing drain layers similar to those shown in Fig. 4-13 [(b) and (c)].

The seepage pressure in a soil mass can be determined by the use of flow net. A flow net is a graphical representation of seepage in soil. As shown in Fig. 4-13 [(a), (b) and (c)], the solid lines (flow lines) represent the direction of flow; the dashed lines are equipotential lines. The pressure head is equal at all points on an equipotential line. These two sets of lines intersect each other at right angles. If the flow net is sketched so that all areas between two sets of flow lines and equipotential lines are approximately square, such as *defg* in Fig. 4-13(a), the flow net can be readily used for determining the seepage pressure (or pore pressure) at any point in the soil mass.

A free water surface, such as line 0 or line 9 in Fig. 4-13(a), is an equipotential line. The back of the wall and other impervious boundary are flow lines. If drain layers are provided, flow lines are directed toward them. To draw a flow net, start with trial lines and keep improving until all requirements are met and all areas are about square. For detailed discussion of sketching flow nets, reference is made to standard text books of soil mechanics. With a little practice, the student should be able to sketch flow nets accurately enough to use in ordinary practical cases.

In Fig. 4-13(a), the equipotential lines are numbered 0, 1, 2, . . ., 9. The pressure is equal to zero on line 0, and equal to $h\gamma_w$ on line 9, where h is the total head differential and γ_w the unit weight of water. On line 2, for example, the pressure is equal to $\frac{2}{9}\,h\gamma_w$.

In the same figure, line *ab* is a trial sliding surface. It intersects equipotential lines at various points. At point j, for example, it intersects equipotential line 2, hence the pressure

$$p_w = \frac{2}{9} h \gamma_w$$

With the pressure at the various intersections known, as shown on the right hand side of Fig. 4-13(a), the total pressure P_w, acting at right angle to the sliding surface ab, can be computed.

In computing the earth pressure during a rain storm, the wedge theory (Sec. 4-4) is used. The computation should include the total hydraulic pressure P_w (pore pressure) acting on the sliding surface; the saturated weight of soil should be used. These forces are shown in Fig. 4-13(d) and the force polygon is shown in Fig. 4-13(e). If the back of wall is vertical, the wall is built on top of a relatively impervious layer, and an adequate drain layer is provided against the back of the wall, the value of total pressure P_w acting on any trial sliding surface can be obtained directly from Fig. 4-13(f).

With the aid of flow net, the hydraulic uplift against the base of the wall can be determined at intersections of equipotential lines.

4-9 Surcharge Load

The lateral pressure introduced on a retaining structure may be classified into four types for convenience of computation.

A. Uniform load. When the ground surface is subjected to a uniformly distributed load, the earth pressure computation is often made by substituting the load by an equivalent surcharge layer. The thickness of this surcharge layer is equal to the distributed load divided by the unit weight of the underlying soil.

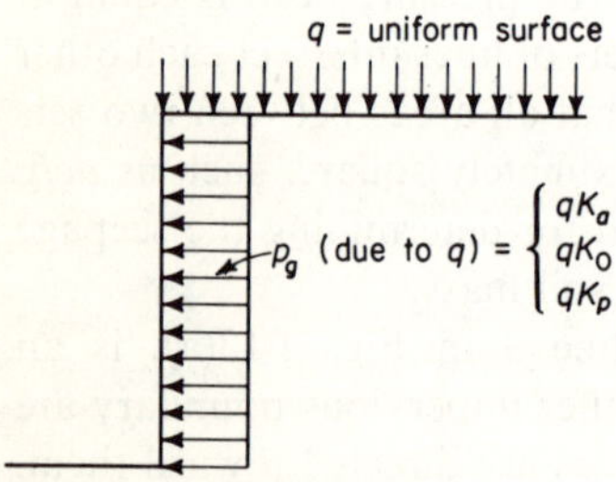

Fig. 4-14 Lateral pressure due to uniform surcharge.

The computation of lateral pressure due to a uniform surcharge is relatively simple. In the case where the wedge theory is used, the surcharge load can be readily included in the computation of the weight of wedges. In the case where Rankine theory is applicable, the pressure caused by a uniform surcharge q is a constant pressure and is equal to

$$p_q = qK_a, \quad p_q = qK_0, \quad \text{or} \quad p_q = qK_p \tag{4-6}$$

where the earth pressure coefficient K_a, K_0, or K_p is a constant within any soil stratum having a constant φ value, Fig. 4-14.

Contrary to the uniform surcharge, the lateral pressure due to a surcharge applied on a limited area of the ground surface is difficult to determine. The effect of a strip load or a line load parallel to a retaining wall may be included in the trial wedge method. In other cases, the theory of elasticity has been widely used. However, the validity of the elastic theory has not been proved in all cases. The results of a limited number of tests (Terzaghi, 1954) have proved that the measured horizontal unit pressure against a vertical rigid wall are about twice as high as those calculated by the elastic theory.

B. Strip load. Highways, railroads and continuous wall footings are strip loads when they are parallel to the retaining structure. Refer to Fig. 4-15 and let:

p_q = horizontal pressure at point a;

β = angle of visibility at point a, in radian;

α = angle between vertical and the bisector of β;

q = strip load, psf.

The actual lateral pressure against a rigid wall is twice the value determined by theory of elasticity:

$$p_q = \frac{2q}{\pi}(\beta + \sin\beta)\sin^2\alpha + \frac{2q}{\pi}(\beta - \sin\alpha)\cos^2\alpha \tag{4-7}$$

as shown in Fig. 4-15, the value of p_q varies with depth.

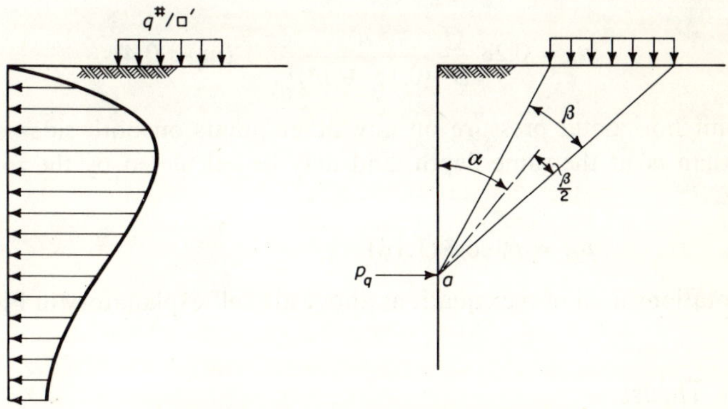

Fig. 4-15 Lateral pressure due to strip load.

C. Line load. A continuous wall footing of narrow width may be taken as a line load when located parallel to the retaining structure. Similar to the case of strip load, the lateral pressure increases from zero at the ground surface to a maximum value at a certain depth and gradually diminishes to zero at a greater depth, Fig. 4-16. The unit horizontal pressure may be computed by the following equation (Terzaghi, 1954):

$$\left.\begin{aligned} p_q &= 1.27q\frac{xz}{R^4} \\ &= 1.27\frac{q}{H}\frac{m^2n}{(m^2+n^2)^2} \end{aligned}\right\} \quad (m > 0.4) \qquad \text{(4-8a)}$$

$$p_q = 0.203\frac{q}{H}\frac{n}{(0.16+n^2)^2} \quad (m < 0.4) \qquad \text{(4-8b)}$$

D. Point load. A wheel load or any load concentrated on a small area may be treated as a point load. The intensity of lateral pressure in this case varies not only with the depth but also with the horizontal distance from the load. The pressure is greatest along the vertical line *ab* closest to the load, Fig. 4-17. Along this line *ab*, the unit horizontal pressure *p* may be computed by the following empirical equations (Terzaghi, 1954):

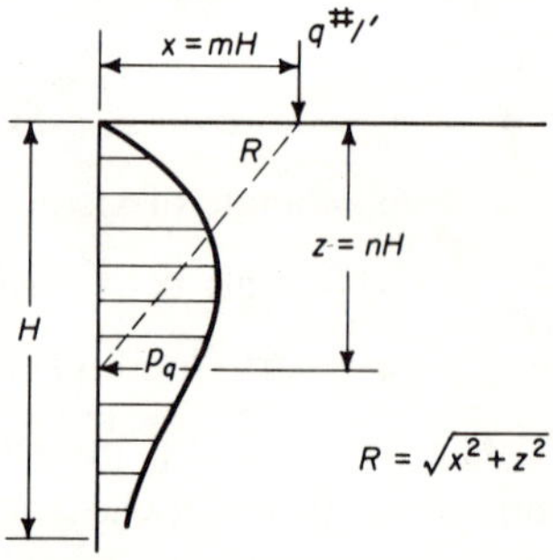

Fig. 4-16 Lateral pressure due to line load.

$$p_1 = 1.77\frac{Q}{H^2}\frac{m^2n^2}{(m^2+n^2)^3} \quad (m > 0.4) \qquad \text{(4-9a)}$$

$$p_1 = 0.28\frac{Q}{H^2}\frac{n^2}{(0.16+n^2)^3} \quad (m < 0.4) \qquad \text{(4-9b)}$$

The unit horizontal pressure on any other points on both sides of *ab* is smaller than p_1 at the same depth, and may be calculated by the following equation.

$$p_Q = p_1 \cos^2(1.1\psi) \qquad \text{(4-9c)}$$

The notations used in the equations above are self-explanatory in Fig. 4-17.

4-10 *Ice Thrust*

Substructures are subjected to ice thrust where the ground water or capillary water is above the frost line (depth of frost penetration). Lateral thrust is caused by the volume expansion of ice upon change in temperature. The magnitude of the thrust is very large, being equal to the buckling or crushing strength of the ice sheet. In practice the horizontal ice thrust acting

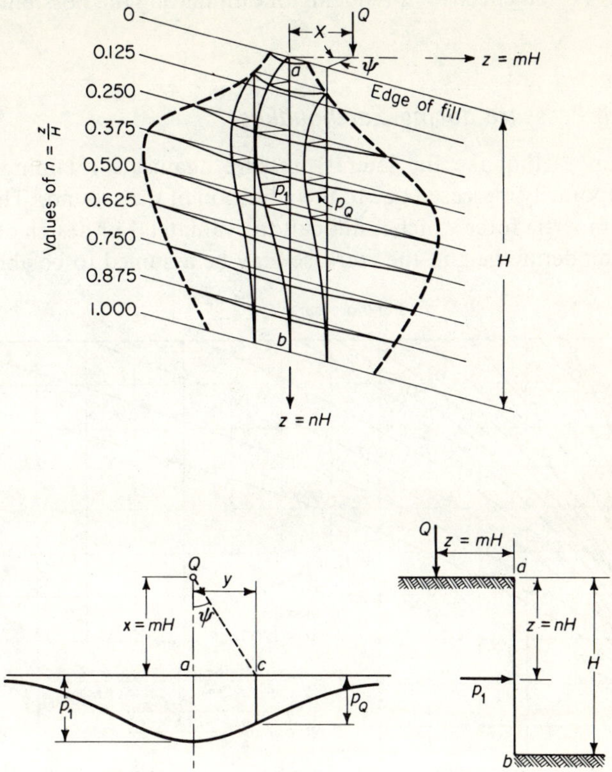

Fig. 4-17 Lateral pressure due to point load.

on a rigid vertical wall against a body of free water is generally taken as 30,000 to 50,000 pounds per linear feet.

Coarse granular soils (sand and gravel) above ground water level are not capable of retaining water in the intergranular spaces. Therefore, no ice or frozen ground will be formed even in severe winter. On the other hand, the voids in the fine grained soils (fine sand, silt and clay) invariably contain some water. The upper layer of such soils is subjected to freezing and consequently, it exerts lateral thrust on any object against the volume expansion of this layer.

In practice the possibility of lateral thrust from ice or frozen ground should be eliminated because the magnitude of this thrust is generally too excessive to be taken into account. Ice thrust may be largely reduced by providing the retaining structure with a sloping surface at the water level. If the wall can yield laterally, there is no problem. Fine-grained soils susceptible to freezing should be replaced with free-draining granular soil above the frost line in the vicinity of the back of the retaining structure. Drainage system installed in the backfill is often effective in reducing or eliminating the possibility of frost formation.

4-11 *Earth Pressure during Earthquakes*

During an earthquake the lateral pressure against a retaining structure may be temporarily increased due to the vibration of the ground. The increase is a result of inertia force which is difficult to evaluate. For design of retaining walls with moderate height, the increase may be assumed to be about 10 per

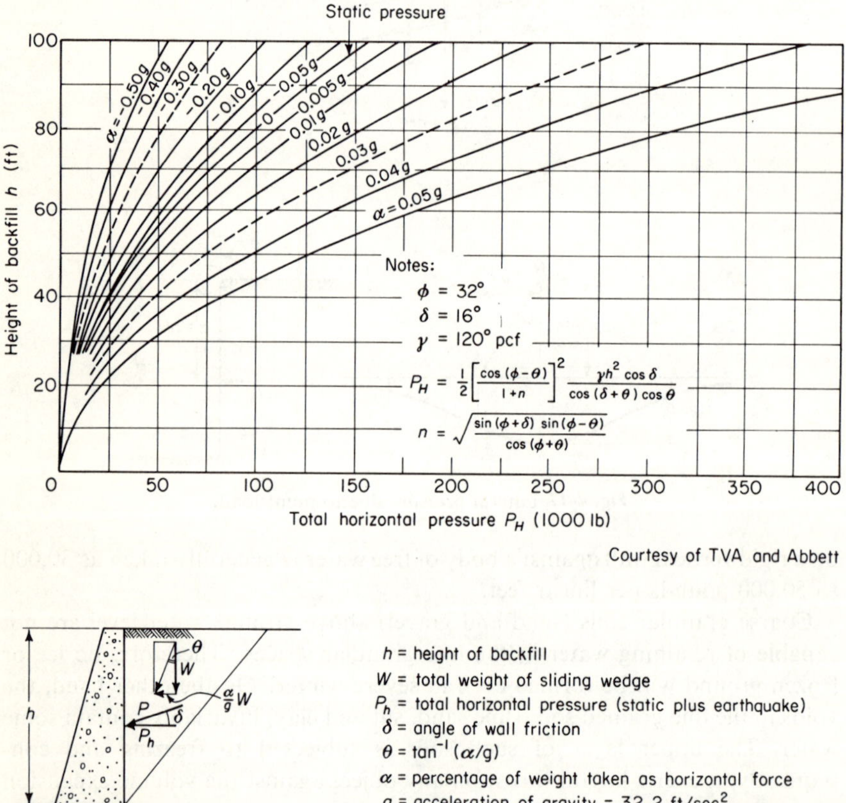

Fig. 4-18 Earth pressure during earthquakes.

cent of the normal design pressure. In the case of high retaining walls, however, combined pressure may be determined approximately by the trial wedge method.

The procedure of analysis is exactly the same as in the normal cases except that the sliding wedge is subjected to a horizontal force in addition to all the other forces. The magnitude of this horizontal force depends upon the fundamental period of the system and the horizontal acceleration during the earthquake, which should be established by examination of the earthquake records of the locality. The TVA engineers assume that the horizontal force is equal to 18 per cent of the weight of the sliding wedge whereas the Japanese engineers use one-third of the gravity. The TVA standard is reproduced in Fig. 4-18.

For the simple case of a uniform soil with a plane surface of ground, the center of static pressure is found to be at about the third point above the base, and the center of dynamic pressure is at about the third point below the top of ground. For this reason the dynamic and static pressures should be differentiated. The static pressure is subtracted from the combined lateral pressure and the difference is considered as the dynamic pressure. The distribution of the static pressure may be hydrostatic or stepped whereas the dynamic pressure may be assumed to vary linearly from zero at the base to a maximum at the top of the soil (TVA).

It should be pointed out that the retaining wall itself may be also subjected to an acceleration during the earthquake and these two forces (acceleration of the sliding wedge and that of the wall) may act simultanuously. Therefore the retaining wall should be designed to resist both these forces.

4-12 *Wave Pressure*

Waves are usually generated by wind, and occasionally by moving vessels, earthquakes, tides, and barometric changes. Structures obstructing the propagation of the wave motion are subjected to lateral pressure.

Waves generated by wind over a large body of water produces an undulating surface consisting of crests and troughs, Fig. 4-19. The horizontal distance from crest to crest, or from trough to trough, is known as the wave length. The vertical distance from crest to trough is known as the wave height. The time required for the wave form to travel a distance equal to the wave length is called the period.

The wave pressure against a continuous vertical surface may be computed by the theories shown in Fig. 4-19. It is seen that the wave pressure depends on the wave length and wave height. In localities where sufficient records are not available to establish the maximum values, the following empirical formulae have been used:

Stevenson: $H = 1.5\sqrt{F}$ $(F > 30)$

$H = 0.15\sqrt{F} + 2.5 - \sqrt[4]{F}$ $(F < 30)$

Molitor: $H = 0.17\sqrt{V_w F}$ $(F > 20)$

$$H = 0.17\sqrt{V_w F} + 2.5 - \sqrt[4]{F} \qquad (F < 20)$$

where H = wave height, ft;

F = fetch = distance over which the wind acts, miles;

V_w = wind velocity, miles per hour.

The wave lengths are, according to Gaillard, 9 to 15 times the wave height for inland lakes and 17 to 33 times the wave height for ocean waves. Molitor derived an empirical formula for the wave length L for inland lakes:

$$L = 840\frac{H}{V_w}$$

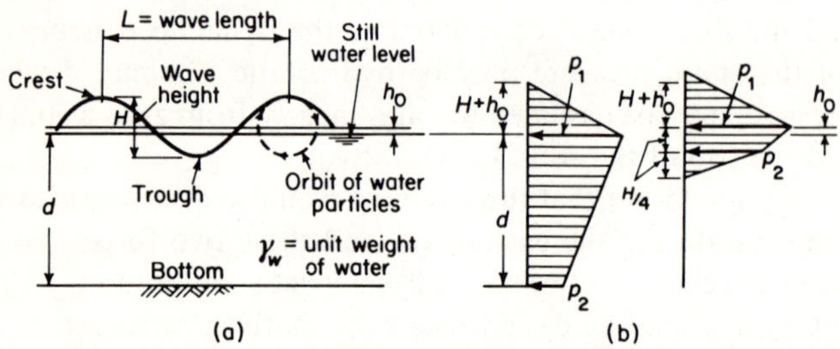

Sainflou theory: $p_1 = (p_2 + \gamma_w d)\dfrac{H+h_o}{d+H+h_o}$

$$p_2 = \frac{\gamma_w H}{\cosh\frac{2\pi d}{L}}$$

$$h_o = \frac{\pi H^2}{L}\coth\frac{2\pi d}{L}$$

Gourret theory: $p_1 = (p_2 + \gamma_w d)\dfrac{H+h_o}{d+H+h_o}$

$$p_2 = \gamma_w\left(\frac{H}{\cosh\frac{2\pi d}{L}} - \frac{\pi H^2}{L}\tanh\frac{2\pi d}{L}\right)$$

$$h_o = \frac{\pi H^2}{L}\left(\tanh\frac{2\pi d}{L} - \coth\frac{4\pi d}{L}\right) + \frac{\pi H^2}{2L}\cdot\frac{\coth\frac{4\pi d}{L}}{\sinh^2\frac{2\pi d}{L}}$$

Molitor rule: $p_1 = \dfrac{k\gamma_w}{2g}(C+V_{oc})^2$

$$p_2 = 0.72\,p_1$$

$$h_o = \frac{2H^2}{L}$$

k = 1.3 for 30 mph wind
1.7 for 70 mph wind
1.8 for ocean storm

$$C = \text{wave velocity} = \sqrt{\frac{gL}{2\pi}\tanh\frac{2\pi d}{L}}$$

$$V_{oc} = \frac{\pi H}{T}\coth\frac{2\pi d}{L}$$

Fig. 4-19 Wave pressure theories. After Hudson, 1953.

A more accurate computation of wave height may be made by the theory of Sverdrup and Munk (Mason, 1953). The relationship between the wave height, wind velocity, and fetch is shown graphically in this reference.

4-13 *Other Lateral Forces*

Substructures and foundations are often subjected to some other lateral forces which must be established in each particular job prior to the design. The following lateral forces are often encountered.

Swelling pressure. Any structure that retains an expansive clay may be subject to a large swelling pressure when the upper layer of the clay absorbs water. The swelling pressure in some cases may exceed the weight of the overlying soil. The amount of swelling pressure is difficult to determine.* In practice, it is almost invariably more economical to eliminate the possibility of swelling pressure against the retaining structure than to design the structure for the full pressure. The most economical way is to use granular backfill. If such material is not available, provision should be made to prevent the surface water from being accumulated adjacent to the retaining structure, and drainage should be provided to direct the seepage water flow, Sec. 4-8.

Thrust due to thermal expansion. When the structure expands or contracts as a result of temperature fluctuation, the substructure or foundation may be subjected to a lateral movement or a lateral thrust. This lateral thrust is eliminated if roller bearings, rockers, or other similar devises, are provided. In the case of a bridge span resting on friction bearings, the lateral thrust is equal to the vertical reaction times the coefficient of friction of the sliding surface.

Traction forces. Traction forces due to moving railway and highway traffic and due to hoist and crane wheels are transmitted to the substructure and foundation. The AREA, AASHO, and AISC Specifications contain information on the magnitude of such traction forces.

Mooring pull. Dock structures are provided with mooring posts for anchoring boats. The magnitude of the mooring pull may be assumed to be equal to the capacity of the winch used on the boat.

Ship impact. The direct impact of a ship collision against a dock or other water front structure is usually too large to be taken into account in the design. Instead, some type of fender systems† or dolphins are generally constructed to reduce the impact to a minimum. Frequently an arbitrary force, such as 25 to greater than 100 tons, is used in design.

---

* For further information, reference is made to *Proceedings of Symposium on Shear Strength* (Colorado School of Mines, 1959).

† See "Docking Fenders Key to Pier Protection," *Engineering News-Record* (May 8, 1958 and June 19, 1958).

Plate Five

Drainage of a Construction Site by Means of Well Points

In the design and construction of any substructure and foundation, the problem of drainage and waterproofing deserves serious consideration. There are a number of methods for dewatering the site. The use of an improper method of dewatering will not only require large expenditure, but may also introduce difficulties and damages to the foundation soil.

Substructures must be dampproofed or waterproofed to prevent moisture or free water from entering the useful floor space. In this chapter, the principles of dewatering and permanent drainage are discussed. The methods of dampproofing and waterproofing are also presented.

5 FOUNDATION DRAINAGE AND WATERPROOFING

5-1 *Introduction*

Prior to the design and construction of a given foundation, the ground water level at the site must be reliably determined. If the ground water is high, some of the following problems are encountered:

Dewatering the site during construction
Foundation drainage
Waterproofing or dampproofing

For each job it is important to determine before hand the method of dewatering, the type of foundation drains, waterproofing, or dampproofing. Sometimes the cost of dewatering the site is excessive and consequently the total construction cost may be high. In such cases, a cost comparison should be made for all feasible types of foundations or schemes of substructures from which the most economical design can be selected. Quite often the amount of free water that will flow into the excavation is difficult to predict due to the erratic pattern of waterbearing layers and pockets. Under such circumstances, the design must be made flexible enough so that it can be adjusted to suit the actual conditions as the excavation proceeds or when the excavation is completed.

5-2 *Methods of Dewatering*

When construction is made below the ground water level, the site must be dewatered for the following purposes (Swiger, 1960):

1. To provide a suitable working surface at the bottom of the excavation.
2. To stabilize the banks of the excavation thus avoiding the hazards of slides and sloughing.
3. To prevent disturbance of the soil at the bottom of excavation caused by boils or piping. Such disturbance may reduce the bearing power of the soil.

The amount of water to be removed from the site varies from a trivial quantity to large volumes, depending upon the height of water head, the permeability of the soil below the water level, and the size of the area to be dewatered. Extensive dewatering is necessary for deep excavation in permeable soils (sand, gravel, or soils containing such seams) whereas little dewatering is required for shallow excavations or excavations in impervious soils (clays).

A successful dewatering job depends upon the proper selection of the method and the constant vigil on the operation. The surface water should be diverted away from the excavation. Furthermore, the possibility of piping or boil should be analyzed, Sec. 5-3. The dewatering may be done in one or a combination of the following methods:

A. Sumps. A sump is merely a hole in the ground from which water is being pumped for the purpose of removing water from the adjoining area. This method is most commonly used for removal of surface water but is also useful where the amount of water to be removed is small.

One sump may be sufficient for a small area, whereas several sumps with ditches leading to them are necessary for dewatering a large area. If the soil

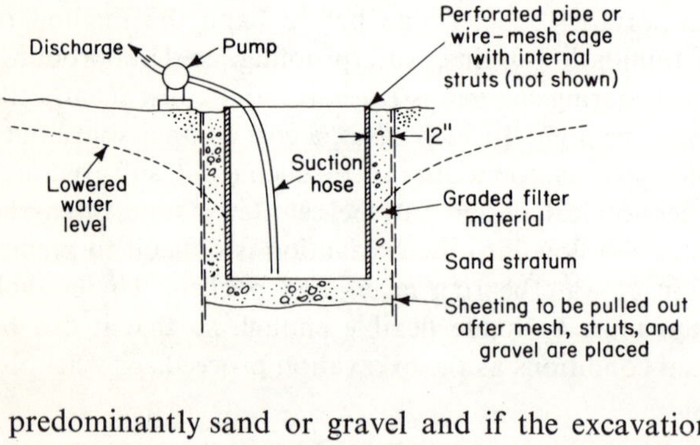

Fig. 5-1 Pump sump.

is predominantly sand or gravel and if the excavation exceeds several feet below the ground water level, the pump sump method may become inadequate, and another method should be used.

In any dewatering operation it is important to guard against the danger of carrying away the fine particles from granular soils. As fine particles are carried away by the flowing water, the bearing capacity of the soil may be impaired. If existing foundations are in the vicinity, pumping may cause settlement of these foundations. To avoid such difficulties, the sump should be lined with a filter material which has grain-size gradations determined by the rules discussed in Sec. 5-5. Generally, the filter material is installed in the following manner, as seen in Fig. 5-1:

1. Drive sheeting around the sump for the full depth of the sump.
2. Install a cage inside the sump. The cage may be made of wire mesh with internal strutting or a perforated pipe.
3. Fill the filter material in the space outside the cage and at the bottom of the cage.
4. Withdraw the sheeting.

The relative amount of soil particles carried away by pumping can be determined by visual examination of the water discharged from the end of the hose. For a prolonged pumping, a bucket of discharged water should be collected periodically, and the water allowed to set for several hours. The amount of soil particles settled at the bottom of the bucket can be observed visually.

B. Well points. A well point is a two to three inch diameter pipe two to four feet long which is perforated and covered with a screen. The lower end of the pipe has a driving head with water holes for jetting. Figure 5-2 shows a photograph of one of the commercially made well points.

Well points are connected to two to three inch diameter pipes known as riser pipes and are inserted into the ground by driving or jetting. The upper ends of riser pipes lead to a header pipe which, in turn, is connected to a pump. The ground water is drawn by the pump into the well points through the header pipe and discharged. The well points are installed with two to five foot spacing, Fig. 5-3. This type of dewatering system is effective in soils constituted primarily of sand fraction or other soils containing seams of such materials. In highly pervious soils such as course gravels, the spacing required to handle the water may be so small that well points become impracticable. They are not useful to draw water out of clays because of the slow process of water seepage. In silt strata, well points may be used if the upper two to three feet of the riser pipes are encased in a tamped clay seal and if pumping is maintained for a period of several weeks. By so doing, a vacuum pressure is created in the silt.

In stratified soils, the screened portion of the well point does not draw water from all the strata above it. In order to facilitate the dewatering in all strata and thus cut down the cost, vertical sand drains may be provided within the influence area of the well points. These sand drains are usually 12 to 24 inches diameter at 15 ft spacings.

The well points can lower a water level to a maximum of 18 ft below the

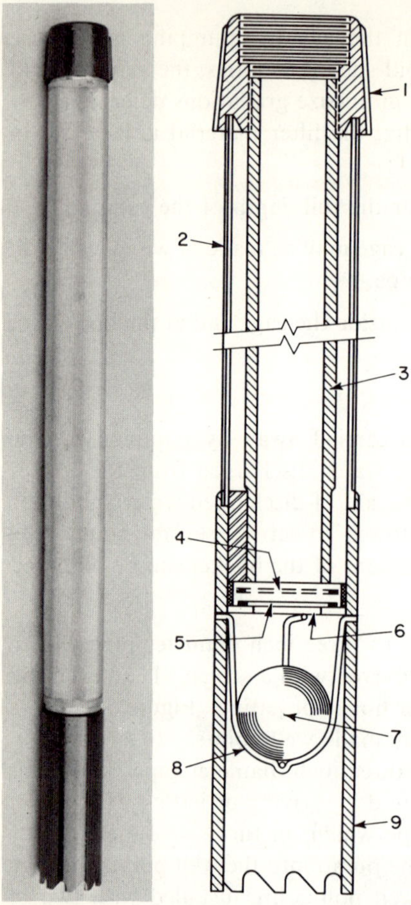

Fig. 5-2 Well point: (1) coupling; (2) screen cylinder – available in tin dipped cold rolled steel or stainless steel; (3) center tube; (4) ring valve – during jetting when small eddy currents, created by the ball seat, raise it to the upper position, closing off the annular space between the tip and the center pipe; (5) ring valve – at position of rest on ball seat when the jetting is shut off; (6) ball seat; (7) ball valve; (8) retainer basket – clamps the ball seat in place and keeps the ball valve in the center of the tip during jetting, providing a steamlined, oversize exit for the full force of the jetting water; (9) jetting tip. Courtesy of the Moretrench Corp.

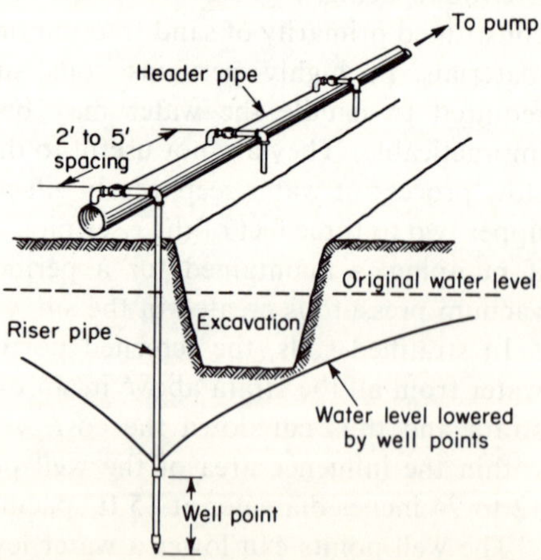

Fig. 5-3 Well point system.

center-line of the header. Under ideal conditions and using special high vacuum equipment, the depth of lowering has been increased to as much as 25 ft (Werblin, 1960). For lowering water level to a greater depth, the multiple stage system of well points must be used which employs two or more tiers of well points. Under average conditions, any number of stages can be used, each stage lowering the water level about 15 ft. A typical setup for a two-stage system is shown in Fig. 5-4. However, multiple stage system requires additional footage of header pipes and additional pumps. It also increases the width of excavation due to the berms required for headers. Therefore, for dewatering a large head of water, other methods should be considered. The selection of dewatering method should be made on the basis of total cost including initial cost and the cost of operation.

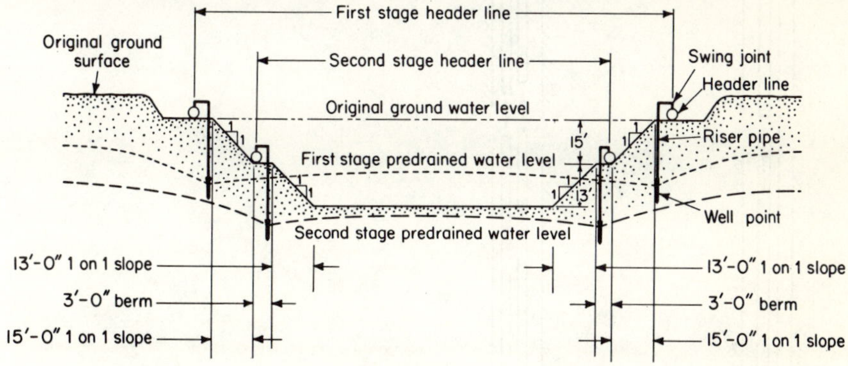

Fig. 5-4 Typical two-stage well point system. After Griffin Wellpoint Corp.

C. Deep well pumps. To install a deep well pump, a well of 6 in. to 2 ft or larger diameter, is bored to the desired depth; then a deep well turbine, a submersible pump or a water ejector is lowered to the bottom of each well. Such wells are capable of lowering a large head of water and are spaced at 25 ft to more than 120 ft apart depending upon the depth of water to be lowered and other conditions. Filter material should be provided in the well to prevent loss of fine particles in the adjacent ground and clogging of the system. The filter material should be selected according to the rules presented in Sec. 5-5.

Although the cost of installation and operation of deep wells is high, this system can be less expensive than the multiple-stage well point system. Furthermore, the wells can be located at some considerable distance from the edge of excavation thus causing very little interference with other construction activities. In large jobs where a number of construction equipment are in use, this factor may be of decisive advantage.

The deep well pumps may be augmented by the use of vacuum, Fig. 5-5(a), which increases the yield of water from the soil to the well, but simultaneously reduces the capacity of the pump due to the reduction of net suction head. When ejectors are used, two parallel headers are required–one for the pressure supply and the other as the collecting main. The motive force in this system is the circulating water under pressure, Fig. 5-5(b). The space between the casing and the return pipe may be used for the pressure supply, hence a single pipe is required inside the well. Single pipe ejectors are made for 2 in. to 6 in. diameter casings and have larger pumping capacity than two pipe units.

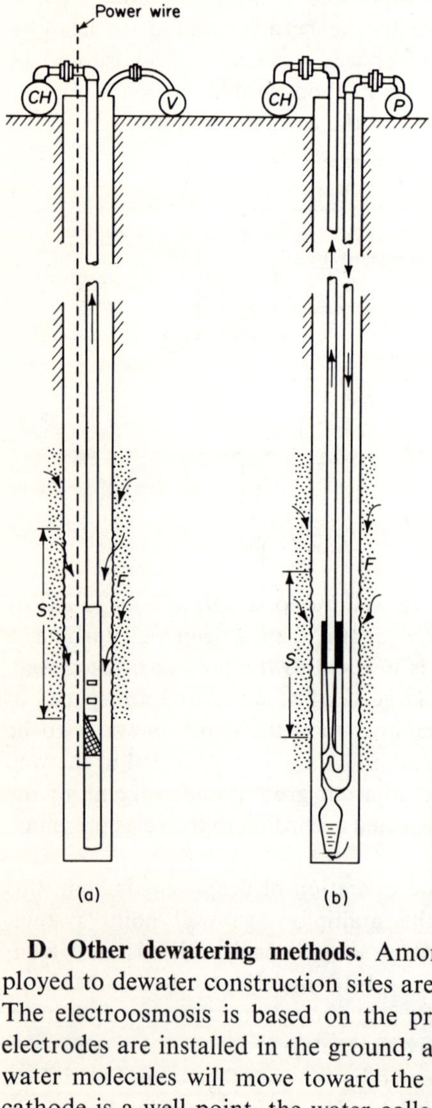

Fig. 5-5 Deep well pumps: (a) submersible turbine pump, 4 to 12 in.; (b) ejector, 2 to 8 in., double pipe. CH = collecting header; P = pump; V = vacuum; F = filter material; S = screen. After B. J. Prugh.

D. Other dewatering methods. Among other methods occasionally employed to dewater construction sites are electroosmosis and ground freezing. The electroosmosis is based on the principle that if positive and negative electrodes are installed in the ground, and an electric potential is set up, the water molecules will move toward the negative electrodes (cathode). If the cathode is a well point, the water collected at the point can be removed by pumping. This method is used advantageously in fine-grained soils (primarily silts) where the efficiency of ordinary well points is low. While the electro-osmosis method has been used occasionally on large dewatering jobs, the freezing method is even less developed. If such methods are contemplated, expert guidance should be employed in the design, installation, and supervision of the system.

The amount of water inflow in the excavation may be reduced by sheetpiling walls or grout-curtains:

1. *Sheetpiling walls.* Sheetpiling may be driven into the ground to form cofferdams to aid in dewatering. This type is used where:

1. The excavation is made in open water, as in the case of bridge piers and abutments;
2. The lowering of water level in the surrounding area may damage the foundations of adjacent structures;
3. The space is not sufficient for open excavation.

If impervious stratum (clay or bedrock) is encountered at a shallow depth, the sheetpiling may be driven to this stratum to cut off the seepage water. Then, the amount of seepage water into the cofferdam is very small. If the pervious soils extends to a large depth below the ground, the amount of inflow decreases with the increasing depth of sheetpiling penetration. In such cases, the rate of inflow should be determined by means of flow nets.* Even when sheet piling is driven to an impervious stratum, there should be provisions for collecting and discharging the seepage and surface water inside the cofferdam by means of well points or sump pumps.

The design of sheetpiling cofferdams is discussed in Chapter 13.

2. *Grout curtain walls.* The amount of seepage water entering the excavation can often be cut off or greatly reduced by a grout curtain wall around the site. Small holes are bored in the ground and grout mix is injected in the holes. Cement, clay, asphalt, chemicals, or a combination of two of these materials may be used as grout material.†

5-3 *Stability of Bottom of Excavation*

In order to provide a dry working surface with an adequate supporting power, the water level in the excavation must be lowered at least to a depth of two to three feet below the bottom of excavation. When sump pumps are

* See any standard textbook of soil mechanics and Sec. 4-8.

† For further information on grouting, reference is made to the "Symposium on Cement and Clay Grouting of Foundations" and "Symposium on Chemical Grouting of Foundations," *Proceedings ASCE* (1958).

used, the danger of boil or piping caused by uplifting pressure due to upward seepage flow in the soil must be avoided. The basic principle of piping in general was presented in Sec. 1-14, and the analysis of piping in cofferdams is discussed in Sec. 13-7.

When well points or deep wells are employed to dewater the site, generally there is no uplifting force causing boil or piping because the seepage water flows essentially in the downward direction towards the screened portion of the wells. However, when well points are used in stratified soils which contain one or more strata of impervious material, Fig. 5-6, the stability of

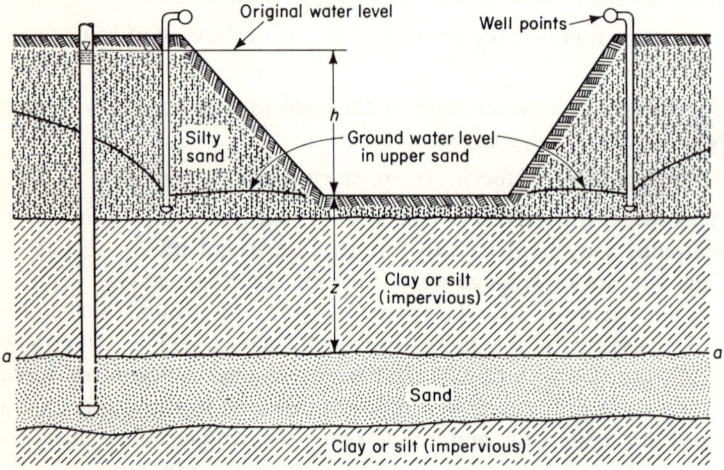

Fig. 5-6 Stability of bottom of excavation due to excess hydrostatic pressure against impervious layer. After W. F. Swiger.

the bottom of excavation should be analyzed. Lowering of water level in the upper water-bearing stratum will not relieve the pressure in the lower water-bearing stratum below the line *aa*. If the pervious layer below the line *aa* is horizontal and is subjected to a hydraulic pressure equal to the height $h + z$, then, the uplift pressure on the soil above the line *aa* at failure becomes equal to the total weight of the material. Thus,

$$62.5\,(h + z) = \gamma z$$

where γ is the saturated unit weight which may be taken as 125 pcf. The equation above leads to a rough rule and that is: to avoid uplift of the bottom of an excavation, the height h should be equal to z divided by a factor of safety, say 1.3. If this rule cannot be satisfied, the impervious layer below the line *aa* should be also drained by well points, deep wells, or by vertical sand drains which permit water to flow from this layer into the excavation (Swiger, 1960).

5-4 Foundation Drainage

When a substructure is carried below groundwater level, provisions must be made to keep the floor area dry. This can be accomplished by one of the following methods:

1. *Foundation drainage*. The water level is lowered to a depth below the floor elevation.
2. *Waterproofing*. The substructure is made watertight by waterproofing.

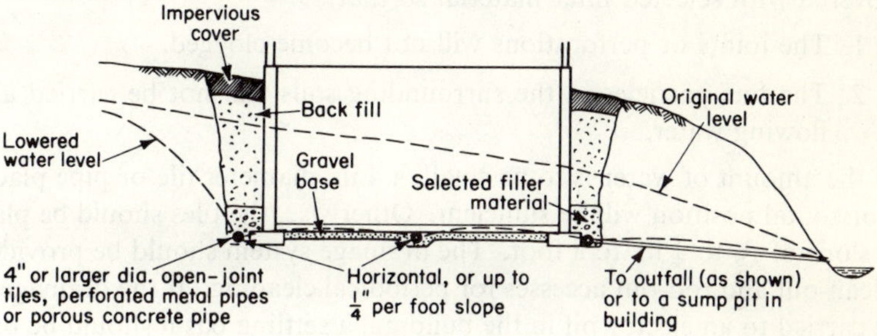

Fig. 5-7 Foundation drains.

Foundation drainage is generally accomplished by installation of drain tiles or drain pipes adjacent to footings and, if necessary, under the floor slabs, Fig. 5-7. Drain tiles are made of clay or concrete, and the sections are placed butting each other with open joints to permit the water entering them.

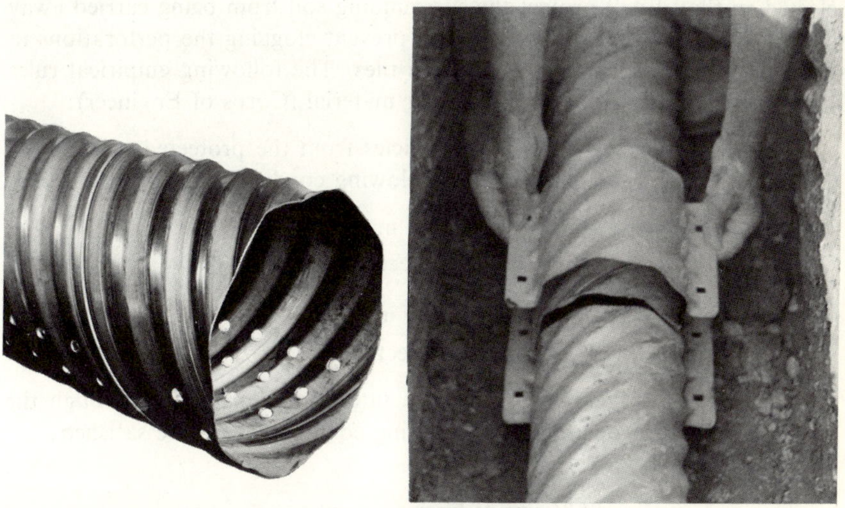

Fig. 5-8 Corrugated perforated pipe and method of splicing. Photographs courtesy of Armco Drainage and Metal Products, Inc.

Drain pipes may be corrugated metal or nonmetallic pipes with $\frac{5}{16}$ in. perforations or porous concrete. Generally two lines of perforations are used, one on each side of the pipe located about 30 degrees down from the horizontal axis. The perforations are spaced about one inch on centers, Fig. 5-8. Drain pipes or tiles are most desirable when the amount of seepage water is small (such as the case of clayey soils) and when the drain pipes can be discharged by gravity into a sewer, ditch, or the like. To avoid the possibility of carrying away fine soil particles, the drain tiles should not be placed below the bottom of the adjacent footings. Also they should be covered with selected filter material so that:

1. The joints or perforations will not become clogged.
2. The fine particles in the surrounding soils will not be carried away by flowing water.

If the amount of water is to be small, a 4 in. diameter tile or pipe placed in a horizontal position will be sufficient. Otherwise, the tiles should be placed on a slope of $\frac{1}{16}$ to $\frac{1}{4}$ in. to a foot. The drainage system should be provided with clean-out and rod-out accesses for periodical cleaning. If the drainage system is carried to an ejector pit in the building, a settling basin should be provided to allow any soil particles to settle down in the basin before going through the pump.

5-5 *Criteria for Selection of Filter Material*

The filter material to be used around the drain pipes should be carefully selected so that it will protect the surrounding soil from being carried away through the drain pipe and that it will prevent clogging the perforations in the drain pipe or openings in the drain tiles. The following empirical rules should be used for selection of the filter material (Corps of Engineer):

1. To prevent the movement of particles from the protected soil into or through the filter material, the following conditions must be satisfied:

$$\frac{15\%\text{ size of filter material}}{85\%\text{ size of protected soil}} \lesseqqgtr 5$$

and

$$\frac{50\%\text{ size of filter material}}{50\%\text{ size of protected soil}} \lesseqqgtr 25$$

2. To prevent clogging the pipe with filter material moving through the perforation or opening, the following conditions must be satisfied:

For slots:

$$\frac{85\%\text{ size of filter material}}{\text{slot width}} > 1.2$$

For circular holes:

$$\frac{85\% \text{ size of filter material}}{\text{hole diameter}} > 1.0$$

For porous concrete pipes, the following criteria may be used:

$$\frac{15\% \text{ size of aggregate in porous pipe}}{85\% \text{ size of filter}} \leqq 5$$

and

$$\frac{15\% \text{ size of filter material}}{15\% \text{ size of protected soil}} \leqq 5$$

The 15 per cent size is the size of particles as shown on the grain size curve, Fig. 1-2, corresponding to 15 per cent finer ordinate. In other words, 15 per cent of the soil particles is finer than that size. Generally, concrete sand (sand satisfying the requirement for fine aggregate for concrete mix) meets the requirements when it is used to protect fine sandy and silty soils.

5-6 *Waterproofing*

With an adequate foundation drainage system, the foundation walls and base slabs are not subjected to hydrostatic pressure. Therefore, the construction cost is generally low. When the soil below water level is pervious, an extensive drainage system may be necessary and consequently very costly. In such cases, the substructure may be designed to resist the permanent hydrostatic pressure, and the portion of the substructure is kept dry by waterproofing.

Waterproofing a substructure may be done in one or a combination of three methods, namely: membrane, hydrolithic and integral waterproofing. Regardless what method is used, all joints below water level should preferably be provided with waterstops, Sec. 5-8.

A. Membrane waterproofing. A waterproofing membrane is a continuous barrier for preventing passage of water, Fig. 5-9 and Fig. 5-10. This barrier

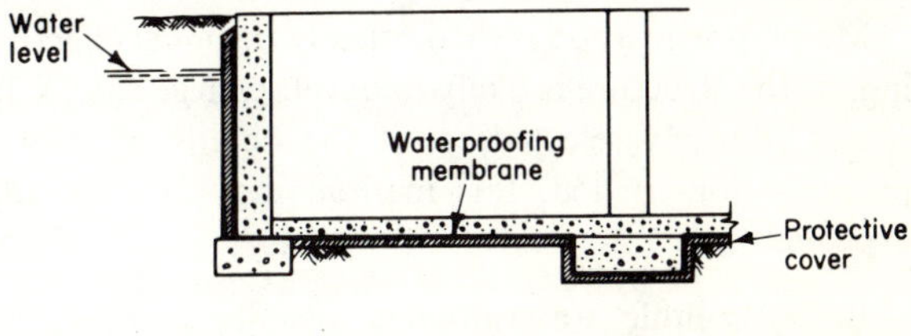

Fig. 5-9 Membrane waterproofing for substructures.

is known as membrane which is made of two or more layers of bitumen-treated cotton fabrics, or two or more layers of bitumen-treated felt with a middle layer of bitumen-treated cotton fabric. The layers are cemented together by bituminous mopping. The fabric or felt should have sufficient tensile strength in order to be able to bridge over cracks in the structure, and the bituminous material should be elastic and self-healing. Both the fabric and bitumen should be of durable type under a prolonged action of free water (AREA, 1953). Coal tar pitch is recognized as more durable than asphalt in this case (Gill, 1959).

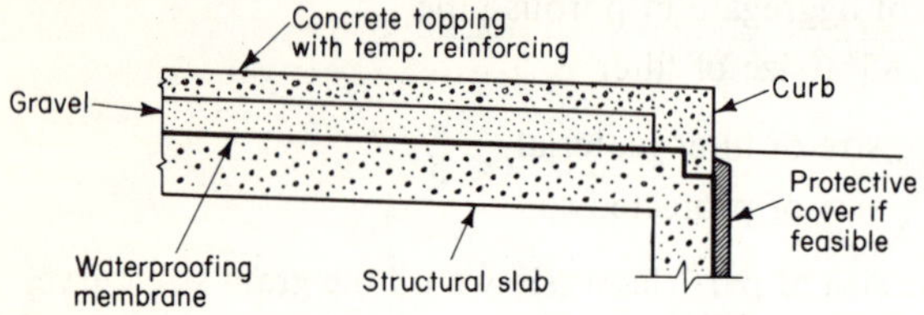

Fig. 5-10 Method of waterproofing roof of sidewalk vault.

In order to provide a continuous water barrier, the successive layers of the membrane should be properly lapped. The lap joints should be staggered, and a minimum two-inch overlap is necessary. The membrane should be extended one or two feet above the highest water level. Caution must be exercised to avoid damaging the membrane during and after it is placed; it should be protected immediately after placing. AREA recognizes the following types of protective covers:

1. A layer of poured-in-place asphalt mastic not less than $1\frac{1}{2}$ in. thick.
2. A layer of asphalt blocks not less than $1\frac{1}{4}$ in thick or a layer of asphalt plank not less than $1\frac{1}{4}$ in. thick, laid in extra heavy mopping of asphalt with joints filled with hot asphalt.
3. A layer of poured-in-place coal tar mastic not less than $1\frac{1}{2}$ in. thick.
4. A layer of adequately reinforced cement mortar or concrete not less than 2 in. thick.
5. A course of hard burned brick not less than $2\frac{1}{2}$ in. thick with joints filled with hot asphalt except when laid as a vertical wall or on a slope, at which time bricks should be laid in cement mortar.

Mastic should not be used on surfaces steeper than $4\frac{1}{2}$ vertical to 12 horizontal.

Membrane method is theoretically the most effective method of waterproofing. If the structure is likely to develop large cracks, if the membrane material is not properly selected, or if the details of joints and terminals are not properly constructed, this method may not provide the watertightness as expected.

B. Hydrolithic waterproofing (ironite method). This method consists of applications of coats of cement plasters containing iron filing to the inside face of the substructure after the shrinkage cracks have developed. Since cracks in masonry substructures are likely to develop in the first one or two years, it may be necessary to apply additional coats from time to time. This method, therefore, is not feasible in cases where the surface is covered with finishing materials. Obviously, this is the best method of waterproofing an existing substructure.

C. Integral waterproofing. A concrete admixture is used to produce a more watertight concrete. There are a variety of admixtures, ranging from lime to commercially made compounds of undefined composition. The purpose of such admixtures is to produce dense concrete but they may not eliminate shrinkage cracks. Therefore, the effectiveness of this type of waterproofing depends largely on the workmanship and on the possibility of complete absence of cracks due to shrinkage or other causes.

5-7 *Dampproofing*

Dampproofing is used to prevent accumulation of surface moisture from condensation. This is achieved by application of asphalt emulsion or coal tar emulsion on the exterior surface of walls and slabs. The surface should be cleaned and the emulsion may be applied by spraying, brushing, or mopping. However, coal tar emulsion should always be sprayed on. Each coat should be allowed to dry before application of the next coat.

Dampproofing is used on the exterior surface of substructures above the high water level. It is not effective in preventing free water under a head, nor effective in preventing moisture from entering cracks and openings.

5-8 *Waterstops*

Construction joints, contraction and expansion joints are possible source of water leakage if not made watertight. Therefore, any such joint below ground water level should be provided with a suitable waterstop. Common waterstops may be classified in three categories: metal, rubber, and mastic types. All waterstops are installed in the joints during construction.

Metal waterstops are continuous strips of thin metal of 6 in. or larger in width. One-half width of the strip is embedded in the concrete on each side of the joint. If some relative movement between the two adjoining parts is anticipated as in the case of expansion and contraction joints, the strip may be bent in a V or bellow shape, Fig. 5-11(a). This type of waterstops may be made of stainless steel, copper, copper-bearing steel, or other durable, corrosion-resisting material. Stainless steel is not commonly used because of its high cost. Copper strips have the disadvantage of being easily damaged during placing concrete. Therefore, metal waterstops are commonly made of copper-bearing steel or ingot iron usually $\frac{1}{8}$ in. thick.

The rubber type waterstops are available in dumbbell (two-bulb), center-bulb (three-bulb), labyrinth, or similar forms, and may be made of rubber,

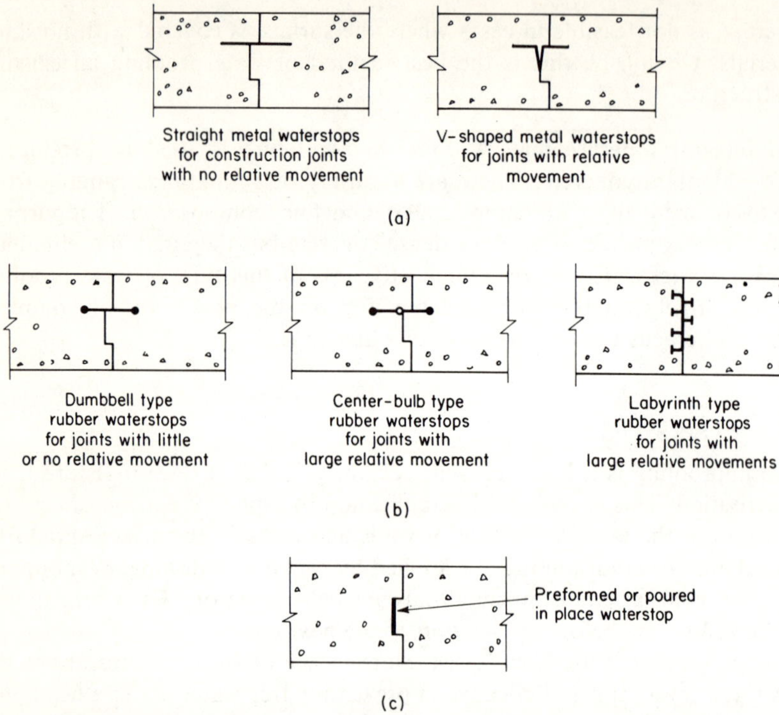

Fig. 5-11 Common types of waterstops: (a) metal waterstops; (b) rubber waterstops; (c) mastic type waterstop.

neoprene, or some forms of plastic material. This type is capable of undergoing displacement because of its high elasticity.

Another type of waterstop is made of preformed or poured-in-place mastic material composed of commercially made compounds. When large movement and large pressure is anticipated, this type may not offer the positive watertightness.

PART 2 FOUNDATIONS

Plate Six

Spread Footings

Spread footings are the most widely used of all types of foundations because of the low cost of construction. Adequately designed spread footings must have ample safety margin against bearing failure and against excessive settlement. They must not overstress any of the underlying soil strata. If they are subjected to eccentric, inclined or uplifting forces, their capacity against such forces must be reassured.

In this chapter, complete procedures of design and analysis of spread footings under conditions encountered in common practice are presented.

6 SPREAD FOOTINGS

6-1 *Use of Spread Footings*

Spread footings are undoubtedly the most widely used type among all foundations because they are usually more economical. Construction of footings requires a least amount of equipment and skill and no heavy or special equipment is necessary. Furthermore, the conditions of the footing and the supporting soil can be readily examined.

Other types of foundations are more favorable than footings when the bearing capacity of the soil is so low that disproportionately large footings would be required, or when the soil immediately under the footing, or at a greater depth, is very compressible that excessive settlement would be resulted if footings are used. Other conditions may require deep foundations instead of footings, for example, bridge piers which must be carried below the depth of scour.

6-2 *Common Types of Footings*

From the point of view of design, footings are classified into four types:

1. Spread footings,
2. Strap footings,
3. Combined footings,
4. Mat foundations.

A spread footing supports one column or one wall. If it supports more than one column or wall, it is called strap footing, combined footing or mat foundation which are discussed in Chapter 7. The common types of footings are shown in Fig. 6-1.

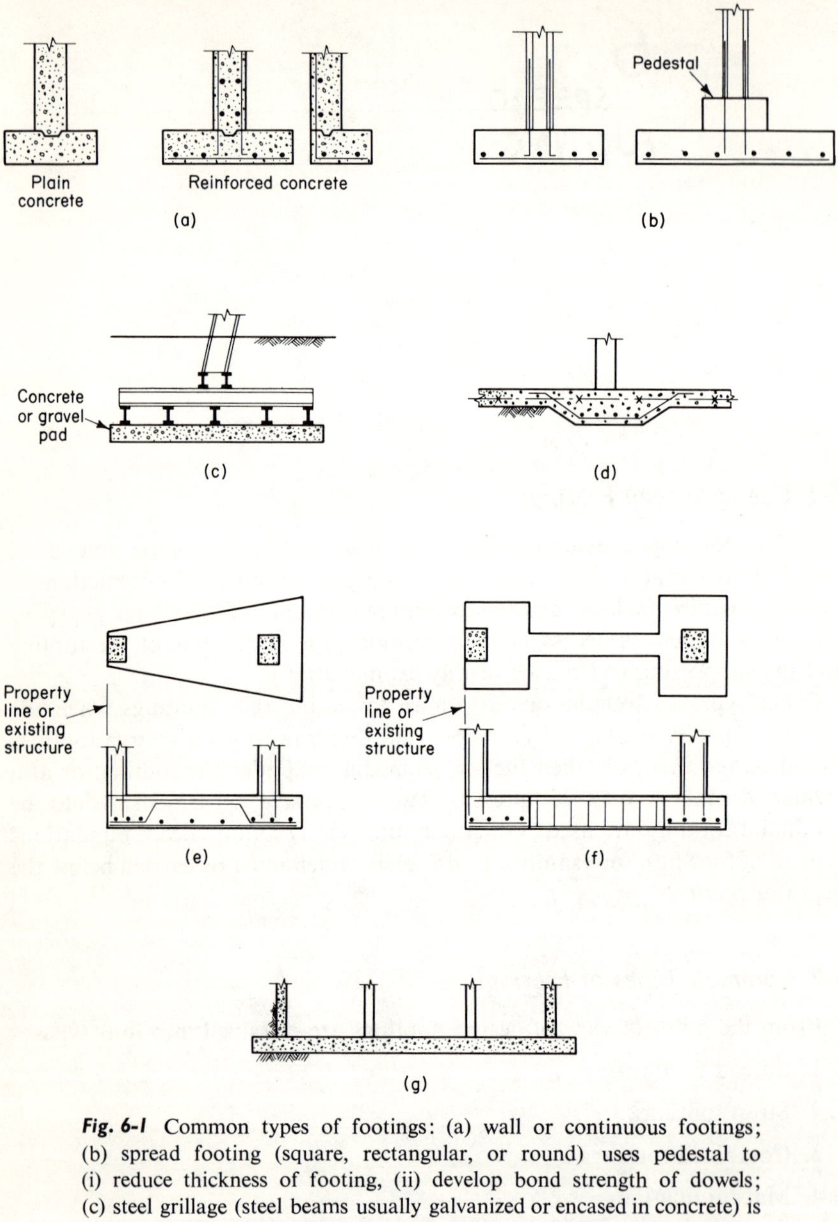

Fig. 6-1 Common types of footings: (a) wall or continuous footings; (b) spread footing (square, rectangular, or round) uses pedestal to (i) reduce thickness of footing, (ii) develop bond strength of dowels; (c) steel grillage (steel beams usually galvanized or encased in concrete) is largely used in transmission tower foundations or under heavy column loads; (d) monolithic footing is used for watertight basement and for resisting uplift pressure; (e) combined footing; (f) strap or pump handle footing; (g) raft or mat foundation is used in soils with low bearing capacity and in soils with extremely erratic characteristics and is very watertight.

6-3 Design of Footings

Footings may be designed by the following procedure:

1. Calculate the loads applied at top of footings. Two types of loads are necessary, one for bearing capacity determination and the other for settlement analysis–Chapter 3.
2. Sketch a soil profile or soil profiles showing the soil stratification at the site. On this profile superimpose an outline of the proposed foundation scheme–Sec. 2-13.
3. Establish the maximum water level–Sec. 2-12.
4. Determine the minimum depth of footings–Sec. 6-4.
5. Determine the bearing capacity of the supporting stratum–Sec. 6-5.
6. Proportion the footing sizes–Sec. 6-6.
7. Check for danger of overstressing the soil strata at greater depths–Sec. 6-7.
8. Predict the total and differential settlements–Sec. 6-8.
9. Check stability against horizontal forces–Sec. 6-10.
10. Check uplift on individual footings and basement slab–Sec. 6-12.
11. Design the footings–Sec. 6-13.
12. Check the need for foundation drains, waterproofing or damp-proofing–Chapter 5.

6-4 Depth of Footings

1. Footings should be carried below the top (organic) soil, miscellaneous fill, abandoned foundation, debris, or muck. If the top soil is too deep, two

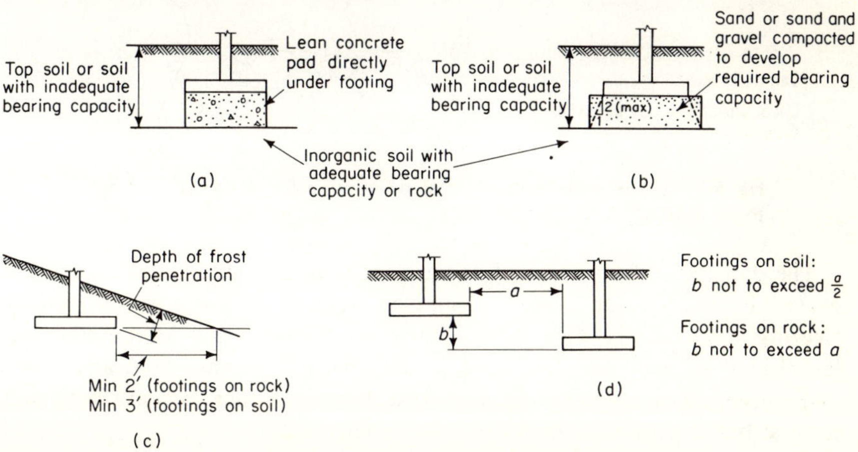

Fig. 6-2 Minimum depth of footings.

alternatives may be used depending upon the relative economy and the time available:

(a) Removing the top soil directly under the footing and replacing it with lean concrete, Fig. 6-2(a);

(b) Removing the top soil in an area larger than the footing and replace it with compacted sand and gravel fill. The area of the compacted fill should be sufficiently large to distribute the footing load as shown in Fig. 6-2(b).

2. Footings should be carried below the depth of frost penetration. In heated buildings, the interior footings are not affected by frost, therefore they may be as high as other requirements permit.

The minimum depths of footings are generally established in large cities and are stipulated in the local building codes. In the absence of such data, reference may be made to the chart shown in Fig. 6-3.

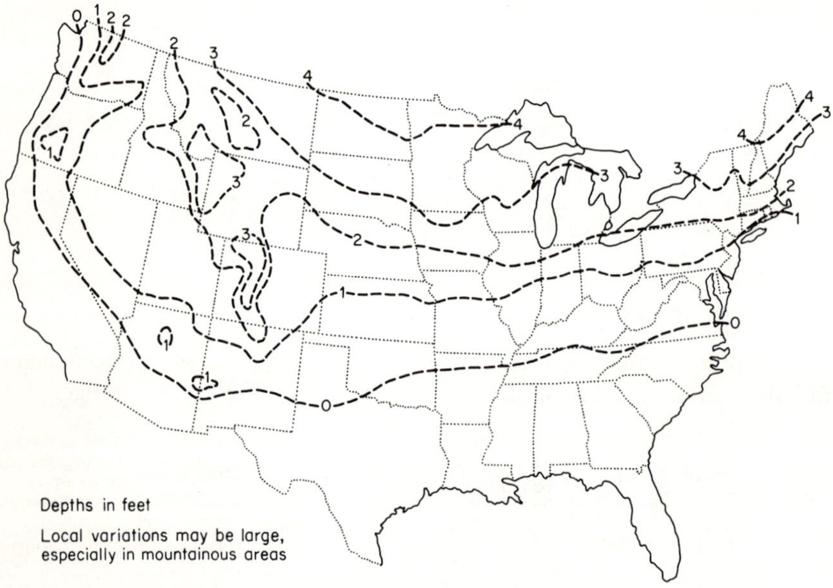

Fig. 6-3 Approximate depth of frost penetration for design of footings. From AREA.

The damage of footings, and of the superstructure, due to frost action is caused by the volume expansion and contraction of water in the soil at freezing temperatures. Gravel and coarse sand above water level containing less than 3 per cent silt, fine sand, or clay particles cannot hold any water and consequently are not subject to frost damage. Other soils are subjected to frost heave within the depth of frost penetration.

3. It is not good practice to place footings on the ground surface even in localities where freezing temperatures do not occur because of the possibility of surface erosion. The minimum depth of footings should be one foot for one and two story dwellings and stores, two feet for heavier construction.

4. Footings on sloping ground should have sufficient edge distance (minimum 2 to 3 ft) as protection against erosion, Fig. 6-2(c).

5. The difference in footing elevations should not be so great as to introduce undesirable overlapping of stresses in soil. This is generally avoided by maintaining the maximum difference in elevation equal to, or equal to one half of the clear distance between two footings, Fig. 6-2(d). This requirement is also necessary to prevent disturbance of soil under the higher footing due to the excavation for the lower footing.

6-5 *Bearing Capacity of Soils under Footings and Mat Foundations*

A. Bearing capacity of granular soils. The bearing capacity of granular soil depends upon the unit weight γ and the angle of internal friction φ of the soil, both of which vary primarily with the relative density of the soil. Compact or dense soils have large γ and φ values and consequently high bearing capacity. Compact soils are naturally not very compressible, therefore cause little settlement. Loose soils, on the other hand, have small γ and φ values, and low bearing power. Even under moderate loads loose soils may suffer large settlement. This is particularly true when the soil is subjected to vibration.

The relative density of granular soils *in situ* is generally determined by standard penetration tests. This test and its limitations are discussed in Sec. 2-7. The relationship between the N values (the standard penetration resistance) and φ values (the angle of internal friction) of granular soils has been established empirically, and is shown in Table 1-1. From this relationship, the bearing capacity can be determined by the Terzaghi theory, Sec. 3-3.

In actual cases the relationship between φ and N values has quite a large scatter. The empirical relationship was established generally on the conservative side. Therefore, the bearing capacity as calculated from this relationship is sometimes considerably lower than the real capacity of the soil. The following quotation from Terzaghi illustrates this point explicitly:

> I consider the [standard penetration] test merely as a means for obtaining preliminary information concerning the degree of homogeneity of the subsoil of . . . sand and as a basis for estimating the upper limiting value for the settlement of the footings. The next step depends on economic considerations. If the building is relatively small it is more economical to design the footings on the basis of the upper limiting value . . . than to make further investigation. On the other hand, if the structure is large and the loads to be carried by the

footings are heavy it is indicated to supplement the results of the . . . penetration data by loading tests to be performed in those locations where the standard penetration tests revealed the presence of the loosest and the densest portion of the subsoil. (Terzaghi, 1957)

In conventional design, the allowable bearing capacity should be taken as the smaller of the following two values.

1. *The allowable bearing pressure based on ultimate capacity.* This allowable pressure is equal to the ultimate bearing capacity divided by an appropriate factor of safety. A factor of safety of 3 is usually used under normal loading conditions and a factor of safety of 2 under combined maximum load. The ultimate bearing capacity of granular soils can be computed by the general equations (3-1a) and (3-1b). However, the following empirical equations are more convenient to use.

For square footing:

$$q_{\text{ult}} = 2N^2BR'_w + 6(100 + N^2)DR_w \tag{6-1a}$$

For very long footings:

$$q_{\text{ult}} = 3N^2BR'_w + 5(100 + N^2)DR_w \tag{6-1b}$$

where q_{ult} = net ultimate bearing pressure, psf;

= pressure at bottom of footing in excess of the pressure at the same level due to the weight of the soil immediately surrounding the footing.

N = standard penetration resistance, number of blows per foot. N values should be adjusted if the penetration test is made at shallow depth, Sec. 2-7.

B = width of footing, ft.

D = depth of footing, ft, measured from ground surface to bottom of footing. If the ground levels on both sides of footing are not equal, D should be measured from the lowest ground level. If $D > B$, use $D = B$ for computation.

R_w and R'_w = correction factors for position of water level, see Fig. 6-4. When the water level is below the bottom of footing, $R_w = 1.0$; and when water level is above the bottom of footing, $R'_w = 0.5$.

2. *The allowable bearing pressure based on tolerable settlement.* This allowable bearing pressure has been established empirically (Terzaghi and Peck, 1948) and may be expressed by the equation:

$$q_a = 720\,(N - 3)\left(\frac{B + 1}{2B}\right)^2 R_w \tag{6-2}$$

where q_a = net allowable bearing pressure in psf for maximum settlement of 1 in. It should be taken as the pressure at the bottom of the footing in excess of the weight of the soil immediately surrounding the footing. The other notations are identical to those in Eq. (6-1). If the maximum tolerable settlement is different from 1 in., Eq. (6-2) may be modified on the assumption

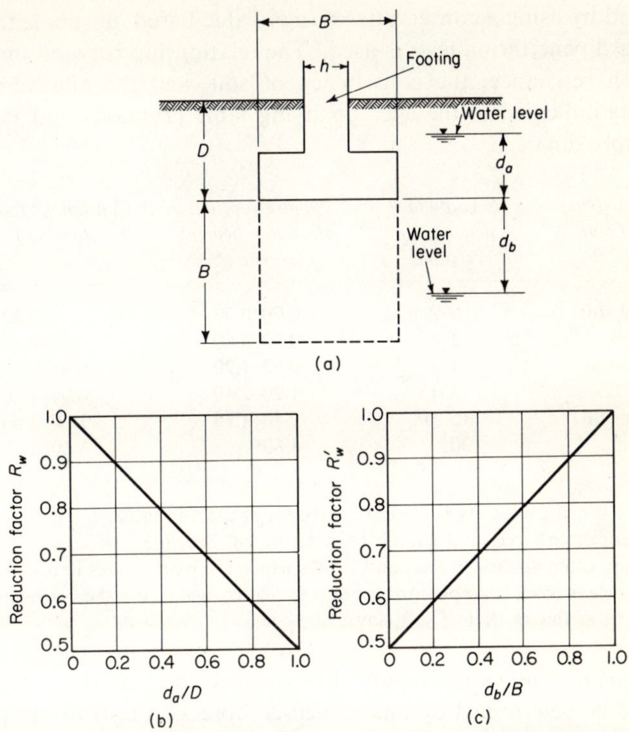

Fig. 6-4 Correction factor for position of water level: (a) depth of water level with respect to dimension of footing; (b) water level above base of footing; (c) water level below base of footing. After AREA.

that settlement is proportional to the bearing pressure. The value of q_a may be increased linearly with depth of footing up to 100 per cent when the depth is equal to the width of the footing. In other words, Eq. (6-2) may be multiplied by the factor $(1 + D/B)$, with a limiting value of 2 when D/B exceeds unity.

The bearing capacity of a footing is largely affected by the characteristics of the volume of soil within a depth equal to about 1 to $1\frac{1}{2}$ times the width of the footing. Unless the soil possesses some cohesion, the upper layer of one to two feet can be easily disturbed and loosened by construction operation. Therefore, it is not advisable to use large bearing capacity for small or narrow footings such as those supporting continuous walls, even if the natural soil is very compact.

B. Bearing capacity of clay and clayey soils. The ultimate bearing capacity of these soils depends primarily upon its consistency (or shear strength). The consistency can be determined by:

1. *Standard penetration tests.* For small jobs where a better economy can be achieved by using a conservative design value based on simple test results, the standard penetration tests is used. The relationship between the standard penetration resistance, the consistency of soil, and the allowable bearing capacity as indicated in the accompanying table (Terzaghi and Peck, 1948) is very approximate.

Consistency of soil	*N (Standard penetration resistance)*	*Square footings Allowable bearing pressure**	*Continuous footings tons/sq ft*
Very soft†	0–2	0.00–0.30	0.22
Soft†	2–4	0.30–0.60	0.22–0.45
Medium†	4–8	0.60–1.20	0.45–0.90
Stiff††	8–15	1.20–2.40	0.90–1.80
Very stiff††	15–30	2.40–4.80	1.80–3.60
Hard	30†	4.80†	3.60†

* Ultimate bearing capacity is equal to three times the allowable.

† Settlement often large, and should be determined, Sec. 6-8.

†† Stiff clays often possess fissures and cracks which are weak planes in resisting shearing forces. Such clays must be kept from being softened by water, the shear strength on these planes may be as low as that of soft clays.

2. *Unconfined compression tests.* For average projects the consistency of soil should be determined by unconfined compression tests of samples taken with thin-walled steel tubing known as Shelby tubes, Sec. 2-8. The ultimate bearing capacity is (Skempton, 1951):

$$q_{\text{ult}} = cN_c \tag{6-3}$$

where q_{ult} = net ultimate bearing capacity,

c = cohesion = $\frac{1}{2}$ unconfined compression strength,

N_c = bearing capacity factor of clay, see Fig. 6-5, which depends on the shape of the footing.

If the clay under the footing (within a depth equal to the width of the footing) consists of two layers, the bearing capacity may be determined by the following equation (Buttons, 1953).

$$q = c_1 N_c' \qquad \text{(6-4}$$

where c_1 = cohesion of the soil in the upper layer, tons/sq ft;

N_c' = bearing capacity of two-layer system, see Fig. 6-6.

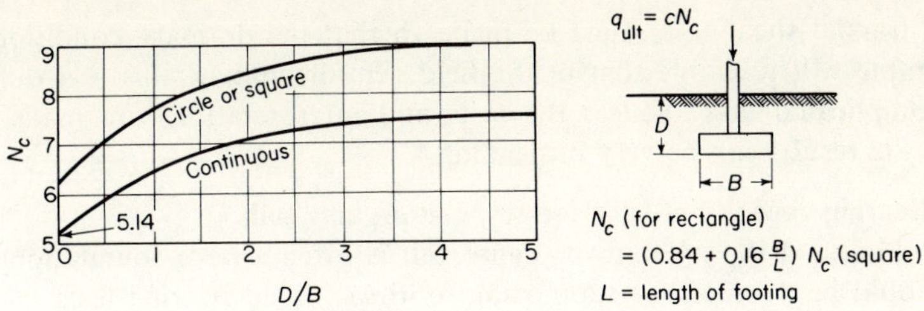

Fig. 6-5 Ultimate bearing capacity of clay. After A. W. Skempton.

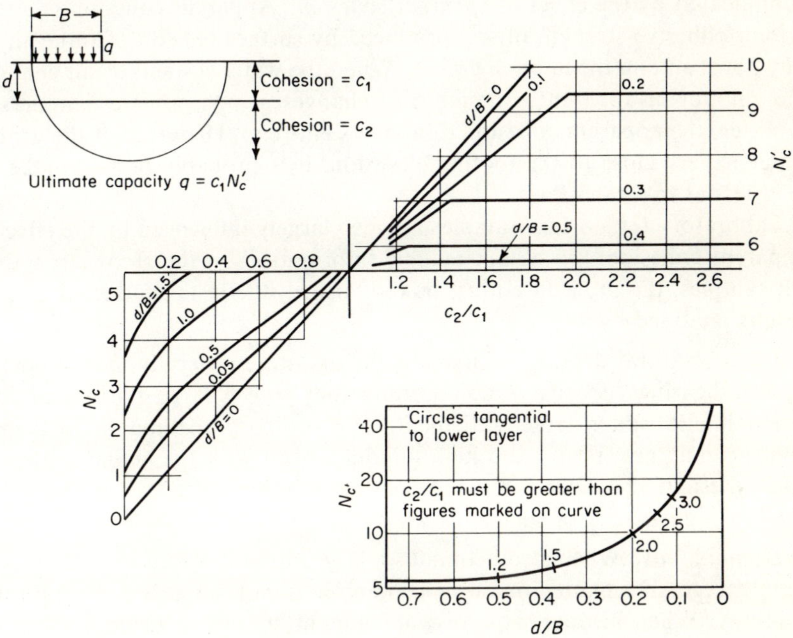

Fig. 6-6 Ultimate bearing capacity of clay (two-layer system). From S. J. Button.

3. *Triaxial tests.* For very large projects, the shear strength may be determined from triaxial tests of undisturbed samples. The test results are plotted in terms of the following parameters (Sec. 1-5):

$$s = c + (\sigma - u) \tan \varphi$$

where s = shear strength,

c = cohesion,

σ = total normal stress,

u = pore water pressure,

φ = angle of internal friction.

The triaxial shear tests must be made so that the drainage conditions in the sample will resemble that in the field. Unfortunately, this is a difficult and complicated task. Unless the tests and interpretations are made correctly, the results can be very misleading.*

C. Bearing capacity of silts, loesses, and organic soils.

Silt. Unless it is very hard or dense, silt is often a poor foundation soil, and should be avoided for supporting footings. Load bearing tests may be resorted to for determining the bearing capacity of silt. In addition to all the shortcomings of load bearing tests, the bearing capacity of silt is further complicated by the effect of *apparent cohesion.* Apparent cohesion is a temporary cohesive strength of soil produced by surface tension of water in the tiny pores among the fine soil grains. When the water content of silt changes, the amount of apparent cohesion also changes. Upon complete immersion in water, the apparent cohesion disappears entirely. Therefore, if the ground water is very close to the footing elevation, it is probably best to make the load test at the water level.

Laboratory tests on silt samples are also largely influenced by the effect of apparent cohesion. To obtain reliable information on the shear strength of silt samples, the procedure must be established, the tests conducted and the results analysed by an expert.

Loess. Natural deposits of loess in the dry state generally have moderate or low bearing capacity. Upon wetting they lose a large portion of their strength and suffer large settlement. The ordinary method of settlement analysis may give results too high in comparison with the actual settlement (Bolognesi, 1957). Therefore, load bearing test should be used in evaluating the bearing capacity of such soils.

Organic soils. When a soil contains a large amount of organic matter with visible vegetable matter or organic odor, it is not suitable for supporting footings. When in doubt, the organic content should be determined in the laboratory. Highly organic soils will settle in due time even under their own weight. Settlement is due partly to the extrusion of water from the soil (consolidation) and partly to chemical reactions in the organic matter (decomposition).

D. Bearing capacity of compacted fills. Soils ranging from plastic clay to sand and gravel have been used for compacted fill to support structures. The bearing capacity of fill depends largely on the type of soil and the degree of compaction. A well compacted sand and/or gravel is capable of supporting large pressure whereas a poorly compacted clay has extremely low bearing

---

* For advanced study, reference is made to the *Proceedings of ASCE Conference on Shear Strength* (1961) and *Proceedings of Conference on Pore Pressure and Suction in Soils*, Institute of Civil Engineers (London, 1960).

capacity. The bearing capacity of compacted fill may be determined before or after the placement of the fill:

1. *Determine the bearing capacity before placement of the compacted fill.* Usually the soil is compacted to 90–100 per cent of the maximum density as determined by the standard or the modified compaction tests. If the soil is cohesive, samples having the desired degree of compaction may be tested to determine the unconfined compression strength (or the c and φ value by triaxial compression tests). From the unconfined strength (or c, φ values) the bearing capacity can be evaluated in the same manner as for natural soil deposits.

If the soil is granular (noncohesive), samples having the desired degree of compaction may be subjected to direct shear tests, triaxial tests, or relative density tests. The purpose of direct shear and triaxial tests is to determine the value of φ from which the bearing capacity can be calculated. Indirectly, the φ value may be estimated from Table 1-1 if the relative density of the fill is determined, Sec. 1-6(a).

2. *Determine the bearing capacity after placement of fill.* Fills not placed under field control should not be used for supporting footings. Otherwise the condition must be thoroughly explored before being used. In order to determine the bearing capacity of the fill in place, soil borings and tests must be made in the same manner as for natural deposits.

E. Bearing capacity of rocks. Almost all rocks can withstand a compressive stress higher than concrete. Following are some of the exceptions:

1. Limestones with cavities and fissures which may be filled with clay or silt.
2. Rocks with bedding planes, folds, faults, or joints at an angle with the bottom of footing.
3. Soft rocks often reduce their strength after wetting. Weathered rocks are very treacherous. Shales may become clay or silt in a matter of hours of soaking.

The common sandstones and limestones have modulus of elasticity from that of a poor concrete to high strength concrete. Very hard igneous and metamorphic rocks exhibit considerably greater value of modulus of elasticity. See Sec. 1-11 to 1-13 for additional discussion.

F. Bearing capacity determined by load test. Load bearing tests give reliable results only when the soil condition is uniform from the bottom of the footing to a depth at least equal to the width of the largest footing. Since settlement in cohesive and partially cohesive soils takes place in a long period of time, load bearing tests on such soils are not very practical. Fortunately, the bearing capacity and the settlement characteristics of such soils can be readily determined by laboratory tests on the relatively undisturbed samples.

The results of load bearing tests on granular soils are useful provided that the test is made with extreme care. The following are some of the factors that should be considered.

1. The test should be made on the loosest area contemplated to support any footing.
2. The depth of ground water in the test case and in the actual cases should be comparable. Avoid making test on a layer affected by capillary water.
3. Each load increment is maintained until no further settlement of significant magnitude takes place.
4. The ground is not frozen during the test.

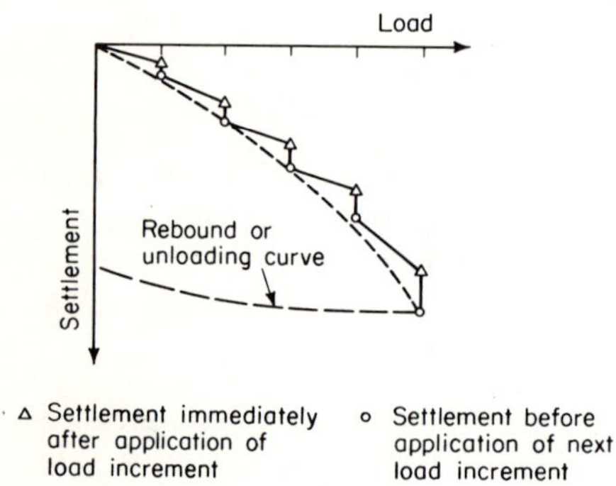

Fig. 6-7 Graphical presentation of results of load bearing test.

There are many other factors which influence the test results. It is advisable to follow the standard test procedure of ASTM Designation D 1194.

The results of load bearing tests should be plotted in a graph similar to one shown in Fig. 6-7. Whenever economically justified, more than one load bearing test should be made. Because of variation in soil characteristics and other factors, two tests made under identical conditions on a presumably uniform soil often have considerably different load-settlement curves. Therefore, results of load bearing tests require careful and expert interpretation.

6-6 *Footing Size Proportions*

Footing sizes determined by allowable bearing pressures are usually satisfactory provided that a settlement analysis is made and that the footing sizes are revised if the analysis indicates excessive settlement. Some engineers intend to minimize the differential settlement due to varying live loads by proportioning the footings in such sizes that all footings will have the same average bearing pressure under the service load. The service load is the actual load expected to act on the foundation during the normal service of the structure. In ordinary buildings, it may be taken as dead load plus one-half live load. A larger percentage of live load should be used in warehouses and other storage floors. This procedure was discussed in Sec. 3-9, and is outlined below for the convenience of the designers.

Let L_{l+d} = live load + dead load for the column which has the largest live load/dead load ratio;

L_s = service load for the same column;

= dead load + $\frac{1}{2}$ live load for ordinary buildings;

q_a = allowable bearing pressure as determined by the principles discussed in Sec. 6-5;

q_d = design pressure for all footings except the one with largest live load/dead load ratio.

Then A = area of footing supporting the column with the largest live load/dead load ratio.

$= L_{l+d}/q_a$

$q_d = L_s/A$

$$\text{Area for other footings} = \frac{\text{Service load}}{q_d}$$

6-7 *Stress on Lower Strata*

1. For stability analysis of footings, the pressure under a footing may be assumed to spread out on a slope of 2 vertical to 1 horizontal. Thus, a load Q acting concentrically on a footing area of $B \times L$ is assumed to be distributed over an area of $(B + Z)(L + Z)$ at a depth Z below the footing, Fig. 6-8. If any stratum of soil is inadequate to sustain this spread-out pressure, the design bearing pressure should be reduced. However, for a two layer system of clays, the procedure described in Fig. 6-11 gives more reliable results.

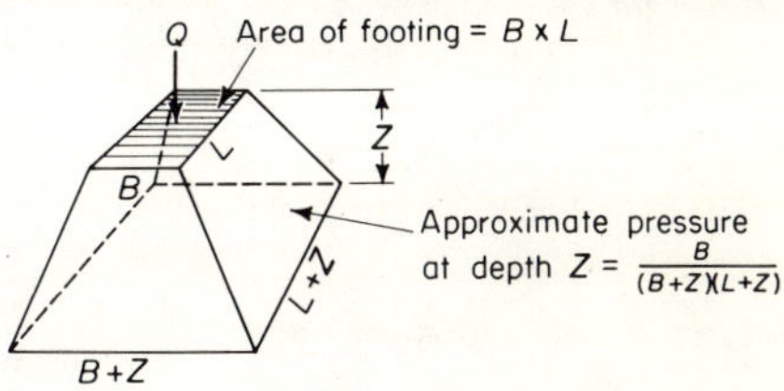

Fig. 6-8 Approximate distribution of vertical pressure under footing.

2. For settlement analysis, the approximation above may not be sufficient, and a more accurate approach based on elastic theory may be required. All elastic methods are developed from the Boussinesq's equation which deals with a single load acting on the surface of a half-space (infinitely large area and depth).

$$q = \frac{3Qz^3}{2\pi R^5} = \frac{3Q}{2\pi z^2} \cos^5\psi \tag{6-5}$$

where q = vertical stress at any given point;

Q = surface load;

z = depth of the given point;

$r = \sqrt{x^2 + y^2 + z^2}$, see Fig. 6-9;

ψ = angle between line R and vertical.

Based on Boussinesq's equation, the vertical stresses under continuous, rectangular and circular footings have been computed. The results are shown in Fig. 6-10. In these figures the magnitude of vertical pressure at various points are given in terms of the bearing pressure q. For example the vertical pressure at any point along the line $0.2q$ is equal to 20

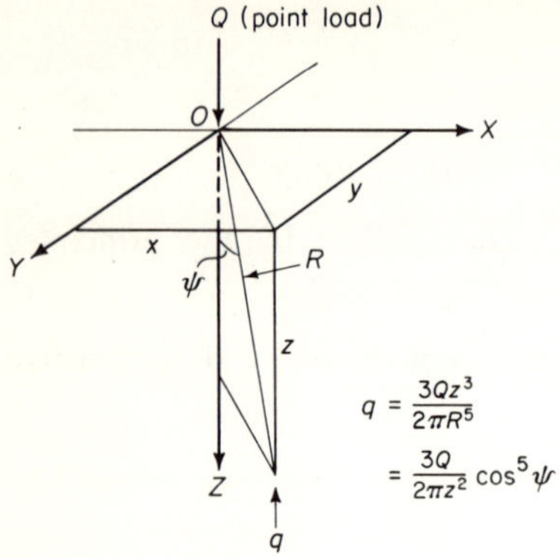

Fig. 6-9 Vertical stress due to a point load.

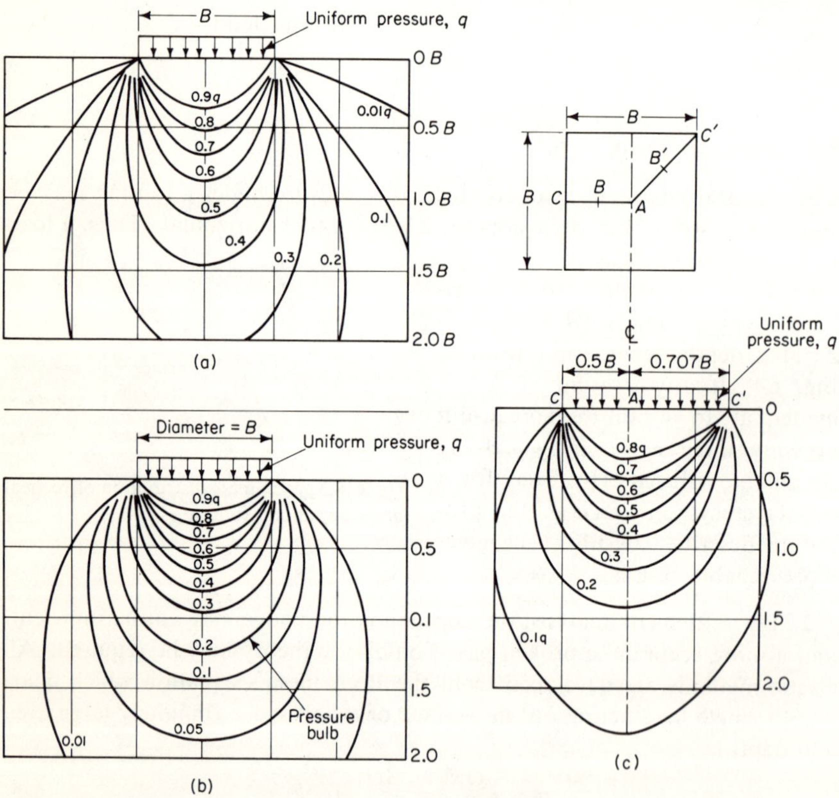

Fig. 6-10 Vertical stresses under footing: (a) under a continuous footing: (b) under a circular footing; (c) under a square footing.

Fig. 6-11 Newmark influence chart for computing vertical pressure. After Corps of Engineers.

per cent of the applied contact pressure. These lines of equal pressure are bulb shaped and consequently are called *pressure bulbs*. The most commonly used pressure bulb is the one for $0.2q$ because in practical cases any stress less than $0.2q$ is often of little consequence. For circular and square footings the pressure bulb is about $1.5B$ wide and $1.5B$ deep, B being the width of the footing.

The computation of vertical pressure by the Boussinesq's equation is a laborious procedure and suitable only for research works. In practice a graphical solution by the Newmark influence chart, Fig. 6-11, is used. The solution is simple, expeditious and can be best illustrated by an example. In this example it is desired to determine the vertical pressure at a depth of 10 ft below point x due to a uniform contact pressure $q = 4500$ psf from a footing shown in Fig. 6-12. The first step is to draw a plan of the footing and the location of point x on a transparent paper in such a scale that the distance AB shown on the influence chart is equal to the depth 10 ft. Then place the plan on top of the influence chart, so that point x lies at the origin of the chart, and count the number of influence areas occupied by the footing. An influence area is an individual area bounded by two adjacent straight lines and two adjacent arcs. The vertical pressure at a depth of 10 ft below point x is equal to the number of influence areas (78) times the intensity of footing pressure (4500) times the influence valve (0.001) which equals 350 psf.

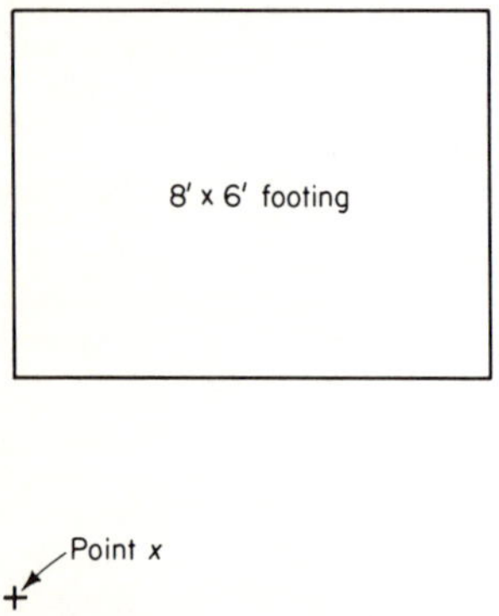

Fig. 6-12 Example illustrating the use of the Newmark influence chart.

Both the Boussinesq's equation and the Newmark influence chart are intended for the case of surface loading. If they are used for computing stresses in the soil due to a deep foundation, the computed stress would be greater than the actual value.

6-8 Settlement of Footings

Footings on granular soils will not suffer detrimental settlement if the smaller value of the two allowable pressures given by Eqs. (6-1) and (6-2) is used. Footings on stiff clay, hard clay, and other firm soils generally require no settlement analysis if the design provides a minimum factor of safety of 3. Soft clay, compressible silt, and other weak soils will settle even under moderate pressure, and therefore settlement analysis is necessary.

The total settlement of a footing on clay may be considered to consist of three parts (Skempton and Bjerrum, 1957):

$$S = S_i + S_c + S_s \tag{6-6}$$

where S = total settlement,

S_i = immediate elastic settlement,

S_c = settlement due to consolidation of clay,

S_s = settlement due to secondary consolidation of clay.

1. *Immediate settlement.* Immediately upon application of load on the footing, elastic compression of the underlying soil takes place causing a settlement of the footing. This amount can be computed by elastic theory. However, it is usually very small and can be neglected for all practical purposes.

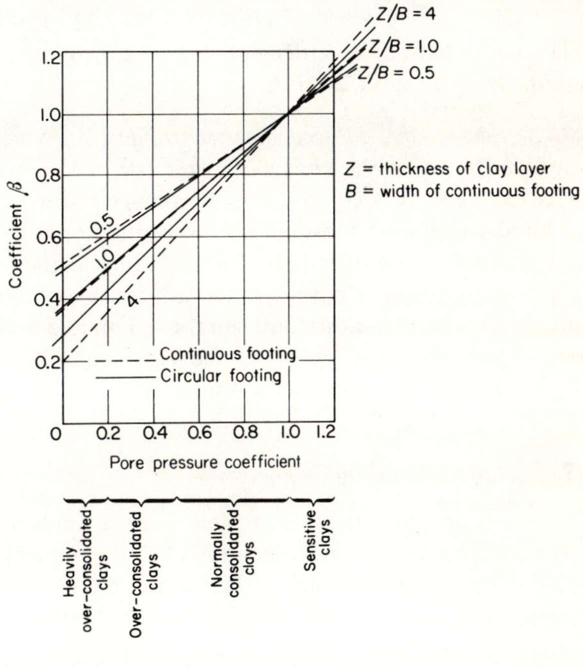

Fig. 6-13 Coefficient β for computing consolidation settlement. From Skempton and Bjerrum.

2. *Settlement due to consolidation.* The settlement caused by consolidation is due to the slow extrusion of water from the pores of the fine particles of clay. The amount of final consolidation settlement S_c can be calculated by the following equation:

$$S_c = S_0\beta \tag{6-7}$$

where β = the coefficient depending on the geometry of the footing and the loading history of the clay. Values of β are shown in Fig. 6-13.

S_0 = settlement calculated by Terzaghi theory of consolidation;

$$= m_v \Delta p H \tag{6-8}$$

$$= \frac{C_c}{1 + e_0} H \log_{10} \frac{p_0 + \Delta p}{p_0} \tag{3-4}$$

where m_v = coefficient of volume compressibility of the clay. This value is determined by consolidation test.

Δp = vertical stress due to load on footing.

H = thickness of the compressible clay. The clay thickness should be divided into several layers to obtain reasonably accurate settlement of a thick layer.

C_c = compression index, also determined by consolidation test.

p_0 = vertical effective pressure due to soil overburden.

The computation of settlement due to consolidation is illustrated in the design example, sheet 2 DE 6.

3. *Settlement due to secondary consolidation.* When an undisturbed soil sample is tested in the consolidometer (or oedometer) the rate of volume decrease checks very closely with the theory. However, when the sample is one hundred per cent consolidated (according to the theory of consolidation) the volume decrease does not stop according to the theory, but instead the sample continues to compress at a reduced and rather constant rate. The amount of consolidation that can be computed by the theory is called primary consolidation; whereas the slow consolidation that takes place afterwards is called secondary consolidation, Sec. 3-5.

6-9 *Eccentric Loading*

Eccentric loading may result from a load applied off the center of the footing or from a concentric load plus a bending moment. For the purpose of determining the pressure under the footing the moment may be removed by shifting the vertical load to a fictitious location with an eccentricity e = moment/vertical load. In the analysis of an eccentrically loaded footing two separate problems are confronted:

1. For the purpose of structural design, the pressure against the bottom of the footing, commonly called contact pressure, is assumed to have a planar distribution. When the load is applied within the kern of the footing area, common flexural formulae are applicable.

$$q = \frac{Q}{A} \pm \frac{M_x}{I_y} x + \frac{M_y}{I_x} y \tag{6-9}$$

where q = contact pressure at a given point (x, y);

Q = vertical load;

A = area of footing;

x and y = coordinates of the point at which the contact pressure is calculated;

M_x, M_y = load Q multiplied by eccentricity parallel to x and y axes, respectively;

I_x, I_y = moment of inertia of the footing area about the x and y axes, respectively.

Equation (6-9) is valid when one of the following conditions exists:

(a) The footing is symmetrical about x and y axes.

(b) The footing is symmetrical about x axis and $e_y = 0$.

(c) The footing is symmetrical about y axis and $e_x = 0$.

For rectangular footings, Eq. (6-9) may be written in a simpler form.

$$q = \frac{Q}{A}\left(1 \pm 6\frac{e_l}{L} \pm 6\frac{e_b}{B}\right) \tag{6-9a}$$

When e_x, e_y or e_b, e_l exceed a certain limit, Eq. (6-9) or (6-9a) gives a negative value of q which indicates tension between the soil and bottom of footing. Unless the footing is weighted down by surcharge loads, the soil cannot be relied upon for bonding to the footing and offering tensile resistance. Therefore, the flexural formulae Eq. (6-9) and (6-9a) are applicable only when the load is applied within a limited area which is known as the kern and is shown shaded in Fig. 6-14(a). The procedure for determination of soil pressure when the load is applied outside the kern is simple in principle but laborious. Cases for rectangular and circular footings have been worked out and the kerns are shown by shaded areas in Fig. 6-14 [(a) and (c)]. For footings of other shapes, the graphical method of successive trials is probably the simplest for practical solutions (Roark, 1954).

The graphical method, similar to any other method, is based on the assumption that the pressure varies linearly with the distance to the neutral axis from zero at the neutral axis to a maximum at the most remote point and on the requirement of statical equilibrium that the resultant of the soil pressure should lie on the line of action of the applied load Q. The procedure is as follows. Draw a trial neutral axis N-N, Fig. 6-14(b) and a line ab perpendicular to N-N, starting from point b which is most remote. The area between point b and N-N is under compression while the area on the other side of N-N is unstressed. The intensity of stress at a given point varies in simple proportion with its perpendicular distance from N-N. The compression area is divided into several narrow strips of uniform width dy, running parallel to

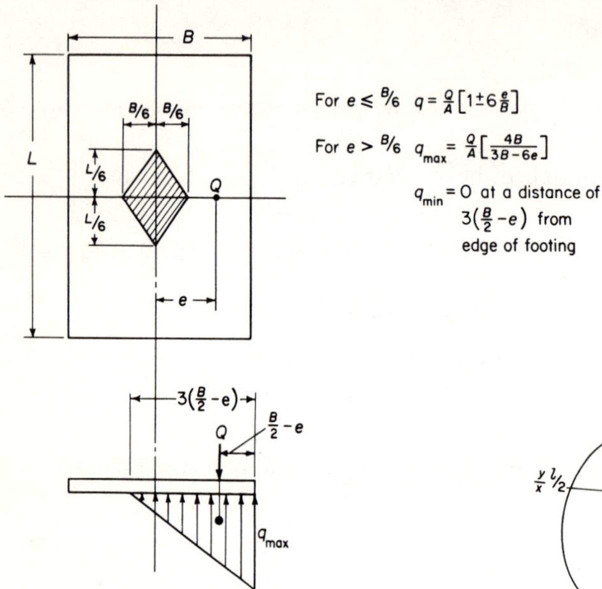

(a) Rectangular footing, load on one of the center lines of footing.

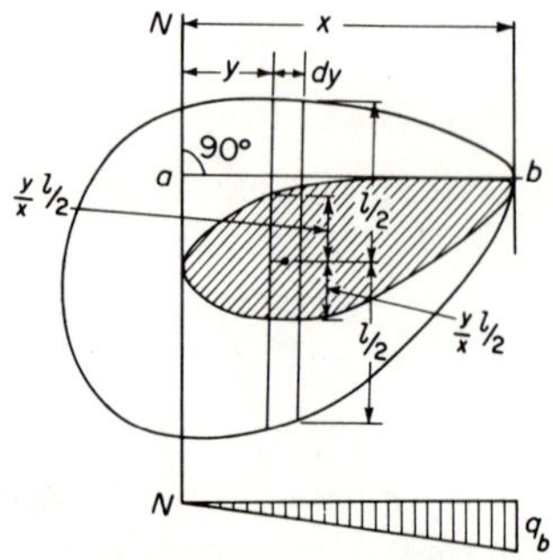

(b) General procedure.

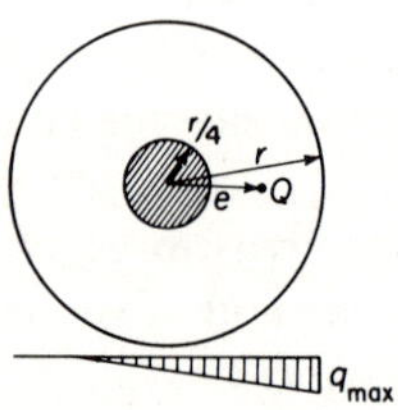

For $e \leq r/4$ $\quad q = \frac{Q}{A}\left[1 \pm 4\frac{e}{r}\right]$

For $e > r/4$ $\quad q_{max} = k\frac{Q}{A}$

$A = \pi r^2$

k values are tabulated below

e/r =	0.25	0.30	0.35	0.40	0.45	0.50	0.55	0.60	0.65	0.70	0.75	0.80	0.90
k =	2.00	2.20	2.43	2.70	3.10	3.55	4.22	4.92	5.90	7.20	9.20	13.0	80.0

(c) Circular footing.

Fig. 6-14 Pressure distribution used for structural design of eccentrically loaded footings.

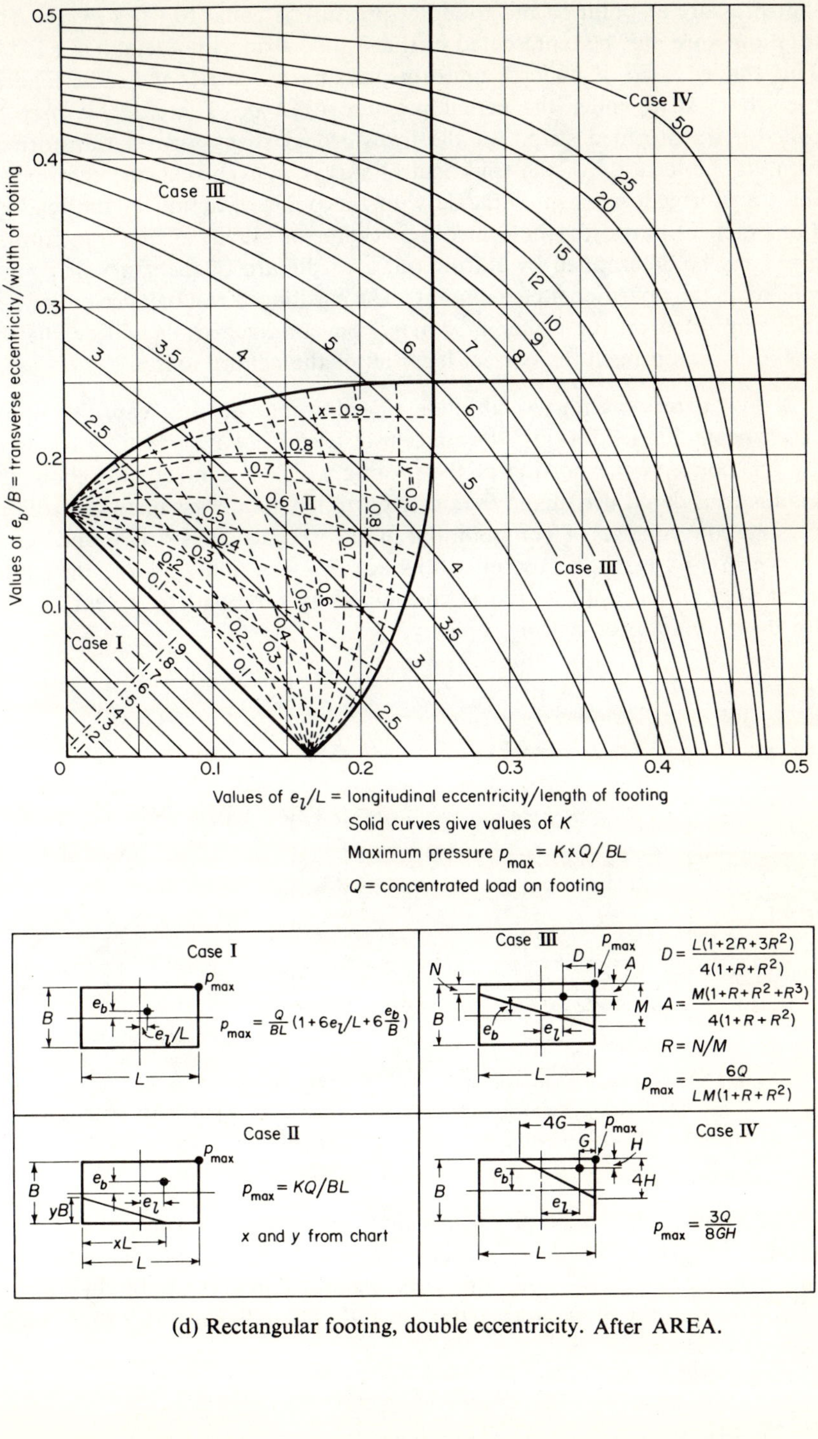

(d) Rectangular footing, double eccentricity. After AREA.

N-N. The unit pressure acting on this strip is equal to $(Y/X)q_b$, where q_b is the unit pressure at point b, and the total pressure is equal to $(Y/X)q_b l dy$. The total pressure may be represented by the shaded strip with a length of $(Y/X)l$. This shaded strip, if under a uniform pressure q_b, carries the same load as the whole strip under the actual pressure $(Y/X)q_b$. Therefore, it may be called a transformed strip. All the transformed strips form a transformed area. If the location of the trial neutral axis N-N is correct, the centroid of the transformed area will coincide with the point of action of the load Q. For practical purposes, the centroid or center of gravity of the transformed area may be determined by cutting out a cardboard of the same shape and balancing the board on a pencil point. The cardboard will balance only when it is supported on the center of gravity. Several such trials will enable the engineer to approach the correct location of the neutral axis.

2. For determination of ultimate or allowable bearing capacity of an eccentrically loaded footing, the concept of *useful width* has been introduced. By this concept, the portion of the footing which is symmetrical about the load is considered useful and the other portion is simply assumed superfluous for the convenience of computation. If the eccentricities are e_l and e_b, as shown in Fig. 6-15, the useful widths are $B - 2e_b$ and $L - 2e_l$, the equivalent area $(B - 2e_b)(L - 2e_l)$ is considered as subjected to a central load for determination of bearing capacity.

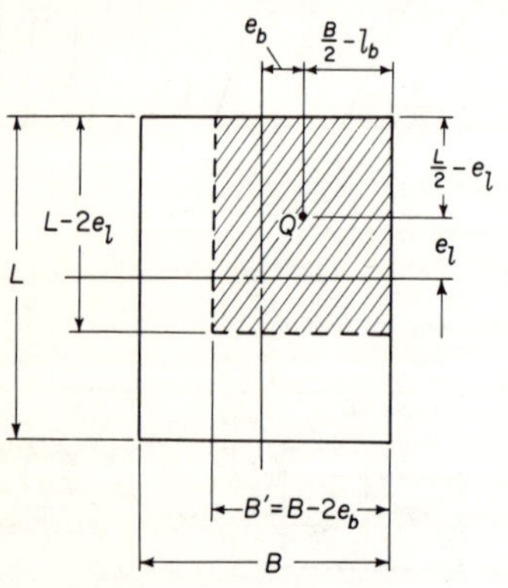

Fig. 6-15 Useful widths for determination of bearing capacity of eccentrically loaded footing on cohesive soils.

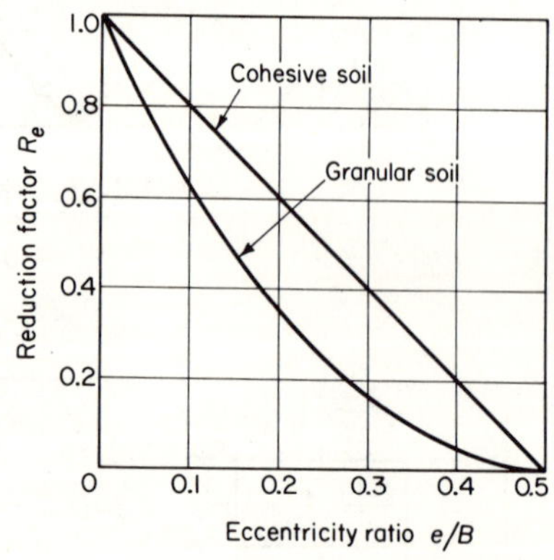

Fig. 6-16 Bearing capacity of eccentrically loaded footing. After AREA.

The concept above simply means that the bearing capacity of a footing decreases linearly with the eccentricity of load, as is shown by a straight line in Fig. 6-16. In cohesive soils, this linear relationship prevails, but in granular soils, however, the reduction is parabolic rather than linear, (Meyerhof, 1953).

Therefore, the reduction factor shown in Fig. 6-16 should be used for design purposes: First the bearing capacity of the footing is determined on the basis that the load is applied at the centroid of the footing. Then, this bearing capacity is corrected by multiplying with the factor shown in Fig. 6-16.

6-10 Inclined Load

The conventional method of stability analysis of footings subjected to inclined loads is as follows: the inclined load Q is resolved into a vertical component Q_v and a horizontal component Q_H. The stability of the footing against ultimate failure under the vertical load is treated by the same principles for footings subjected to vertical load only, and the effect of the horizontal component is ignored. Then, the stability of the footing against the horizontal force is analysed by calculating the factor of safety against sliding which is defined as the ratio between the total horizontal resistance and the horizontal force. The total horizontal resistance in general consists of a passive resistance of soil, P_p, and a frictional resistance R, Fig. 6-17. The value of P_p can be

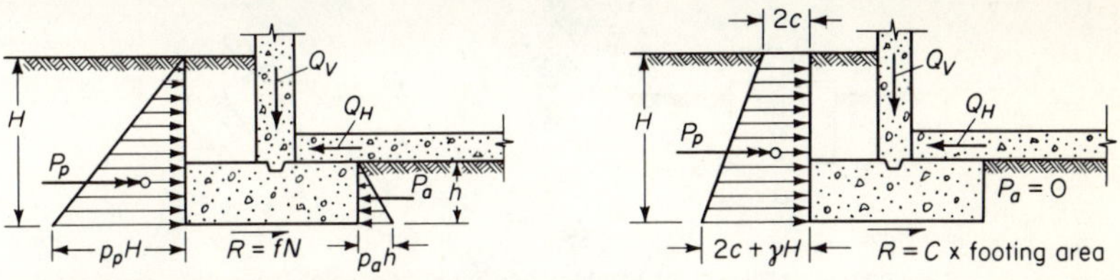

N = total vertical force acting on the base of footing

Factor of safety against sliding $= \dfrac{P_p - P_a + R}{Q_H}$

Granular soils

Type of Soil	P_p psf		Coef. of Friction, f
	submerged	dry or moist	
Sand and/or gravel with less than 5% silt	210	350	0.55
Sand and/or gravel with 5% or more silt	180	250	0.45
Silt or soils containing more than 30% silt	120	150	0.35

Cohesive soils

Type of Soil	Cohesive Strength c = psf	Unit Weight, r pcf
Very soft clay	200	110
Soft clay	400	120
Medium, stiff, and hard clay	600	125

The values above may be used in small jobs. Backfill must be well compacted to insure the design passive pressure

Fig. 6-17 Conventional method of analysis of footings subjected to inclined loads.

determined by the principles discussed in Chapter 4. However, for smaller projects, conservative values such as those shown in the figure may be used. It should be emphasized that high values of passive earth pressure P_p may not be realized in granular soils unless it is backfilled and well compacted in layers.

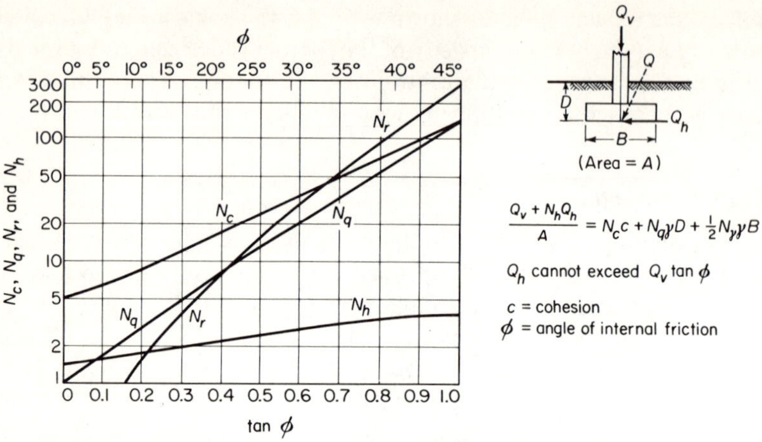

Fig. 6-18 General formula for bearing capacity of continuous footing subjected to inclined load. After N. Janbu.

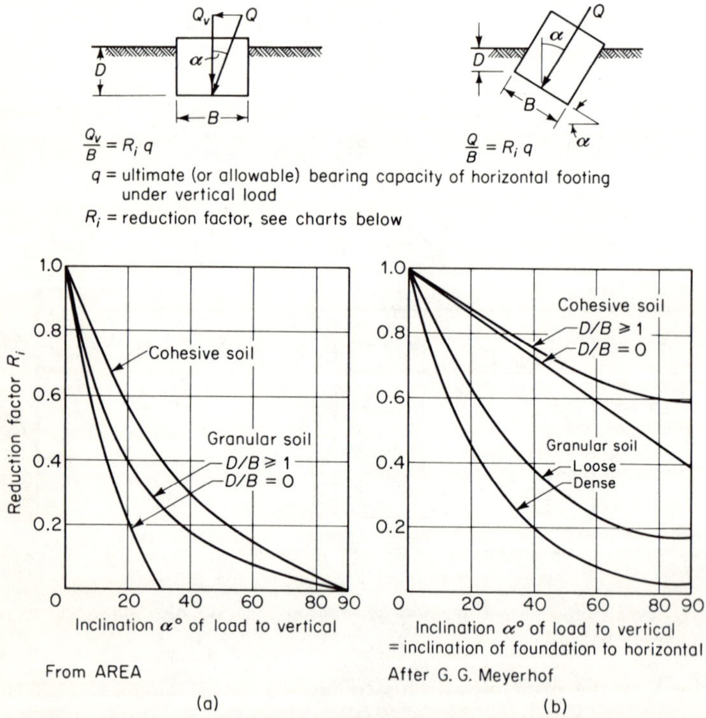

Fig. 6-19 Bearing capacity of footing subjected to inclined load: (a) horizontal foundation; (b) inclined foundation (after G. G. Meyerhof and AREA).

Research in soil mechanics has extended the bearing capacity theory into the case of inclined loading (Meyerhof, 1953; Janbu, 1957). Janbu's analysis is a direct extension of Terzaghi theory with a factor N_h in addition to the Terzaghi bearing capacity factors N_c, N_γ, and N_q.

$$\frac{Q + N_h Q_h}{A} = N_c c + N_q \gamma D + \frac{1}{2} N_\gamma \gamma B \qquad (6\text{-}10)$$

The notations and values of N_c, N_q, N_γ, and N_h are shown in Fig. 6-18.

Meyerhof has calculated ultimate bearing capacity of footings subjected to inclined loading and published the results in graphical form. They have been constructed in convenient charts shown in Fig. 6-19. The load is assumed to apply vertically and the bearing capacity is determined by the normal procedure. Then it is corrected by the factor R_i shown in Fig. 6-19.

6-11 Footings on Slopes

The bearing capacity of footings on sloping ground may be determined by the following equation (Meyerhof, 1957):

$$q = cN_{cq} + \tfrac{1}{2}\gamma B N_{\gamma q} \qquad (6\text{-}11)$$

The values of the bearing capacity factors N_{cq} and $N_{\gamma q}$ for continuous footings are shown in Fig. 6-20. These factors vary with the slope of the ground, the relative position of the footing and the angle of internal friction of the soil.

Before construction of footings on sloping ground, the stability of the slope itself must be investigated. Footings should not be constructed on slopes which are unstable. They should also be avoided on slopes where slow creep of the superficial material takes place. The stability of a stable slope may be endangered by the addition of footings.

6-12 Uplift of Footings

The resistance of a footing against uplift is derived from the weight of the footing and the weight of soil above it. For soil below ground water level the submerged weight should be used.

As a footing is being uplifted, a prism of soil is carried by the footing, Fig. 6-21(a). The shape of the prism depends upon the characteristics of soil above the footing. Due to lack of conclusive data, no rational design rules have been developed. However, conventional method assuming a 60 degree cone, Fig. 6-21(a) may lead to unsafe results. For footings subjected to a small uplift, the method shown in Fig. 6-21(b) may be used. If a large number of footings are subjected to high uplift forces, some model tests or full-sized field pull-out tests may be economically justified.

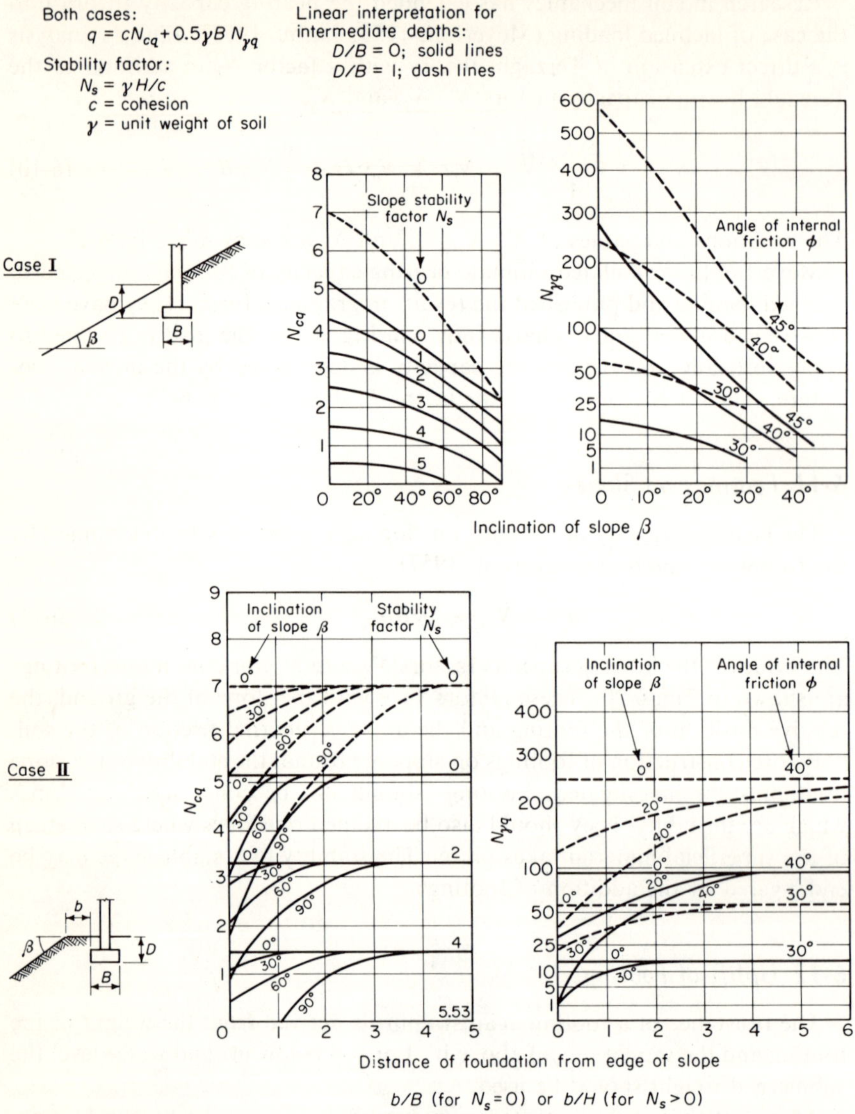

Fig. 6-20 Ultimate bearing capacity of continuous footings on slopes. After G. G. Meyerhof.

6-13 *Structural Design of Footings*

In practice all individual and wall footings are designed on the assumption that the distribution of the soil pressure against the bottom of the footing is straight-line or planar. When the load is applied at the centroid of the footing area, the unit pressure is equal to the total load divided by the footing area. In case of eccentric load, the pressure may be calculated by the procedure described in Sec. 6-9.

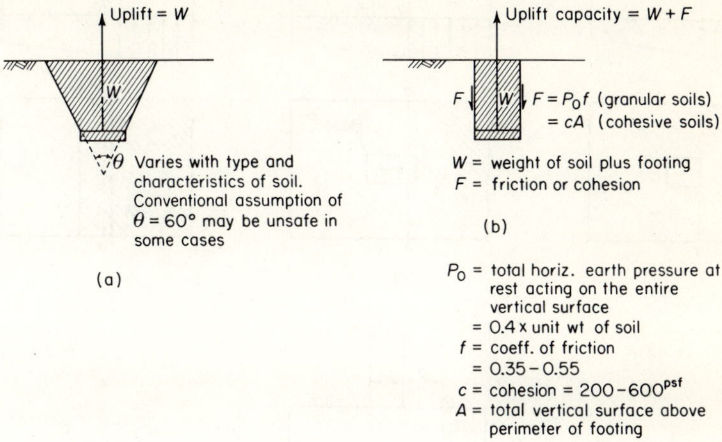

Fig. 6-21 Uplift capacity of footing: (a) probable uplift capacity; (b) minimum theoretical uplift.

By far the majority of footings are constructed of concrete, and the design of such footings should follow the concrete codes.* The design criteria used in the current American practice are shown in Fig. 6-22.

If a pedestal is so proportioned that its height is at least equal to twice its width beyond the face of column, Fig. 6-23, the critical sections for computing bending, bond, and shear stresses are as shown in Fig. 6-22, and there is no need to analyse the stresses in the pedestal. For pedestals having smaller depth/width ratio, the stresses in the pedestal must be analysed. The analysis may be made on the assumption that the bond stress along the entire embedment of dowels below the top of the pedestal is uniformly distributed. Based on this assumption, the total stress acting on the bottom of the pedestal is equal to the total stress in the concrete of the column plus the amount of stress in the column vertical reinforcement transmitted through bond within the depth of the pedestal. Fig. 6-23 illustrates the stresses acting on each element of the footing.

The members in a steel grillage are designed as cantilever beams subjected to uniformly distributed soil pressure.

---

* American Concrete Institute, American Association of State Highway Officials, American Railroad Engineers Association, Canadian National Code, British Code of Practice, or the local building codes.

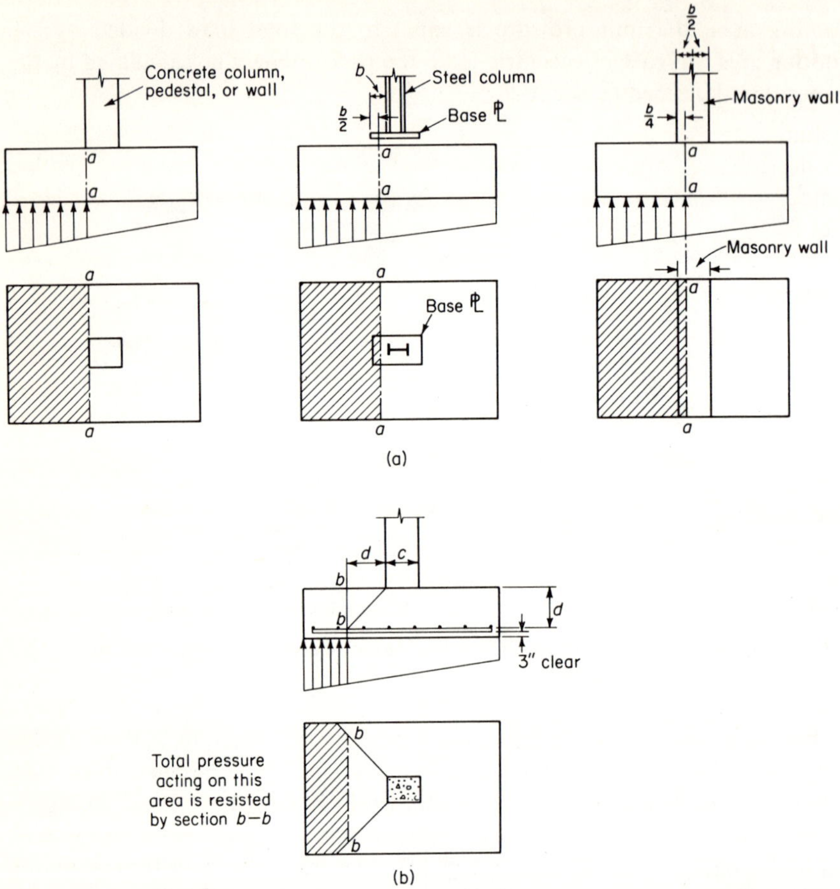

Fig. 6-22 Criteria for design of concrete footings: (a) critical section (*a-a*) and load area for computing bond and bending stresses; (b) critical section (*b-b*) and load area for computing shear stress.

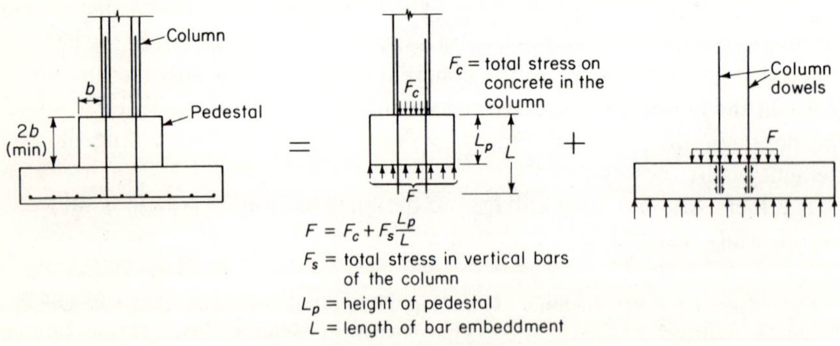

Fig. 6-23 Stresses in pedestaled footings.

6-14 *Fixity of Column Base and Rotation of Footing*

The engineer is sometimes confronted with the question of whether the column bases should be fixed or free to rotate. At other times he is compelled to design the footings for a central load and a moment, and for a limited amount of rotation. Therefore an understanding of the rotation characteristics of the column base and the footing is essential.

When the lower end of a column is subjected to a bending moment, the joint between the column and the footing must be strong enough to transfer the stresses. In the case of concrete columns, this can be readily done by embedding the dowels in the footing, and the column may be considered fully fixed to the footing. The lower end of steel columns may be fixed to the footings by means of anchor bolts. When the anchor bolts are required to resist tension due to column bending, they are generally pretensioned to a desired minimum stress. Unless this stress is exceeded from actual bending of the column, there is no elongation of the anchor bolts. Again, the column bases may be considered as fixed to the footings. In either case, the rotation of column base is caused only by the elastic deformation due to the greater compression at the toe of the base, which is generally small and insignificant. In the event when the anchor bolts are not pretensioned and the rotation of the base is critical, reference is made to a paper by Salmon, Schenker, and Johnston (1957).

Regardless of the degree of fixity between the column base and the footing, a movement from the column will cause unsymmetrical soil pressure. The soil pressure is assumed to have straight line or planar distribution, Sec. 6-9A. Unfortunately the pressure distribution is not likely to be planar and cannot be determined quantitatively. Therefore, the rotation of a footing acted by a moment or an eccentric loading can only be estimated on the basis of some simple computations guided by good engineering judgement. For example, small and shallow footings on sand are prone to rotation because the sand readily runs out from under the toe of the footing. If the footing is located at a greater depth, the sand is subjected to a confining pressure due to the weight of the overlying soil. The relative effect of the edge condition diminishes as the size of the footing increases. It becomes apparent that small and shallow footings on granular soils should not be relied upon for providing fixity to the column bases.

Contrary to sand, clay and clayey soils resemble elastic material and are capable of resisting a concentrated stress at the edge. Furthermore, since a large portion of the settlement of footings on clay is due to consolidation, over a long period of time bending moment acting only in short durations would not cause significant rotation.

As an example of the type of approach to this problem, the rotation of footings subjected to moment or eccentric load may be estimated by the following analysis. The toe of the footing will probably not settle more than the amount S_1 which is the average settlement if the entire footing is subjected to the maximum toe pressure; the heel of the footing probably not more than the amount S_2 which is the average settlement if the entire footing is subjected to the minimum pressure at the heel. The maximum and minimum pressures are computed on the assumption of straight line or planar distribution. The probable amount of rotation, therefore, is equal to or less than $(S_1 - S_2)$ divided by the width (or length) of the footing.

6-15 *Construction*

Footings are the simplest type of foundation in so far as the construction procedure is concerned. In addition to the normal exercise of precaution there are relatively few points that require special attention, namely: the inspection of subsoil conditions, the relative depth of footings, and the dewatering of the excavation when necessary.

The construction of footings for buildings is usually started after the general grading work is completed at which time the ground is leveled to an elevation at, or 6 in. below, the bottom of the lowest floor slab. Then the area is excavated by simple or power operated hand tools. The bottom of the excavation is carefully excavated to the required depth, the form work for the sides of footing is placed and held by stakes, and the reinforcement is placed on cement block supports (and high chairs if top bars are used). Before placing the concrete, anchor bolts or column dowels must be accurately secured on the form work. Short and straight dowels of small diameter may be placed by hand immediately after the concrete is poured. The form work for the sides may not be necessary and the concrete may be poured against the vertical sides of the excavation if the soil does not slough in.

A. Inspection of subsoil conditions. Natural soil deposits are seldom truly uniform. An apparently uniform soil stratum often contains pockets or lenses of material having somewhat different engineering properties. It is impractical and almost impossible to ascertain the soil condition under each footing by ordinary soil boring program. Therefore, it is the responsibility of the engineer to evaluate the average soil condition based on the soil boring results, and often he has to make conservative generalizations. Before the foundation is finally constructed, he must check the actual conditions in the field. If the soil conditions at certain footing locations are not as good as he has assumed, the footing must be either lowered to a stratum having sufficient bearing power or enlarged to reduce the pressure to suit the bearing capacity of the soil. The choice between these two methods depends upon the relative economy, the time, or other factors involved.

The method for checking the soil conditions at the footing excavations must be simple and expeditious. The soil conditions should be inspected after the excavation but before concreting. For clays or clayey soils, soil samples may be taken by a hand auger or shovel, and the approximate strength may be determined by a simple portable unconfined compression tester or by a pocket size penetrometer. In most cases, the shear strength can be estimated by the simple thumb test described in Table 1-2, Sec. 1-8A.

For sand or gravel, some simple penetration tests may be used for comparison of soil density at various locations. The penetration test may be simply the counting of blows required to drive a certain size reinforcing rod with a specific weight dropping a given height. (For example $\frac{7}{8}$ in. diam rod driven by a 7 lb hammer falling 18 in.). Such tests should be made first at locations where the soil density (and consequently the bearing capacity) is known from the soil borings or tests, and the results should be used as a basis for comparison. If further tests made at any other footing locations encounter smaller resistance, the adequacy of the soil for sustaining the design pressure must be carefully investigated by more accurate tests or load bearing tests.

B. Relative depth of footings. Any adjacent footings should not be constructed at such different levels that the construction of the lower footing would disturb the soil supporting the upper footing, and that the pressure from the upper footing would not introduce undue additional stress to the soil under the lower footing. This difficulty is generally avoided by keeping the difference in footing elevations not greater than one-half the clear distance between the footings. For this reason it is always a good practice to construct the lower footings first, and when necessary to construct the lower footing at a greater depth than contemplated, the elevation of the upper footing can be adjusted accordingly.

Sometimes the adjacent footings must be constructed at largely different levels, for example, when a new basement is constructed adjacent to footings under an existing first floor. Sheeting may be used to retain the adjacent ground when excavation is made.

The problem of footings at two different levels is illustrated in Fig. 6-24 where a wall footing at the first floor adjoins a basement wall. It is the common practice to lower the first floor footing in gradual steps down to the level of the basement footing as shown in Fig. 6-24(a). By so doing the natural state of the subsoil is considered unaltered. An alternative method is shown in Fig. 6-24(b). In order to construct the basement, an excavation larger than the basement floor must be made. After the basement wall is matured, the overexcavated area is backfilled with suitable soil. If the original soil is sand or gravel, and the backfill consists of the same material which is compacted in layers (6 to 9 in.) to a density equal to or greater than that of the original soil, the footing at the first floor may be supported on the

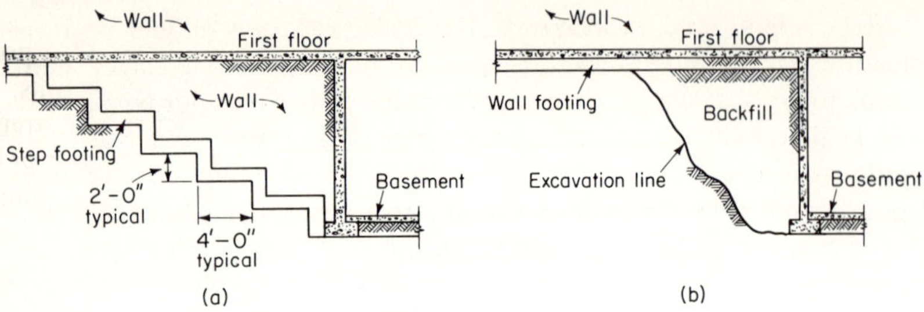

Fig. 6-24 Wall footings at different levels.

backfill. If there is any doubt of the bearing capacity of the backfill, the wall footing should be stepped down as shown in Fig. 6-24(a) or else the wall itself should be designed to span between the basement wall and a point on the original ground at several feet from the excavation line.

C. Dewatering. The excavation should be kept dry during the construction period because free water is objectionable for several reasons. In clay or clayey soils, free water tends to soften the upper portion of the soil and causes settlement of footings. The soil conditions under water cannot be readily inspected. Excavation in water is expensive and not satisfactory. Furthermore, the quality of concrete placed in water is questionable, particularly when the water is not stagnant.

To avoid the difficulties mentioned above, excavations below ground water level are kept dry by various methods discussed in Chapter 5.

6-16 Design Example

On sheet 1, Plate DE 6, the column loads, walls loads, floor loads, and pertinent soil data are shown. On the right-hand side of the soil profile, results of the standard penetration tests are shown for the granular soils, and the unconfined compression strength q_u, natural void ratio e_0, and compression index C_c for the soft clay are also indicated. The water level was 25 ft below the finished grade.

The first step was to determine the bearing capacity of the upper sand layer. The N value was adjusted in accordance with Eq. (2-1). Because the adjusted value exceeds 2 times the test value (N') a reduction factor of 2 was used. This gives $N = 28$.

The stress on the layer of loose sand was analysed by the approximate method discussed in Sec. 6-7.

The total settlement of the footings consisted of three components, namely the settlement due to each of the three layers of soil above the hardpan. The hardpan itself contributes practically no settlement, as indicated by the local experience. The bedrock lies immediately under the hardpan.

According to Eq. (6-2) the medium-dense sand layer will settle 1 in. at a pressure q_2 = 8600 lb per sq ft. Since an allowable value of 5000 lb per sq ft was used in design, the approximate settlement is equal to 5000/8600 = 0.58 in. The same procedure is used for settlement calculation for the loose sand layer. The consolidation settlement of the soft clay was computed by Eq. (3-4), with the values of e_0 and C_c determined by laboratory tests.

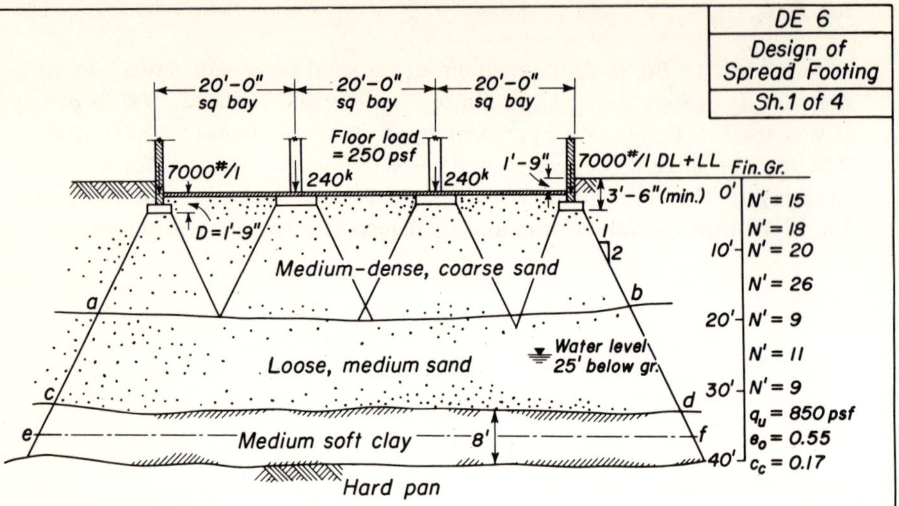

Bearing Capacity Assume borings were taken from finish grade. At depth of 8 ft, $p = 8 \times 110/144 = 6$

Eq. (2-1) $N = 18\left(\frac{50}{6+10}\right) = 56$

$\times \frac{1}{2} = 28$ (use)

Col. footing: Eq. (6-1a), $q_1 = \frac{1}{3}q_{ult} = \frac{2}{3}N^2B + 2(100 + N^2)D$

$= \frac{2}{3}(28)^2B + 2(100 + 28^2)2.5 = 520B + 4400$

$D = 2.5$ (estimated)

$q_1 = 7500$ psf (trial $B = 6'$)

Eq. (6-2), $q_2 = 720(28-3)\left(\frac{6+1}{12}\right)^2 = 6100$ psf

$\times\left(1 + \frac{D}{B}\right) = 8600$ psf

Because of cumulative settlement, q_2 must be reduced to 5000 psf, as will be seen in settlement analysis

Wall footing: Eq. (6-1b), $q_1 = \frac{1}{3}q_{ult} = N^2B + 1.67(100 + N^2)D$

$= 780B + 1470D$

$= 4130$ psf ($D = 1{:}75'$, trial $B = 2'$)

Eq. (6-2), $q_2 = 720(28-3)\left(\frac{2+1}{2\times 2}\right)^2 = 10{,}100$ psf

Use 2'-0" wide wall footing

Try $q = 5$ ksf for col. footing

Stress on Lower Strata

Since the 2:1 stress distribution lines intersect practically at the top of the layer of loose sand, the stress on this and the lower layer may be approximated by assuming a uniform distribution

Average total structural load $= (7000 + \frac{240{,}000}{20})/31'$

$= 613$ psf

Floor load $= \underline{250}$

863

Weight of soil removed $= 1.75 \times 110 = \underline{193}$

Net $q = 670$ psf

(1) On top of layer of loose sand,

$$q = 670 \times \frac{62'}{77'} = 540 \text{ psf}$$

(77' = length ab)

(2) At mid-height of clay layer,

$$q = 670 \times \frac{62'}{95'} = 437 \text{ psf}$$

(95' = length ef)

Both of these stresses are very low. No problem in shear failure.

Settlement Analysis

(1) Medium-dense sand layer:

$$S = \frac{q}{q_2} = \frac{5000}{8600} = 0.58 \text{ in.}$$

(2) Loose sand layer:

$$q_2 = 720(9-3)\left(\frac{77+1}{2 \times 77}\right)^2 0.5 = 540$$

($0.5 = R_w$) $\times 2$ (due to benefit of depth) $= 1080$ psf

$$S = \frac{q}{q_2} = \frac{540}{1080} = 0.50 \text{ in.}$$

(3) Clay: Eq. (3-7), $S = \frac{C_c}{1+e_o} H \log_{10}\left(\frac{p_o + \Delta p}{p_o}\right)$

$C_c = 0.17$, $e_o = 0.55$ } From laboratory tests

Δp = Increase in vertical pressure at mid-height of layer = 437 psf

p_o = Original vertical pressure
= 20' moist medium dense sand = 20 x 115 = 2300
5' moist loose sand = 5 x 105 = 525
8' submerged loose sand = 8 x 60 = 480
4' clay, submerged = 4 x 60 = 240
3545 psf

$$S = \frac{0.17}{1.55}(96'')\log\left(1+\frac{437}{3545}\right) = 0.53 \text{ in.}$$

Layer	Total settlement		Probable differential settlement
(1)	0.58"	x 3/4 =	0.44"
(2)	0.50"	x 1/2 =	0.25"
(3)	0.53"	x 1/2 =	0.27"
	1.61"		0.96"

Settlements are acceptable for warehouse structure, although they are higher than normally allowed. 1/2 total settlement is taken as possible differential settlement at greater depths.

Horizontal Thrust Wind = 25 psf
x 48' height = 1200#/' of building

$$\text{Passive } P = \tfrac{1}{2}p_pH^2 \geq \tfrac{1}{2}(200)(3.5)^2 = 1225\#/' \quad \text{OK}$$

Uplift None

Structural Design of Col. Footing

For the purpose of illustration, the footing is assumed to be limited to a maximum of 6 ft width

A = 240/5 = 48 sq ft 8' x 6' footing

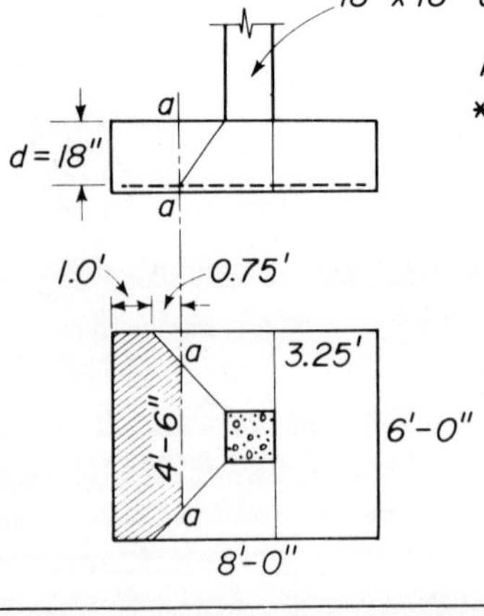

ACI Code is used in this design.
* indicates 1956 Code value

$f_c' = 3000$ psi
$f_s = 20{,}000$ psi
v = 75 psi*
u = 240 psi*

Bearing q = 240/48 = 5 ksf

Try $d = 18''$ $D = 22''$

Long way: Total shear area (shaded area) $= \frac{6 + 4.5}{2} \times 0.75 + 6' \times 1 = 9.94^{sf}$

V (for shear) $= 9.94 \times 5 = 49.7^{k}$

$$v = \frac{49,700}{54(7/8)*18} = 59 \text{ psi} \qquad < 75 \text{ OK}$$

V (for bond) $= 6.0' \times 3.25' \times 5 = 97.5^{k}$

$$\Sigma O = \frac{97,500}{240(7/8)*18} = 26.0'' \text{ req'd.}$$

$$M = 97.5 \times \frac{3.25}{2} = 159^{lk}$$

$$As = \frac{159}{1.44(18)} \times 0.85 = 5.2 \,\square'' \text{ req'd.}$$

Use 12–#6 ($As = 5.28 \,\square''$, $\Sigma O = 28.8''$)

Short way:

V (for bond) $= 8.0 \times 2.25' \times 5 = 90^{k}$

$$\Sigma O = \frac{90,000}{240(7/8)*18} = 24.0'' \text{ req'd.}$$

$$M = 90 \times \frac{2.25}{2} = 101^{lk}$$

$$As = \frac{101}{1.44(18)} \times 0.85 = 3.21 \,\square'' \text{ req'd.}$$

Use 12–#5 ($As = 3.72 \,\square''$, $\Sigma O = 23.9''$)

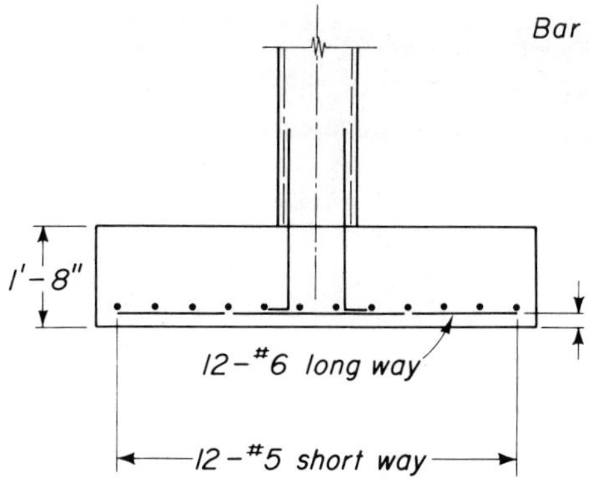

Bar placement:

$$\frac{2}{8/6 + 1} = 0.86 \times 12 = 11$$

Require 11 short bars in central 6'-0" width

Uniform spacing OK

Plate Seven

A Mat Foundation under Construction

The design of combined footings and mat foundations is a difficult problem in two ways. First the structure is so highly indeterminate that rigorous analysis based on elastic theory is not available. Second, the foundation soil is not elastic and the reaction against combined footings and mat foundations is difficult to predetermine. Because of these difficulties, such foundations are commonly designed by simplified procedures and modified with conservative and experienced judgement.

Progress has been made in the structural analysis and the understanding of the subgrade reaction. These analyses are developed from the advanced structural theories. Unlike the conventional method they are more complex. However, they provide information regarding the stresses in various parts of the foundation. Again, they must be also used with experienced judgement. They are presented, together with the conventional method, in this chapter for the benefit of advanced students.

7 STRAP FOOTINGS, COMBINED FOOTINGS, AND MAT FOUNDATIONS

7-1 Uses of Strap Footings, Combined Footings, and Mat Foundations

Square footings are most economical for supporting square and round columns. Under rectangular columns or walls, rectangular footings are appropriate. These shapes should be used wherever the conditions permit.

However, when a column is near or right next to a property limit, a square or rectangular footing concentrically located under the column would extend into the adjoining property. If the adjoining property is a public sidewalk or alley, local building codes may permit such footings to project into public property. Before doing so the engineer should consult the local codes. But when the adjoining property is privately owned, the footings must be constructed within the property. In such cases, there are three alternatives which are illustrated in a schematic plan shown in Fig. 7-1. These alternatives are:

1. *Strap footing*. A strap footing comprises two or more footings connected by a beam called a strap. This type is also known as a cantilever footing or a pump-handle foundation.
2. *Combined footing*. A combined footing is a long footing supporting two or more columns in one row.
3. *Mat or raft foundation*. A mat foundation is a large footing, usually supporting several columns in two or more rows.

The choice between these types depends primarily upon the relative cost. As a rule the strap footing is more economical than the combined footing where the subsoil has large bearing capacity. However, if the required strap becomes large and deep, the combined footing may be less expensive.

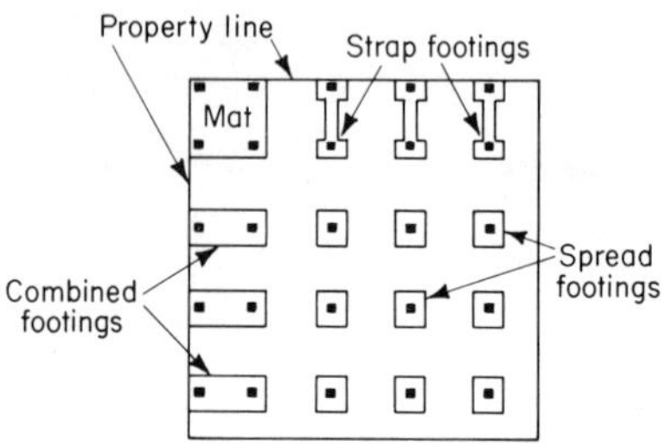

Fig. 7-1 Schematic plan illustrating common uses of various types of footings.

In the majority of the cases, mat foundations are used where the soil has low bearing capacity. By combining all individual footings into one big mat, not only the unit pressure on the subsoil is reduced but also the bearing capacity is often increased. Figure 7-2 illustrates the latter point. In the case of individual footings, the depth of foundation is the dimension from top of the base slab to the bottom of the footing, whereas the depth of foundation of the mat is measured from the exterior ground surface to the bottom of the mat. Since the bearing capacity increases with increasing depth and width of foundation, Sec. 6-5A, and the settlement decreases with the increasing depth of foundation, the advantage of mat foundation is two-fold.

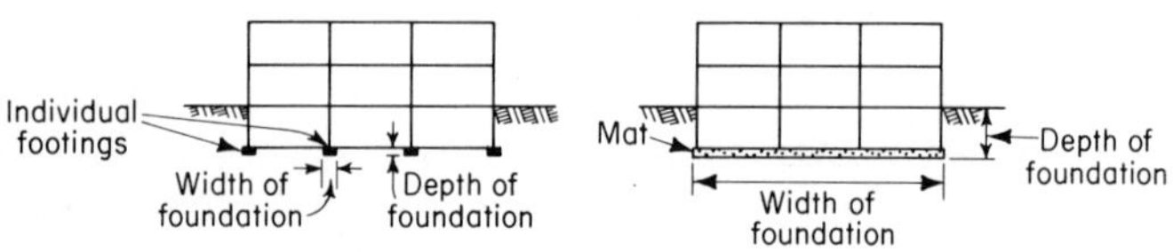

Fig. 7-2 Depth and width of foundation for individual footings and mat foundations.

In localities where the subsoil is very compressible and extends to a great depth, the so-called compensated design is used to the best advantage, Fig. 7-3. In this design, a deeper basement is made under the higher portion of the superstructure so that the net pressure (the total building load minus the weight of soil replaced by basement) at any depth in the subsoil is relatively uniform, therefore large differential settlements are avoided.

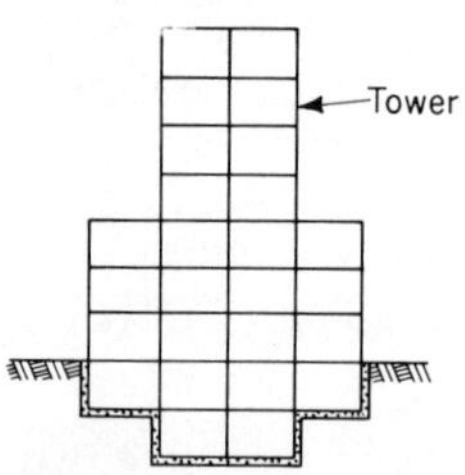

Fig. 7-3 Compensated foundation.

Some structures naturally lend themselves to the use of mat foundations; silos, chimneys, and large machineries being notable examples. Vibratory machinery is often supported on a massive pad to reduce the settlement. When large hydrostatic pressure is encountered, a mat is most desirable because of its structural strength and its possibility of being watertight.

7-2 *Common Types and Arrangement of Strap Footings, Combined Footings, and Mat Foundations*

Straps may be arranged in a variety of ways, Fig. 7-4, and their choice depends on the physical conditions of each specific case. A strap may be connected to a footing, to a wall, or to a column above the footing. In either case a strap should not be designed to withstand only the bending moment and shear stresses but also to transfer the end shear or reaction to the footings or columns at both ends. It should be so arranged that it does not require an unusual construction procedure.

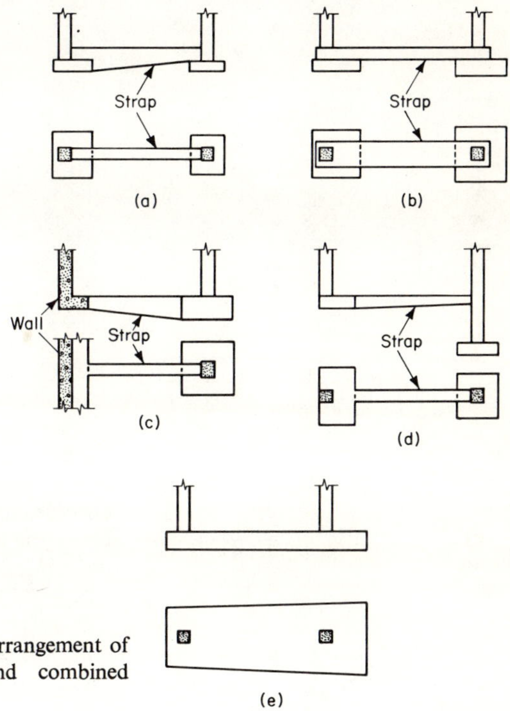

Fig. 7-4 Common arrangement of strap footings and combined footings.

A combined footing is limited to a trapezoidal, rectangular, and other simple shapes. When two columns must be extended to different levels, strap footings should be used, Fig. 7-4(d).

A true mat is a flat concrete slab with uniform thickness throughout the entire area Fig. 7-5(a). This type is most suitable where the column spacing is fairly small and uniform and the column loads relatively small. For large column loads a portion of the slab under the column may be thickened, Fig. 7-5(b), to provide sufficient strength for negative moment and shear (diagonal tension). This may also be accomplished by providing a pedestal under each column Fig. 7-4(d). If bending stresses become large because of large column spacing and unequal column loads, thickened bands may be used along the column lines in both directions Fig. 7-5(c). The empty cells are formed by the use of sheet metal or paper domes.

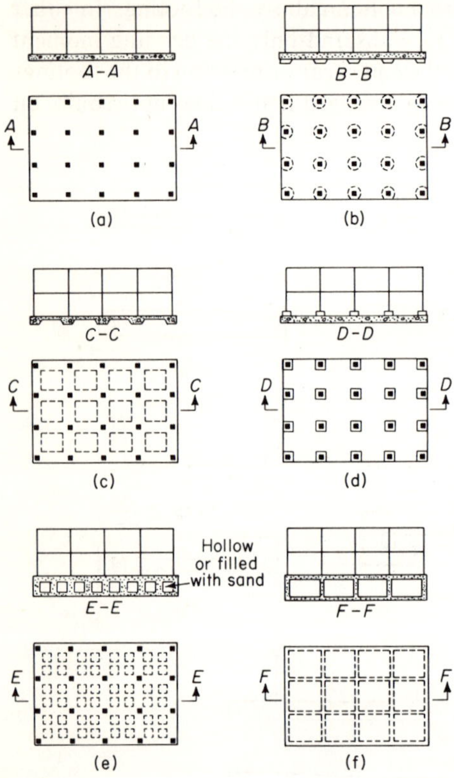

Fig. 7-5 Common types of mat foundations: (a) flat plate; (b) flat plate thickened under columns; (c) two way beam and slab; (d) flat plate with pedestals; (e) cellular construction; (f) basement walls as rigid frame.

Under extremely heavy column loads, a two-way grid structure made of cellular construction [Fig. 7-5(e)] and of intersecting structural steel trusses has been used (Teng, 1949). Basement walls have been also used as ribs or deep beams, Fig. 7-5(f).

A mat often rests directly on soil or rock. However it may rest on piles just as well.

7-3 Design of Strap Footings

Strap footings are designed on the basis of the following assumptions:

1. The strap is infinitely stiff. It serves to transfer the column loads onto the soil with equal and uniform soil pressure under both footings.

2. The strap is a pure flexural member and does not take soil reaction. To avoid bearing on the bottom of the strap, several inches of the underlying soil may be loosened up prior to the placement of the concrete.

With the assumptions above, the design of strap footing is a simple procedure. It begins with a trial value of e, Fig. 7-6. Then the reactions R_1 and R_2 are computed by the principles of statics. The tentative footing areas are equal to the reactions R_1 and R_2 divided by the allowable bearing pressure q. With tentative footing sizes, the value of e is computed. These steps are repeated until the trial value of e is identical with the final one. The

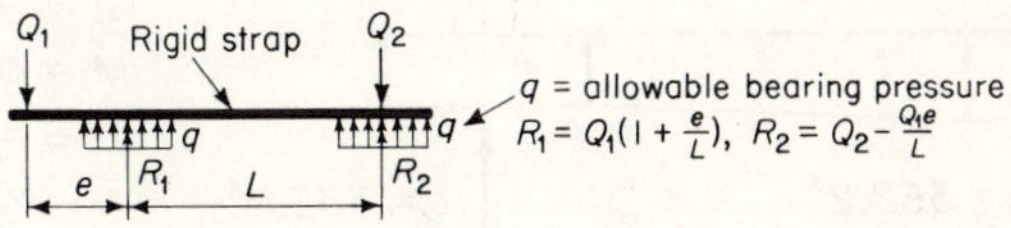

Fig. 7-6 **Principle of strap footing design.**

shear and moment in the strap are determined, and the strap is designed to withstand the shear and moments. The footings are assumed to be subjected to uniform soil pressure and designed as simple spread footings. Under the assumptions given above, the resultant of the column loads Q_1 and Q_2, would coincide with the center of gravity of the two footing areas. Theoretically, the bearing pressure would be uniform under both the footings. However, it is possible that sometimes the full design live load (after reduction according to the building codes) acts upon one of the columns while the other may be subjected to little live load. In such a case, the full reduction of column load from Q_2 to R_2 may not be realized. It seems justified then that in designing the footing under column Q_2 only the dead load or dead load plus reduced live load should be used on column Q_1.

An example of strap footing design is given in Plate DE 7-1.

Strap Footing

$f'_c = 3750$ psi
$f_s = 20{,}000$ psi
$q = 8000$ psf
Allow. bearing

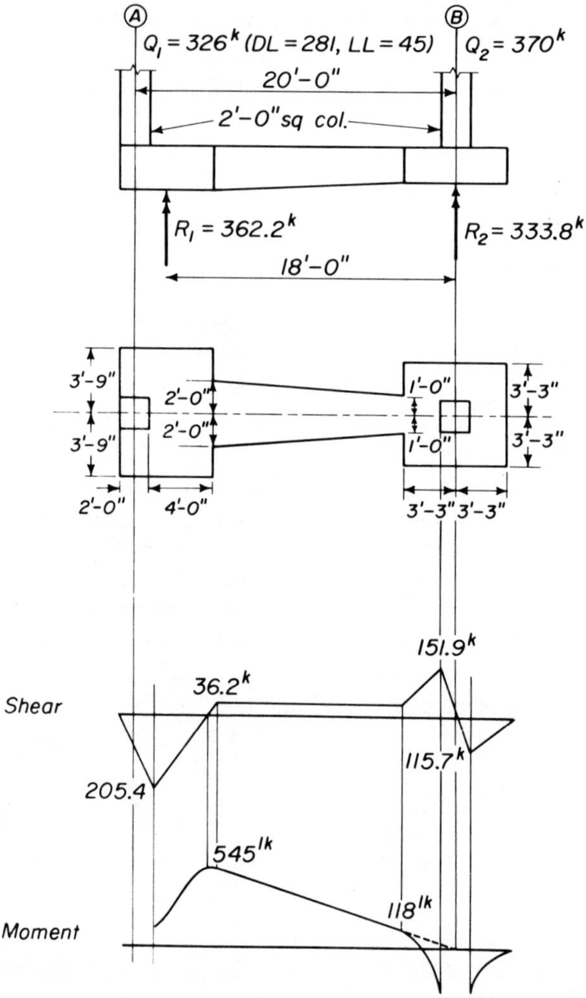

Try 6′ cantilever footing

$e = 2'$

$M = Q_1 e = 652^{lk}$

$V = \frac{652}{18} = 36.2^k$

$$R_1 = \begin{array}{r}326\\ 36.2\\ \hline 362.2^k\end{array} \qquad R_2 = \begin{array}{r}370\\ 36.2\\ \hline 333.8\end{array}$$

$\frac{362.2}{8} = 45.3^{\square'}$

Use 6′0″ x 7′ 6″

$\frac{333.8}{8} = 41.6^{\square'}$

Use 6′-6″sq
e = 2′ OK

$362.2/6 = 60.3^{k}/'$
$333.8/6.5 = 51.3^{k}/'$
$326 - 2(60.3) = 205.4^k$

$$\begin{array}{r}2.25' \times 51.3 = 115.7\\ 36.2\\ \hline 151.9^k\end{array}$$

$M = 36.2^k X$
$= 118^{lk}$ $(X = 3.25')$
$= 545^{lk}$ $(X = 15')$

DE 7-1
Strap Footing
Sh. 2 of 3

ACI Code is used in this design ✱ indicates 1956 Code value

$v = 75$ psi*

Top bars $u = 245$ psi*

Bottom bars $u = 280$ psi*

Footing at Col Ⓑ

$R'_2 = 370 - \frac{281 \times 2}{18} = 339^k$ (Designed as spreading footing)

$D = 2'\text{-}2''$

$+d = 22''$

12 – #7 bott. each way

Footing at Col Ⓐ

Try $D = 2'\text{-}8''$ $+d = 32 - 4 = 28''$ $-d = 32 - 3 = 29''$

Transv. bending:

$+M = \frac{1}{2}(8)(2.75)^2 = 30.3^{'k/'} \times 6 = 182^{'k}$

$+As = \frac{182}{1.44(28)} = 4.5^{\square''}$

$V = 8(2.75)6 = 132^k$

$\Sigma O = \frac{132}{(0.280)\frac{7}{8}^*(28)} = 19.4''$ Use 8–#7 bot.

Long. bending:

Extend top bars from strap Use 6–#6 bot.

Strap

Left end:

Try 4'-0" width Req'd. $d = \sqrt{\frac{545}{(0.298)4}} = 21.4''$

$D = 32''$ OK

$As = \frac{545}{1.44(29)} = 13.0^{\square''}$

$\Sigma O = \frac{36.2}{(0.245)\frac{7}{8}^*(29)} = 5.9''$ Use 9–#11 top.

Right end:

Try 2'-0" width Req'd. $d = \sqrt{\frac{118}{(0.298)4}} = 14''$ $D = 26''$ OK

$As = \frac{118}{1.44(23)} = 3.6^{\square''}$

$\Sigma O = \frac{36.2}{(0.245)\frac{7}{8}^*(23)} = 7.4''$ Use 3–#11 top.

DE 7–1
Strap Footing
Sh. 3 of 3

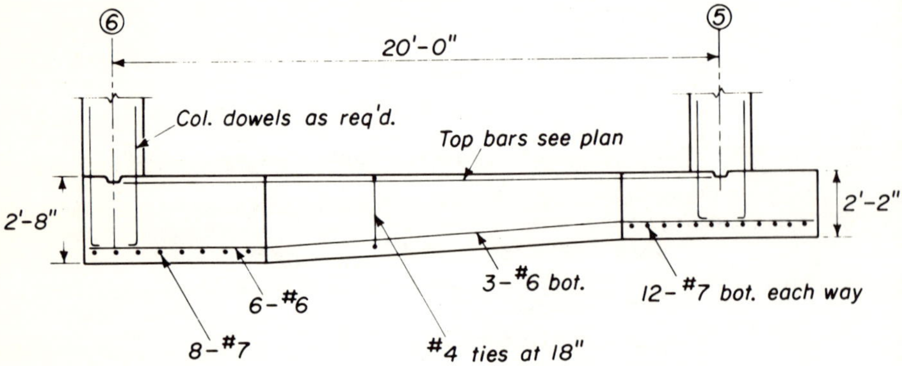

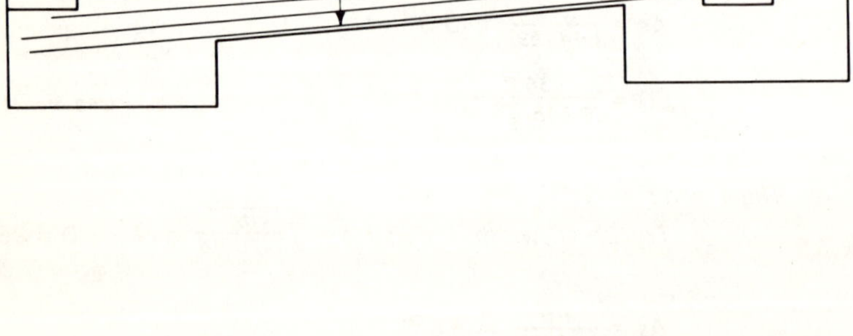

7-4 *Design Methods for Combined Footings and Mat Foundations*

The methods of design of combined footings and mat foundations may be classified according to the assumptions used, as follows.

A. Rigid method: conventional method. The conventional method of design assumes that:

1. The footing or mat is infinitely rigid, and therefore, the deflection of the footing or mat does not influence the pressure distribution.
2. The soil pressure is distributed in a straight line or a plane surface such that the centroid of the soil pressure coincides with the line of action of the resultant force of all the loads acting on the foundation.

In the case of piled foundations, the piles are assumed to be perfectly elastic, Fig. 7-7(a).

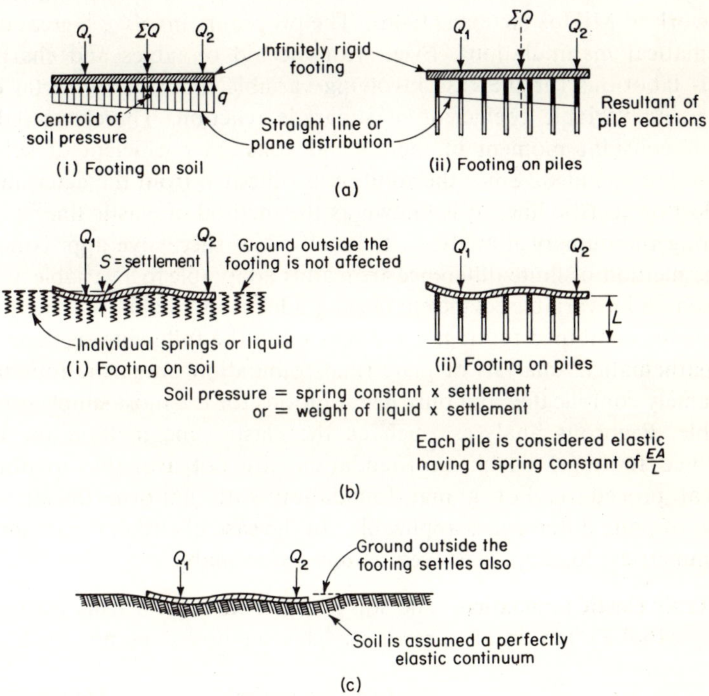

Fig. 7-7 Design methods for combined footings and mat foundations: (a) rigid method (conventional); (b) elastic method (simplified elastic foundation); (c) elastic method (true elastic continuum).

B. Simplified elastic foundation. This method assumes that the soil behaves like an infinite number of individual coil springs each of which is not affected by others, Fig. 7-7(b). The elastic constant of the springs is equal to the coefficient of subgrade reaction of the soil. Further, the springs are assumed to be able to resist tension or compression. This assumption was first used by E. Winkler and therefore this type of foundation is referred to as a Winkler foundation. It is often compared to foundation supported on a dense liquid whose unit weight is equal to the coefficient of subgrade reaction.

The coefficient of subgrade reaction of a soil is the unit pressure required to produce a unit settlement and is discussed in the latter part of this chapter. In the case of piled foundations, each pile is considered as a coil spring with an elastic constant equal to EA/L, where E is the modulus of elasticity of the pile material, A the average cross-section of the pile, and L the effective length of the pile. The effective length may be taken as the full length of point-bearing piles and about one-half length of skin-friction piles.

A number of procedures have been developed for the analysis of beams on the simplified elastic foundation concept. The most complete and rigorous is the work of Miklos Hetenyi (1946). The procedure involves a great deal of mathematical manipulations. Even with the aid of tables and charts, the work is laborious for the case involving variable moment in inertia of the footing and variable coefficient of subgrade reaction. Therefore, only the case for constant moment of inertia and constant coefficient of subgrade reaction are included. Since the solution is obtained from the determination of deflection (elastic line), it is known as the method of elastic line.

Among the numerical analyses, the method of successive approximations and the method of finite difference are readily adaptable to a variable moment of inertia and a variable coefficient of subgrade reaction. All three methods of analyses for combined footings are discussed in the following sections.

A mathematical analysis of plate (mat foundation on elastic foundation) is extremely complicated and only few solutions for the most simple cases are available. Rigorous analyses, such as the elastic line method for beams (combined footings) on elastic foundation, are not available to obtain a practical procedure. For a mat foundation with uniform thickness, the method of finite difference is applicable. In the case of circular mats supporting symmetrical loading, elastic equations are available.

C. Truly elastic foundation. The soil is assumed to be a truly elastic solid obeying Hooke's law in all directions. This method does not readily lend

	Combined footings	*Mat foundations*
Conventional method	Rigid method	Rigid method
Simplified elastic foundation	Method of elastic line Method of successive approximation Method of finite difference	Method of finite difference
Truly elastic foundation	—	—

itself to engineering applications because it is extremely difficult and solutions are available for only few extremely simple cases.

The accompanying table is a summary of the available procedures for analysis of combined footings and mat foundations.

7-5 Design of Combined Footings

A. Design of combined footings by conventional (rigid) method. The conventional design is based on the assumption that the footing is infinitely stiff and that subgrade (or soil) reaction has a straight-line distribution. The procedure is simple and is described step by step in Fig. 7-8.*

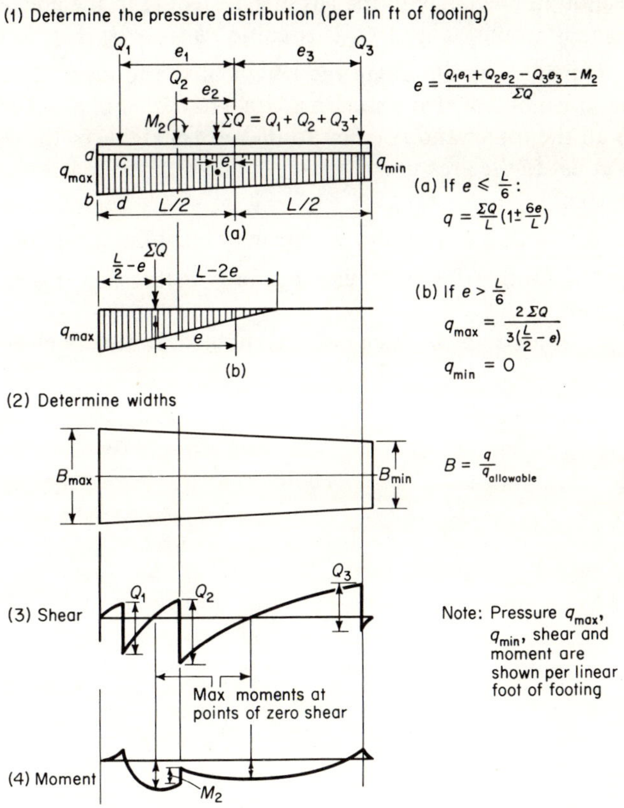

Fig. 7-8 Procedure for design of combined footing (conventional method).

---

* With the column loads and the subgrade reactions known, the problem is statically determinate. Hence, it should not be confused with the problem of indeterminate structures, such as the moment distribution method.

1. Determine the total column loads

$$\sum Q = Q_1 + Q_2 + Q_3 \ldots$$

   and the location of the line of action of the resultant ΣQ. If any column is subjected to bending moment, the effect of the moment should be taken into account.
2. Determine the pressure distribution, per lineal foot of footing.
3. Determine the width of the footing.
4. Draw shear diagram along the length of footing. By definition, the shear at any section along the beam is equal to the summation of all vertical forces, to the left or right of the section. For example, the shear at a section immediately to the left of Q_1 is equal to the area *abcd*, and immediately to the right of Q_1 is equal to *abcd* $- Q_1$, Fig. 7-8.
5. Draw moment diagram along the length of the footing. By definition the bending moment at any section is equal to the summation of moment due to all the forces and reaction to the left (or right) of the section. It is also equal to the area under the shear diagram to the left (or right) of the section.
6. Design the footing as a continuous beam to resist the shear and moment.
7. Design the footing for transverse bending in the same manner as for spread footings.

An example of a complete design of a combined footing is given in Plate DE 7-2.

DE 7-2
Combined Footing
Sh. 1 of 2

$f'_c = 3750$ psi
$f_s = 20000$ psi

Allow. pressure = 8.5^{kSF}

* indicates 1956 ACI Code value

$DL = 1050^k$, $LL = 325$, total 1375^k $\qquad$ $DL = 1210^k$, $LL = 615$, total 1825^k $\qquad$ $1375 + 1825 = 3200^k$

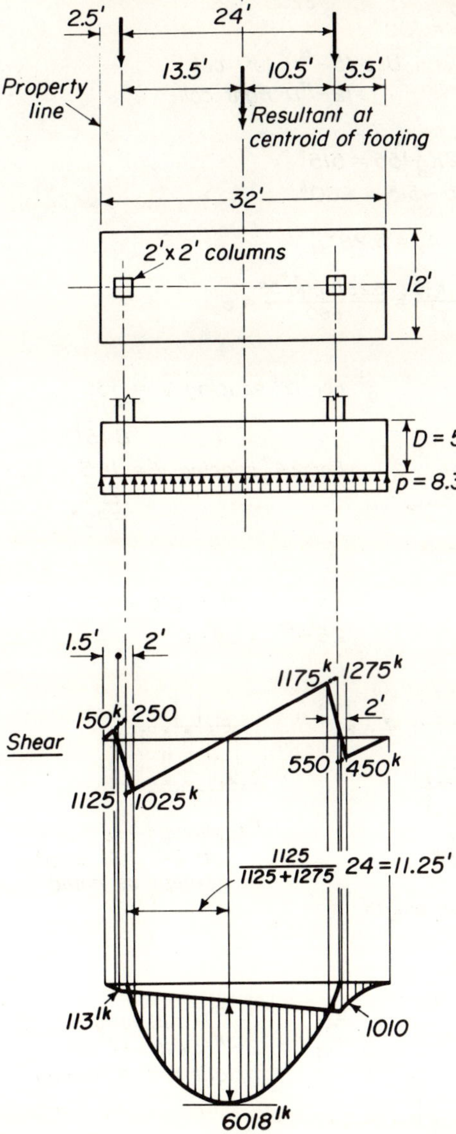

$$\bar{X} = \frac{1825}{3200} \times 24 = 13.5 \pm$$
$$+2.5 = 16.0'$$

Length = 2 x 16.0 = 32.0'
$- 26.5'$
$= 5.5'$

$$\text{Width} = \frac{3200}{32 \times 8.5} = 11.8 \qquad \text{Use } 12'$$

$$p = 8.33^{kSF} \times 12 = 100^{k}/_{1}$$

$$\text{Req'd } d = \sqrt{\frac{6018}{0.298 \times 12}} = 42''$$

Try $D = 5'\text{-}0'' = 60''$

$d = 54''$

$$v = \frac{1175000}{12(12)\frac{7}{8}^{*}(54)} = 175 \text{ psi} \quad \text{OK}$$

Longitudinal steel:

$$+A_s = \frac{6018}{1.44 \times 54} = 79^{\square''}$$

$$\Sigma O = \frac{1175000}{280(\frac{7}{8})^{*}54} = 90''$$

Use 52 – #11 in 2 layers

$$-A_s = \frac{1010}{1.44(57)} = 12.3^{\square''}$$

Use 9 – #11 at col, extend 7 – #11 full length

Combined Footing

Transverse steel: $M = 5' \times 32' \times 8.33^{KSF} \times 2.5 = 3330^{lk}$

$$As = \frac{3330}{1.44(55)} = 41.6^{\square''} \quad \times \frac{1825}{3200} = 23.8^{\square''}$$

$$\times \frac{1375}{3200} = 17.8^{\square''}$$

$$\Sigma O = \frac{5 \times 32 \times 8330}{280(\frac{7}{8})^{*}55} = 100''$$

Use 15–#11 left col.
12–#11 right col.

Web reinforcing: $V_c = 0.075(12)(12)(\frac{7}{8})^{*}55 = 515^{k}$

$$V' = V - V_c = 1175 - 515 = 660^{k}$$

6 sets of #6 ⊔ $A_v = 5.28^{\square''}$

$$\text{Spacing} = s = \frac{A_v f_v jd}{V'} = \frac{5.28(16)(\frac{7}{8})^{*}55}{660} = 6''$$

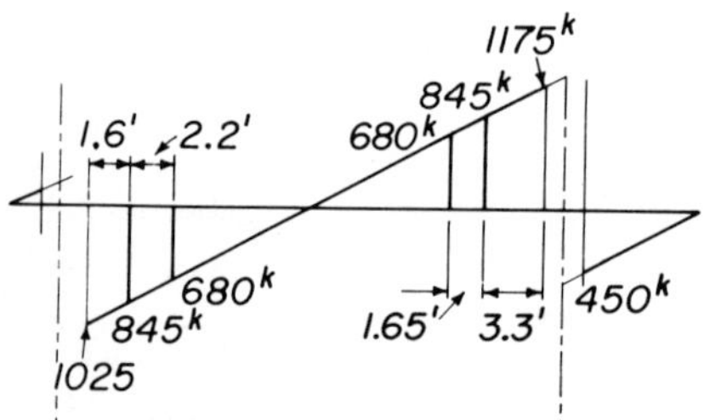

For 12" spacing $V' = 330 + 515 = 845^{k}$

For 24" spacing $V' = 165 + 515 = 680^{k}$

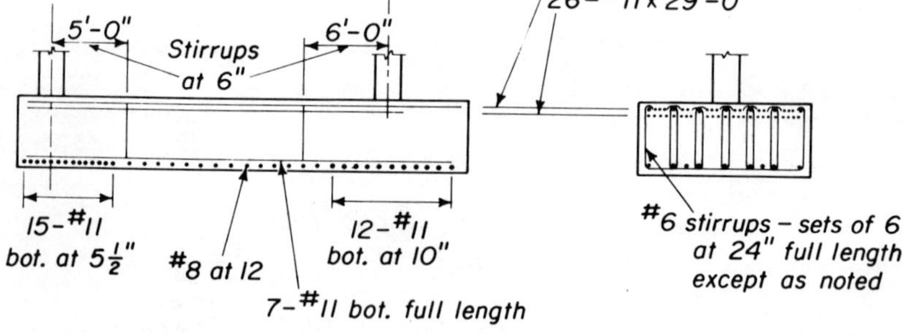

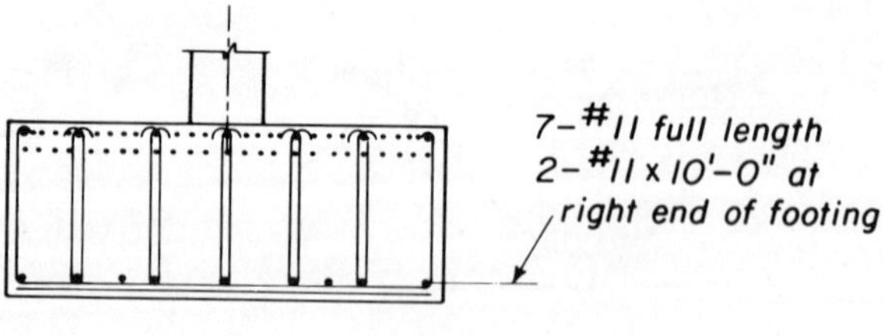

B. Design of combined footings by method of elastic line. It is well known to the structural engineers that (see any test on strength of material):

$$\frac{M}{EI} = \frac{d^2y}{dx^2}$$

$$\frac{d}{dx}\left(\frac{M}{EI}\right) = V$$

$$\frac{d}{dx}V = q$$

where M = bending moment at any given point on the beam or footing,

x = coordinates along the length of the footing,

y = deflection, in this case, settlement of footing,

V = shear,

q = reaction at a given point on the elastic foundation,

E and I = modulus of elasticity and moment of inertia of the footing.

From the basic assumption of elastic foundation:

$$q = -yBk$$

where B = width of footing,

k = coefficient of subgrade reaction.

$$EI\frac{d^4y}{dx^4} = -yBk \qquad (7\text{-}1)$$

The general solution of the equation above is available (Hetenyi, 1946). With the deflection along the length of footing (elastic line) known, the shear and moment can be determined.

The solutions of the elastic line method for footings with constant EI and acted by a concentrated load are shown in Fig. 7-9. Each curve corresponds to a footing acted by a load at the left end, $L/12$, $L/6$, . . . from the left end. The deflection is shown in terms of y_0 (y_0 = average deflection of the footing). The moment is shown in terms of M_0 ($M_0 = Q/4\lambda$ where Q = the magnitude of the concentrated load, and $\lambda = \sqrt[4]{Bk/4EI}$).

These curves are useful in practical problems. Their uses are illustrated in Plate DE 7-3. First the coefficient of subgrade reaction k is adjusted according to Eq. (7-10) (see Sec. 7-8). The value of λ is computed. It is found that L/λ is equal to $2/\lambda$ approximately, the curves shown in Fig. 7-9(a) are applicable. The location of concentrated loads are approximately $L/12$ and $L/6$. By superimposing the two moment curves for $L/2$ and $L/6$, the maximum moments are found to be $0.63M_0$ and $0.32M_0$ respectively.

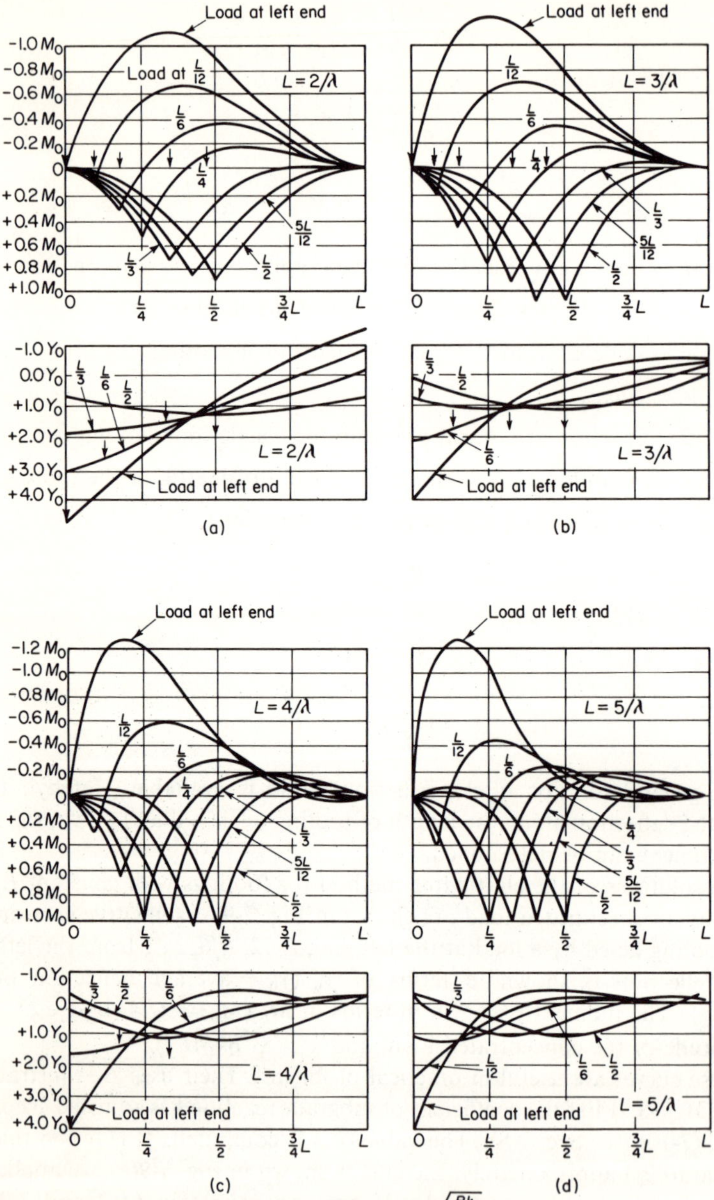

Note: $\lambda = \sqrt[4]{\frac{Bk}{4EI}}$

Fig. 7-9 Elastic lines for footings of finite length acted by a concentrated load. From Seely and Smith.

DE 7–3
Combined Footing
Sh. 1 of 1

Analysis by Elastic Line Method

Given: 12′ wide x 5′ deep footing
32′ long
$E = 4,000,000$ psi
$I = \frac{1}{12}(144)(60)^3 = 2,592,000$
For dense sand, Table 7–1:
$K_1 = 600$ pci

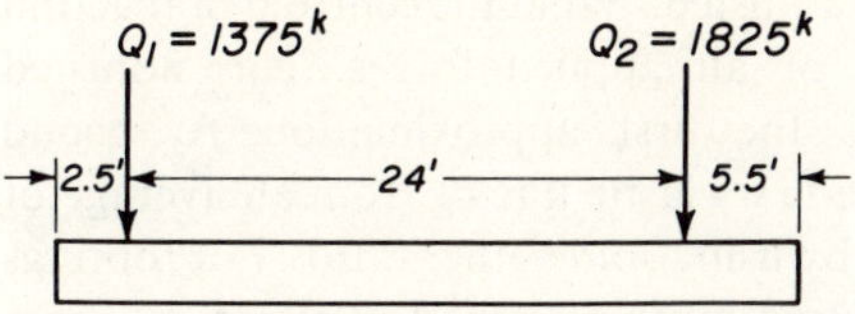

Eq. (7–10) $k = k_1(1 + 2 \times \frac{5}{12})(\frac{12+1}{2 \times 12})^2 = 320$

$$\lambda = \sqrt[4]{\frac{Bk}{4EI}} = \sqrt[4]{\frac{144 \times 320}{4(4,000,000)2,592,000}} = 0.0049$$

$L = 32' = 384''$,

$\therefore L = \frac{2}{\lambda}$ Use Fig. 7–9 (a)

$2.5' \approx \frac{L}{12}$, $5.5 \approx \frac{L}{6}$, superpose curves for $\frac{L}{12}$ and $\frac{L}{6}$

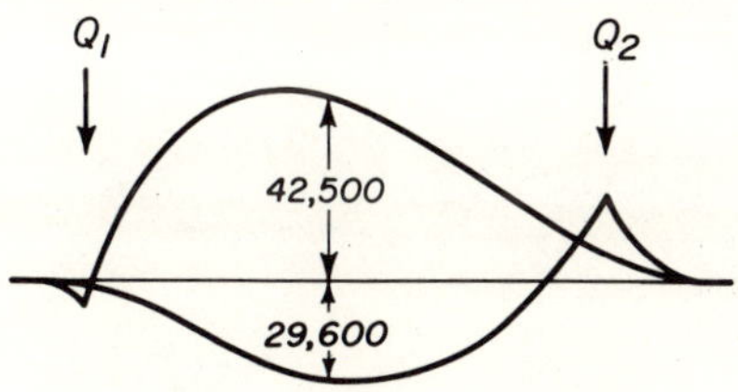

$$0.63\, M_0 = 0.63 \frac{Q_1}{4\lambda} = 42,500$$

$$0.32\, M_0' = 0.32 \frac{Q_2}{4\lambda} = \frac{29,600}{72,100}{}''^k = 6,018'^k$$

Note: $6,300'^k$ in plate DE7–2

C. Design of combined footing by method of successive approximations (Popov, 1951). To start the solution, a certain pressure distribution is assumed. Generally, the footing is assumed to be infinitely rigid and the pressure distribution is assumed planar. On the basis of this pressure distribution, the deflection (or elastic line) of the footing is determined by the moment-area or elastic-weight method. By "locking" the footing in this deflected shape, the footing position is adjusted so that the centroid of reaction ($\Sigma\, yk$) coincides with the resultant of all applied forces. This adjusted deflection or elastic line represents the first approximation. A second approximation may be made from the first elastic line or from an average of the straight line and the first curve, or by using some other ratio. For footings with length smaller than π/λ, the solution by this method may not converge and the results are in error.

In determining the elastic line, the footing is divided into a number of segments. A larger number of segments gives a greater accuracy. For ordinary cases, four to six segments between two columns give sufficiently accurate results. The values of E, I, and k may be different between segments. The total pressure acting against each segment is taken as a concentrated force. The elastic line is then determined by the moment area method: deflection at any point is equal to the bending moment on the conjugate beam with M/EI as loading. The ordinates of the elastic line may be determined arithmetically or graphically (by force polygon and funicular polygon) or by a numerical procedure.

An example of this method is shown in Plate DE 7-4. The procedure is explained step by step.

Analysis by Method of Successive Approximation

Given

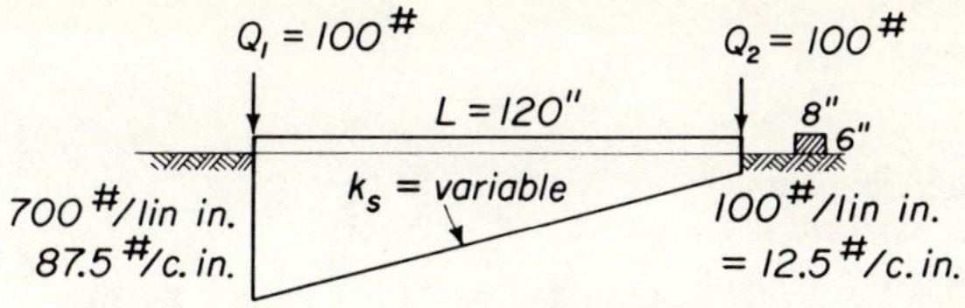

$E = 2.5(10)^6$ #/s.I.
$I = 144 \text{ in}^4$

Solution

(1) Determine transformed area for uniformed $k_s = 12.5$ #/c.in.

$56'' = 8'' \times \frac{87.5}{12.5}$

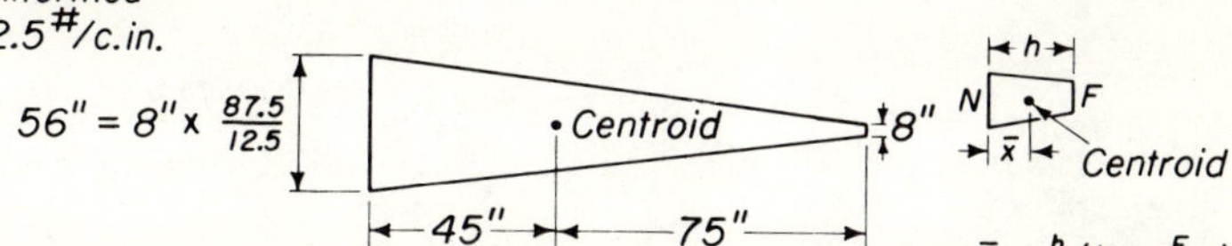

$\bar{x} = \frac{h}{3}(1 + \frac{F}{N+F})$

(2) Determine soil pressure by rigid method:

$q = \frac{\Sigma Q}{A} \pm \frac{M_c}{I'_0}$

At $c = 45''$

$q = \frac{200}{3840} - \frac{3000(45)}{3,744,000}$

$= 0.016$ psi

$\times 56 = 0.90$ #/lin. in

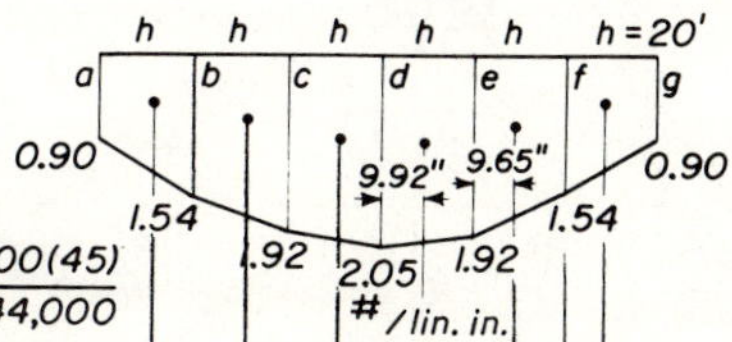

$\Sigma Q = 200$#

$A = \frac{1}{2}(8+56)120$

$= 3840 \text{ in}^2$

$M = 100(75) - 100(45)$

$= 3000$ in.#

I'_0 (transformed area)

$= \frac{120^3[(56+8)^2 - 2(56)8]}{36(56+8)}$

$= 3,744,000 \text{ in.}^4$

(3) Substitute soil pressure by concentrated reactions

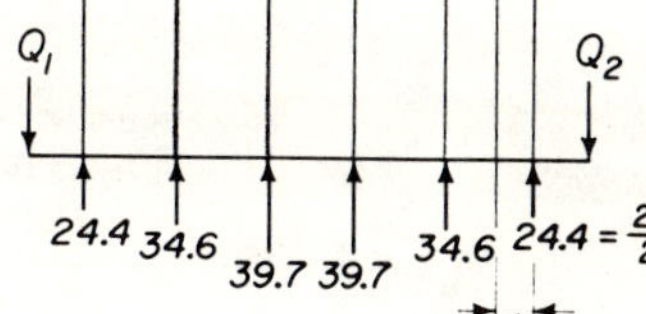

$24.4 = \frac{20}{2}(1.54 + 0.90)$

$\bar{x} = \frac{20}{3}(1 + \frac{0.90}{1.54 + 0.90}) = 9.14''$

(4) Calculate bending mom. at each panel point

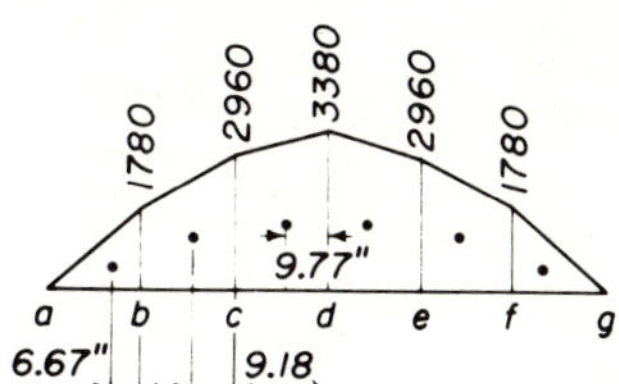

$= \frac{20}{3}(1 + \frac{1780}{1780 + 2960})$

(5) Conjugate beam under concentrated $\frac{M}{EI}$ loading (elastic weight)

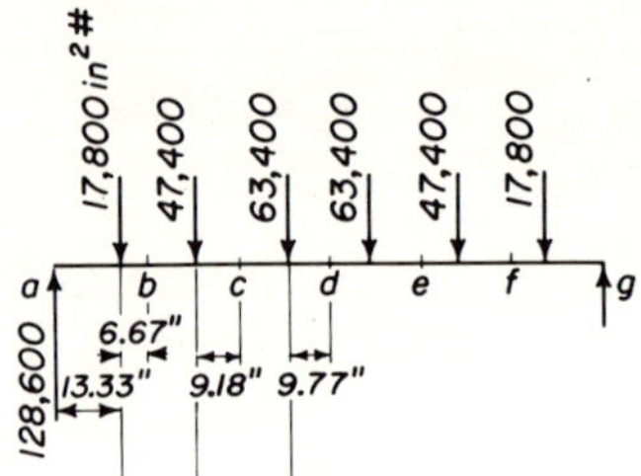

$EI = 2.5(10)^6(144)$
$= 3.6(10)^8$

(6) Deflection = mom. under elastic wt.

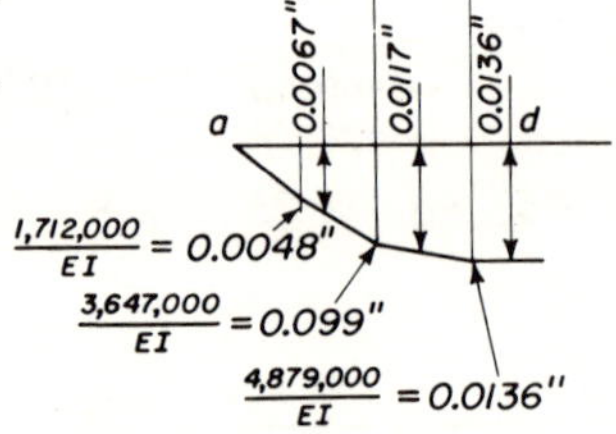

128,600 x 13.33 =	1,712,000
17,800	
110,800 x (6.67 + 10.82) =	1,935,000
	3,647,000
47,400	
63,400 x (9.18 + 10.23) =	1,232,000
	4,879,000

(7) k_s values

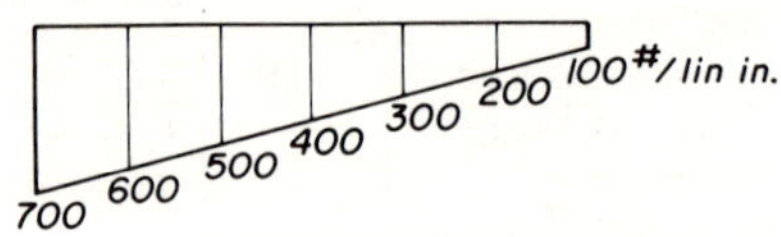

(8) Reaction = (6) x (7) and determine centroid of total soil reaction

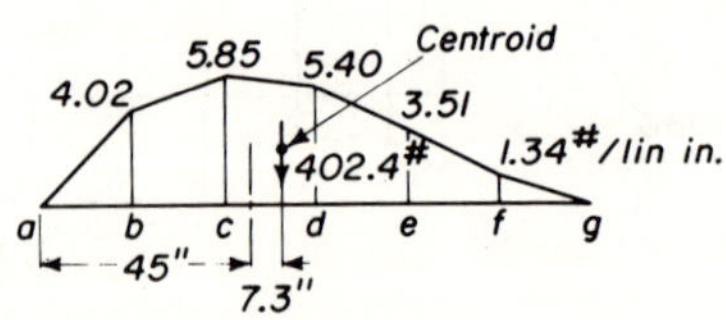

(9) Corrective force = $\Sigma Q + 402.4 = 602.4$
Corrective mom. = $402.4 \times 7.3 + M$ (Step 2) $= 2930 + 3000 = 5930''\#$

Soil pressure due to corrective F and M:
$q = \frac{602.4}{A} \pm \frac{5930c}{I_0'}$

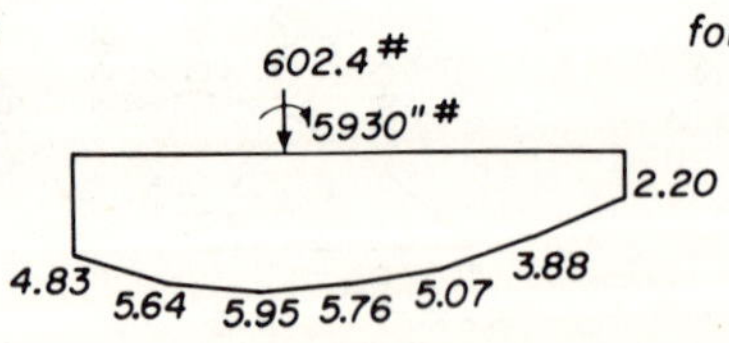

for $c = 45''$
$q = 0.086\#/\text{s.in.}$
$\times 56 = 4.83$

(10) Soil reaction for
$\Sigma Q = 200^{\#}$
$M = 3000''^{\#}$
$= (9) - (8)$

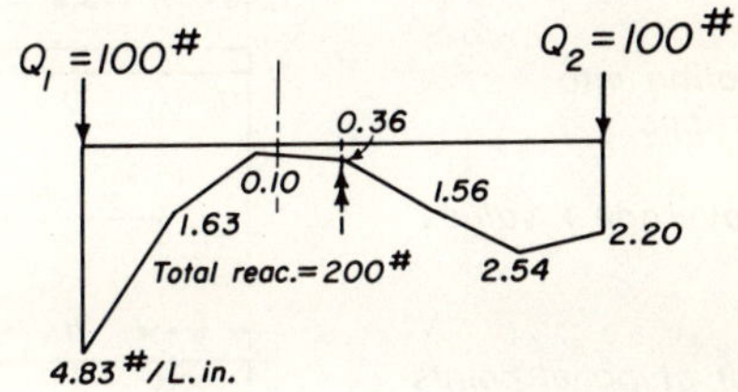

This is the result of 1st trial

(11) Take average value of (10) and (2) (shown solid line)

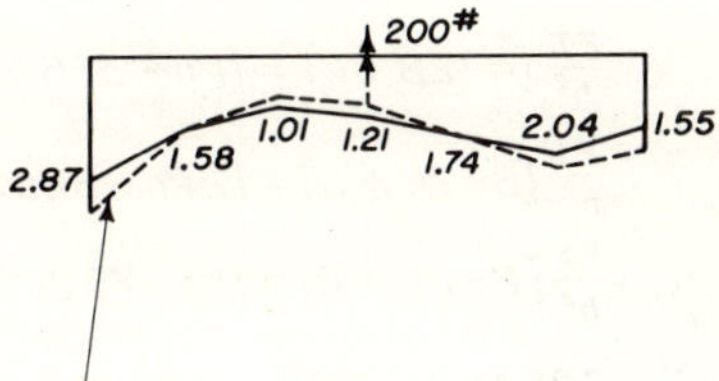

(12) Obtain 2nd approximation by starting with the average pressure (solid line) and repeat steps (2) to (10)

After Popov

Analysis by Method of Finite Differences

Given:

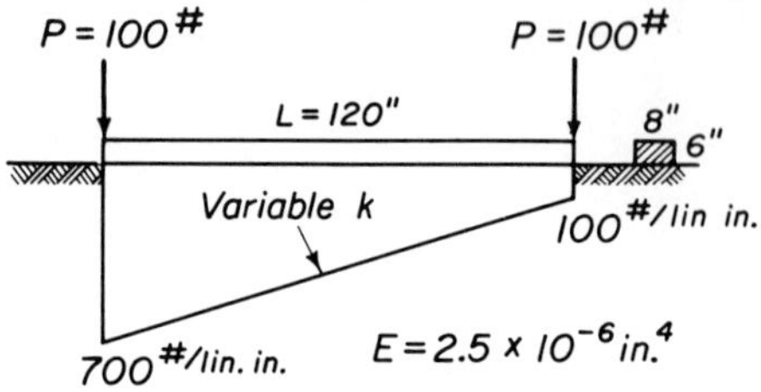

Solution:

Divide footing into equal sections

Assume average k values

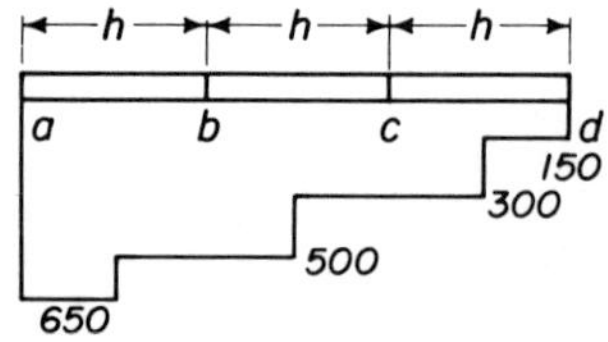

h = 40"

Settlement at panel points are unknowns (A, B, C and D)

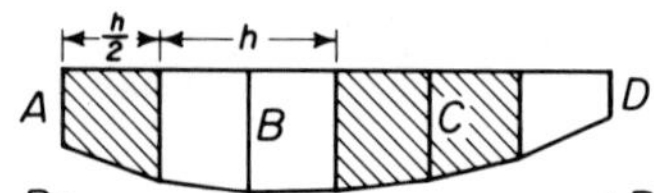

Equivalent concentrated forces (approximate)

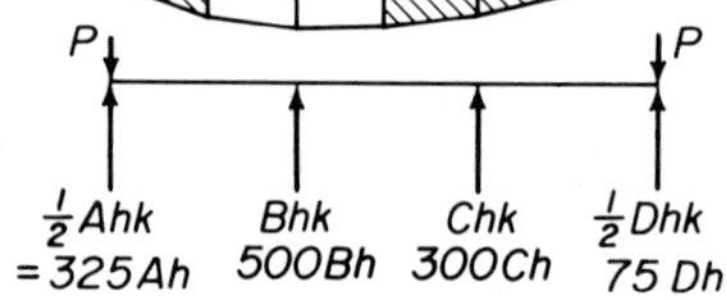

M at b, left	$\frac{EI}{h^2}[A-2B+C] = [Ph - 325Ah^2]$
M at c, left	$\frac{EI}{h^2}[B-2C+D] = [2Ph - 325Ah(2h) - 500Bh(h)]$
M at c, right	$\frac{EI}{h^2}[B-2C+D] = [Ph - 75Dh^2]$
$\Sigma V = 0$	$325Ah + 500Bh + 300Ch + 75Dh = 2P$

Solving for A, B, C and D:

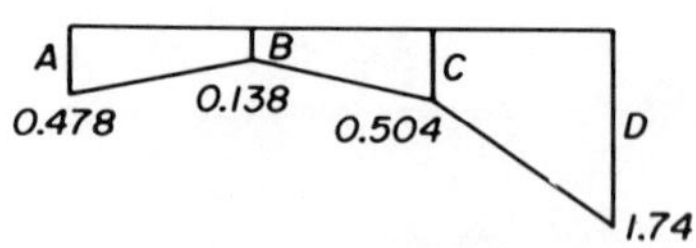

($\times 10^{-2}$) inch

Pressure = kA, kB....

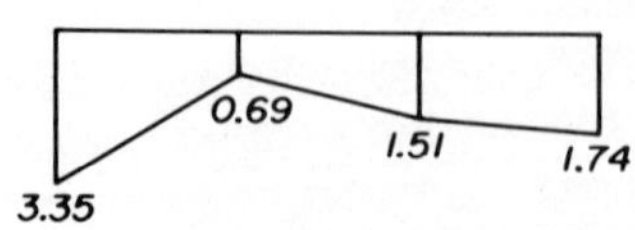

#/lin in.

D. Design of combined footing by method of finite difference. The finite difference method (Malter, 1958) treats the footing as a flexural member consisting of sections, usually of equal length h. Instead of being supported on a continuous soil pressure, each section is supported by equivalent concentrated reactions R_a, R_b, . . ., at the panel points a, b, . . ., Plate DE 7-5. The forces and the reactions, R_a, R_b, . . ., should behave according to well-known relationships for flexural members where:

$$\text{Deflection} = y$$

and

$$\text{Moment} = EI\frac{d^2y}{dx^2}$$

The equations above are substituted with finite difference operators:

$$\text{Deflection at } a, b, c, \ldots = A, B, C, \ldots$$

$$\text{Moment at } b = \left(\frac{\Delta^2}{dx^2}\right)_b EI = \frac{(A - 2B + C)EI}{h^2} \tag{7-2}$$

Analysis of combined footing by means of the difference equation above is illustrated by a numerical example shown in Plate DE 7-5.

1. Divide the footing into 4 to 6 equal lengths $= h$.

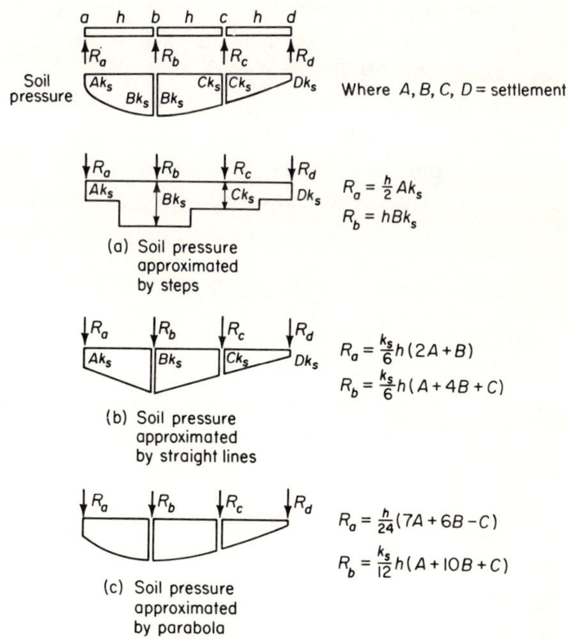

Fig. 7-10 Methods for calculating approximate equivalent concentrated reactions.

2. Let A, B, . . ., be the settlement, or deflection, at points a, b, ... The soil reaction at a, b, . . ., is Ak, Bk, ...

3. Replace the continuous soil reaction by equivalent concentrated reactions R_a, R_b, ... There are three methods of converting the distributed reaction into concentrated reactions; they are explained in Fig. 7-10. The simplest method (a) is used in the example.
4. The footing under the applied loads and the equivalent reactions should satisfy the equations for equilibrium, i.e. $\Sigma M = 0$ and $\Sigma V = 0$. Write equations for the $\Sigma M = 0$ at any panel points and $\Sigma V = 0$ for the whole system, in terms of A, B, C, ...
5. Solve the simultaneous equations for values of A, B, C, ...

It is seen that this method requires very little labor. The only tedious work is the solution of simultaneous equations. With the advent of electronic digital computers, this is no longer a lengthy procedure. However, this method, as the method of successive approximation, may not converge.

7-6 *Allowable Bearing Pressure for Mat Foundations*

The procedure for determining the allowable bearing pressure under footings and mat foundations was described in Chapter 6. However, a structure supported on a mat of 20 ft $\times$ 20 ft or larger can withstand greater settlements than one supported on spread footings. Mat foundation tends to bridge over irregularities or heterogeneity of the soil and the average settlement does not approach the extreme values of spread footings. Therefore, the allowable bearing pressure for mat foundation can be greater than that for footings on the soil insofar as the settlement is concerned. For mat foundations on granular soils, an increase of 100 per cent has been used (Terzaghi and Peck, 1948). The following allowable pressures are applicable for design of mat foundations on sand or gravel.

$$q_1 = \tfrac{4}{3}N^2BR_w + 4(100 + N^2)\,DR'_w \tag{7-3}$$

$$q_2 = 360(N - 3)\,R'_w \tag{7-4}$$

where q_1 and q_2 = allowable soil pressure under mat foundation, psf;
N = number of blows per foot in standard penetration test;
B = smaller dimension of the mat, ft;
D = depth of foundation, ft;
R_w and R'_w = reduction factors for water level, see Fig. 6-4, Sec. 6-3B.

The smaller of q_1 and q_2 should be used.

7-7 *Design of Mat Foundations*

A. Design of mat foundation by conventional rigid method. In the conventional method it is assumed that the mat is infinitely rigid and that the bearing pressure against the bottom of the mat follows a planar distribution where the centroid of the bearing pressure coincides with the line of action of the resultant force of all loads acting on the mat. The procedure of design by this method is described as follows.

The maximum column and wall loads are computed using a live load reduction in accordance with local building codes. The line of action of the resultant of all these loads is determined. The weight of the mat, however, may not be included in the structural design of the mat because every point of the mat is supported by the soil directly under it, causing no flexural stresses.

After the resultant force is located the pressure distribution can be readily determined by the following formula:

$$q = \frac{\Sigma Q}{A} \pm \sum Q \frac{e_y}{I_x} y + \sum Q \frac{e_x}{I_y} x$$

where ΣQ = total loads on the mat,

A = total area of the mat,

x, y = coordinates of any given point on the mat with respect to the x and y axes passing through the centroid of the area of the mat,

e_x, e_y = coordinates of the resultant force,

I_x, I_y = moment of inertia of the area of the mat with respect to the x and y axes respectively.

For a more detailed discussion on pressure distribution of eccentrically loaded footings see Sec. 6-9.

If the mat rests on a soft soil, an eccentricity of loading may cause largely different settlement at extreme corners. In such cases, a computation of settlement should be made for the corners. The vertical pressure in any soil stratum under each corner of the mat may be determined by means of Newmark's influence chart (Fig. 6-11). Since a mat occupies the entire area of the building, it is often unfeasible and uneconomical to proportion the mat so that the centroid of the mat coincides with, or is close to, the line of action of the resultant force.

The mat is analysed as a whole in each of two perpendicular directions. Thus, the total shear force acting on any section cutting across the entire mat is equal to the arithmetic sum of all forces and reactions (bearing pressure) to the left, or right of the section. The total bending moment acting on such section is equal to the sum of all moments to the left, or right, of this section.

Although the total shear and moment forces can be determined by the principles of simple structures, the stress distribution along this section is a problem of a highly indeterminate nature. No simple practical procedure is available to solve this problem. If the column loads and spacing are about equal, an approximate idea as to how the moment and shear are distributed along each section may be arrived. In most cases, however, the variation of moment and shear is often far different from the average value. This point can be illustrated by a simple example shown in Fig. 7-11. The total bending moment on section *a-a* is equal to the difference between the positive moment (tension on the bottom of slab) due to the soil reaction and the negative moment due to the column load Q. Let us say the net total moment is positive. Then, the average bending moment on section *a-a* is equal to this net moment divided by the length of section *a-a*. But it is obvious in this case that the strip b is subjected to a positive moment and the strip c is subjected to a negative moment. The average moment is not indicative of the sign and the magnitude of the bending moments in the individual strips.

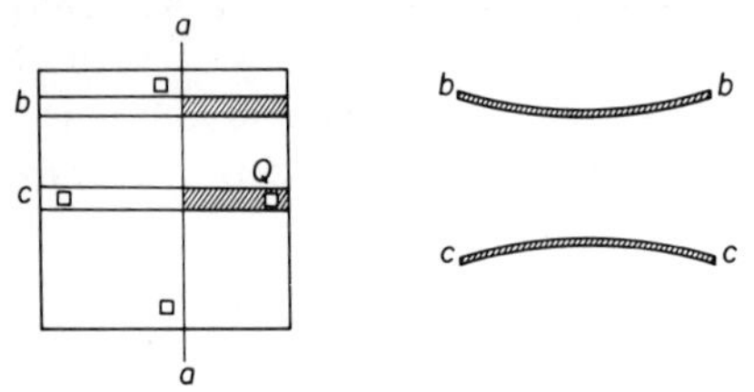

Fig. 7-11 Example illustrating variation of bending moment in mat.

In order to obtain some idea as to the upper limit of stresses, each strip bounded by center lines of column bays may be analysed as independent, continuous, or combined footings. Full column loads are used and the soil reaction under each strip is determined without reference to the planar distribution determined with the mat as a whole. This method undoubtedly gives very high stresses because it ignores the two-way action of the mat. Therefore certain arbitrary reduction in stresses (for example 15 per cent, 25 per cent or sometimes greater than $33\frac{1}{3}$ per cent) are used.

A numerical example of design by the conventional method is given in Plate DE 7-6. The column loads, wall loads, and allowable bearing pressure are assumed already known, and are shown on sheet 1 of Plate DE 7-6. The centroid of all column loads and wall loads ΣQ, $\bar{x}$, and $\bar{y}$ are first determined. Then a trial size for the mat is assumed. The centroid of the mat e_x and e_y is readily located.

The soil pressure under the mat is determined by the general equation:

$$q = \frac{\Sigma Q}{BL}\left(1 \pm 6\frac{e_b}{B} \pm 6\frac{e_l}{L}\right)$$

With the soil pressure determined, the mat is analysed as individual bands along column center lines. In this analysis, moment coefficient $\frac{1}{10}$ is used.

Design by Conventional Method

$f'_c = 3000$ psi $\quad f_s = 20{,}000$ psi
Allowable bearing pressure $q_a = 5^{tsf}$

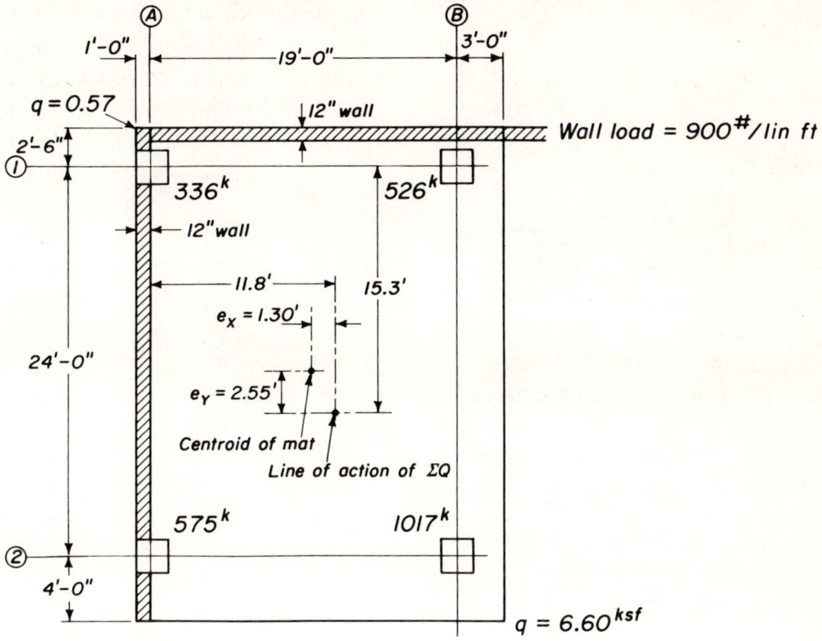

$$\Sigma V = 336 + 526 + 575 + 1017 + \underset{27.5}{0.9 \times (2.5 + 24 + 4)} + \underset{20}{0.9(19+3)} = 2502^k$$

$$\bar{X} = \frac{(526+1017)19 + 20(11) - 27.5(0.5)}{2502} = 11.8'$$

$$\bar{Y} = \frac{(575+1017)24 + 27.5(13) - 20(2.0)}{2502} = 15.3'$$

Try mat as shown

$$e_X = 11.8 + 1.0 - \frac{23}{2} = 1.30' \qquad I_X = \frac{1}{12}(23)(30.5)^3 = 54{,}400 \text{ ft}^4$$

$$e_Y = 15.3 + 2.5 - \frac{30.5}{2} = 2.55' \qquad I_Y = \frac{1}{12}(30.5)(23)^3 = 30{,}800 \text{ ft}^4$$

$$q = \frac{2502}{23 \times 30.5}\left(1 \pm 6\frac{1.30}{23} \pm 6\frac{2.55}{30.5}\right) = 6.60^{ksf\ max} < 5^{tsf} \quad OK$$
$$= 0.57^{ksf\ min}$$

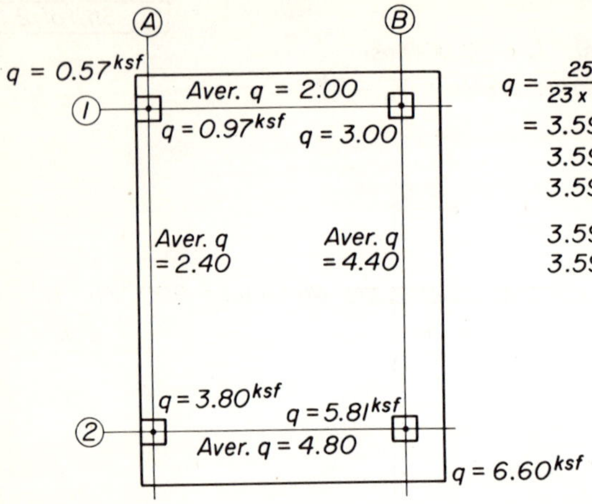

$$q = \frac{2502}{23 \times 30.5} \pm \frac{2502(1.3)}{30,800} X + \frac{2502(2.55)}{54,400} Y$$

$$= 3.59 \pm 0.106X \pm 0.118Y$$

$$3.59 - 0.106(10.5) + 0.118(12.75) = 0.97$$

$$3.59 - 0.106(10.5) + 0.118(11.25) = 3.80$$

$$3.59 + 0.106(8.5) - 0.118(12.75) = 3.00$$

$$3.59 + 0.106(8.5) + 0.118(11.25) = 5.81$$

Since each band is approximately a simple span, use moment coef.

Approx. M, Line ① $= \frac{1}{10}(2)(19)^2 = 72^{\text{lk}/'}$

Line ② $= \frac{1}{10}(4.80)(19)^2 = 173^{\text{lk}/'}$

Line Ⓐ $= \frac{1}{10}(2.4)(24)^2 = 138^{\text{lk}/'}$

Line Ⓑ $= \frac{1}{10}(4.40)(24)^2 = 253^{\text{lk}/'}$

$d = \sqrt{\frac{253}{0.236}} = 33''$ req'd.

Try $D = 4'\text{-}6''$

$d = 50$

$As = \frac{253}{1.44(50)} = 3.52$ □"/'

#11 at 5	Top on Line Ⓑ	12'-6" wide band
#11 at $9\frac{1}{2}$	Top on Line Ⓐ	10'-6" wide band
#11 at 16	Top on Line ①	14'-6" wide band
#11 at $7\frac{1}{2}$	Top on Line ②	16'-0" wide band

Chech Shear $V = 4.4 \times 12 = 52.8^k$ (approx.)

$v = \frac{52,800}{12(7/8)50} = 101$ psi — As shear in beam ok

(7/8 ← not in 1962 Code) — (this case ok)

Shear on periphery of col.

B. Design of mat foundation by method of finite difference. This method is based on the assumption that the subgrade can be substituted by a bed of uniformly distributed coil springs with a spring constant (coefficient of subgrade reaction) k.

The differential equation for deflection of such a mat foundation is

$$\nabla^4 w = \frac{q - kw}{D}$$

where $\nabla^4 w = \dfrac{\delta^4 w}{\delta x^4} + 2\dfrac{\delta^4 w}{\delta^2 x \delta^2 y} + \dfrac{\delta^4 w}{\delta^4 y}$

q = subgrade reaction per unit area of mat,

k = coefficient of subgrade reaction,

w = deflection,

D = rigidity of the mat,

$$= \frac{Et^3}{12(1 - \mu^2)}$$

E = modulus of elasticity,

t = thickness of mat,

μ = Poisson's ratio.

Equation (7-5) corresponds to Eq. (7-1) in Sec. 7-5B. To solve the differential equation above by finite differences, the mat is divided into square areas ($h \times h$). In the case of an interior point, point a in Fig. 7-12(a), the deflection at point a can be expressed by a difference equation in terms of deflections at the adjacent points to the right, left, top, and bottom. This difference equation is

$$20w_a - 8(w_t + w_b + w_r + w_l) + 2(w_{tl} + w_{tr} + w_{bl} + w_{br}) + (w_{tt} + w_{bb} + w_{ll} + w_{rr}) = \frac{qh^4}{D} + \frac{Qh^2}{D}$$

The notations w_a, w_t, . . ., represent deflection at points a, t, ... The difference equation above is diagrammatically represented by Fig. 7-12(b). The difference equations for other points of the network are shown in a similar manner.

For a given mat foundation, one difference equation can be written for each point of intersection. By solving these simultaneous equations, the deflections at all points are determined. The equations can be solved rapidly with a digital computer.

Notations w_a, w_t .... = deflection at point a, t, ...
h = size of network
Q = concentrated load at point a
EI = modulus of elasticity and moment of inertia of mat, constant throughout the mat.
μ = Poisson's ratio
$D = \frac{Et^3}{12(1-\mu^2)}$, t = thickness of mat

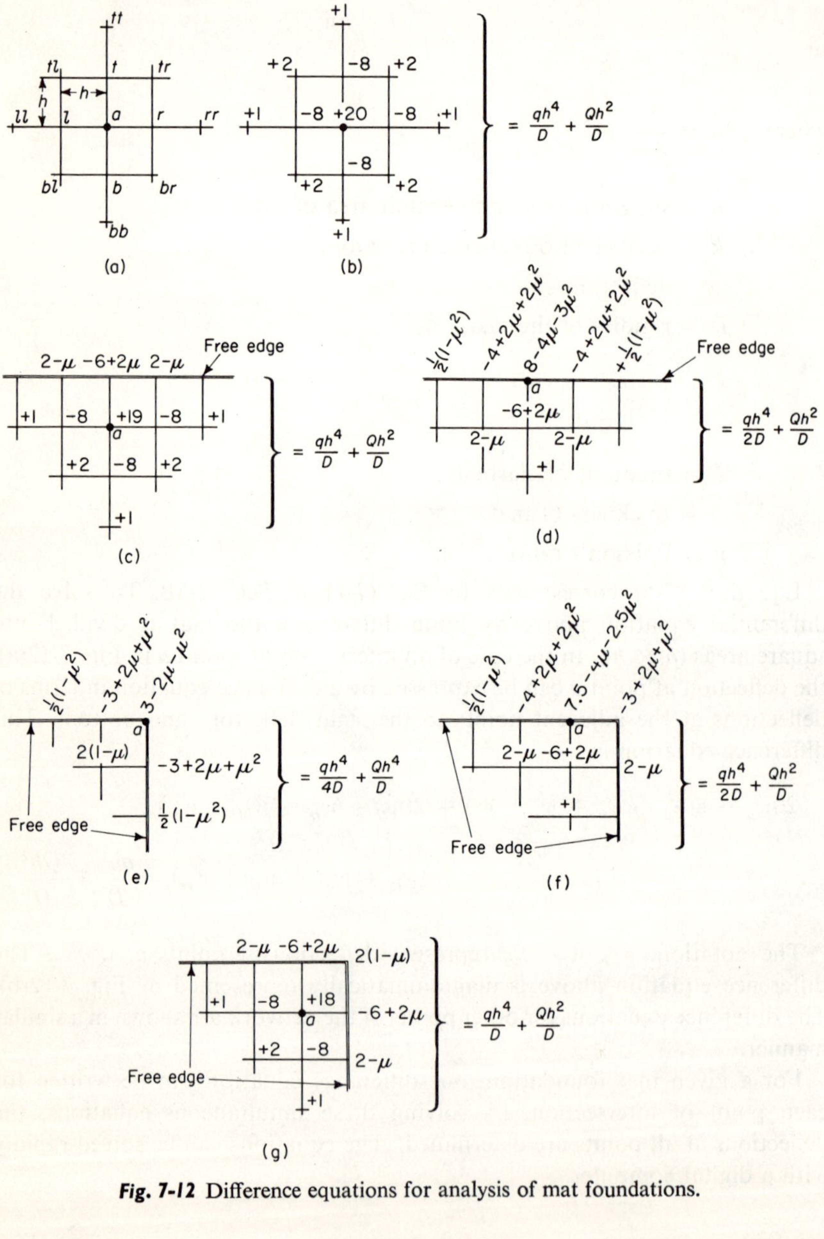

Fig. 7-12 Difference equations for analysis of mat foundations.

After the deflections are known, the bending moment at any point in each direction can be determined. From theory of elasticity, it is known that

$$M_x = M'_x + \mu M'_w$$

where M_x = bending moment per unit strip in x-direction,

M'_x = bending moment in the x-direction not including the influence of the bending moment in the y-direction,

M'_w = bending moment in the y-direction not including the influence of the bending moment in the x-direction.

By using the finite difference operators, Eq. (7-2), the total bending moment on a strip in the $l - r$ direction can be expressed for an interior point

$$M_{l-r} = -\frac{D}{h^2}[(w_l - 2w_a + w_r) + \mu(w_t - 2w_a + w_b)]$$

The procedure of analysing a mat foundation by the method of finite differences is illustrated by a simple example shown in Plate DE 7-7. A square footing 10 ft $\times$ 10 ft is subjected to a central load $Q = 100^k$.* This example is worked out by Rijhsinghani (1961). Due to symmetry in two directions, there are only 15 unknowns in the network of 64 points. The equations are arranged and tabulated on sheet 2 of Plate DE 7-7. The solution for deflections are shown at the bottom of the table. With these values, the bending moment at any point can be calculated. For example, at point 2, bending moment

$$M_2 = -\frac{D}{h^2}[(w_1 - 2w_2 + w_3) + \mu(w_6 - 2w_2 + w_6)]$$

$$= 7370 \text{ in.-lb/in.}$$

The results of bending moment along the center line of this footing is shown in a graph on sheet 1 of Plate DE 7-7.

The accuracy of the finite method depends upon the number of networks divided. When the squares are considerably larger than the size of columns, the results are very unreliable adjacent to the columns. This difficulty can be overcome, however, by introducing subdivisions of the network adjacent to the columns. For difference equations involving subdivisions, reference is made to textbooks of numerical methods, for example, Marcus, Newmark, and Austin (1954). Unfortunately, extremely large numbers of equations are required in practical problems even if subdivisions are employed.

---

* In order to obtain a larger deflection, a relatively thin footing, 12 in., is used in this example.

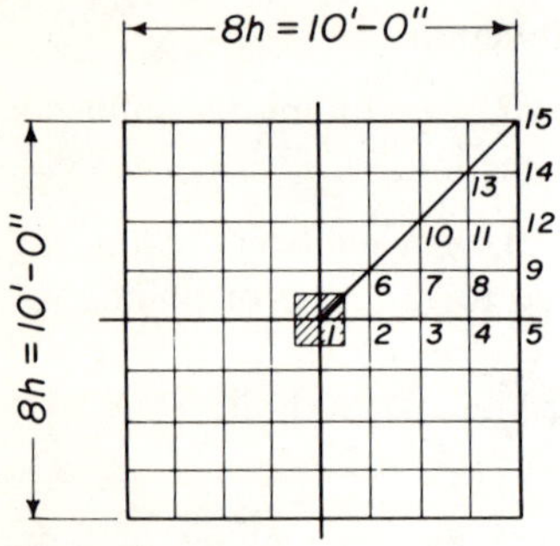

Given

A square footing 10'-0" x 10'-0" subjected to a central load Q = 100 kips

$$t = 12''$$
$$E = 3 \times 10^6 \text{psi}$$
$$\mu = 0.15$$
$$h = 15''$$
$$k = 100 \text{ lbs/cu in.}$$
$$D = \frac{Et^3}{12(1-\mu^2)} = 4.41944 \times 10^8 \text{lb in.}$$
$$qh^4/D = 0.01145$$
$$Qh^2/D = 5.091075 \times 10^{-2} \text{in.}$$

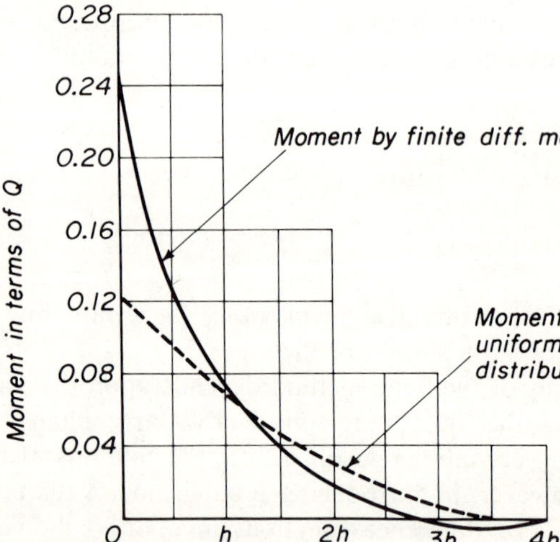

Moment along center line of footing

DE 7-7

Mat Foundation

Finite Difference Method

Sh. 2 of 2

Difference Equations for Points 1,2,3.....15

y / Eq.	ω_1	ω_2	ω_3	ω_4	ω_5	ω_6	ω_7	ω_8	ω_9	ω_{10}	ω_{11}	ω_{12}	ω_{13}	ω_{14}	ω_{15}	Constant
1	20.01145	-32.0	4.	0	0	8.	0	0	0	0	0	0	0	0	0	0.05091075
2	-8.	25.01145	-8.	1.	0	-16.	6.	0	0	0	0	0	0	0	0	0
3	1.	-8.	20.01145	-8.	1.	4.	-16.	4.	0	2.	0	0	0	0	0	0
4	0	1.	-8.	19.01145	-5.7	0	4.	-16.	3.7	0	2.	0	0	0	0	0
5	0	0	1.	-5.7	7.338226	0	0	3.7	-7.31	0	0	0.9775	0	0	0	0
6	2.	-16.	4.	0·	0	22.1145	-16.	2.	0	2.	0	0	0	0	0	0
7	0	3.	-8.	2.	0	-8.	23.01145	-8.	3.	0	0	0	0	0	0	0
8	0	0	2.	-8.	1.85	1.	-8.	20.01145	-5.7	2.	-8.	1.85	1.	0	0	0
9	0	0	0	1.85	-3.655	0	1.	-5.7	7.826976	0	1.85	-3.655	0	0.48875	0	0
10	0	0	2.	0	0	2.	-16.	4.	0	20.01145	-16.	2.	2.	0	0	0
11	0	0	0	1	0	0	3.	-8.	1.85	-8.	21.01145	-5.7	-8.	2.85	0	0
12	0	0	0	0	0.48875	0	0	1.85	-3.655	1.	-5.7	7.338826	1.85	-3.655	0.48875	0
13	0	0	0	0	0	0	0	2.	0	2.	-16.	3.7	18.01145	-11.4	1.7	0
14	0	0	0	0	0	0	0	0	0.48875	0	2.85	-3.655	-5.7	8.69948	-2.6775	0
15	0	0	0	0	0	0	0	0	0	0	0	0.9775	1.7	-5.356	2.68036	0
Defl.	9.3254997	8.7651519	7.9388617	7.0937881	6.3127221	8.4004645	7.6998976	6.9168705	6.1638721	7.1470927	6.4645359	5.7667868	5.8637645	5.2163497	4.5994236	$\times 10^{-2}$ in.

C. Mat foundation for circular structures. Large circular or cylindrical structures such as chimneys, silos, water tanks, etc., are best supported on a circular or octagonal mat formation. The latter, being very close to a circle, may be treated as a true circle for all practical purposes. The stresses in a circular mat of uniform thickness may be computated by the formulae (Beyer, 1956) shown in Fig. 7-13.

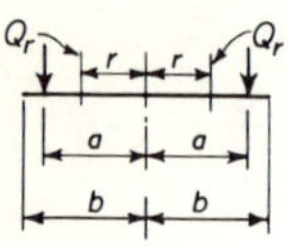

$\rho = \frac{r}{a} \qquad \beta = \frac{b}{a}$

μ = Poisson's ratio

$\phi_0 = 1-\rho^4 \qquad \phi_1 = 1-\rho^2 \qquad \phi_2 = \rho^2 \log_e \rho \qquad \phi_3 = \log_e \rho \qquad \phi_4 = \frac{1}{\rho^2} - 1$

q = uniform bearing pressure

M_r = radial moment $\qquad M_t$ = tangential moment

Q_r = shear per linear foot of circumference

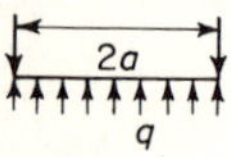

$$M_r = \frac{qa^2}{16}(3+\mu)\phi_1 \qquad M_t = \frac{qa^2}{16}[2(1-\mu) + (1+3\mu)\phi_1]$$

$$Q_r = q\frac{a}{2}\rho$$

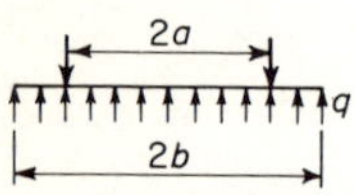

$$k_1 = 2(1-\mu) + (1+3\mu)\beta^2 - 4(1+\mu)\beta^2 \log_e \beta$$

$$k_2 = 2(1-\mu) - (3+\mu)\beta^2 - 4(1+\mu)\beta^2 \log_e \beta$$

For $\rho \leq 1$

$$M_r = \frac{qa^2}{16}[k_1 - (3+\mu) + (3+\mu)\phi_1]$$

$$M_t = \frac{qa^2}{16}[k_1 - (1+3\mu) + (1+3\mu)\phi_1]$$

$$Q_r = q\frac{a}{2}\rho$$

For $\rho \geq 1$

$$M_r = \frac{qa^2}{16}[k_1 - (3+\mu) + (3+\mu)\phi_1 - 2(1-\mu)\beta^2\phi_4 + 4(1+\mu)\beta^2\phi_3]$$

$$M_t = \frac{qa^2}{16}[k_1 - (1+3\mu) + (1+3\mu)\phi_1 + 2(1-\mu)\beta^2\phi_4 + 4(1+\mu)\beta^2\phi_3]$$

$$Q_r = \frac{q}{2}a\left(\frac{\beta^2}{\rho} - \rho\right) \qquad (\phi_0 \text{ and } \phi_1 \text{ are negative when } \rho > 1)$$

Fig. 7-13 Equations for analysis of circular foundations.

7-8 *Coefficient of Subgrade Reaction*

The coefficient of subgrade reaction (modulus of foundation, subgrade modulus) is defined as the ratio between the pressure against the footing or mat and the settlement at a given point.

$$k = \frac{q}{S} \tag{7-5}$$

where k = coefficient of subgrade reaction, lb per cu in.;

q = pressure against the footing or mat at a given point, psi;

S = settlement of the same point of the footing or mat, in.

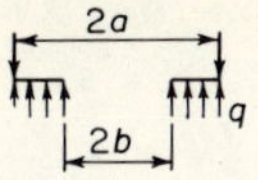

$$k_1 = (3+\mu) + 4(1+\mu)\frac{\beta^2}{1-\beta^2}\log_e \beta$$
$$k_2 = (3+\mu) - 4(1+\mu)\frac{\beta^2}{1-\beta^2}\log_e \beta$$

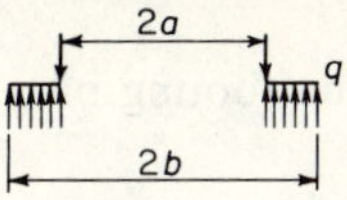

$$M_r = \frac{qa^2}{16}\left[(3+\mu)\phi_1 - \beta^2 k_1\phi_4 + 4(1+\mu)\beta^2\phi_3\right]$$
$$M_t = \frac{qa^2}{16}\left\{(1+3\mu)\phi_1 + \beta^2 k_1\phi_4 + 4(1+\mu)\beta^2\phi_3 + 2(1-\mu) - 2\beta^2\left[2(1-\mu) - k_1\right]\right\}$$
$$Q_r = q\frac{a}{2}\left(\rho - \frac{\beta^2}{\rho}\right)$$

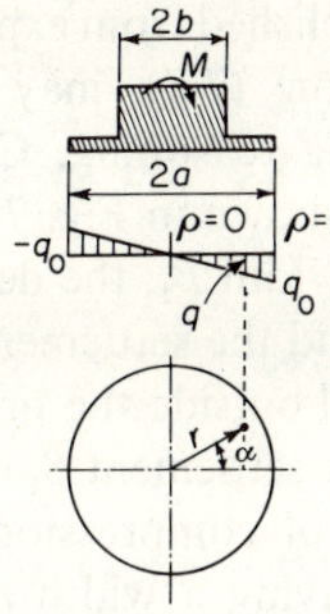

$$k_1 = 3 + \mu + (1-\mu)\beta^4$$
$$k_2 = 4(2+\mu) + (1-\mu)(3+\beta^4)\beta^2$$
$$k_3 = 4(2+\mu)\beta^4 - (3+\mu)(3+\beta^4)\beta^2$$

$$\bar{q}_0 = \frac{4M}{\pi a^3}$$

$$M_r = \frac{\bar{q}_0 a^2}{48k_1}\left\{(5+\mu)k_1\rho^3 - (3+\mu)k_2\rho + 3(1+\mu)k_1\rho^{-1} - (1-\mu)k_3\rho^{-3}\right\}\cos\alpha$$
$$M_\alpha = \frac{\bar{q}_0 a^2}{48k_1}\left\{(1+5\mu)k_1\rho^3 - (1+3\mu)k_2\rho + 3(1+\mu)k_1\rho^{-1} + (1-\mu)k_3\rho^{-3}\right\}\cos\alpha$$
$$M_{r\alpha} = \frac{\bar{q}_0 a^2}{48k_1}(1-\mu)\left\{k_1\rho^3 - k_3\rho + 3k_1\rho^{-1} + k_3\rho^{-3}\right\}\sin\alpha$$
$$Q_r = \frac{\bar{q}_0 a}{24}\left(9\rho^2 - 2\frac{k_2}{k_1} - 3\rho^{-2}\right)\cos\alpha$$
$$Q_\alpha = -\frac{\bar{q}_0 a}{24}\left(3\rho^3 - 2\frac{k_2}{k_1}\rho + 3\rho^{-1}\right)\sin\alpha$$

Fig. 7-13 Equations for analysis of circular foundations (cont).

In other words, the coefficient of subgrade reaction is the unit pressure required to produce a unit settlement. In clayey soils, settlement under the load takes place over a long period of time and the coefficient should be determined on the basis of the final settlement. On purely granular soils, settlement takes place shortly after load application.

Equation (7-5) is based on two simplifying assumptions:

1. The value k is independent of the magnitude of pressure.
2. The value k has the same value for every point of the surface of footing or mat.

Actually a number of factors affect the value of coefficient of subgrade reactions. They are discussed in the following (Terzaghi, 1955).

1. *Effect of size.* For the same soil, the value of k decreases with increasing width of the footing. The following relationship applies to very long footings (Terzaghi, 1955):

$$k = k_1\left(\frac{B+1}{2B}\right) \qquad \text{(granular soil)} \tag{7-6}$$

$$k = \frac{k_1}{B} \qquad \text{(cohesive soil)} \tag{7-7}$$

where k = coefficient of subgrade reaction for a very long footing of width B, B in ft.

k_1 = coefficient of subgrade reaction for a very long footing of one foot width.

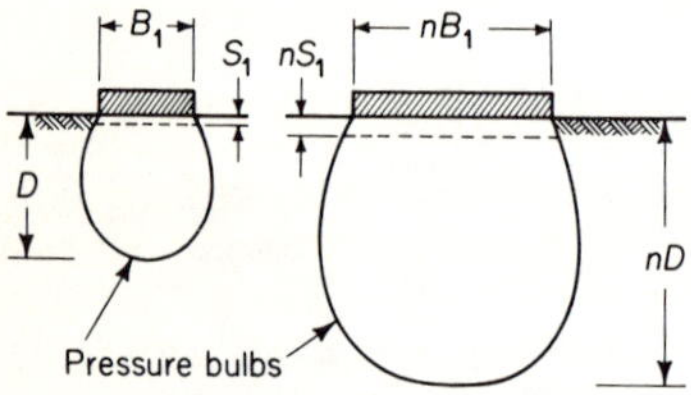

Fig. 7-14 Effect of size of footing.

Equation (7-6) is established from experimental results. Equation (7-7) may be derived by the following reasoning. Consider the pressure bulbs shown in Fig. 7-14. For a footing having a width B_1, the depth of pressure bulb is D, and the settlement is S_1. The stress in the soil outside the pressure bulb is small, and the settlement S_1 may be thought as a result of compression of the soil within the pressure bulb only. For a footing having a width nB_1, and under the same unit load q, the depth of pressure bulb is nD, and consequently the settlement is nS_1. That is

$$S = nS_1$$

Therefore,

$$k = \frac{q}{s} = \frac{1}{n} \cdot \frac{q}{S_1} = \frac{k_1}{B}$$

2. *Effect of shape.* For footings having the same width B under the same uniform load q and supported on the same soil, the value of k decreases with increasing length L of the footing (Terzaghi, 1955).

$$k = \frac{k_s(1 + B/L)}{1.5} \tag{7-8}$$

where k = coefficient of subgrade reaction for rectangular footing having a length L and width B,

k_s = coefficient of subgrade reaction for square footing ($B \times B$).

This equation indicates that k_s value for an infinitely long footing is equal to $\frac{2}{3}$ that for a square footing.

3. *Effect of depth.* The modulus of elasticity* of sand increases with the depth and it may be expressed by the following equation

$$E_{av} = C\gamma z$$

where C = constant, depending on the property of sand;

γ = density of sand;

z = depth.

Consider the sketch of a footing shown in Fig. 7-15. Again, the settlement of the footing may be considered as a result of compression of the soil contained in the pressure bulb. An examination of the pressure bulb indicates that the approximate average pressure is equal to $\frac{1}{2}q$. If the pressure bulb is further simplified by substitution of a cube $B \times B \times B$, the average depth z would be equal to $D + B/2$. Substitute these values to the equation above,

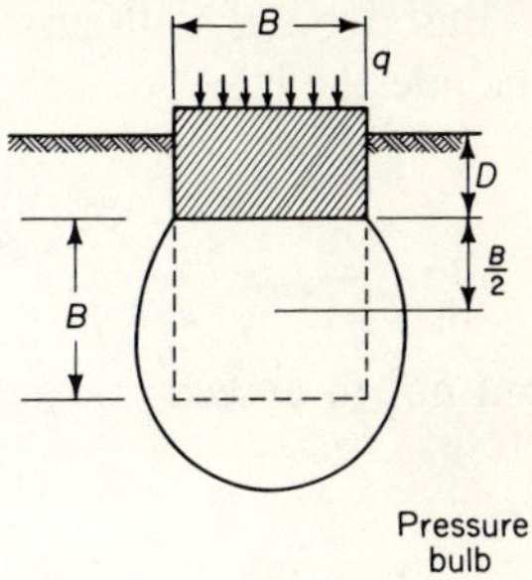

Fig. 7-15 Effect of depth of footing supported on granular soil.

$$E_{av} = \frac{\text{Average stress}}{\text{Average strain}} = \frac{\frac{1}{2}q}{\dfrac{S}{\text{Depth of bulb}}} = C\gamma\left(D + \frac{B}{2}\right)$$

where S = total settlement of footing,

D = depth of foundation.

Therefore,

$$k' = \frac{q}{S} = C\gamma\left(1 + 2\frac{D}{B}\right)$$

where k' = coefficient of subgrade reaction of a foundation at depth D. For a footing on surface of ground,

$$k' = C\gamma$$

Therefore,

$$k' = k\left(1 + 2\frac{D}{B}\right) \qquad (k' \leqq 2k) \tag{7-9}$$

This equation indicates that the settlement of a footing is reduced to one-half if it is lowered from ground surface to a depth equal to one-half of the width of the footing.

---

* In the strict sense the term modulus of elasticity should not be used since sand is not truly elastic. Some researchers prefer the term average modulus of stress and strain.

Due to the loosening of the soil during construction, the settlement of footings is usually greater than predicted on the basis of undisturbed soils. In fact, for narrow wall footings less than two feet wide the design should be based on the loosened condition. For the purpose of design, the modulus of subgrade reaction for granular soils should not be taken as greater than two times that at the ground surface.

Based on the discussions above, a general equation may be written to include the effect of size and depth for square footings on granular soils

$$k = k_1\left(\frac{B+1}{2B}\right)^2\left(1 + \frac{2D}{B}\right) \tag{7-10}$$

but not to exceed

$$2k_1\left(\frac{B+1}{2B}\right)^2$$

The modulus of elasticity for a purely cohesive soil with uniform properties from the ground to a great depth is practically constant throughout the depth. Therefore, the depth has no effect on the value of modulus of foundation.

Table 7-1 COEFFICIENT OF SUBGRADE REACTION k_1 FOR 1 FT $\times$ 1 FT PLATES OR LONG FOOTINGS OF 1 FT WIDTH* (lb per cu in.)

	Granular soils		
Relative density	*Loose*	*Medium*	*Dense*
Dry or moist	45 (Range = 20–70)	150 (Range = 70–350)	600 (Range = 350–1200)
Submerged	30	90	350

		Cohesive soils		
Consistancy	*Soft and very soft*	*Stiff*	*Very stiff*	*Hard*
Unconfined strength tons/sq ft	0–1	1–2	2–4	4 and over
	Design as if foundation is perfectly rigid	85 (Range = 60–120)	175 (Range = 120–230)	350 (230 and over)

* After Terzaghi (1955).

However, for stiff and hard clays, the soil is actually partly cohesive and partly frictional. In such cases a general equation may be expressed to include the effect of depth and size of footing.

$$k = k_a\left(\frac{B+1}{2B}\right)^2\left(1 + \frac{2D}{B}\right) + \frac{k_b}{B} \tag{7-11}$$

where the values of k_a and k_b must be evaluated by at least two tests using two different sizes, say 1 ft sq and 2 ft sq.

7-9 General Considerations in Design of Mat Foundations

When a mat is supported on strong bedrock, the column loads are transmitted to the rock on a relatively small area directly under the columns. The pressure distribution is similar to the one shown in Fig. 7-16(a). If it rests on stiff or compact soils, the column loads are distributed to the subsoil in larger areas, Fig. 7-16(b). On soft soils, the pressure against the bottom of mat

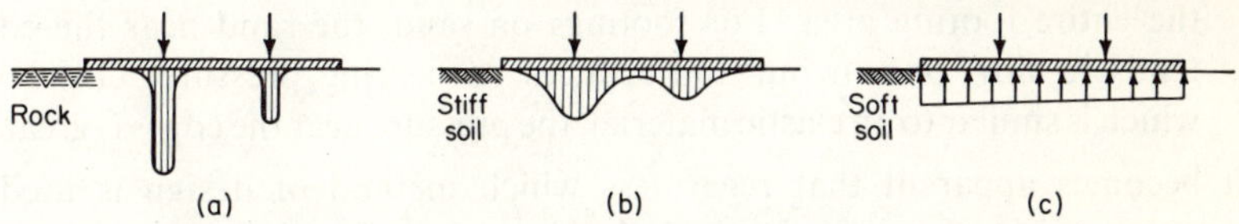

Fig. 7-16 Types of pressure distribution under mat foundation.

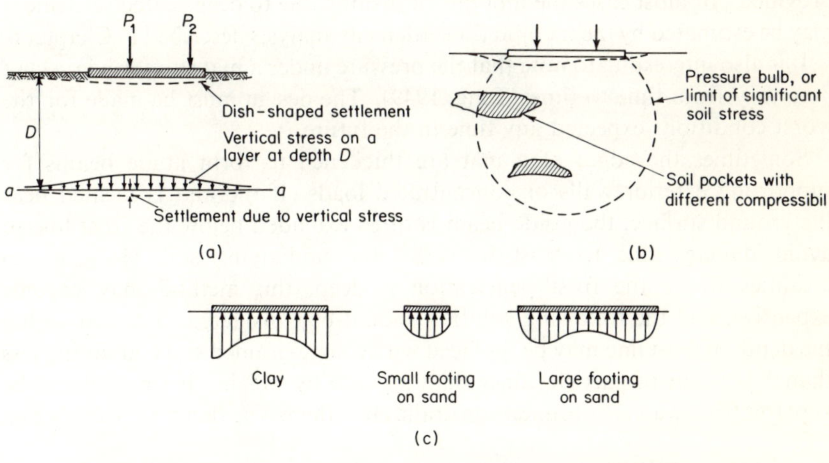

Fig. 7-17 Some additional factors affecting soil pressure and settlement of large footings and mat foundations: (a) dish-shaped settlement under a large footing or raft as a result of compression of deep soil strata; (b) variable soil conditions under large footing or raft; (c) soil pressure varies with type of soil.

foundation approaches planar distribution, Fig. 7-16(c). Therefore it is fully justified to design a mat on mud, soft clay, medium clay, peat, and organic soils by the conventional rigid method.

On stiff and compact soils the stresses in the mat resulting from the pressure distribution shown in Fig. 7-16[(a) and (b)] are smaller than those computed on the basis of planar distribution. Consequently, a greater economy may be achieved by designing the mat by elastic methods.

Regardless of the method used in the design, the stresses may be different from the calculated values because of the following reasons (Fig. 7-17):

1. If there is a layer of compressible soil extending to a relatively great depth, or if there is a compressible layer at a lower depth, a greater compression is produced in the soil under the center of the footing. Consequently it causes a dish-shaped ground settlement.
2. In apparently uniform soil deposits, there are always lenses and pockets of soil having characteristics different from the typical or average. Such pockets introduce unequal settlement.
3. Footings and mat foundations are not subjected to equal pressures over the entire footing area. For footings on sand, the sand near the edge tends to flow or run out and reduces its bearing pressure. On clays, which is similar to an elastic material, the pressure near the edges is greater.

It becomes apparent that regardless which method of design is used a generous amount of reinforcing is desirable. A certain minimum amount of reinforcing running both ways on the top and bottom of slab should be provided. In most cases the amount of dishing due to deep seated settlement may be estimated by the method for settlement analysis described in Chapter 6.

It is also interesting to note that the pressure under a mat supported on clay may vary from time to time (Teng, 1949). The design must be made for the worst conditions expected any time in the future.

Sometimes the edges of a mat are thickened to form grade beams for supporting exterior walls or concentrated loads. If the mat is located near the ground surface, the grade beam is often extended below the frost line to avoid damage due to frost heave of the underlying soil. However, in localities where the frost penetration is deep, this method may become expensive, and the following solution should be considered. The soil within the depth of frost line may be replaced with coarse-grained soil containing less than 3 per cent of particles finer than .02 mm by weight. If this soil can be kept free from water (by foundation drain, etc.), there is no danger of frost heave.

7-10 Construction of Mat Foundation

Mat foundations are almost invariably constructed of reinforced concrete. To avoid excessive shrinkage cracks mats are poured in small areas, commonly in the vicinity of 30 ft $\times$ 30 ft. Construction joints should be carefully located at sections of low shear stress. The common practice is to locate them along the center lines between columns. An elapse of at least 24 hours is desired between pours of adjacent areas. Reinforcing bars should be continuous across the joints. If bar splicing is needed, a lap of 24 bar diameter should be provided. The concrete should be strong enough to transfer the shear stress across the joint. This is commonly done by providing a shear key along the joint. The shear key, usually occupies the middle third of the thickness of the mat and should be designed for the maximum shear stress. If necessary, the mat may be thickened to provide sufficient strength in the joints.

Plate Eight

A Pile Driver Driving Batter Piles

Piles may be made of timber, concrete, or steel, and a large variety of each are available. In cast-in-place concrete piles alone, there are quite a number of commercial types. From the engineer's point of view, piles are classified as friction piles, point-bearing piles, compaction piles, etc. The procedures of design of a pile foundation include the selection of the material and the type of piles, the determination of bearing capacity, the pile length, and the pile spacing. The stresses in the lower soil strata must also be checked. In addition, adequate provision must be made for any lateral or uplifting force. The complete procedures for design and analysis of pile foundations are presented in this chapter.

8 PILE FOUNDATIONS

8-1 *Use of Piles*

Piles may be used for the following purposes:

1. To transfer loads through water or soft soil to a suitable bearing stratum by means of end bearing of the piles (*end-bearing or point-bearing piles*).
2. To transfer loads to a depth of a relatively weak soil by means of "skin friction" along the length of the piles (*friction piles*).
3. To compact granular soils, thus increasing their bearing capacity (*compaction piles*).
4. To carry the foundation through the depth of scour to provide safety in the event the soil is eroded away.
5. To anchor down the structures subjected to uplift due to hydrostatic pressure or overturning moment (*tension piles or uplift piles*).
6. To provide anchorage against horizontal pull from sheetpiling walls or other pulling forces (*anchor piles*).
7. To protect water front structures against impact from ships or other floating objects (*fender piles* and *dolphins*).
8. To resist large horizontal or inclined forces (*batter piles*).

8-2 *Types of Piles*

Piles may be classified according to their composition or function.

Table 8-1 GENERAL CHARACTERISTICS AND USE OF PILES

Type of pile	*Usual maximum design load, tons per pile*	*Advantages*	*Disadvantages*	*Common use*
Timber	25	Low cost per foot of pile Timber is a resilient material suitable for impact absorption	Small bearing capacity Untreated piles above ground water may last more than 25 years but are not permanent Prone to damage by hard driving, should not be driven through hard stratum or boulders	Foundation for structures with moderate load Protection of waterfront structures from damage by floating objects and crafts Timber trestles and bents Foundation for temporary structures
Precast concrete	80, except for large prestressed piles	Relatively large bearing capacity Permanent Can be treated for sea water installation	Must be reinforced to withstand handling stresses Requires space for casting and storage Takes time to set and cure before installation Requires heavy equipment for handling and driving Incurs large cost in cutting off extra lengths or adding more lengths (this is often the case since exact pile lengths can be determined in advance only in exceptional cases)	Trestles and bridge bents Water front installations (docks, piers, bulkheads, etc.) Foundation for bridges Prestressed piles of large bearing capacity are advantageous in bridge foundations

Table 8-1 (cont.)

Type of pile	*Usual maximum design load tons per pile*	*Advantages*	*Disadvantages*	*Common use*
Cast-in-place	75, except for compacted pedestal piles	Relatively large bearing capacity Permanent Can be treated for sea water installation Easy to alter pile lengths Damage due to handling or driving can be eliminated May be installed by pre-excavation thus eliminating vibration due to driving		Foundations for buildings, bridges, etc., of moderate to heavy loads
Composite	Governed by the weaker of the two parts	Relatively low cost Permanent	Small bearing capacity Joint between two parts constitutes a weak point prone to damage due to driving	Foundations for buildings, bridges, etc., of moderate loads where upper part of piles is above ground water level
Steel	100	Large bearing capacity Can penetrate through stiff layer or boulders Small volume displacement of soil Can stand rough handling	Possibility of damage from corrosion and electrolysis Relatively expensive unless the bearing stratum can develop large pile capacity Less effective as friction pile or compaction pile	Foundations for large structures of heavy loads Trestles and bridge bents

Composition of Pile:

Timber: Plain
Treated with preservative

Concrete: Precast
Cast-in-place–a number of commercial varieties are available

Composite: Commonly timber or steel (lower portion of the pile) and concrete (upper portion of the pile)

Steel: H-pile
Pipe pile
Sheet pile

Function of Pile:

Point-bearing pile, friction pile, compaction pile, uplift pile, anchor pile, fender pile, and dolphin

The functions of piles are defined in Sec. 8-1. The significant characteristics and the common uses of all types of piles are summarized in Table 8-1.

8-3 Timber Piles

A timber pile is the trunk of a tree, trimmed of branches. It must satisfy the minimum requirements before being qualified as a foundation pile. The quality, the treatment, and the constructional characteristics of timber piles are discussed below.

A. Quality and classification of timber piles. A timber pile should be of sound quality and free of defects. It should be straight and have a uniform taper. A straight line drawn between the center of the butt to the center of the tip should be contained entirely within the pile. The general requirements of timber piles are described in the ASCE Manual No. 17 in the section entitled "Timber Piles and Construction Timber."

According to this manual, timber piles are divided into three classes on the basis of the quality of timber and the dimensions of the pile:

Class A pile: To be used for heavy loads or large unsupported length.

Class B pile: To be used for medium loads.

Class C pile: To be used below permanent water level or for temporary works.

The quality of piles is determined by the amount or lack of defects (decay, splits, twist of wood grains, . . .), size of knots, holes, etc. The dimensions of a given class of pile depends upon the length of pile and species. In any case, the minimum butt diameter is 14 in. for class A piles, 12 in. and 13 in. for class B piles, and 12 in. for class C piles. Bark on class C piles may not be peeled.

B. Preservative treatment. Untreated piles entirely embedded below ground water table are considered permanent, provided that marine borers are not present. When projecting above water, the timber piles are subjected to decay by fungi and attack by insects and borers. Therefore, building codes usually prohibit the use of untreated timber piles above water table to support permanent structures.

The most effective and common method for prevention of decay and animal and plant attacks is treatment of piles with preservatives (creosote oil being universally used). When a sufficient amount of creosote is impregnated properly in the piles, (16 lb per cu ft of timber for piles in fresh water, and 22 lb in sea water) the protection against decay and attack is excellent, with the exception of few borers.

After driven to final depth, all pile heads, treated or untreated, should be sawed square to sound undamaged wood to receive the pile cap. But before concrete for the pile cap is poured, the heads of the treated piles should be protected by zinc coat, lead paint, or by wrapping the pile heads with fabric upon which hot pitch is applied.

The treated piles should be handled with care. Cutting, framing, and drilling should be done before treatment as much as possible. Cuts, abrasions, etc., should be covered with coats of creosote and pitch. Bolt holes should also be treated with creosote in the field.

C. Splices of timber piles. Timber piles should preferably be driven the full length without splicing. Splices of timber piles should be avoided when piles are subjected to uplift or lateral forces.

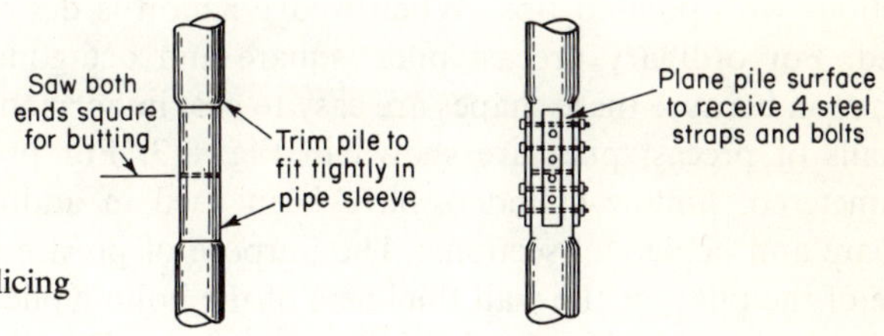

Fig. 8-1 Details of splicing timber piles.

If necessary, sections of piles may be spliced by either one of the two methods shown in Fig. 8-1. One method uses a section of a pipe sleave with a length of about 4 to 5 times the diameter of the pile. The butting ends of the pile are sawn square to ensure full contact after being driven, and the spliced portions are trimmed smooth and are fitted tightly into the pipe sleave. This type of splice is simple to make, but it lacks ability to transmit uplift force.

The other method of splicing utilizes steel straps and bolts. The butting ends are sawn square, and the four sides are planed flat to receive the splicing straps. This type of splice can resist some uplift and lateral force.

D. Overdriving of timber piles. One of the most significant drawbacks of timber piles is the possibility of damage due to overdriving. Piles may be damaged at the tips or above, Fig. 8-2. Providing a metal driving shoe on the tip does not materially reduce the possibility of such damages. Therefore, the design capacity of timber piles is limited empirically to about 25 tons in order to avoid the possibility of damages due to the necessity of hard driving. Furthermore, the behavior of the pile and the blows per foot of penetration during the pile driving operation should be carefully observed. If there is any doubt as to the possibility of such damage, the pile driving should be stopped immediately, and if necessary, one or two driven piles may be pulled out for visual examination.

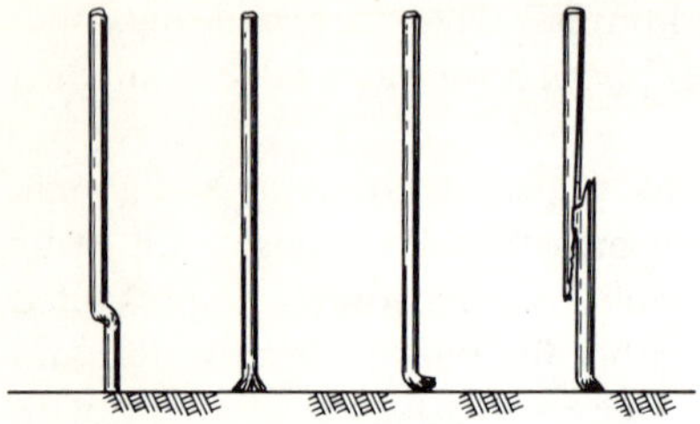

Fig. 8-2 Damage of timber piles by overdriving. After Chellis.

8-4 Precast Concrete Piles

Precast concrete piles are cast and cured in a casting yard, then shipped to the site for driving. Sometimes they are cast at the site if space is available. The information concerning the design and construction of precast piles are discussed below.

A. Design of precast piles. Precast piles are often made of uniform sections with pointed tips. When wedge action is desired, tapered piles are used. For ordinary precast piles, square and octagonal sections are most common because these shapes are easy to cast in horizontal position. Typical details of precast piles are shown in Fig. 8-3. For prestressed piles, large diametered, hollow cylinders have been used in addition to the ordinary square and octagonal sections. The purpose of prestressing is to reduce the size of the piles, or the wall thickness of the hollow pile, and consequently it facilitates the handling and driving procedure. When precast piles project above ground to support the bridge deck, the large cylindrical type has the desired rigidity and is relatively light in weight (Lin and Talbot, 1961).

The cross-section and the reinforcing of precast piles are usually governed by the handling stresses. These stresses depend upon the method of lifting and the location of points of support. For piles up to 25 ft long, one point of support is sufficient. For longer piles more points (as much as 5 or more points) are used to reduce the handling stresses, thus resulting in a more economical section. The supporting points should be clearly painted on the piles. They should be so located to ensure equal reactions and equal bending stresses at points of support.

In the design of precast piles for handling stresses, the concrete and the reinforcing steel may be allowed to an overstress up to 50 per cent. However, it is a good practice to provide a minimum of reinforcing equal to 1 per cent to 4 per cent of the average cross-section area of the pile, regardless of the calculated lifting stresses. Closely spaced ties or spirals should be provided near the top and bottom of the pile to avoid damage due to high impact stresses. Driving shoes may be cast at the tips of the piles in the event hard driving is anticipated.

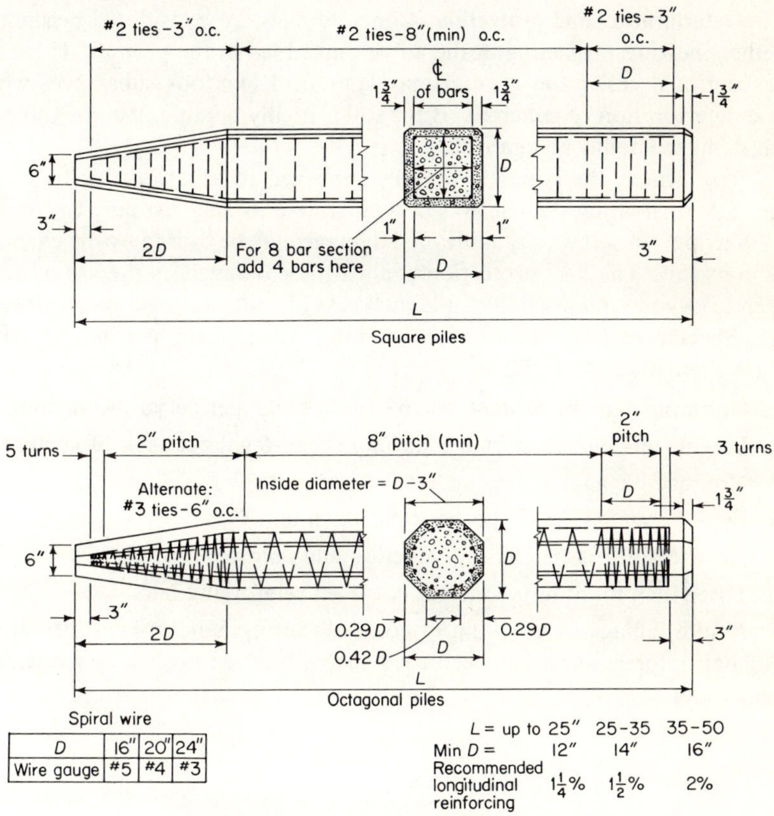

Fig. 8-3 Typical details of precast piles.

B. Cut-offs and splices of precast piles. After the piles are driven "home" (to the required depth), pile heads are cut off or spliced to the desired elevation for proper embedment in the pile caps. Any portion of the pile shattered or cracked due to driving or cutting off should be removed and spliced with fresh concrete. Since the exact pile lengths usually vary considerably even within a small area of the site, cut-offs or splices are inevitable.

To cut off a precast pile, a groove is cut by a chisel around the pile at the cut-off level; the reinforcing bars are exposed; and the bars projecting above the desired height are cut with a torch. Then the extra length of the pile above the groove is snapped off. If the pile heads are below the anticipated level, they may be spliced by adding a new length. In this case, a section of the piles must be cut off first so that a sufficient length of the reinforcing bars will project into the new section. If conditions permit, the splicing may be avoided by lowering the pile caps and extending the columns above by means of additional pedestals.

C. Deterioration and protection. Concrete piles are considered permanent whether they are projecting in the air or embedded in the ground. However, in exceptional cases, the ground may contain deleterious substances which cause deterioration of concrete. If the soil is highly organic, the possibility of such damage should be analized.

In fresh water, the concrete may be subjected to abrasion due to waves, sand, ice, or floating debris, and in temperate and cold regions, to freezing and thawing. In salt water, additional damage may be caused by the chemical action of salt. The best protection against these damages is the use of dense concrete of good quality and ample thickness of concrete cover on reinforcing bars. Specific recommendations concerning concrete for marine structures are (Wakeman et al., 1958):

1. Optimum cement content–use $6\frac{1}{2}$ to $7\frac{1}{2}$ sacks per cu yd of concrete.
2. Low water-cement ratio–use no more than 6 gal per sack of cement.
3. Nonreactive aggregates.
4. Type V cement, having 5 per cent maximum C_3A.
5. Air entrained concrete–in temperate and cold regions.
6. Three inch minimum concrete cover on reinforcing bars.
7. Avoid damage or cracking of concrete during handling and driving.

Asphalt impregnation has been used successfully on precast concrete piles in salt water.

8-5 *Cast-in-place Concrete Piles*

A cast-in-place pile is constructed by making a hole in the ground to required depth and then fill it with concrete. This type of pile is widely used because of a number of advantages as indicated in Table 8-1. Consequently, a variety of cast-in-place piles are available in the United States, and each bears the name of the manufacturer. The basic characteristics of these different varieties of cast-in-place piles are summarized in Fig. 8-4.

In general, cast-in-place piles may be classified into three types, namely shell type (cased type), shell-less type (uncased type), and pedestal type. The shell type is made by driving a steel shell into the ground with the aid of a mandrel (or core) inserted into the shell. The mandrel is withdrawn and the concrete is placed in the shell. The shell is made of corrugated and reinforced thin sheet steel (Raymond and Western Piles), thin fluted steel (Monotube Piles), or pipes (Armco welded pipes or common seamless pipes). The shell-less type is formed by withdrawing the shell while the concrete is being placed.

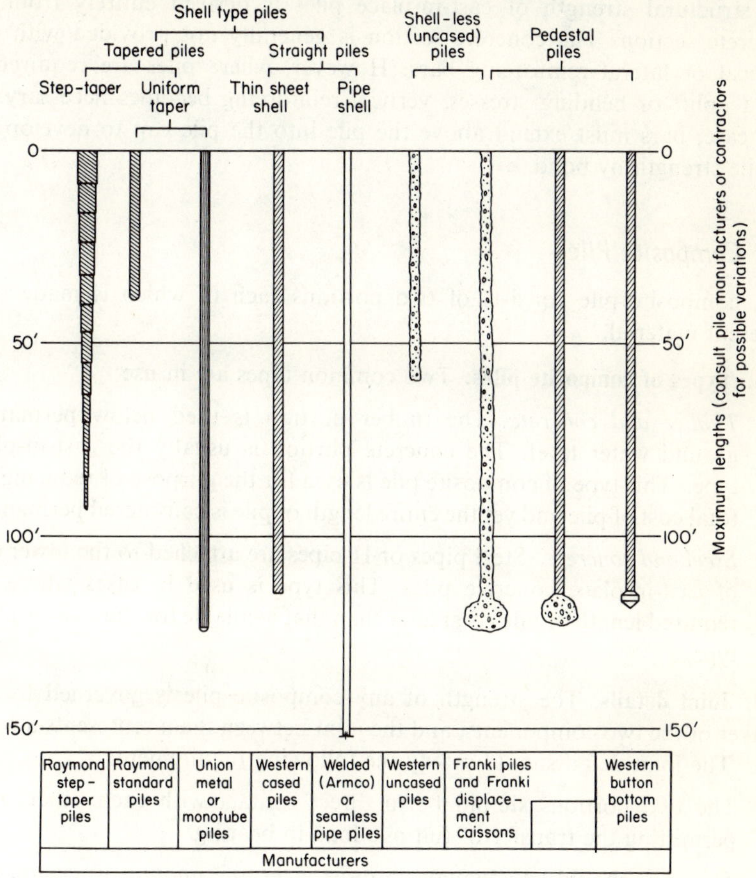

Fig. 8-4 Common types of cast-in-place piles.

Usually the fresh concrete is expounded in short sections by a heavy weight to eliminate the possibility of void space left in the pile or between the pile and the surrounding soil. Either the cased or the uncased piles may have an enlarged bulb or pedestal at the bottom. The pedestal may be a precast concrete cone (Western Piles) or an expanded concrete formed by impact of heavy drop weight (Franki Piles).

The steel shells may be tapered or uniform in cross section throughout the full length, except at the tip of the pile. The tapered piles are advantageous for friction piles and compaction piles in granular soils. Piles of uniform section are used for the point-bearing type driven to bedrock or very hard stratum. Where the bearing stratum is not too strong or not too thick, the pedestal type may be advantageous.

With the exception of pipe piles where thick metal is used to carry stresses, the structural strength of cast-in-place piles is derived entirely from the concrete section. The concrete section is generally not provided with any vertical or lateral reinforcing bars. However, where piles are required to resist uplift or bending stresses, vertical reinforcing becomes necessary. In this case, bars must extend above the pile into the pile cap to develop the tensile strength by bond.

8-6 *Composite Piles*

A composite pile consists of two portions each of which is made of a different material.

A. Types of composite piles. Two common types are in use:

1. *Timber and concrete.* The timber portion is used below permanent ground water level. The concrete portion is usually the cast-in-place type. This type of composite pile is used for the purpose of reducing the total cost of pile and yet the entire length of pile is considered permanent.
2. *Steel and concrete.* Steel pipes or H-pipes are attached to the lower end of cast-in-place concrete piles. This type is used in cases where the required length of piles is greater than that available for the cast-in-place type.

B. Joint details. The strength of any composite pile is governed by the weaker of the two components, and the joint between them represents a weak link. The joint detail should satisfy the following requirements:

1. The two portions should be in direct contact with each other, thus permitting the transfer of full pile load in bearing.
2. The joint should be adequate to resist a certain amount of tension and bending. In the case of piles subjected to calculated uplift, the joint should be strong enough to develop $1\frac{1}{2}$ to 2 times this uplift force. Even for those piles which are not subjected to calculated uplifting and bending stresses, often there is a certain amount of such stresses due to ground heaving and lateral movement resulted from pile driving.
3. The joint should be simple and easy to make in the field.

A variety of methods have been used to connect the two portions of timber-concrete type composite piles. The more commonly used methods are shown in Fig. 8-5 and discussed below.

1. *Wood tenon joint.* The upper end of the timber portion is trimmed into a tenon about 9 in. diameter × 18 in. long. The concrete portion is cast around the tenon in the form of a socket. To ensure good concrete in the socket, a steel sealing ring is welded to the shell of the concrete portion, thus keeping the joint water tight. The socket portion may be plain concrete or reinforced with vertical bars and horizontal ties.

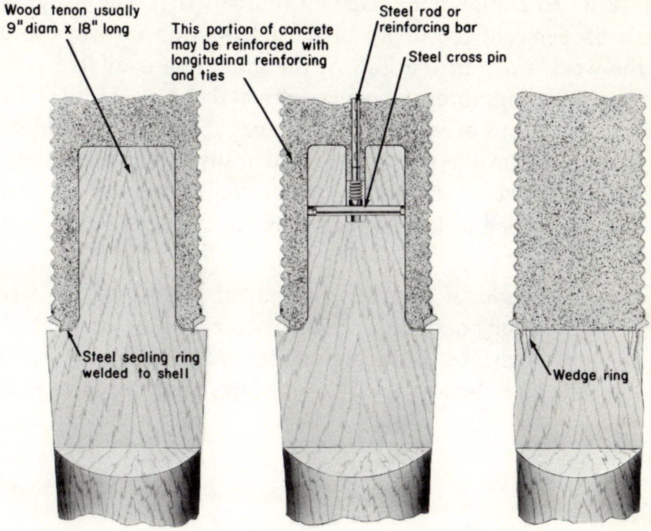

Fig. 8-5 Joint details commonly used in composite piles. Courtesy of Raymond Concrete Pile Co.

2. *Steel rod joint.* A $\frac{3}{4}$ in. or larger diameter pin or reinforcing rod is used to connect the two portions together. The steel pin is usually driven into the timber portion whereas the reinforcing rod is anchored in the timber by means of a steel cross pin. In either case, the lower end of the shell for the concrete portion is welded to a wedge ring which is forced into the timber pile in order to form a watertight joint.
   When additional strength is desired, the reinforcing rod and cross pin may be used on top of the wood tenon, thus a combination of the two methods are used.
3. *Wedged joint.* The head of the timber portion is wedged tight inside the steel sealing ring which is welded to the shell above. The joint is relatively weak.

8-7 *Steel Piles*

A steel pile may be a rolled section, a fabricated shape, or a piece of sheet pile. Two or more sections of sheet piles may be connected together in a box shape and driven as one pile.

A. Common types of steel piles. H-piles and pipe piles are the most commonly used types.

1. H-piles are proportioned especially to withstand the large impact stresses during hard driving. The flanges and the web are rolled with equal thickness in order to eliminate damage on thinner part. The flange width is made at least 85 per cent the depth of the pile section in order to provide rigidity in the weak axis. In the U.S.A. the minimum available size of the H-pile is 8 in. and the minimum thickness of metal is $\frac{3}{8}$ in. H-piles are rolled in mild steel equivalent to other structural shapes. Steel plates welded on top of H-piles had been considered necessary to transfer the pile loads to the concrete pile cap. However, tests* have indicated that if the pile heads are cut square and embedded at least 6 in. in the concrete cap, there is no need of such plates.

2. Pipe piles are made of seamless or welded pipes and are frequently filled with concrete. They may be driven closed-ended or open-ended. The open-end piles may be driven to the desired depth and the soil inside the pipe is cleaned out. The closed-end piles are formed by fixing a driving point to the

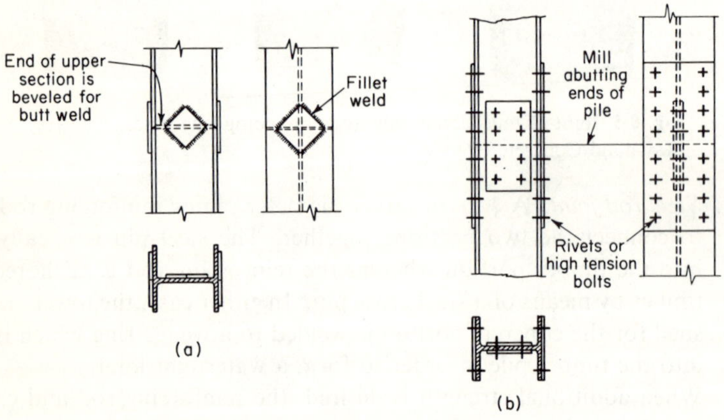

Fig. 8-6 Splice of H-piles: (a) welded splice; (b) riveted or bolted splice.

tip of the pile. For major columns the pipe piles should be at least 10 in. in diameter, and the thickness of wall should not be less than $\frac{5}{16}$ in. The choice between the open-end and the closed-end types depends upon the soil

* *Investigation of the Strength of the Connection between a Concrete and the Embedded End of the Steel H-piles*, Research Report No. 1 (Columbus, Ohio: Department of Highways).

conditions at the site. In some cases an open-end pipe can be driven to a greater depth since the soil inside the pipe can be cleaned as the driving progresses. If boulders or other obstructions are encountered before reaching the desired depth, they may be removed by means of a chopping bit, or, in the case of large diameter pipes, by blasting. In other cases where cleaning of soil inside the pipes is difficult, closed-end pipe should be used.

B. Splices of steel piles. H-piles are spliced in the same manner as steel columns, Fig. 8-6. Welded, riveted or bolted (high tension bolts) splice may be used, depending primarily on the preference of the engineer or the available equipment. If the splice is located above the ground surface, or if large bending moment will act upon the pile, the splice should be designed to resist such stresses. In normal cases, even when the piles are not subjected to horizontal thrust or bending moment, there is a certain amount of bending stress in the pile (The pile is never perfectly straight, and may be bent due to driving, etc.)

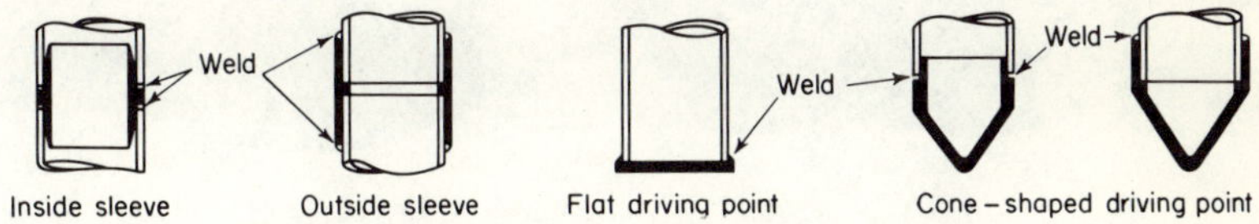

Fig. 8-7 Splice and driving shoe for pipe piles.

It is customary to design the splice to resist a moment equal to one-third to one-half of the moment capacity of the H-section. Some building codes require the splice to develop full strength of the H or pipe section.

The pipe piles may be spliced by butt welding, or by the use of a sleeve, Fig. 8-7.

C. Driving points. Open-end pipe piles are usually driven with a square cut end without other aid. The common types of driving points (shoes) for closed-end pipe piles are shown in Figs. 8-7, 8-8, and 8-9.

The selection of driving points for steel piles depends largely on the soil conditions. When driven to shale or other soft rocks, a square cut tip will penetrate the rock and seat itself in a secure position. Base plates welded on to the bottom of H-piles have been used to bear on compact sand and gravel layers. Such base plates are useful for reducing the depth of penetration in the hard stratum when this stratum is thin and is underlain by a soft soil.

When steel piles are driven to hard rock, the problem of slippage of pile tips on the uneven or sloping surface should be considered. Lateral shifting or sliding of a pile point may exceed several inches before it is seated in the rock. Slippage is even more severe in the case of batter piles. To overcome this difficulty the Norwegian engineers have used a special driving point known as the "Oslo point" (Bjerrum, 1957). It is made of a steel rod 3 to 4 in. diameter. The bottom of the rods are hollow-ground to provide sharp cutting

Fig. 8-8 Pipe piles with flat driving points. Photograph courtesy of Armco Drainage and Metal Products, Inc.

Fig. 8-9 Pipe piles with cone-shaped driving points. Photograph courtesy of Armco Drainage and Metal Products, Inc.

edge on the perimeter. For best result, the lower 4 in. of the rods should be hardened to a Brinnell hardness of 400 to 600. The web of the pile is slotted and the rod is welded to the pile, Fig. 8-10.

To insure good seating, the following driving procedure is followed. As soon as the pile tip is in contact with bed rock (this can readily be detected if

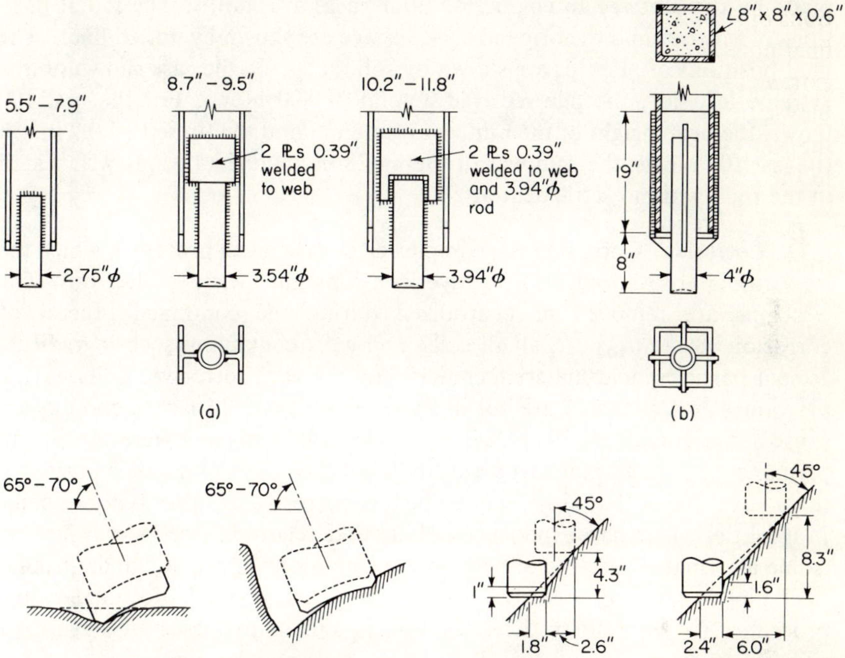

Number of blows	Height of ram, in.	Settlement, in.
100	4 – 8	0.2
100	8	0.55
100	8 – 10	0.24
100	8 – 10	0.12
100	8 – 10	0.04
100	8 – 10	0.04
100	28	0.08
100	28	0
800		1.27

Numb. of blows	Height of ram, in.	Settlem., in.
100	4 – 6	3.5
100	12 – 14	0.5
25	24 – 28	0.08
5	47	0.04
230		4.12

Numb. of blows	Height of ram, in.	Settlem., in.
10	10	3.1
10	10	?

Numb. of blows	Height of ram, in.	Settlem., in.
10	10	4
40	10	?
20	10	0
5	10	0

(c)

Fig. 8-10 Driving points ("Oslo points") for steel piles to hard rock: (a) driving point for H-piles; (b) driving point for box piles (for difficult foundation conditions); (c) results of test driving of piles to hard rock (all driving points 4 in. ϕ "Oslo points"). From L. Bjerrum.

the soil above the bed rock is soft; otherwise, it is difficult to know whether the pile has reached bed rock or still in the hard overburden) the pile driving is stopped, and the elevation of the pile head is measured by a level. Pile driving is then continued using a very small heights of drop of the ram (6 to

8 in.). The pile penetration is measured for every 20 blows of driving. When the penetration is practically zero after a series of 20 blows, the drop heights may be doubled. This procedure of increasing the drop heights by steps is repeated until a satisfactory contact area is chiseled out in the bedrock. A large number of blows is necessary for a pile point to seat on a satisfactory contact area on hard rock. Fig. 8-10(c) shows the results of test driving piles made by the Norwegian engineers. Four cases are shown. The initial positions of pile tips and the original rock surface are shown by dotted lines. The final positions of pile tips are shown by solid lines. In the case shown on the extreme left side, the pile received a total of 800 blows. For the first 100 blows, the drop height of the ram was 4 to 8 in., and the tip settled 0.2 in. In the last 100 blows, the ram was dropping 28 in., and the tip was well seated in the rock with no settlement at all.

D. Corrosion. Corrosion is a complex electrochemical process in which the iron atoms are ionized into positive iron ions and negative electrons. The electrons are removed or neutralized during the continued process of corrosion. Swamps, peat, alkali soils, and soils containing such ingredients as coal particles, acid material, cinders, etc. are very corrosive. Soils having pH values greater than 7 are not likely very corrosive. Moisture and oxygen cause active corrosion. Previously steel piles driven in clays were considered permanent because of the low permeability of clay. However, some investigations have shown that clays may be just as corrosive as other types of soils. In localities where past experience of long time record of steel piles is lacking (some information may be of help on the conditions of existing underground steel pipes), the soil corrosivity should be determined. Unfortunately the procedure for corrosivity determination is still in the research stage. For ordinary projects, one of the following methods may be used.

1. *Additional thickness.* This method, employed by most codes, requires that the pile section contains a certain extra thickness (usually $\frac{1}{16}$ in.) in excess of the sectional area as required by strength. This may be accomplished by either deducting $\frac{1}{16}$ in. from the actual metal thickness while computing the pile capacity, or splicing another piece of steel ($\frac{1}{16}$ in. thick) to the pile length where corrosive action is anticipated.

2. *Removal of corrosive soils.* Organic soils, cinder fill, unburned carbon or industrial waste near the ground surface may be removed and replaced by noncorrosive soils. This method is economical only when such soils are at shallow depth.

3. *Concrete encasement.* Near the ground surface where moisture and oxygen are abundant, the piles may be protected very effectively by encasing with concrete. Concrete encasement extending to a great depth is a costly method.

4. *Cathodic protection.* Since corrosion is a continuing process of removing electrons (electric current) from the piles, a method of preventing such a removal would be effective. This is done by cathodic method whereby a slow current is introduced towards the steel piles, instead of away from them. This may be effected by either supplying a low-voltage direct current to an iron or graphite anode buried near the piles or depressing the potential of the piles by grounding them, or a combination of both. It should be pointed out here that painting or other types of coating applied prior to driving are of doubtful value since the coating in the embedded portion is likely to be damaged from abrasion or handling. The real value of copper-bearing steel (usually 0.02 per cent copper) against corrosion has not generally been accepted. Piles projecting into atmosphere should be painted periodically. Piles projecting in polluted water should be protected with concrete encasement or coal tar paint. In fresh water, the section near the water surface should be protected.

8-8 Design of Pile Foundations

The complete procedure for the design of pile foundation may be summarized by the following steps:

1. Calculate the loads — two types: one for bearing capacity and the other for settlement analysis, see Chapter 3. The total load acting on the piles should include the weight of the pile cap and the soil above it. If the ground is newly filled or will be filled in the future, the additional load on piles due to negative skin friction should be included, Sec. 8-18.
2. Sketch a soil profile or profiles showing the soil stratification, and on this profile, superimpose the outline of the proposed foundation and substructure (Sec. 2-13).
3. Establish the permanent water level. If untreated timber piles are used, the possibility of ground water lowering in the future should be studied. Ground water may be lowered by pumping water from deep wells for water supply, industrial use, or air conditioning purposes. In any case, the permanent water level should not be assumed higher than the invert elevation of the sewer pipes and the low water level in the adjacent wells, rivers, or lakes. In the case of bridge foundations, the depth of scour should be established.
4. Determine type and length of piles (Sec. 8-9).
5. Determine pile capacity (Sec. 8-10).
6. Establish pile spacing (Sec. 8-11).
7. Check stresses in lower strata (Sec. 8-12).

8. Analyze settlement (Sec. 8-13).
9. Design the pile cap (Sec. 8-14).
10. Check uplift and lateral load (Sec. 8-15, 8-16, and 8-17).
11. Establish pile load test criteria (Sec. 8-19).

8-9 *Determination of Type and Length of Piles*

The behavior of a single pile and in particular a group of piles is extremely complex. There is no available theory which predetermines the exact length and bearing capacity of a given type of pile driven in a given soil. The selection of the type, length, and capacity is usually made from estimation based on the soil conditions and the magnitude of load. In large cities where the soil conditions are well known and where a large number of pile foundations have been constructed, the experience gained in the past is extremely useful. Generally the foundation design is made on the preliminary estimated values. Before the actual construction begins, pile load tests must be made to verify the design values. In the event the test results are significantly different from the design values, the foundation design must be revised according to the test results. Occasionally pile load tests can be made before the foundation design, therefore, more reliable and reasonable pile length and pile capacity can be ascertained.

In selecting the type of piles to be used for a given job, the engineer should first familiarize himself with the characteristics of all the piles available. The information shown in Table 8-1 and Fig. 8-4 are useful. The following discussion will aid him in selecting the type and length of the piles.

Point-bearing piles may be used if a hard stratum or the bedrock is within a reasonable depth and if there is no soil below this stratum which is too soft to impair the supporting capacity of the hard stratum. The length of piles driven to bear on the bedrock can be estimated with fair accuracy if sufficient borings are made to delineate the surface of bedrock. The point-bearing piles on hard stratum may be driven several feet in this stratum to develop the capacity or else pedestal-type piles are carried on top of it. In any one of these cases, the pile length can be estimated within a few feet.

Friction piles are used when hard stratum or bedrock is deep which would require very long point-bearing piles. The length of friction piles varies with the characteristics of the soil, the magnitude of loads, and the size of the pile. The theoretical values given in the next article are only approximate.

Compaction piles are used to increase the relative density of granular soil near the ground surface. The length of such piles is very difficult to predict, and therefore, test-driving is usually necessary. In general this type of piles are relatively short because there is no need to compact the soil to a depth greater than required.

8-10 Pile Capacity

The bearing capacity of a single pile is controlled by the structural strength of the pile and the supporting strength of the soil, and the smaller of the two values should be used for the design.

A. Pile capacity as determined by the structural strength of pile.

1. To avoid damage due to overdriving, timber piles are limited commonly to a design capacity not exceeding 25 tons per pile, and concrete piles are limited to 65 tons per pile. If properly protected with driving helmet and cushion, steel piles may approach the full value of the allowable stress. However, severe damage due to overdriving should be avoided.
2. To avoid overstress in the pile under design loads, the allowable pile stresses are governed by the values in the codes. In Table 8-2, the pile stresses allowed by several building codes are assembled for reference.
3. To avoid buckling failure of the pile, the total load should not exceed the critical buckling load divided by an appropriate factor of safety.

Table 8-2 ALLOWABLE STRESS IN PILES

Type of pile	*New York City Codes 1948*	*Boston City Codes 1958*	*Uniform Building Code 1961*	*AASHO 1957*	*AREA 1951*
Timber	600–800 psi 20^t (6″ ϕ tip) 25^t (8″ ϕ tip)	16^t–30^t depending on species	60% basic stress for clear material but not exceeding 1000 psi	18^t (10″ ϕ) 20^t (12″ ϕ) 24^t (14″ ϕ) 28^t (16″ ϕ)	
Concrete*	0.25 fc' but not exceeding 1000 psi	0.25 fc' but not exceeding 900 psi	0.225 fc'	20^t (10″ ϕ) 24^t (12″ ϕ) 28^t (14″ ϕ) 32^t (16″ ϕ) 40^t (20″ ϕ) 50^t (24″ ϕ)	0.25 fc' (friction piles) 0.20 fc' (point-bearing piles)
Steel	9000 psi (uncased) 12000 psi (cased)	8000 psi (concrete filled pipe piles) 7000 psi (H-piles) 15000 psi (core)	9000 psi	6000 psi	12000† psi (if splice sleeves are welded) 9000† psi (if splice sleeves are not welded)

* Note fc' = concrete cylinder strength at 28 days.
† Deduct $\frac{1}{16}$ in. for corrosion.

However, theory and experience have proven that the lateral restrain required to prevent pile buckling is very small. Therefore, in ordinary cases when the unit stress in the pile is kept below the allowable value, there is no need for checking the buckling capacity. Buckling enters the design problem when piles are projecting above ground surface. In such cases, approximate analysis are generally made on the assumption that the piles are laterally supported at 10 ft below ground surface in soft soils, and 5 ft in firm soils. However, in extremely soft mud, the pile may not obtain any lateral support throughout the full depth of the mud. In such extreme cases, pile load tests are desirable.

B. Pile capacity as determined by the supporting strength of soil.

1. *Point bearing piles.* When piles are driven to bedrock, the full structural strength often can be utilized, provided that the pile tip is well seated on the rock. The capacity is impaired if lateral shifting or uplifting of the pile take place due to pile driving operation (Sec. 8-20).

The bearing capacity of piles driven on, or into, a hard strata or soft rock is very difficult to determine, and usually it can only be estimated by approximate formulae. This estimated value must be verified by a full-scale loading test. Among a number of such so-called static pile formulae, the basic bearing capacity equation (3-1b) has given more reliable results.

$$Q_{\text{ult}} = \pi R^2(1.3cN_c + \gamma DN_q + 0.6\gamma RN_\gamma) \tag{3-1b}$$

If the pile tip is embedded in a granular soil for a depth of h, the bearing capacity may be estimated.

$$Q_{\text{ult}} = \pi R^2(\gamma DN_q + 0.6\gamma RN_\gamma) + 2\pi Rhf_s \tag{8-1}$$

where Q_{ult} = ultimate bearing capacity of single pile;

R = radius of pile tip;

c = cohesion of soil;

D = total penetration of pile, from ground surface to pile tip;

γ = unit weight of soil above the supporting stratum, use buoyant weight for the portion below ground water;

N_c, N_q, N_γ = bearing capacity factors, depending on the φ value of the soil, see Sec. 3-3;

h = depth of penetration of pile tip in the supporting layer of granular soil;

$f_s = \gamma(D - h/2) \tan \varphi$, and φ = angle of internal friction of the supporting soil.

When selecting the proper values of φ and c for estimating the bearing capacity it is customary to assume that the supporting soil is not affected by pile driving. The assumption is on the conservative side because actually within a distance of about three diameters below the pile tip and a lateral distance of about two diameters from the pile the granular soil is highly compacted.

2. *Pedestal type bearing piles.* The bearing capacity of a point-bearing pile can be increased by the use of an enlarged base or a pedestal. This base may be made of a precast concrete cone attached to the pile shell and driven in place, or it may be more effectively formed after the shell is driven. In the latter case, stiff concrete mix (zero slump) is placed in the hole and driven by a heavy ram, thus a pedestal is formed by displacement of the surrounding soil (granular type only). Therefore, the capacity of the pile is increased as a result of the increased base area and the increased density of the soil.

3. *Friction piles in granular soils.* The bearing capacity of a single friction pile driven in sand and gravel may be approximated by the formula proposed by Ireland (1957).

$$Q_{\text{ult}} = 2\pi RL(\gamma Z + q)\,K \tan \varphi \tag{8-2}$$

where Q_{ult} = ultimate bearing capacity;
R = average radius of pile;
L = total length of embedment of pile;
γ = unit weight of soil, taking average for full length of pile; buoyant weight should be used below ground water;
Z = depth of center of gravity of the embedded portion of the pile;
q = permanent surcharge load;
K = coefficient of lateral earth pressure; for step taper concrete piles, $K = 1.75$; use smaller values for steel piles;
φ = angle of internal friction of soil.

4. *Friction piles in clays.* A friction pile in clay is supported by adhesion between the pile and the soil. The bearing capacity of such a pile is approximately equal to the unit adhesion multiplied by the embedded area of the pile.

Pile driving in soft clays tends to disturb the clay around the pile. Consequently the clay loses a very large portion of its strength. Fortunately the disturbed clay begins to consolidate and to gain strength rapidly and immediately after driving. In most cases, the full strength of the virgin clay is likely to be regained one month or less after driving. The soft soil immediately surrounding the pile finally becomes stronger than the undisturbed soil. In ordinary projects, the temporary reduction in shear strength does not affect the pile capacity because piles will be loaded several months after driving. In the rare cases where the piles are required to sustain load immediately after driving, this effect must be considered. In such cases, laboratory tests should be made to determine the disturbed strength and the rate of regaining the strength after remolding (Seed and Reese, 1957).

Soft clays tend to come into direct contact with the pile due to their high plasticity. Therefore, the adhesion is equal to the cohesive strength of the clay. Pile driving in stiff clays not only disturbs the surrounding soil but it may also create a small open space between the pile and the clay. Consequently, the adhesion is always smaller than the cohesive or shear strength of the soil. Unless proven by pile load test, the values in the accompanying table should be used (Tomlinson, 1957).

Material of pile	*Cohesive strength psf*		*Adhesion to pile psf*
Concrete and timber	Soft	0– 750	0– 700
	Medium	750–1500	700– 900
	Stiff	1500–3000	900–1300
Steel	Soft	0– 750	0–600
	Medium	750–1500	600–750
	Stiff	1500–3000	?

5. *Friction piles in silt.* If the silt has a large amount of cohesive strength, the criteria for piles in clays may be used. If the silt is purely granular, the feasibility of friction piles depends primarily upon the relative density of the silt. Driving displacement piles (compact or friction piles) in soft saturated (or near saturated) silt will not compact it as in the case of sand. Instead, silt may be squeezed laterally and upwards, or liquified into a quick condition which requires a period of several days before they start to regain part of their strength. In short, loose or soft silts are unsuitable for supporting friction piles. If piles are desired, they should be carried to a lower stratum.

Compact silts may support spread footings directly or supporting the piles by friction. The length and capacity of such piles cannot be determined by static or dynamic formula. Pile load tests are necessary.

6. *Friction piles in loess.* Results of a thorough investigation of piles in loess deposit in Nebraska may be quoted as reference (Holtz and Gibbs, 1953).

1. When the dry density of loess is less than 80 lb per cu ft, friction piles are not recommended. Piles should be driven to a lower, more competent stratum.
2. When the dry density of the loess is 80 to 90 lb per cu ft, friction piles are satisfactory. The loess deposit should be prewetted or water jets should be used. Driving friction piles in dry loess will not get good results.
3. When the dry density of the loess is larger than 90 lb per cu ft, the foundation may be supported on footing without piles.

In any event, the bearing capacity and the required length of piles can be determined only by pile load test. Dynamic driving formulae (Engineering-News formula, etc.) generally will give misleading results.

7. *Friction piles in several soil strata.* Frequently piles are driven through several layers of soil. Such piles derive their bearing capacity from the supporting strength of each of these layers. When one layer is considerably stronger than the others the bearing capacity of the pile may be evaluated on the assumption that only this layer offers support to the pile. In the cases where the soil profile consists of two or more layers each contributing significant support to the pile, the length of pile can only be determined by test driving and the bearing capacity by load tests.

8. *Combined point-bearing and friction piles.* In many cases a pile derives part of its support from point-bearing and part from skin friction. The ultimate capacity of the pile is the sum of these two values.

C. Pile capacity as determined by pile driving formula. Qualitatively speaking a pile exerting a greater resistance against driving is capable of sustaining a greater load. For decades engineers have endeavored to equate the pile capacity in terms of driving energy and a large number of such formulae have been developed. Among them the Engineering-News formula is the simplest and most widely used.

When a pile hammer hits the pile, the total driving energy is equal to the weight of hammer times the height of drop or stroke. This energy is consumed by the work done in penetrating the pile and by certain losses. To express it algebraically:

$$\text{Driving energy} = \text{Work of pile penetration} + \text{Loss of energy}$$

or

$$12E = RS + L$$

where E = driving energy = weight × stroke of hammer, in case of drop hammers or single acting hammers, ft-lb;

R = resistance of soil, lb;

S = pile penetration per blow, in.;

L = loss of energy including loss in impact, in driving cap, in pile and in soil.

If L is assumed to be proportionate to the pile resistance, it can be written

$$L = RC$$

where C = empirical constant = 0.1 for steam hammers and 1.0 for drop hammers.

$$12E = RS + RC$$

$$R = \frac{12E}{S + C}$$

With a factor of safety of 6,

$$R = \frac{2E}{S + C} \qquad \text{(Engineering-News formula)} \qquad (8\text{-}3)$$

A dynamic formula is based on the premise that the soil resistance remains constant during and after driving operation. This is true in coarse-grained soils where free and fast movement of water in the voids is possible. In soils containing a fair amount of fine particles (silt, clay, fine sand) water can not escape readily during driving, and the excess water tends to reduce the frictional resistance along the periphery of the pile. The driving operation also disturbs the surrounding soil and reduces the shear strength to a fraction of the original value. Immediately after driving, however, the soil regains its strength rapidly. Therefore, dynamic formula cannot be relied upon for determination of pile capacity in soils containing a fair amount of fine grains.

Even in coarse-grained soils dynamic formulae give varying results. The Engineering-News formula, Eq. (8-3) indicates overly conservative values in some cases and unsafe values in other cases. Hence no dynamic formula should be used for determination of pile capacity in general. However, there are cases where dynamic formula may be used. These cases are:

1. In small jobs where piles are driven to coarse-grained soil and where the cost of pile load test is disproportionate to the job cost. The dynamic formula should be used conservatively and cautiously.
2. In localities where sufficient knowledge exists as to the reliability of the formula when piles are driven to certain layer under certain conditions.
3. As a criterion as to when to stop driving if the bearing capacity is expected to be the same as that of the test pile and of other piles driven in the same subsoil.

The penetration resistance should be used also as a guide against the possibility of damage of pile due to overdriving.

8-11 *Pile Spacing and Group Action*

Piles can never be driven perfectly straight and at the exact desired locations. A certain amount of variation does occur. Even if the piles are located accurately true in position, there is some bending moment transmitted to the pile cap by the column. Therefore, building codes do not permit the use of less than three piles to support a major column and less than two piles to support a foundation wall, unless the substructure is so framed as to bring the load concentrically to the centroid of piles. Consequently, piles are seldom installed singularly, but instead, they are always in groups or clusters. The bearing capacity and settlement of pile groups are the end results needed for the design of the foundation.

When a single point-bearing pile is acted upon by an axial load Q, the soil below the pile tip is stressed. According to the theory of elasticity the stress in the soil at the level of the pile tip is as shown in Fig. 8-11(a). If a group of piles are arranged at ordinary pile spacing and each is acted upon by an axial load Q, the stress in the soil overlaps. The total stress at any given point is equal to the sum of stresses introduced by each of these piles. This total stress, shown by a heavy line in Fig. 8-11(b), may be several times greater than that under a single individual pile.

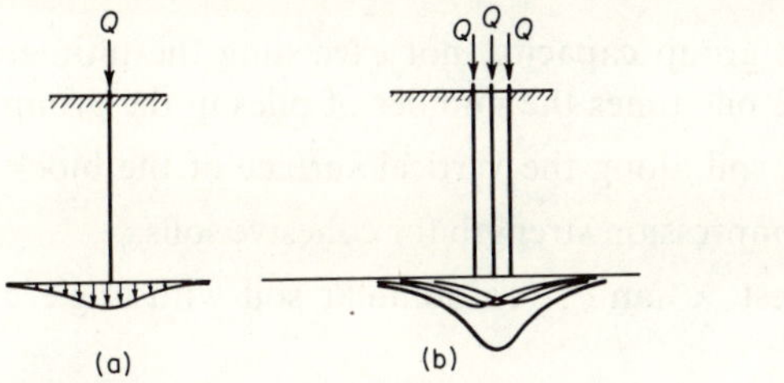

Fig. 8-11 Vertical stress at the level of pile tips.

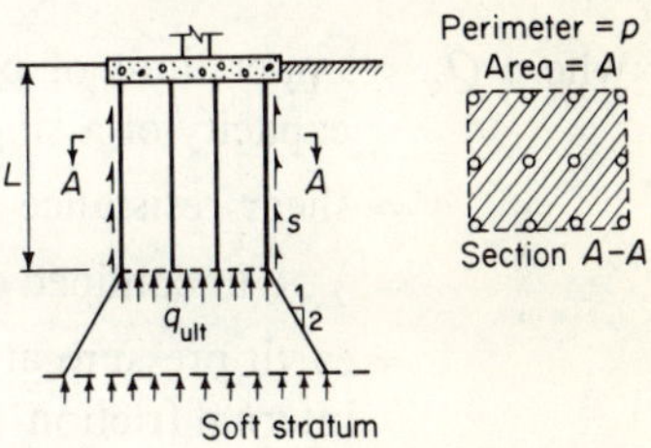

Fig. 8-12 Maximum capacity of pile group.

Similarly the stress overlapping takes place in a group of friction piles. Unless the piles in a pile group are spaced at great distance so that there is no significant overlapping of stress, which is highly impractical, the bearing capacity of a pile group is usually less than the sum of the individual pile capacities.* In practice the pile spacings given in the accompanying table are

Function of piles	*Minimim pile spacing, center to center of piles*
Point-bearing piles in hard stratum	2 to 2½ × butt diameter, or 2 ft 6 in.
Point-bearing piles on hard bedrock	2 × butt diameter, or 2 ft 0 in.
Fricton piles	3 to 5 × butt diameter, or 3 ft 6 in.

used to keep the cost of the pile cap to a minimum. These pile spacings should be used only as a guide, not as a hard and fast rule. In dense sand, plastic soil, saturated silt, etc., a close spacing may cause objectionable upheaval or lateral displacement of ground, whereas in loose sand, smaller spacing may be desired because of the benefit of compaction. Smaller spacing may also be advantageous where negative skin friction is a significant

* Except when a large number of piles will compact the soil, and thus, the bearing capacity is greater than one single pile.

factor. In any event the pile spacing should be checked for group capacity and for settlement.

The bearing capacity of a pile group may be analysed by the following method (Terzaghi and Peck, 1948). In this method two assumptions are made:

1. The pile cap is perfectly rigid;
2. The soil contained within the periphery circumscribing all the piles behaves as a solid block.

Therefore, the entire block may be visualized as one deep footing, Fig. 8-12. The ultimate bearing capacity is then

$$Q_g = sLp + q_{\text{ult}}A - \gamma LA \tag{8-4}$$

where Q_g = upper limit of pile group capacity, not exceeding the ultimate capacity of a single pile times the number of piles in the group;

s = shear resistance of soil along the vertical surface of the block;

= $\frac{1}{2}$ × unconfined compression strength for cohesive soils,

= earth pressure at rest × tan φ, for granular soil with angle of internal friction φ;

L = length of pile embedment in soil;

p = perimeter of area enclosing all the piles in the group;

q_{ult} = ultimate bearing capacity of soil at the level of pile tip;

A = area enclosing all the piles in the group;

γ = unit weight of soil within the block $L \times A$.

If the group capacity Q_p is smaller than nQ (where n = the number of piles and Q = the ultimate bearing capacity of an individual pile), the bearing capacity is controlled by Q_p. This case indicates that the pile spacing is too small to utilize the full capacity of the individual piles. On the other hand, with a large pile spacing the value of Q_p may become greater than nQ. In this case, the individual pile capacity nQ governs, and perhaps the spacing could be reduced to the limit where $Q_p = nQ$ in order to minimize the size of pile cap.

Usually the value of Q_p for friction piles in sand is greater than nQ when the common pile spacing is used, due to the benefit of compaction. However, the reverse is true for friction piles in soft clays.

8-12 *Stress on Lower Strata*

Since the use of elastic theory for determination of vertical stresses in the soil surrounding and below the pile tips is extremely laborious for practical cases, several approximate methods have been proposed. Among them the

60-degree rule has been accepted by a number of building codes. These methods are discussed below.

1. For point-bearing piles the pile group is substituted (for the purpose of determination of stresses in lower strata only) by a perfectly rigid footing whose periphery circumscribes the entire pile group. The total load on the pile group is assumed to spread out uniformly at a 60-degree angle, or 2:1 slope, from this fictitious footing as shown in Fig. 8-13(a).

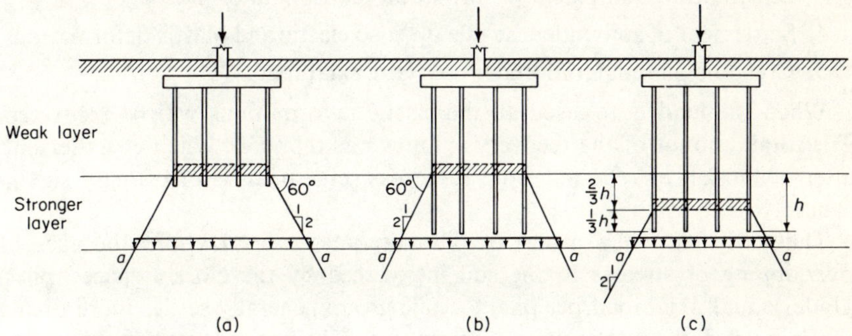

Fig. 8-13 Approximate methods for determination of stresses on lower strata.

2. For friction piles the total load on the group is assumed to spread out uniformly at a 60-degree angle starting at the top of the layer of soil which offers support to the piles. This is equivalent to assuming a rigid footing at the top of the layer, Fig. 8-13(b).

3. For friction piles, an alternate method (Peck, et al., 1953) may be used whereby the load is assumed to spread out from a depth of $L/3$ measuring from the pile tip where L = the length of the embedment of piles in the layer of soil offering skin friction to the piles, Fig. 8-13(c). A comparison between Fig. 8-13(a) and Fig. 8-13(b) indicates that both cases would give equal stress intensity at any level below the top of the stronger layer. But it is known that greater penetration would bring the stresses to a lower level. Method 3 appears more reasonable.

All these methods are very approximate and must be used with caution. In some cases, the settlements calculated by means of these methods are considerably smaller than the measured values. However, these simple approximations gives sufficient information for determination of the supporting strength of the lower strata. If the calculated stress exceeds the ultimate bearing capacity of the layer, the design should be revised by reducing the pile load, by increasing the pile spacing, or by extending the piles to another layer below.

8-13 Settlement Analysis

The total settlement of a single pile under axial load is complex because it may consist of a number of components:

1. Elastic compression of the pile.
2. Movement (or slip) of pile relative to the surrounding soil.
3. Settlement of surrounding soil due to pile load; this comprises elastic deformation and plastic deformation (consolidation).
4. Settlement of soil under the pile tip–also elastic and plastic deformations.
5. Creep of pile material under constant axial load.

When the load is released all the elastic deformations will be recovered. The total amount of the recovery is known as rebound. The net settlement, after deduction of rebound from the gross settlement, is of primary significance.

The settlement of a group of piles becomes more complex because of overlapping of stresses in the soil introduced by the closely spaced piles. Under equal axial load per pile, the pile group generally settles more than a single pile due to the stress overlapping. The procedures for settlement analysis varies with the type of piles and the soil conditions. Each case is discussed separately below.

A. Point-bearing pile on bedrock. If the pile tips are well seated on the rock and if the rock is not soft, the net settlement of a test pile should not be more than a fraction of an inch after deduction of rebound. However, well-designed and constructed buildings supported on piles driven to solid hard rock have been subjected to total settlements several times as much as the net settlement of the test pile. The larger settlements are thought to be the result of one or more of the following factors:

1. Small uplift of piles due to driving of adjacent piles.
2. Long time creep of pile material under constant load.
3. Overlapping of stress in soil.
4. Negative skin friction as a result of disturbance of clay due to pile driving.

Even under the influence of all these factors, the total net settlement of well-designed and constructed pile foundations on bedrock generally will not be so large as to cause special concern or analysis, unless the bedrock is soft. In this case, quantitative analysis is very difficult and in practice it can only be estimated by judging from the characteristics of the rock core sample. Local experience, if available, should be relied upon as guidance.

B. Point-bearing piles in sand and gravel. The load-settlement relationship of a single pile driven in granular soils can be readily determined by load test since the settlement of such pervious soils takes place shortly after load application. However, the test must be so made to differentiate or eliminate the skin friction, as discussed in Sec. 8-19.

The settlement of a pile group is considerably greater than that of a single point-bearing pile as discussed in the previous paragraphs. For a group of piles having the customary pile spacing, the tentative relationship shown in Fig. 8-14 may be used. For example a pile group 10 ft wide would settle, according to this figure about five times that of a single test pile. Pile groups closely located should be considered one large cluster even if they are connected to different pile caps.

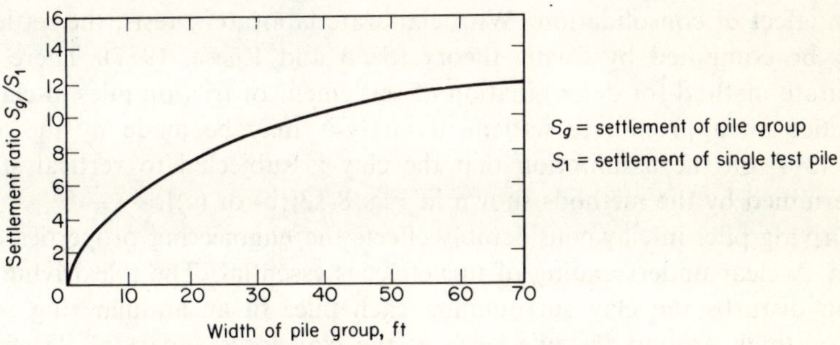

Fig. 8-14 Settlement of group of point-bearing piles in sand. After Skempton.

Engineers should never overlook the possibility of additional settlement due to layer or layers of weak soil below the compact or hard layer which supports the pile tips. The stress in lower layers may be approximated by the 60-degree distribution method as shown in Fig. 8-13.

C. Point-bearing piles on hard clay. Settlement of a single pile or a group of piles driven to hard clay cannot be determined by practical means. A very rough idea as to the order of magnitude of settlement may be obtained by laboratory tests on the basis of broad assumptions, approximations, and simplifications. In localities where records of existing pile foundations are available, these records constitute the most valuable information.

Load tests on such piles usually give optimistic results because an extremely long time is necessary for consolidation. The process of consolidation may be accelerated if the tip of the test pile is provided with porous material and the excess water is drained out during the test. Even so, a load test may require weeks' time to complete the consolidation. Such lengthy load test is not feasible in almost all construction projects. Therefore, if pile load tests are conducted for the purpose of determining the ultimate bearing capacity, the load settlement relationship as established by the tests must be interpreted with extreme caution.

D. Friction piles in sand and gravel. The load-settlement relationship of a single friction pile in granular soils can be determined reliably by pile load test. The time required for the test is relatively brief since settlement in such soils takes place shortly after load application.

If the settlement of the test pile is acceptable, the settlement of a pile group in such soils will be of no concern. This is because the granular soil between the piles is compacted by displacement of the piles and becomes locked in between as a dense mass. The settlement probably approaches that of a pier foundation having a depth and base area equal to those of the pile group.

E. Friction piles in clays. The settlement of a single friction pile in clay cannot be determined within a practicable length of time because of the long term effect of consolidation. With elaborate laboratory tests, the settlement may be computed by elastic theory (Seed and Reese, 1957). There is no accurate method for determination of settlement of friction piles in clays. In practice an approximate settlement analysis may be made by the aid of Eq. (3-4) on the assumption that the clay is subjected to vertical stresses determined by the methods shown in Fig. 8-13[(b) or (c)].

Driving piles in clay considerably effects the engineering properties of the clay. A clear understanding of this effect is essential. The pile driving operation disturbs the clay surrounding each pile. In an annular ring several inches thick around the pile surface, the clay loses a part of its strength due to the disturbance. (The disturbance beyond about one pile diameter is moderate). Immediately the disturbed clay begins to lose water due to the stresses set up by volume displacement. If the clay in this disturbed zone is very sensitive to disturbance, it may lose a large part of its strength and becomes unable even to support the soil above, hence, it begins to consolidate under its own weight. The strength is usually regained very rapidly as a result of water expulsion. In Fig. 8-15 an example is given to illustrate the speed at which this particular soil regains its strength. In about 30 to 50 days, 90 per cent or more of its original undisturbed strength is regained. Eventually the clay in the disturbed zone may become stronger than the virgin soil. In practice, the full load is never applied until several months after pile driving, and hence, the damage due to driving disturbance does not affect the useful strength of the foundation. However, there are doubtful cases where the clay may not regain its full strength or may not regain it rapidly enough. In either case laboratory tests of the undisturbed and disturbed samples should be made, as already discussed in Sec. 8-10(b).

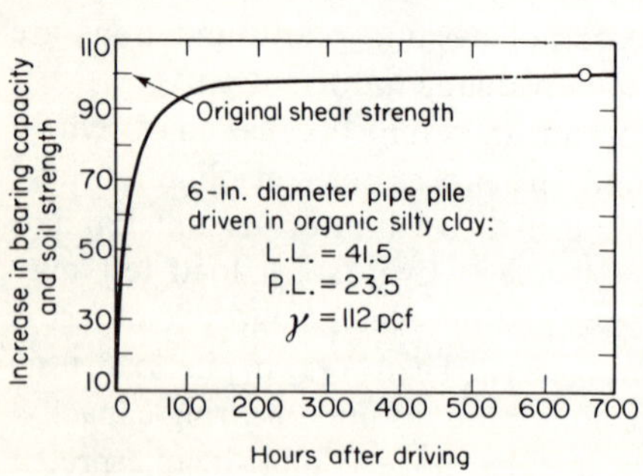

Fig. 8-15 Example of regain of clay strength after pile driving. After Seed and Reese.

8-14 Design of Pile Caps

Pile caps are almost invariably made of reinforced concrete and are designed as individual footings subjected to the column loads plus the weight of the pile cap and the soil above the cap. Under a concentric load, all piles in the same group are assumed to take equal axial loads.* The soil under the pile cap is not assumed to offer any support. Wherever the conditions permit, the piles should be arranged in the most compact geometric form in order to keep the stresses in the pile cap to a minimum. These geometric forms are shown in Fig. 8-16. The criteria for the structural design of pile caps are summarized in Fig. 8-17.

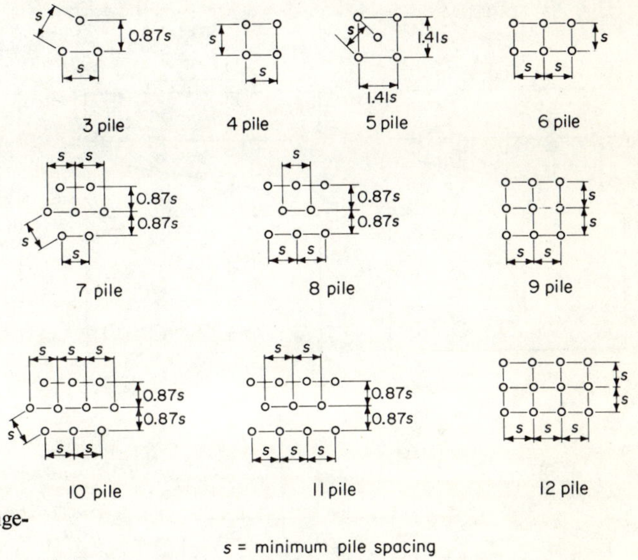

Fig. 8-16 Typical arrangement of pile groups.

Under an eccentric loading or a concentric loading plus a moment, the pile cap is designed in accordance with the following assumptions:

1. Pile cap is perfectly rigid.
2. Pile heads are hinged to the pile cap, therefore, no bending moment is transmitted from the pile cap to the piles.
3. Piles are short and elastic columns, therefore, the deformations and the stress distribution are planar.

These assumptions permit the use of elastic theory for calculation of the pile loads and the stresses in the pile cap. The elastic equation is

---

* Theory, model tests, and field measurements have proved that piles in one group do not take equal reaction. Instead, the center piles take less than the outer piles and the corner piles are subjected to the greatest reaction.

$$Q_m = \frac{Q}{n} \pm \frac{M_y x}{\Sigma (x^2)} \pm \frac{M_x y}{\Sigma (y^2)}$$

where Q_m = axial load on any given pile m,

Q = total vertical load acting at the centroid of the pile group,

n = number of piles in the group,

M_x, M_y = moment with respect to the X and Y axes respectively,

x, y = distance of pile from Y and X axes respectively.

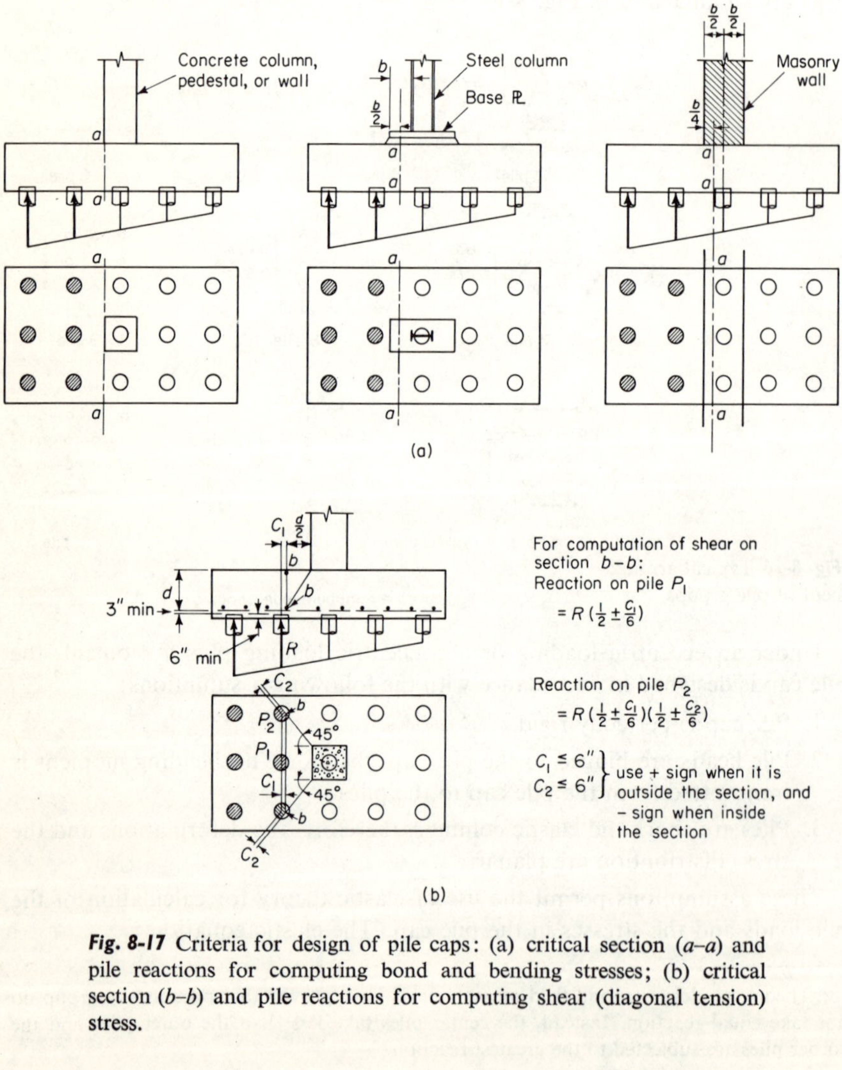

Fig. 8-17 Criteria for design of pile caps: (a) critical section (*a–a*) and pile reactions for computing bond and bending stresses; (b) critical section (*b–b*) and pile reactions for computing shear (diagonal tension) stress.

X and Y axes are two perpendicular axes passing through the centroid of the pile group.

In the cases where piles are arranged in one row at uniform pile spacing s, the following equation may be used.

$$\sum (x^2) = \frac{1}{12} s^2 n(n - 1)$$

where n = number of piles in the row.

Pile caps, similar to spread footings, may have pedestals, stepped or sloping tops. One cap may also support more than two columns, and in this case, the design principles are discussed in Chapter 7.

Pile caps should be large enough to have a minimum edge distance of 4 in. to 6 in. of concrete beyond the outside face of the exterior piles. In difficult driving conditions where the actual locations of piles may deviate considerably from the required, the edge distance should be increased to provide for such field variations. Ordinarily the piles are embedded at least 6 in. in the cap and the reinforcing bars are placed at a clear distance of 3 in. above the pile head. Therefore the effective depth d of a pile cap is generally about 10 in. less than the total depth D of the pile cap.

8-15 Uplift

When piles are required to resist uplift force in excess of the dead load of the structure, the following steps must be taken:

1. The piles must be anchored sufficiently into the cap, the cap tied to the column, and the cap designed for the uplift stresses.

   *Timber piles** embedment in concrete: bond value = 40 psi (concrete placed in water). Expanding the pile head by wedges and driving spikes into the pile heads were found of no significant benefit.

   Concrete piles. Tension reinforcing must be extended into the pile cap to develop the bar strength by bond or by means of other anchorage.

   Steel piles. Reinforcing bars or strap anchors may be welded to the pile heads and extended into the cap for embedment.

2. Concrete piles must be reinforced with longitudinal steel for the full net uplift. Splices in all types of piles should be designed to the full uplift.

3. Uplift resistance of a pile is not necessarily a function of its bearing capacity under compression. For friction piles in soft clay, the capacities against compression and uplift are about equal. A friction pile in

---

* Public Roads, IX (Nov., 1928) p. 169.

granular soils may not have an uplift resistance approaching its bearing capacity. Except for friction piles in soft clays, the uplift capacity of the pile must be determined by pull tests. When large uplift forces are anticipated, the pedestal type piles may be of merit.

The total uplift resistance of a pile group is the smaller of the following two values:

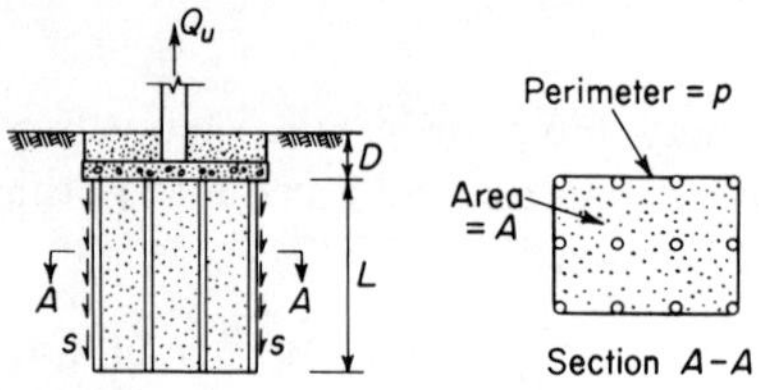

Fig. 8-18 Uplift capacity of pile group.

1. Uplift resistance of a single pile times the number of piles in the group;
2. Uplift resistance of the entire group as a block, Fig. 8-18.

The ultimate uplift capacity, Q_u, as a block is the sum of the following components:

1. The weight of pile cap plus the weight of soil above it.
2. The weight of the block of soil $= \gamma LA$, where γ is the average unit weight of soil, use buoyant weight for the portion below the water level. L and A are as illustrated in the figure.
3. The frictional resistance along the perimeter of the block = approximately $sp(L + D)$, where p, L, and D are shown in the figure, s = shear resistance $= c + \sigma \tan \varphi$. σ is the normal stress on the surface of the block, and may be taken as the earth pressure at rest. c and φ are the cohesion and angle of internal friction of the soil.

8-16 Lateral Load

Any structure is inevitably subjected to a certain amount of horizontal force due to wind, earth pressure, etc., or traction forces from automobiles or trains. It is unsafe to assume that frictional resistance exists between the bottom of the pile cap and the soil because in this type of foundation the vertical load is transmitted through the piles to the lower stratum, not to the soil immediately below the pile cap. In extreme cases, the soil may even settle away from under the pile caps and leaving a small space in between. Unless the structure is supported laterally by other means, the piles should be designed to resist such lateral loads.

The amount of lateral force carried by each pile or pile group depends upon the structural framing which brings the lateral forces down to the bottom of the columns. It is common to utilize the basement floor or tie beams to distribute the horizontal load to a large number of piles. Batter piles are generally resorted to when the lateral force exceeds the allowable amount.

The allowable lateral load on a given pile foundation depends on the type of pile, the type of soil, the embedment of pile head, the nature of the force, and the amount of lateral movement considered acceptable. The lateral movement is generally limited to $\frac{1}{4}$ in. for building and $\frac{1}{2}$ in. or larger for miscellaneous structures such as transmission towers. Based on a number of full size tests, allowable horizontal loads have been proposed and are shown in Table 8-3.

Table 8-3 ALLOWABLE HORIZONTAL LOAD ON VERTICAL PILES*
(for $\frac{1}{4}$ in. lateral movement)

Type of pile	*Pile head*	*Type of soil*	*Allowable load (lb per pile)*
Timber (12 in. diam)	Free-end	Sand	1500
		Medium clay	1500
	Fixed-end†	Sand	4500
		Medium clay	4000
Concrete (16 in. diam)	Free-end or fixed-end†	Medium sand	7000
		Fine sand	5500
		Medium clay	5000

* From McNulty (1956).

† Fixed-end condition may be attained by embedding the pile head at least 24 in. in the concrete cap.

It should be noted that vertical piles driven in a deep bed of soft or very soft clays and silts should not be relied upon for resisting lateral force unless it is very small, say 1000 lb. Concrete piles subjected to lateral forces in excess of about 1000 lb each pile should be reinforced to withstand the flexural stress.

8-17 Batter Piles

When piles are subjected to lateral load in excess of about 1000 lb per pile, it might be more economical and desirable to provide batter piles. Common batter varies from 1 horizontal : 12 vertical to 5 horizontal : 12 vertical. When batter exceeds 3 horizontal : 12 vertical, special driving equipment is necessary. Therefore, before using large amount of batter, pile contractors should be consulted as to the cost involved.

The usual assumption in design of batter piles is that the pile is capable of resisting the same axial load as a vertical pile of the same type and size and driven to the same stratum. There are several methods for analysis of pile foundations involving batter piles. Even the most complicated "elastic method" involves rather unrealistic assumptions.

1. The most crude and conservative method is one that provides enough batter piles to resist all the horizontal force. The capacity of a batter pile in resisting a horizontal force is assumed, in this case, to be equal to the horizontal component of the pile capacity along the direction of batter. This method certainly requires more piles than necessary for a given condition and is not commonly used.

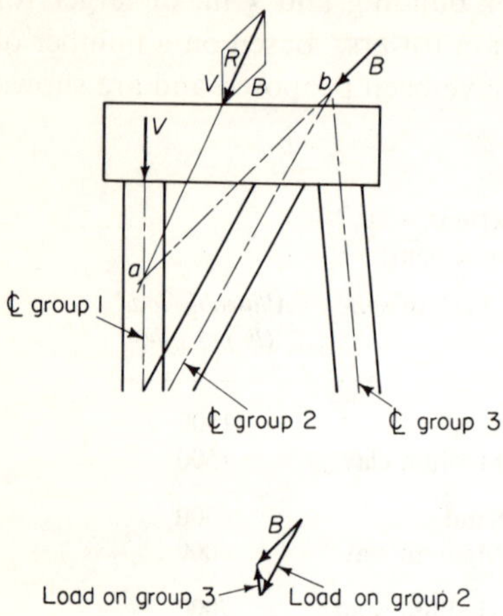

Fig. 8-19 Graphical method (Culmann's method).

2. A commonly used method of analysis is known as Culmann's graphical method, Fig. 8-19. Piles are grouped according to their slopes. It assumes that all piles are subjected to axial load only and that piles in each group are subjected to equal axial load. Based on these assumptions the center of reactions can be located. Culmann's method may be described step by step as follows:

(a) Sketch a profile of the pile foundation and locate the center line of each group of parallel piles.

(b) Draw the resultant R of all external forces applied on the pile foundation. R intersects the center line of the pile group 1 (vertical piles) at point a.

(c) Intersect center line of group 2 and center line of group 3 at b. Connect ab.

(d) Resolve R into components V and B. V is vertical and B is parallel to line ab.

(e) Group 1 is subjected to total axial force V. Group 2 and group 3 are subjected to force B.

(f) Resolve force B into axial loads along center line of group 2 and center line of group 3.

In the cases where piles are arranged in not more than three directions, the solution by this method is statically determinate. For cases where piles are driven in more than three directions, this method may be used if further compounding of pile reactions is made.

An example of analysis by Culmann's method is given in Plate DE 8-1. Five batter piles are used to support a 3 ft section of a long retaining wall. It is assumed that the resultant force in this section due to lateral earth pressure, H the vertical load V, including the weight of the retaining wall, and the weight of the soil above the base, is already determined. The magnitude, slope, and point of application of this force R are shown.

The solution of this problem can be found by following the steps outlined above. The procedure for determination of lateral movement is discussed later in this section.

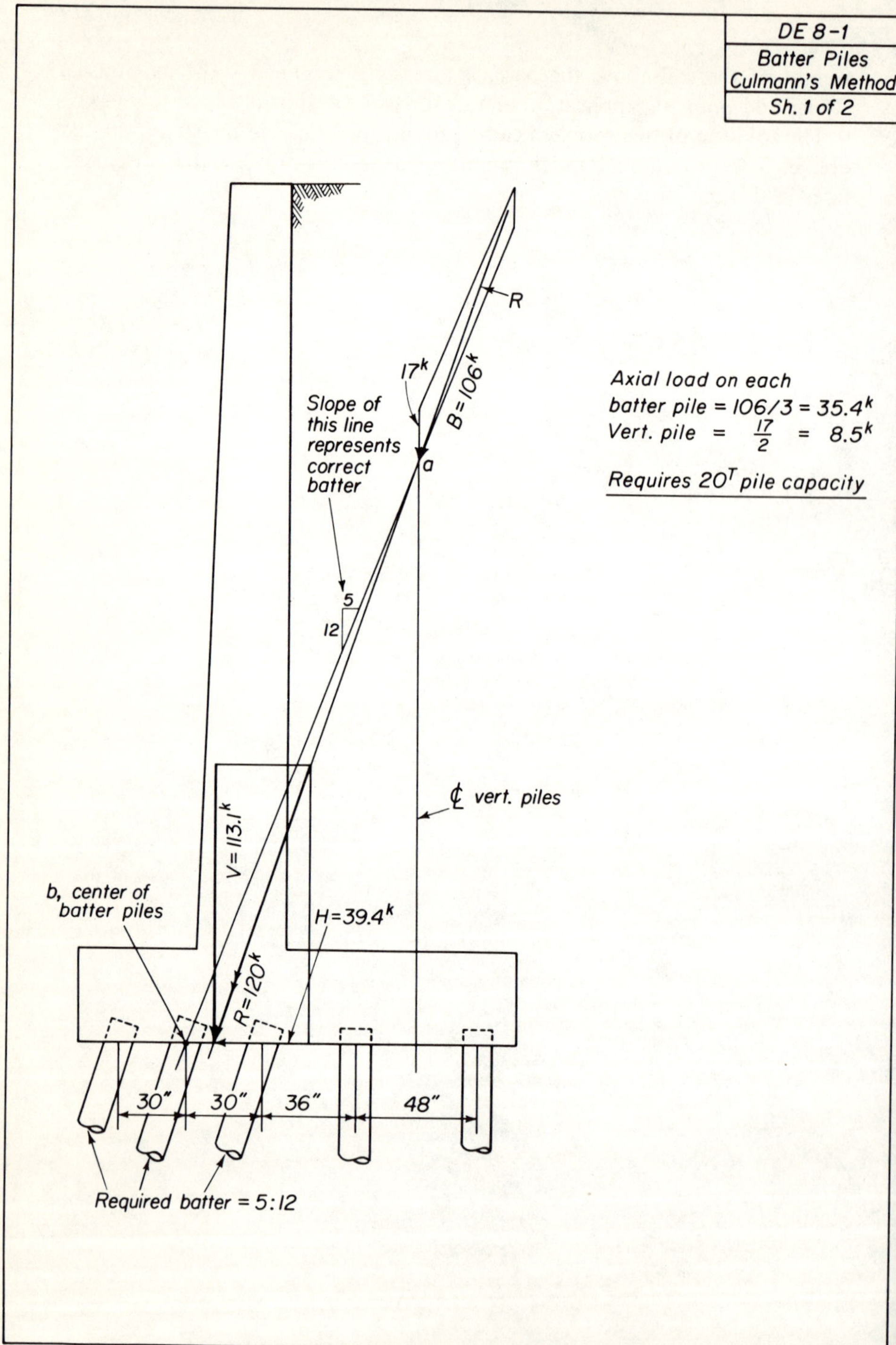
DE 8-1
Batter Piles
Culmann's Method
Sh. 1 of 2
R
17^k
B = 106^k
Slope of this line represents correct batter
a
5
12
Axial load on each batter pile = 106/3 = 35.4^k
Vert. pile = 17/2 = 8.5^k
Requires 20^T pile capacity
℄ vert. piles
V = 113.1^k
b, center of batter piles
H = 39.4^k
R = 120^k
30"
30"
36"
48"
Required batter = 5:12

DE 8-1
Batter Piles Culmann's Method
Sh. 2 of 2

Determine lateral movement based on pile loads calculated by Culmann's Method:

Pile settles 1″ at 265^k axial load

$$\Delta_1 = \frac{35.4}{265} = 0.133''$$

$$\Delta_2 = \frac{8.5}{265} = 0.032''$$

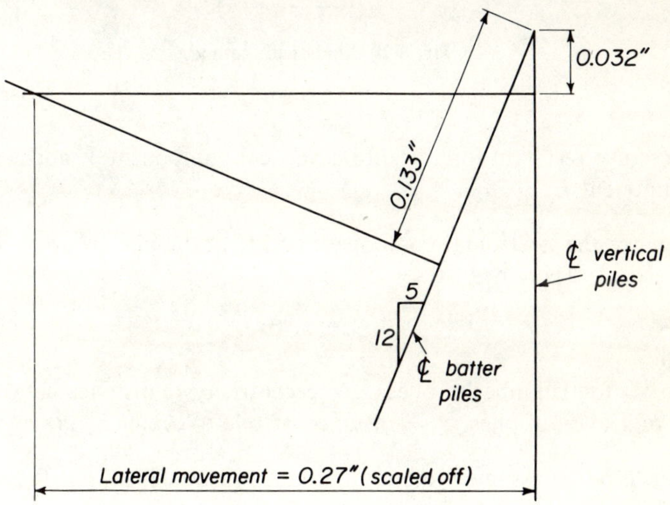

3. Another simple and commonly used method is referred to as the analytical method and is shown here step by step, Fig. 8-20. This method is based on the same assumptions as the Culmann's method.

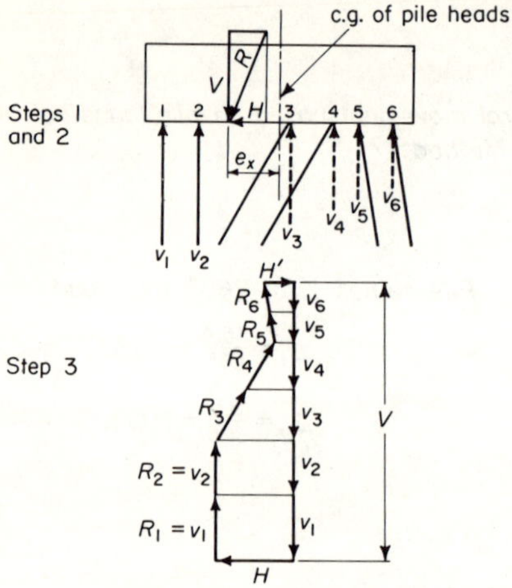

Fig. 8-20 Analytical method.

1. Resolve resultant force R into a vertical component V and a horizontal component H.

2. Ignore the horizontal component and treat the pile group as if all piles were vertical. Therefore,

$$\text{Pile load } v_n = \frac{V}{n} \pm \frac{Ve_x}{\Sigma(x^2)}x$$

where n = total number of piles, e_x = eccentricity = distance between V and center of gravity of piles, x = distance of pile to center of gravity of piles.

3. Each pile is assumed to be subjected to an axial load, R_1, R_2, . . ., whose vertical component is equal to the corresponding vertical reaction determined in step 2. This can be done analytically:

$$R_n = \frac{v_n}{\cos\theta}$$

where θ = the angle between the pile and the vertical.

Hypothetically, the batter piles should have slopes such that the force polygon should close up (i.e. $H' = 0$), Fig. 8-20. However, it is considered acceptable if H' is less than 1,000 lb per pile.

An example of analysis by this method is given in Plate DE 8-2.

DE 8-2
Batter Piles Analytical Method
Sh. 1 of 1

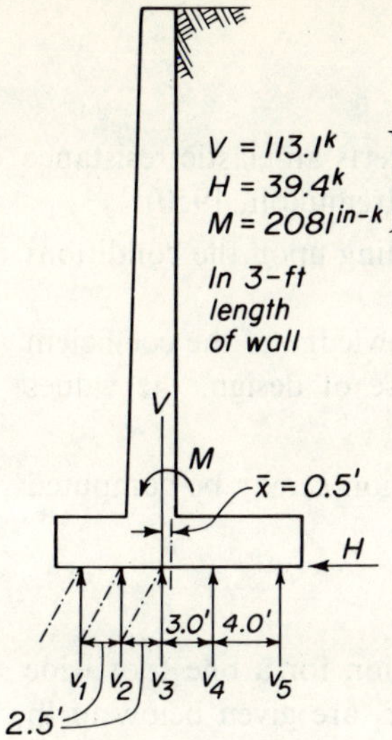

$V = 113.1^k$
$H = 39.4^k$
$M = 2081^{in\text{-}k}$
In 3-ft length of wall

Longitudinal pile spacing

3'-0" c/c

$$\bar{x} = \frac{(3.0) + (3.0 + 4.0) - 2.5 - (2.5 + 2.5)}{5} = 0.5'$$

$$\Sigma(d^2) = 5.5^2 + 3.0^2 + 2.5^2 + 6.5^2 = 87.75 \text{ ft}^2$$

$$M' = 2081 + 113.1 \times 6'' = 2759.6''^k = 230'^k$$

$$v = \frac{113.1}{5} \pm \frac{230}{87.75} d = 22.62^k \pm 2.62\, d$$

$d_1 = 5.5$	$v_1 = 37.02$
$d_2 = 3.0$	$v_2 = 30.48^k$
$d_3 = 0.5$	$v_3 = 23.91^k$
$d_4 = -2.5$	$v_4 = 16.07^k$
$d_5 = -6.5$	$v_5 = 5.57^k$

$v_1 + v_2 + v_3 = 91.41^k$

Req'd. batter

$$= \frac{39.4}{91.41} = 0.43 = \frac{5}{12}+$$

Try $\frac{5}{12}$ batter

$$91.41 \times \frac{5}{12} = 38.0 \qquad 39.4 - 38.0 = 1.4^k = 700 \text{ lb/Pile OK}$$

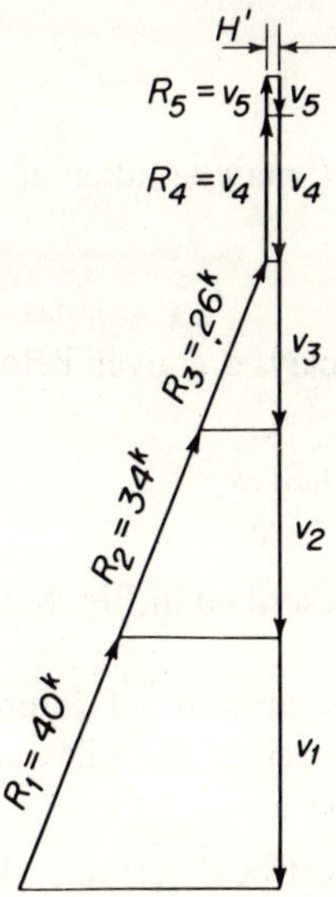

Requires 20^T pile capacity

4. A number of elastic methods have been proposed among which the method based on the theory of beams on elastic foundation is most rigorous. This method assumes that:

(a) The pile cap is perfectly rigid.

(b) The piles are elastic under axial loads.

(c) The soil surrounding the pile is elastic and exerts an elastic resistance against rotation and translation of the pile (Hrennikoff, 1950).

The pile heads may be assumed fixed or free depending upon the conditions of embedment.

This method is quite laborious and requires a knowledge of the coefficient of horizontal subgrade reaction k. For the purpose of design, the values given by Terzaghi (1955) may be used:

Sand. The coefficient of horizontal subgrade reaction k may be computed from

$$k = k_1 \frac{Z}{B}$$

where k_1 = coefficient of horizontal subgrade reaction for a one foot wide pile at one foot depth. The values of k_1 are given below in lb per cu in.

	Loose sand	*Medium sand*	*Dense sand*
Dry or moist	8	24	65
Submerged	4	16	40

Z = depth, ft.

B = width of pile, ft.

Clay. The coefficient of horizontal subgrade reaction k may be taken as

$$k = \frac{1}{1.5B} k_1$$

where k_1 and B are defined as above, the values of k_1 for clays are given below in lb per cu in.

Soft clay	*Stiff clay*	*Very stiff clay*	*Hard clay*
0–50	85	170	350

The procedure for analyzing a batter pile foundation is described in Fig. 8-21. A numerical example is given in Plate DE 8-3.

In practice it is frequently desired to determine the amount of lateral movement under the design loads. If the elastic properties of the piles are known, the following simple method, Fig. 8-22, may be used:

(a) Settlement of single pile subjected to axial load must be determined by load test (or by experience on similar conditions):

δ_Q = settlement under compression load Q.

δ_T = upward movement under tension T.

1. Determine the elastic constants for a single pile:

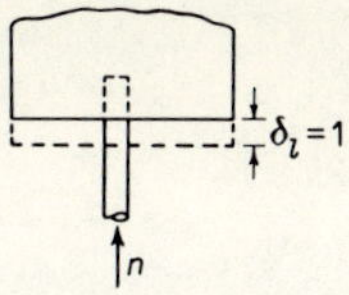

A unit longitudinal displacement ($\delta_l = 1$) produces an axial load, n, from the pile.
This value should be determined from pile load test.
If only elastic shortening is included, the value will be misleading and excessive.

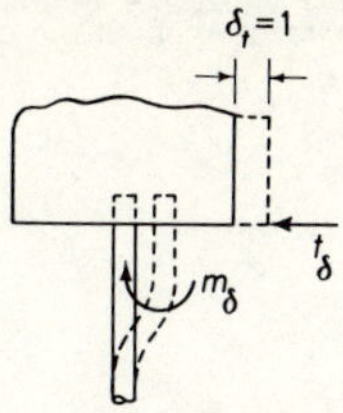

A unit transverse displacement ($\delta_t = 1$) produces a transverse resistance, t_d, and a fixed-end moment m_δ
From theory of elasticity:

$$\left.\begin{aligned} t_\delta &= \frac{k}{\beta} \\ m_\delta &= \frac{k}{2\beta^2} \end{aligned}\right\} \text{Pile head fixed} \qquad \left.\begin{aligned} t_\delta &= \frac{k}{2\beta} \\ m_\delta &= 0 \end{aligned}\right\} \text{Pile head pinned}$$

Where k is the elastic constant of soil known as coefficient of horizontal subgrade reaction

$$\beta = \sqrt[4]{\frac{k}{4EI}}$$

E = modulus of elasticity of pile material
I = moment of inertia of pile cross-section

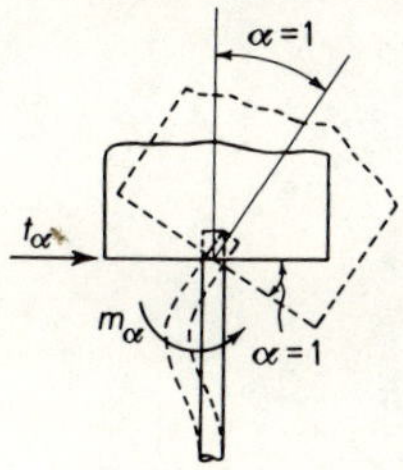

A unit rotation ($\alpha = 1$ radian) produces a transverse resistance t_α and a moment m_α.
From theory of elasticity:

$$m_\alpha = \frac{k}{2\beta^3} \qquad m_\alpha = t_\alpha = 0 \text{ for free pile head}$$

$$t_\alpha = m_\delta$$

2. Determine the elastic constants of the foundation as a whole (assuming pile footing infinitely rigid)

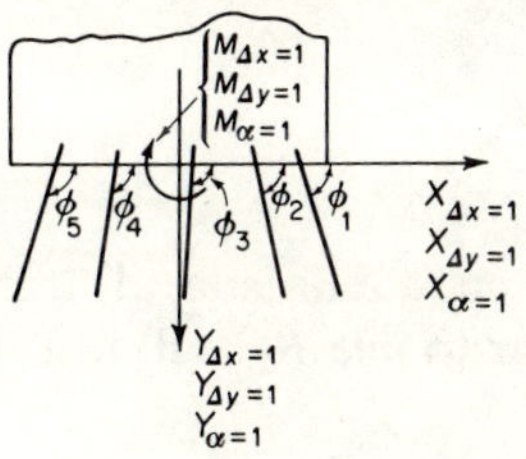

A unit horizontal displacement of the pile footing ($\Delta x = 1$) produces a horizontal resisting force $X_{\Delta x=1}$, a vertical resisting force $Y_{\Delta x=1}$, and a resisting moment $M_{\Delta x=1}$
A unit vertical displacement and a unit rotation produce similar resistances as shown
From theory of structures:

$$Y_{\Delta x=1} = X_{\Delta y=1}$$
$$M_{\Delta x=1} = X_{\alpha=1}$$
$$M_{\Delta y=1} = Y_{\alpha=1}$$

By means of simple geometry and statics, all the constants above can be expressed in terms of the elastic constants for individual piles determined in step 1:

$$X_{\Delta x=1} = -\Sigma(n\cos^2\phi + t_\delta \sin^2\phi)$$
$$Y_{\Delta x=1} = -\tfrac{1}{2}(n - t_\delta)\Sigma(\sin 2\phi)$$
$$M_{\Delta x=1} = -\tfrac{1}{2}(n - t_\delta)\Sigma(X\sin 2\phi) + m_\delta\Sigma(\sin\phi)$$

where X = coordinate of pile head from an arbitrarily chosen origin
Pile heads to the left of center have negative X

$$Y_{\Delta y=1} = -\Sigma(n\sin^2\phi + t_\delta\cos^2\phi)$$
$$M_{\Delta y=1} = -\Sigma(n\sin^2\phi + t_\delta\cos^2\phi)X - m_\delta\Sigma(\cos\phi)$$
$$M_{\alpha=1} = -\Sigma\left[(n\sin^2\phi + t_\delta\cos^2\phi)X^2\right] - 2m_\delta\Sigma(X\cos\phi) - Nm_\alpha$$

where N = total number of piles in the foundation

Fig. 8-21

3. Solve the following equations for Δx, Δy, and α

$$X_{\Delta x=1}\ \Delta x + X_{\Delta y=1}\ \Delta y + X_{\alpha=1}\ \alpha + H = 0$$
$$X_{\Delta y=1}\ \Delta x + Y_{\Delta y=1}\ \Delta y + Y_{\alpha=1}\ \alpha + V = 0$$
$$X_{\alpha=1}\ \Delta x + Y_{\alpha=1}\ \Delta y + M_{\alpha=1}\ \alpha + M = 0$$

where Δx, Δy, and α are the pile foundation displacement under the action of a force H along the X-axis, V along the Y-axis and a moment M about the center of the pile group.

4. Determine the pile displacements

By simple geometry the following equations are derived for any individual pile:

$$\delta_l = \text{longitudinal displacement (compression)}$$
$$= \Delta x \cos\phi + \Delta y \sin\phi + \alpha X \sin\phi$$

Where X = coordinate of pile head

$$\delta_t = \text{transverse displacement, to the right}$$
$$= \Delta x \sin\phi - \Delta y \cos\phi - \alpha X \cos\phi$$

5. Determine the pile forces

P_l = longitudinal force (compression) = $n\delta_l$

P_t = transverse force (acting on the pile footing to the right) = $-t_\delta \delta_t + M_\delta \alpha$

P_m = moment (acting on the footing clockwise) = $M_\delta \delta_t - M_\alpha \alpha$

Fig. 8-21 Procedure of elastic analysis of batter piles. After A. Hrennikoff.

(b) $\Delta_1 = \dfrac{\delta_Q R_1}{Q} \qquad \Delta_2 = -\dfrac{\delta_T R_2}{T}$

(c) Draw line *ce* perpendicular to pile R_1 and at a distance Δ_1 from point of intersection *b*. Draw *de* perpendicular to pile R_2 and at a distance Δ_2 from point of intersection.

(d) Intersect *ce* and *de* at point *e*. Horizontal distance between points *b* and *e* is the lateral movement.

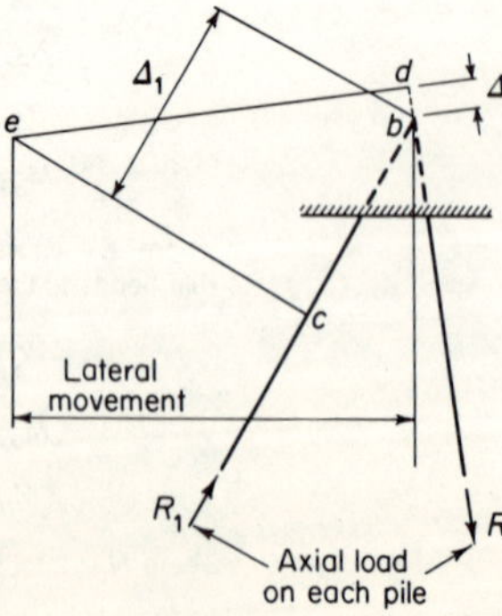

Fig. 8-22 Elastic method for determination of lateral movement of batter piles.

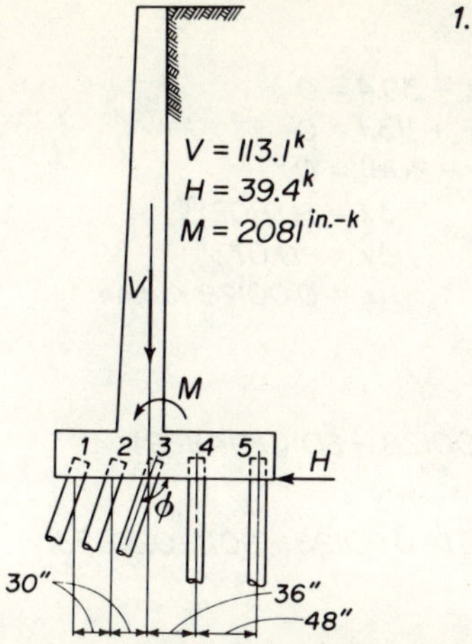

All batter piles 3:1 ($\phi = 108°26'$)

Pile spacing 3'-6" c/c

1. Elastic constants for single pile:

Wood pile driven to rock; assume no slipping at pile tip

$$n = \frac{AE}{L} = 265^{k*}$$

A, for aver. 9" dia. = 63.5 s.i.

E = 1,500,000 psi

L = 360 in.

k : for very soft soil
= 3 psi

$$\beta = \sqrt[4]{\frac{k}{4EI}} = \sqrt[4]{\frac{3}{4 \times 1500000 \times 322}} = 0.00634$$

$$t_\delta = \frac{k}{\beta} = 0.493\ ^{k}/_{in.}$$

$$m_\delta = \frac{k}{2\beta^2} = 38.9\ ^{in.-k}/_{in.}$$

$$m_\alpha = \frac{k}{2\beta^3} = 6140\ ^{in.-k}/_{radian}$$

2. Elastic constants for the whole pile foundation:

$\phi = 108°26'$	$\phi = 90°$
$\sin\phi = 0.94869 \quad \sin^2\phi = 0.9$	$\sin\phi = 1.0$
$\cos\phi = -0.31623 \quad \cos\phi = 0.1$	$\cos\phi = 0$
$\sin 2\phi = -0.6$	$\sin 2\phi = 0$

$$X_{\Delta x=1} = -[(265 \times 0.1 + 0.493 \times 0.9)3 + (0.493)2] = -82^k$$

$$X_{\Delta y=1} = -\tfrac{1}{2}(265 - 0.493)[(-0.6)3 + 0] = +238^k$$

$$M_{\Delta x=1} = -\tfrac{1}{2}(265-0.493)(30+60)(-0.6) + 38.9(3 \times 0.94869 + 2 \times 1.0) = -6920^{in-k}$$

$$Y_{\Delta y=1} = -[265(0.9)3 + 265(1.0)2 + 0.493(0.1)3] = -1245^k$$

$$M_{\Delta y=1} = -[(265 \times 0.9 + 0.493 \times 0.1)(-60-30+0) + 265(1.0)(36+84)] - 38.9(-0.31623 \times 3) = -10320^{in.-k}$$

$$M_{\alpha=1} = -[(265 \times 0.9 + 0.493 \times 0.1)(60^2 + 30^2 + 0) + 265(1.0)(36^2 + 84^2)] - 2(38.9)[-0.31623(-60-30+0)] - 5 \times 6{,}140 = -3{,}320{,}000\ in^2k$$

* Values in this example are taken from "Analysis of pile foundations with batter piles" by A. Hrennikoff, Trans. ASCE Vol. 115, 1950
The value of 265^k for woodpile is excessive.

DE 8-3
Batter Piles Elastic Analysis
Sh. 2 of 2

3. Solve Simultaneous Equations:

$$-82\Delta x + 238\Delta y - 6{,}920\alpha - 39.4 = 0$$
$$238\Delta x - 1245\Delta y - 10{,}320\alpha + 113.1 = 0$$
$$-6920\Delta x - 10320\Delta y - 3{,}320{,}000\alpha - 2081 = 0$$

$$\Delta x = -0.802^{in.}$$
$$\Delta y = -0.073^{in.}$$
$$\alpha = 0.00128 \text{ radian}$$

4. For Pile No. 1:

$$\delta_l = (-0.802)(0.31623) + (-0.073)(1.0) + 0.00128(-60)(0.94869) = 0.112^{in.}$$

$$\delta_t = -0.802(0.94869) - (-0.073)(-0.31632) - 0.00128(-60)(-0.31623) = -0.811^{in.}$$

5. For Pile No. 1:

$$P_l = 0.112 \times 265 = 29.7^k \text{ (comp.)}$$
$$P_t = 0.493(-0.811) + 38.9(0.00128) = 0.46^k \text{ (transverse)}$$
$$P_m = 38.9(-0.811) - 6140(0.00128) = -39.3^{in.-k}$$

Procedure for calculating stresses for other piles is not shown here. Final pile stresses are tabulated below:

	Pile No. 1	Pile No. 2	Pile No. 3	Pile No. 4	Pile No. 5
P_l	29.70^k	39.40^k	49.00^k	-7.10^k	$+9.10^k$
P_t	0.46^k	0.44^k	0.44^k	0.45^k	$+0.45^k$
P_m	$-39.30^{in.-k}$	$-39.00^{in.-k}$	$-38.00^{in.-k}$	$-39.00^{in.-k}$	$-39.00^{in.-k}$

Conclusions: Requires 25^{ton} woodpile at 1:3 batter
Lateral movement = 0.802 in.

8-18 Negative Skin Friction

Negative skin friction is a downward drag acting on the piles due to relative movement between the piles and the surrounding soil. When piles are driven through compressible soils, and the site has a newly placed fill or will be filled in the future, the possibility of negative skin friction should be investigated. Soft to medium clays, soft silt, peat, mud, etc., are compressible soils. Lowering of ground water level in such compressible soils may also bring about negative skin friction. If there is any doubt as to the compressibility of the soil, consolidation tests should be made. If the results indicate that the layer will settle under the weight of the fill, even only a small amount, the pile capacity should be reduced to compensate the drag due to negative skin friction.

The conditions which cause negative skin friction on piles or other types of deep foundation are shown in Fig. 8-23. Under the weight of the new fill, or due to the lowering of ground water to a level below the compressible layer,

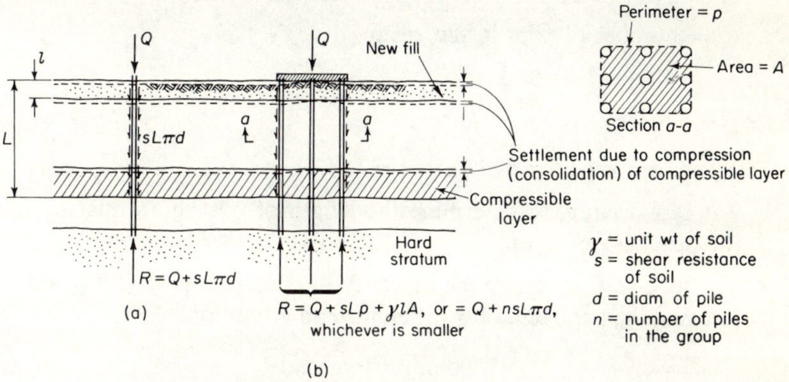

Fig. 8-23 Negative skin friction.

the compressible layer settles gradually for a long period of time (several months to many years). Consequently all soil strata above this layer settle with it. However, piles driven to a lower depth are prevented from downward movement by the resistance offered by the lower stratum. Hence, there is a relative movement between the piles and the surrounding soil. The downward movement of the soil tends to drag the piles by shear along the pile surfaces. For a single pile this downward force is equal to the shear resistance times the surface area of the pile. For a pile group, the soil between the piles may be held by friction or adhesion on the pile surface, and hence, there is no relative movement between the piles and the soil in between. In this case, the total negative skin friction acting on the entire pile group is equal to the weight of this block of soil held in between the piles plus the shear along the periphery of the pile group as a whole. Therefore the negative skin friction, *N.S.*, on a pile group is

$$N.S. = sLp + \gamma LA \quad \text{(whichever is smaller)}$$
$$= nsL\pi d$$

where s = shear resistance of the soil;
= cohesion = $\frac{1}{2}$ unconfined compression strength of cohesive soils,
= $N \tan \varphi$ for cohesionless soils (N may be taken as the earth pressure at rest, and φ is the angle of internal friction),
L = length of pile embedment above the bottom of compressible layer;
A = area enclosing all the piles;
p = perimeter of the area A;
γ = unit weight of soil, average for length L (use buoyant weight for the portion below ground water level),
d = average diameter of the pile in the length L,
n = number of piles in the group.

8-19 Test Piles

Test piles are used for two purposes:

1. For test driving, to determine the length of pile before placing orders for purchase of piles.
2. For load test, to determine or verify the design capacity of piles. Load test may be conducted for different load conditions:

Compression load
Uplift load
Lateral load

Generally load tests are made to determine the bearing capacity and to establish the load settlement relationship under compression load. Local building codes generally require such tests unless the pile capacity is kept under certain limits when driven to a certain depth under certain driving requirements which is proven safe in the locality. When piles are required to resist large uplift or lateral force, corresponding tests should be made.

Usually the pile foundations are designed on an estimated capacity which is arrived from a thorough soil study. At the beginning of construction, load tests are made for the purpose of verifying the adequacy of the design capacity. If the test results show an inadequate factor of safety or excessive settlement, the design must be revised before construction is under way.

Occasionally, time and money may be available for pre-design tests. In such cases, the tests can be carried out more extensively to determine the most economical design capacity, thus resulting in large savings in the total construction cost.

In any event, pile load tests are not inexpensive and the results may be misleading if not conducted and interpreted properly. The procedures of conducting pile load tests are discussed here.

1. *Driving test piles.* Test piles should be driven at a location near a soil boring where the soil characteristics are known and are representative of the most unfavorable conditions at the site. The test piles should be exactly the same (type of pile, dimensions, etc.) as the piles to be used in the construction and should be installed by the same equipment and manner of driving. A penetration record should be kept for every foot of penetration for the entire length of pile. This record may be used as a guide for future driving for this project. Any pile encountering a smaller resistance will be considered as having a smaller bearing capacity than the test pile.

If the test pile or piles are selected from the piles driven for the final construction, which often is the case, it may be as well to select the pile not only at the location with unfavorable soil conditions but also the pile whose verticality and penetration into the bearing stratum is questionable. For example, if the pile tip is broomed, the test may indicate unsatisfactory results.

A rest period should be allowed before application of loads on test piles in clays or silts. This period should preferably be at least a week but in no case less than three days in order to permit the disturbed soil to regain most of its strength. Otherwise the test results are erroneously low.

2. *Methods and procedures of loading.* Detailed information concerning the procedures for conducting pile load test under compression are contained in the ASTM Procedures book (Designation D1194) to which the reader is referred. However, several pertinent points should be discussed here. Load tests give reliable results for piles in granular soils, and frequently the bearing capacity cannot be established other than by load tests. In such cases, load tests should be conducted. In fine-grained soils (silts, clays, or soils containing a fair amount of silt or clay particles) load test may be used to determine the ultimate bearing capacity but the results do not represent the true load settlement relationship. This is due to the longtime effect of consolidation. Even if the test procedures are strictly followed, the time of loading is still too short to include the settlement due to consolidation.

Load tests on point-bearing piles driven through silt or clay require special attention. The test load is carried partly by point-bearing and partly by skin friction, and test results would indicate large bearing capacity and small settlement. However, in the actual case where the load is carried on the pile for long periods of time, the stress in the clay incurred by skin friction compresses the soil. In a long period of time, the clay is compressed and the entire load is shifted to the pile point. Since the shifting of load takes place after a long period of loading, it is impracticable to carry such longtime test. Therefore, tests should be made in such a manner that the effect of skin friction can be eliminated or differentiated.

To eliminate the effect of skin friction on test piles, sleeves may be used for the entire length of the pile above the pile tip. To differentiate the effect of skin friction, two test piles should be used adjacent to each other. One of the

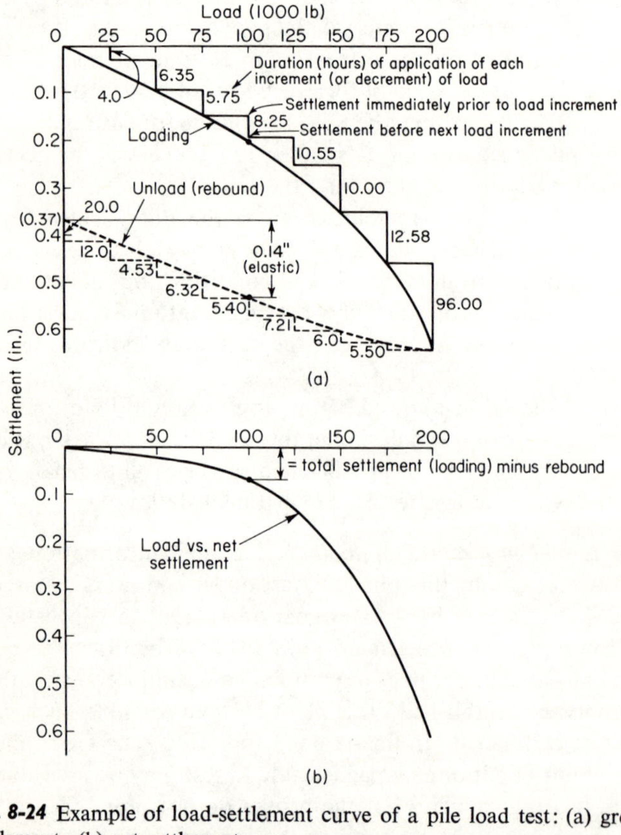

Fig. 8-24 Example of load-settlement curve of a pile load test: (a) gross settlement; (b) net settlement.

test piles is driven home as contemplated for the working piles whereas the other is carried 2 or 3 ft short of the supporting layer. Therefore, one pile derives its resistance from skin friction and point-bearing, and the other is only a friction pile. The difference between these two is the net capacity due to point-bearing.

3. *Test results.* The results of pile load test are usually reported in a tabulated form from which a load-settlement graph is plotted, Fig. 8-24. First the settlement at the beginning and at the end of each load increment is plotted, thus, the load-settlement curve is a stepped, broken line. On this curve, the duration of application of each increment is deciphered. Based on these broken lines, the true curve of loading and unloading curves may be interpreted by connecting the low points.

The most important feature of a pile load test is the relationship between the load and the net settlement. The net settlement of a pile is the permanent settlement excluding the elastic deformation of the pile and the supporting soil. All piles being of approximately the same length and under the same axial stress, this elastic deformation may be considered approximately equal for all piles. In practice, the elastic deformation is not included in the permanent pile settlement.

This elastic deformation of a test pile can be readily determined by measuring the rebound upon unloading (see unloading curve). In the example given in Fig. 8-24, the net settlement under a total load of 100,000 lb is 0.20 − 0.14 = 0.06 in.

4. *Bearing capacity.* The bearing capacity of a pile must satisfy two requirements:

(a) The factor of safety against ultimate failure should not be less than 2 or 3.

(b) The load will not incur an objectionable amount of settlement. For building foundations, it is often limited to 0.01 in. per ton of pile load at the test load (2 × design load). Greater settlement may be justified in some other foundations.

When the load test is carried to failure, the bearing capacity can be readily determined according to the definition given above, Unfortunately, most pile load tests are carried to only 150 per cent or 200 per cent of the design loads. Such test results do not provide information as to the actual factor of safety, although it is greater than 1.5 or 2.0. Therefore, various procedures have been used to determine the design capacity. The Uniform Building Code uses the following procedures:

1. Fifty per cent of the yield point–a yield point is defined as the pile load at which an increase in load produces a disproportionate increase in settlement.
2. One-half of the load which causes a net settlement of 0.01 in. per ton of test load.
3. One-half of the load under which no additional settlement takes place during a 40-hour period.

In practice, piles often penetrate a soft layer or layers before they encounter a stiff soil. An analysis must be made to determine the possibility of loss of skin friction in the soft soil as a result of consolidation. If necessary, the design capacity should exclude the total skin friction of pile in such layer or layers.

8-20 *Construction of Pile Foundations*

The construction of pile foundations involves two separate steps, namely the installation of piles and the making of pile caps. The second step is a relatively simple process similar to the construction of spread footings. Procedures and equipment required for installation of piles are discussed in the following paragraphs. Other problems involved in the construction of pile foundations are discussed in Sec. 8-21.

A. Pile driving equipment. Piles are installed by a special pile driving device known as a pile hammer. The hammer may be suspended from the boom of a crawler crane, supported on a large frame called a pile driver or carried on a barge for construction in water. In all cases, the hammer is guided between two parallel steel members known as leads. The leads may be adjusted at various angles for driving vertical and batter piles.

Several types of hammers are in use and each of which has different sizes. The hammer types are:

1. *Drop hammer.* The drop hammer consists of a heavy ram in between the leads. The ram is hoisted up to a certain height and released to drop on the pile. This type is slow and therefore not in common use. It is used in the cases where only a small number of piles are driven.
2. *Single-acting hammer.* The single-acting hammer is a long stroke machine consisting of a heavy ram. The ram is lifted by steam or compressed air but dropped by its own weight. The energy of a single-acting hammer is equal to the weight of the ram times the height of fall. A photograph and cross-sections of this type of hammer are shown in Fig. 8-25.
3. *Double-acting hammer.** The double-acting hammer employs steam or air for lifting the ram and for accelerating the downward stroke. It operates with a succession of rapid blows. A photograph and sections of a double-acting hammer is shown in Fig. 8-26.
4. *Diesel hammer.* The diesel hammer is a small, lightweight, self-contained and self-activating type using gasoline for fuel, Fig. 8-27. The total driving energy is the sum of the impact of the ram plus the energy delivered by explosion. In Fig. 8-28, the basic principle of operation of this type of hammer is described.

* A differential-acting hammer is also double-acting but uses a different steam cycle for raising and driving the ram.

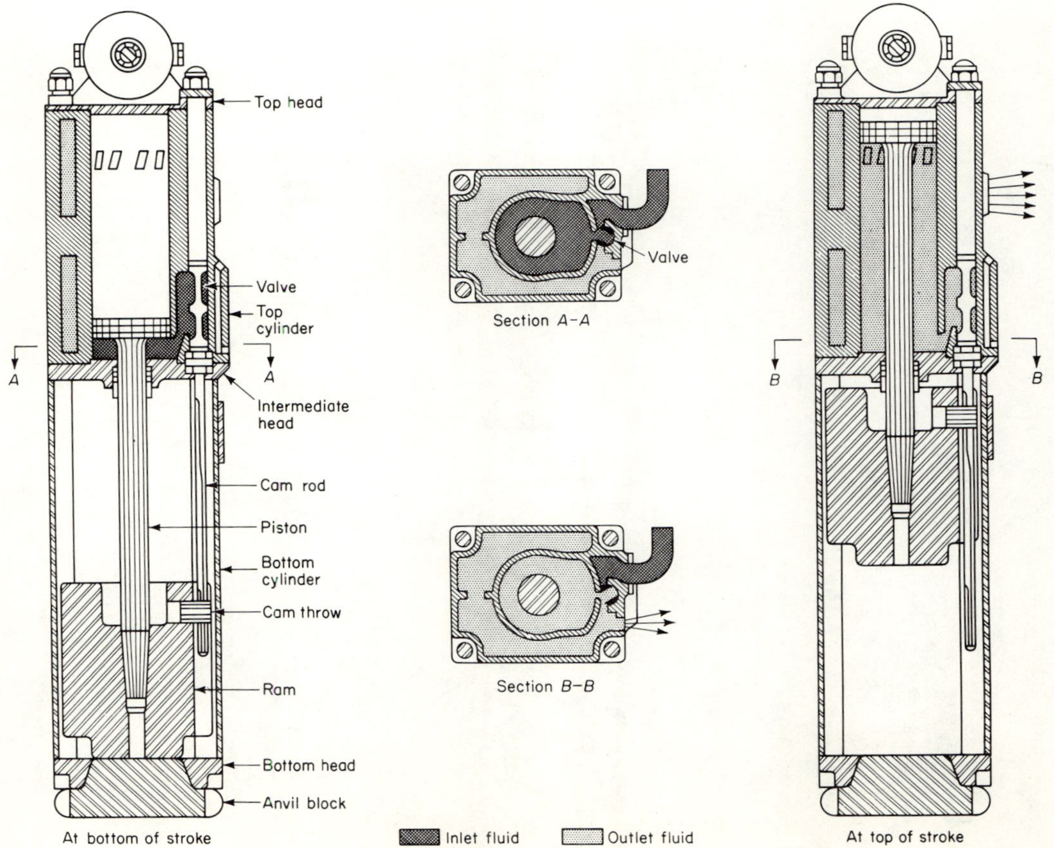

Fig. 8-25 Single-acting hammer. McKiernan-Terry Corp.

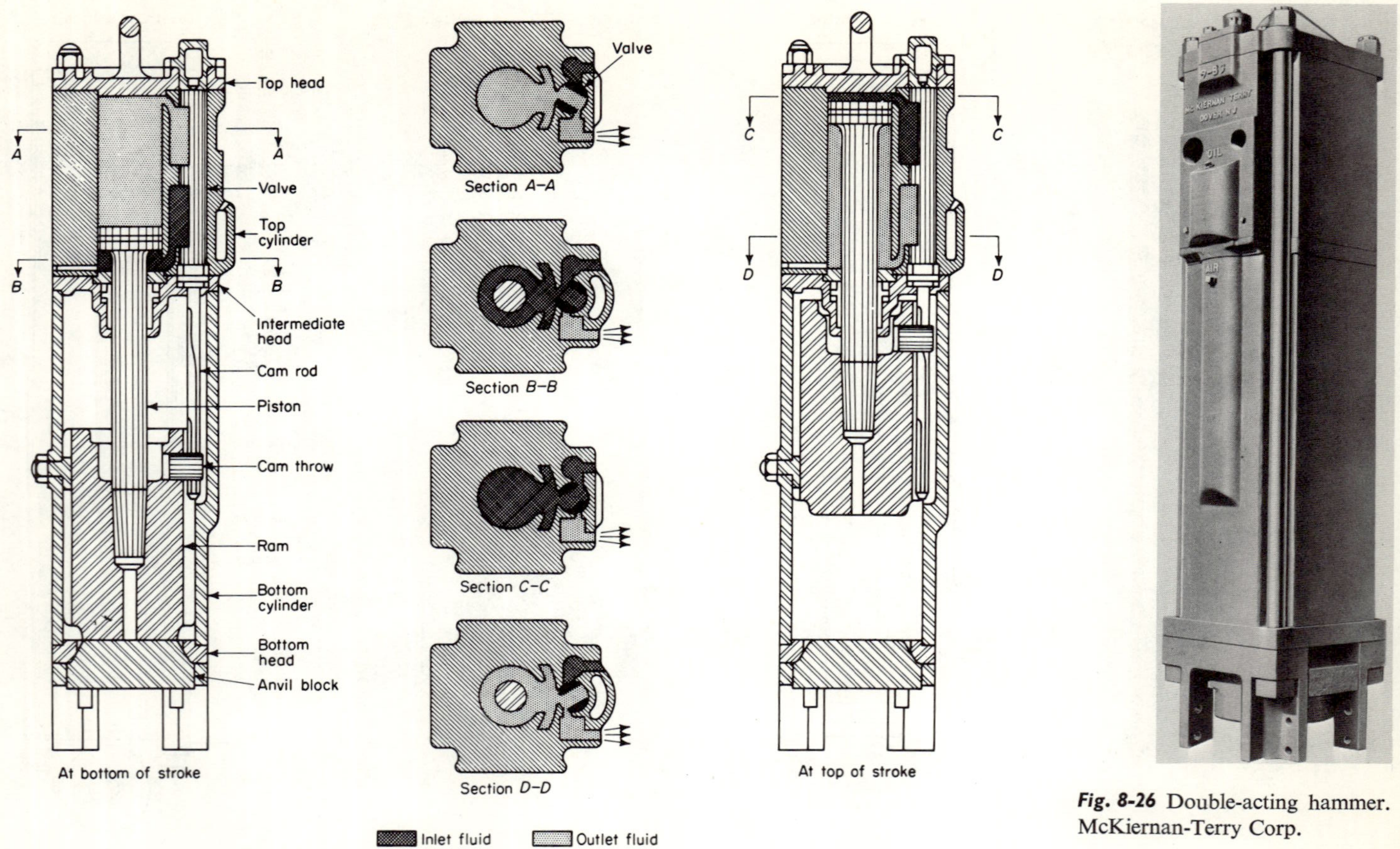

Fig. 8-26 Double-acting hammer. McKiernan-Terry Corp.

5. *Vibratory hammer.* The driving unit vibrates at high frequency. This method has been used in Europe, Asia, and the U.S.A. (Engineering News-Record, 1961).

In construction of pile foundations several parts and devices are frequently used. For convenient reference, they are defined as follows:

Anvil is the part at the bottom of a power hammer which receives the impact of the ram and transmits it to the pile head.

Drive cap or *helmet* is made of cast steel to be placed on top of pile to prevent the pile head from damage during driving and to maintain the axis of pile in line with the axis of hammer.

Cushion is a pad of resilient material, hardwood or rope, placed between the cap and the top of pile to protect the pile head.

Follower is an extension used between the pile and the hammer when the pile head is below the guides and is out of reach of the hammer. The use of the follower introduces an uncertainty in terms of dynamic driving formula.

Leads is a steel frame with two parallel members for guiding the piles in correct alignment.

Ram is the moving part of the hammer. It consists of a piston and a driving head.

B. Selection of pile hammer. Pile hammers are available in a wide range of weights, types, and driving energy. The choice of hammer type and size for a given job depends upon many factors. Generally the size is a more important factor than the type. A heavy pile should be driven by a heavy hammer delivering large energy. Preferably the weight of a hammer should be at least one-half the total weight of the pile, and the driving energy should be at least one foot-pound for each pound of pile weight. Data for selection of driving energy used by TVA engineers are shown in Table 8-4.

Fig. 8-27 Diesel hammer.

Each type of hammer has its use under suitable conditions. The selection also depends upon the availability of hammer, availability of steam or air pressure, headroom clearance, batter piles, etc. The advantages and disadvantages of each type are summarized as follows:

Single-acting hammers are advantageous when driving heavy piles in compact or hard soils; the heavy ram striking at low velocity produces least

Table 8-4 DATA FOR SELECTION OF PILE HAMMERS FOR DRIVING CONCRETE, TIMBER, AND STEEL SHEETPILING UNDER AVERAGE AND HEAVY DRIVING CONDITIONS*

Length of pile (ft)	*Depth of penetration (per cent)*	*Sheet pile†*			*Timber pile*		*Concrete pile*	
		Light	*Medium (ft-lb per blow)*	*Heavy*	*Light (ft-lb per blow)*	*Heavy*	*Light (ft-lb per blow)*	*Heavy*
1. Driving through Earth, Sand, Loose Gravel–Normal Frictional Resistance								
25	50	1000–1800	1000–1800	1800–2500	3600–4200	3600–7250	7250–8750	8750–15000
	100	1000–3600	1800–3600	1800–3600	3600–7250	3600–8750	7250–8750	13000–15000
50	50	1800–3600	1800–3600	3600–4200	3600–8750	7250–8750	8750–15000	13000–25000
	100	3600–4200	3600–4200	3600–7500	7250–8750	7250–15000	13000–15000	15000–25000
75	50		3600–7500	3600–8750		13000–15000		19000–36000
	100			3600–8750		15000–19000		19000–36000
2. Driving through Stiff Clay, Compacted Gravel–Very Resistant								
25	50	1800–2500	1800–2500	1800– 4200	7250– 8750	7250– 8750	7250– 8750	8750–15000
	100	1800–3600	1800–3600	1800– 4200	7250– 8750	7250– 8750	7250–15000	13000–15000
50	50	1800–4200	3600–4200	3600– 8750	7250–15000	7250–15000	13000–15000	13000–25000
	100		3600–8750	3600–13000		13000–15000		19000–36000
75	50		3600–8750	3600–13000		13000–15000		19000–36000
	100			7500–19000		15000–25000		19000–36000
Weight (per lin. ft)		20 lb	30 lb	40 lb	30 lb	60 lb	150 lb	400 lb
Pile size (approx)		15 in.	15 in.	15 in.	13 in. diam	18 in. diam	12 in.^{2}	20 in.^{2}

* Tennessee Valley Authority.

† Energy required in driving single sheet pile. Double these when driving two piles at a time.

damage due to impact. The disadvantages are low driving speed and large headroom requirement.

Double-acting hammers are generally used to drive piles of light or moderate weight in soils of average resistance against driving. This type of hammer can drive piles at fast speed, requires less headroom, and some models can be used to extract piles.

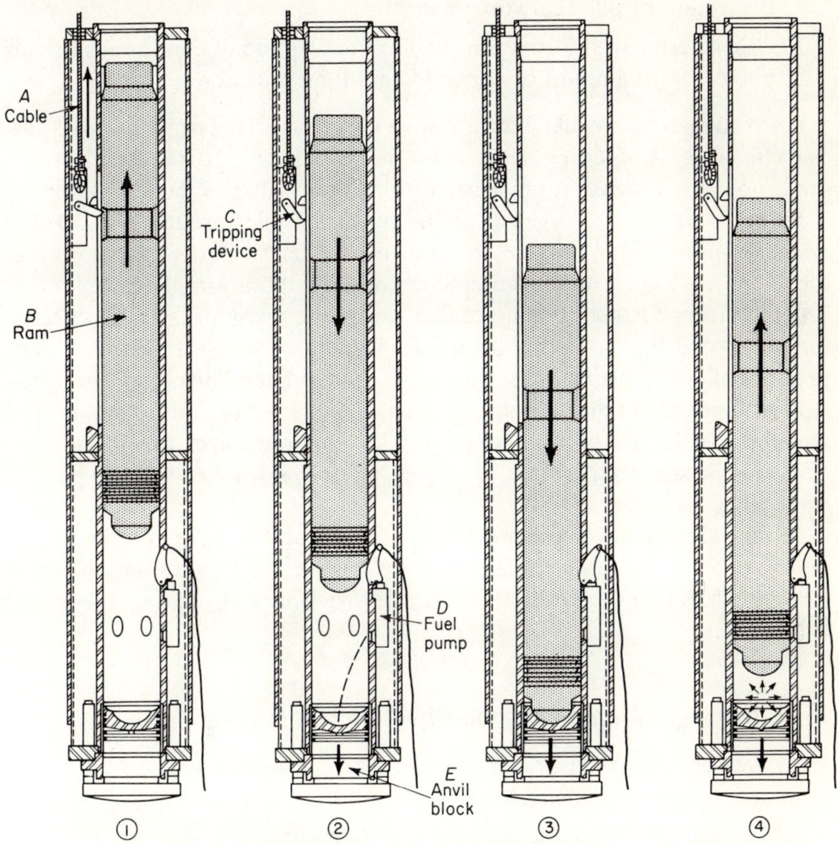

Fig. 8-28 Principle of diesel hammer. After *Engineering-News Record* (May, 1958).

Diesel hammers are similar in application as double-acting hammers, but driving may become difficult in extremely soft ground.

C. Pile driving record. In any job, large or small, the information concerning the pile driving should be kept in an orderly form. It should include:

1. The type and make of hammer and its stroke or rated energy, the other driving equipment including water jet, driving cap, cushion, etc.;

2. The pile size and length, location of the pile in the pile group, and the location or designation of the group;
3. The sequence of driving in the group;
4. The number of blows per foot for the entire length, and the set for last 10 blows;
5. The final tip and head elevation;
6. Inspection of pile shaft for verticality;
7. Other pertinent information such as interruption of continuous driving, pile damage, amount of concrete, and time of concrete placing.

D. Water jetting. Water jetting is often used to assist the penetration of piles. The effectiveness of jetting depends on the type of soil. It is most effective in sand, not too effective in gravel, questionable in silt and ineffective in clay.

Jetting water is discharged near the tip of the pile through 2 in. or 3 in. diameter pipes, with a nozzle $\frac{3}{4}$ in. to $1\frac{1}{4}$ in. diameter. Under a large pressure the water jets loosen up the soil and create a condition similar to quicksand. The pile drops through this quicksand with little resistance. To be effective, the jet should be supplied with ample volume of water flow and a pump pressure of 100 lb per sq in. or greater. To avoid lateral displacement of pile, two jet pipes are preferred, one on each side of the pile. Although jet pipes may be cast in the precast concrete piles, separate pipes are found more effective. Separate pipes may be moved up and down and thus keeping the sand "live."

Jetting should be discontinued at a depth about three feet before the anticipated penetration, and the last several feet of penetration should be made by driving only. If the adjacent piles are affected by jetting, they should be redriven.

8-21 *Damage, Alignment, and Effect of Pile Driving*

A. Damage due to improper driving. Damages commonly resulted from improper driving can be avoided if precaution is exercised to:

1. Deliver the driving blows square and on the axis of the pile.
2. Properly protect the pile head by the use of drive caps and cushions.
3. Stop driving as soon as the penetration reaches the desired resistance. The following resistance is commonly used:

   Timber piles: 4 blows per in. (driving energy, 15,000 ft lb.)

   Concrete: 6–8 blows per in.

   Steel: 10 or more blows per in.

In order to stop driving in time before the pile is damaged by overdriving, the behavior of the pile during the entire period of driving should be observed. It is time to stop driving a timber pile when the following phenomena are observed.

1. The pile shivers and springs near the ground surface.
2. The pile hammer bounces.
3. The pile head shows distress under moderate driving.

Pile may have been already damaged if the following behavior is noticed:

1. Penetration suddenly increases or becomes irregular, whereas the soil formation cannot account for it.
2. Pile suddenly changes direction.

B. Alignment. Piles can never be driven absolutely vertical and true to position. Even in ideal conditions the center of a pile head must be allowed to deviate a certain amount from the required location, and the pile at lower depth to vary from the required vertical or batter line. Therefore, every precaution should be exercised to maintain the piles in position. The general procedure for determining the pile alignment and elevations is as follows:

1. Measure the elevation at top of piles immediately after driving of each and check the final elevations after the adjacent piles are driven or at the completion of all pile driving. If point-bearing piles are uplifted, they should be redriven. A small amount of uplift of friction piles is not harmful.
2. Check the location of all piles after the adjacent piles are driven or at the completion of all pile driving. In ordinary soil conditions a 3 in. tolerance is considered reasonable. Piles which are driven at greater variation may throw a greater load to some piles in the group, and in such cases, the pile reaction and cap design must be checked.
3. Inspect the pile shaft for verticality or required batter. For cast-in-place piles, the general practice is to lower an electric light into the shaft before placing concrete. If the light can not be seen from the top, the pile is rejected for poor verticality. In the case of heavily loaded piles, such crude procedure may not be sufficient, and measurement must be made by specially devised instruments, unless load test is made with piles having questionable verticality.

C. Defective piles. A pile may be considered defective if:

1. It is damaged by driving;
2. It is driven out of position, is bent or bowed along its length.

To avoid damage to fresh concrete in a cast-in-place pile by driving of adjacent piles, the pile should not be concreted until all piles within a certain radius are driven. The radius depends upon the soil condition, the length and size of pile, and the pile spacing. If past experience of pile driving is lacking in the locality, test driving may be necessary to determine the ground movement due to driving.

A defective pile may be withdrawn and replaced by another pile. It may be left in place and another pile be driven adjacent to it, sometimes the damaged part of the pile can be removed and new length of pile spliced in. Unless the remaining portion of the pile is proven to be absolutely intact, the method is not desirable.

D. Effect of pile driving. Pile driving may introduce some of the following effects on the ground.

1. *Subsidence.* Vibration due to pile driving in loose sand may cause compaction of the sand, consequently, the area may settle, and adjacent structures may be affected. In saturated fine sand and silt, the shock may introduce catastrophical subsidence.
2. *Heave.* Pile driving in clays and dense sand is commonly associated with surface heave, and sometimes with lateral displacement. The upheaval may well exceed a foot in plastic soils. The heave of clay is followed by settlement immediately after driving. Piles uplifted by ground heave should be redriven. To avoid heave and lateral movement, pile driving should be started from the center of the ground and proceed outwards.
3. *Compaction.* Sand and gravel within a lateral distance of about 3 diameters of the pile and 2 diameters below the tip is largely compacted due to the displacement of pile, consequently, a pile group in sand behaves as a rigid block of compacted soil.
4. *Disturbance.* Clayey soil surrounding the pile is greatly disturbed due to the displacement of pile. The disturbance may extend to a large lateral distance and the strength of clay is largely reduced. However, in ordinary cases it starts to regain its strength and in 30 to 50 days, 90 per cent or more of its strength may be regained.

Plate Nine

A Cutout View of a Drilled Caisson in Clay before Placement of Concrete

Advancement in the construction equipment and techniques has made the drilled caisson one of the most commonly used types of foundations. This chapter deals with the procedure of design and construction including a brief description of the caisson excavating machines.

9 DRILLED CAISSONS

9-1 Use of Drilled Caissons

The terms *caisson*, *foundation pier*, and *sub-pier* are interchangeably used by engineers to denote a cylindrical foundation with or without steel reinforcing and with or without enlarged bottom which is concreted in place after excavation and which is utilized to transfer structural load to the bedrock or a hard stratum. To avoid confusion with caissons, which are sunk in place (Chapter 10), the term *drilled caisson* is used, although it may also be excavated by hand instead of by drilling machines.

A drilled caisson is largely a compression member subjected to an axial load at the top and a reaction at the bottom. When properly designed, it can resist bending moment. In comparison with other types of deep foundations, a drilled caisson has the following advantages:

1. It can be carried through soils that prevent penetration of piles. Machine drilled caissons have been constructed to a depth greater than 100 ft in very dense sand and gravel deposits with standard penetration resistance greater than 100 blows per foot.

2. It does not cause heaving or vibration of the ground as is often the case for driven piles. This may be a decisive factor when the adjacent structure is on spread footings or short piles.

3. It does not displace any volume of soil, therefore, the problem of shifting and lifting of piles is eliminated.

4. It requires comparatively mobile and light construction equipment which is simple to operate, consequently, the cost of construction is

generally low. Further economy is achieved because it does not require concrete cap for pile group.

5. The soil or rock conditions at the bottom of the caisson can often be inspected visually and tested physically.
6. The construction equipment is less noisy and therefore suitable for areas near hospitals and similar institutions.

Because of economical and other advantages, the drilled caisson has become one of the most popular types of foundations in recent years. Tall buildings, bridges and highway interchanges, wharves, and many other heavy structures have been supported on hand excavated or drilled caissons. In the case where a layer of top soil, miscellaneous fill, or other unsuitable soil extends to a moderate depth, ordinary spread footings would have to be carried below this layer to a suitable soil and would require extra cost in excavation and backfilling. The cost in moving pile driving equipment to and from the site is large and often exceeds the total foundation cost for small jobs. Therefore, footing and pile foundations are not competitive in such cases, and many stores, warehouses, etc., are supported on short drilled caissons.

Because of the advantages stated above, drilled caissons also have been extensively used in underpinning works.

Although drilled caissons are advantageous in several ways, one major drawback must be kept in mind. That is the danger of caving of the shaft, and particularly, the bell during construction. Clays and other soils possessing some cohesive strength will stand up on a cylindrical surface. Compact sand and gravel above ground water may not cave in. Excavation of caisson in granular soils (even with a small amount of cohesion) below water level is difficult because of the hazard of cave-ins and may increase the construction cost by several fold. This is almost invariably so if the bells must be excavated in saturated granular soils.

9-2 *Types of Drilled Caissons*

The drilled caissons may be classified according to the materials used, namely concrete, concrete in steel shell and concrete plus steel core in steel shell, Fig. 9-1. For small jobs plain concrete caissons are commonly used. The caisson may be reinforced with steel bars in the upper portion or in full length of the shaft, depending upon the loading and soil conditions as discussed later. This type is most common because it is simple to construct and it is often inexpensive. When construction of the bell in saturated soil becomes difficult and expensive, the caisson with steel shell may be carried down to, or into, the bedrock.

When a caisson consists of a straight shaft without a bell bottom, it is virtually a pre-excavated cast-in-place pile. A caisson that consists of a steel shell is the same as a pipe pile. The pipe may be driven before or after the excavation of the soil inside it. The difference of terminology in such cases has no engineering significance. It is generally called a pile when driven in a group of two or more to support a column and a caisson when constructed

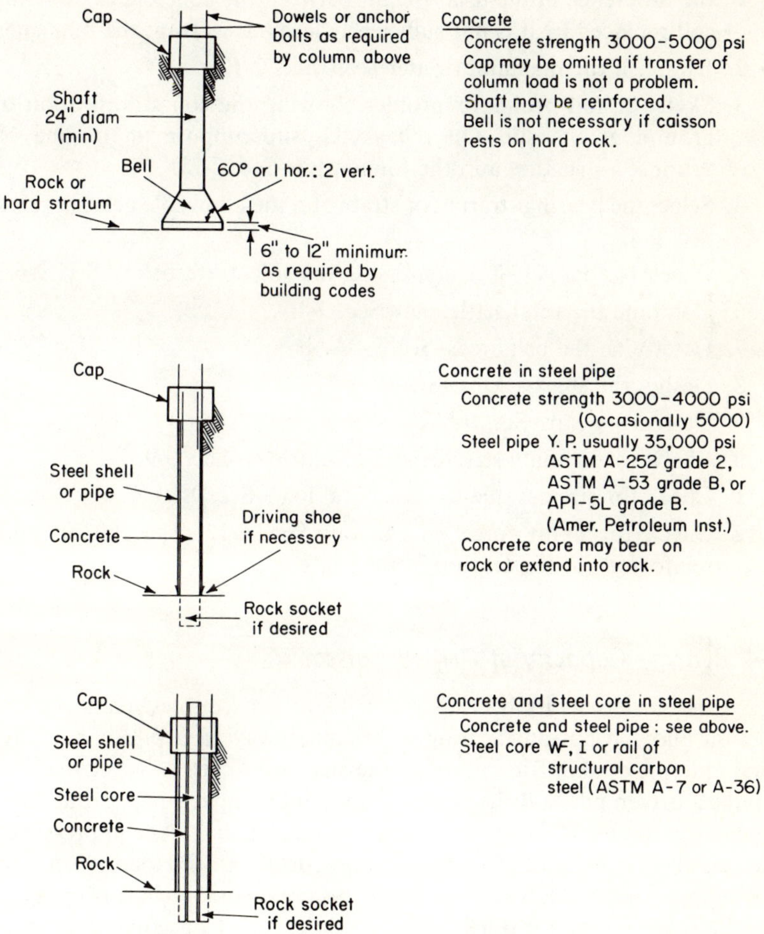

Fig. 9-1 Common types of drilled caissons.

concentrically with the column and when the shaft is sufficiently large to permit a workman descending in it. However, in exceptional cases, two or more caissons are used under one column.

Caissons are called Chicago wells, Gow caissons, etc., depending on the method of construction as discussed later.

9-3 Design of Drilled Caissons

Drilled caissons may be designed by the following steps:

1. Calculate the foundation loads to top of caissons–Chapter 3. The weight of the caisson is usually not included in the design loads because the difference in the total weight between the concrete caisson and the soil replaced by it is not sufficiently large to warrant the refinement.
2. Establish the maximum water level–Sec. 2-12.
3. Sketch a soil profile or profiles showing the soil stratification of the ground at the site. On the sketch superimpose an outline of the proposed structure and the foundation–Sec. 2-13.
4. Select the bearing stratum or strata and the allowable bearing pressure–Sec. 9-4.
5. Check for danger of overstressing the soil strata below–Sec. 9-6.
6. Compute the total settlement–Sec. 9-7.
7. Determine the bell areas–Sec. 9-8A.
8. Design the shafts–Sec. 9-8B.
9. Design the caps–Sec. 9-8C.
10. Check for bending stresses and eccentricity–Sec. 9-9.
11. Check for stresses due to horizontal load–Sec. 9-10.
12. Check for uplift force on caissons–Sec. 6-12. The caisson must be reinforced for the tension in the shaft.

9-4 Bearing Capacity of Drilled Caissons

A caisson derives its bearing power from two sources, namely, the skin friction and the bottom bearing in a similar way to a pile. Generally, the total amount of skin friction along the shaft of a caisson is relatively small. Unlike a driven pile, a drilled caisson does not compact the surrounding soil. Since a caisson usually rests on a hard stratum which is less compressible than the soil above, the hard stratum takes practically all the load. Furthermore, the surface area available for skin friction on a caisson shaft is considerably smaller than the total surface area in a pile group. Consequently, the benefit of skin friction is generally neglected and the caisson is designed as a compression member subjected to a load on top and an equal reaction at the bottom. Occasionally skin friction on caisson shaft may be an important factor.

Another factor that must be taken into consideration is the so-called negative skin friction. A negative skin friction is a dragging force along the shaft of a caisson when the soil surrounding the shaft is settling, whereas, the caisson resting on a hard stratum stays stationary. The surrounding soil may settle as a result of dewatering of the site, compression from a surcharge load (additional fill, for example), or deterioration of organic content of the soil. As long as there is a downward movement of the soil with respect to the caisson shaft, there is a drag which tends to exert an additional load to the bottom of the caisson.

The principle and procedure for determination of negative skin friction discussed in Sec. 8-18 are applicable to caissons. In the numerical example, Sec. 9-13, the effect of negative skin friction is included.

A. Drilled caissons on cohesive soils. According to Skempton (1951) the ultimate bearing capacity of a circular footing or caisson is

$$q_{\text{ult}} = cN_c$$

where c = cohesion which may be taken as one-half of the unconfined compression strength. The value of N_c depends on the depth/diameter ratio of the caisson.

Depth/diam caisson bottom	0	0.5	1.0	1.5	2.0	2.5	3.0	4.0 and over
Value of N_c	6.2	7.1	7.7	8.1	8.4	8.6	8.8	9.0

Since a caisson is drilled in the ground, the depth/diameter ratio seldom becomes less than 1 and the value of N ranges from 7.7 to 9.0. Taking an average of 8.4 and using a factor of safety of 3 for normal loading, the allowable bearing pressure may be expressed as

$$q_a = 1.4q_u \tag{9-1}$$

where q_u = unconfined compressive strength of the clay. It may be noted that the value of q_a varies within 10 per cent when the depth/diameter ratio changes from 1 to infinity. Therefore, it may be considered as a general equation for all caissons on cohesive soils.

With the exception of very small jobs where column loads are moderate, the value of q_u or c should always be determined by laboratory tests.

B. Drilled caissons on sand and gravel. Equations (6-1a) and (6-2) may be applied to drilled caissons. They are

$$q_1 = \frac{2}{3}N^2BR_w + 2(100 + N^2)DR'_w \tag{9-2a}$$

$$q_2 = 1440(N - 3)\left(\frac{B + 1}{2B}\right)^2 R'_w \tag{9-2b}$$

where q_1 = allowable bearing pressure, psf = $\frac{1}{3}$ ultimate bearing pressure.

q_2 = allowable bearing pressure, psf, for a maximum settlement of one inch; this equation is modified from Eq. (6-2) on the assumption that on sand and gravel a deep foundation settles half as much as a foundation at the ground surface;

N = number of blows per ft in standard penetration tests;

B = diameter of caisson bottom, ft;

D = depth of caisson, ft; If $D > B$, use the value of B in calculation;

R_w, R'_w = reduction factors, see Fig. 6-4.

C. Drilled caissons on bedrock. Local building codes usually stipulate safe bearing values for concrete and steel pile on bedrock. These values are based on past experience and are generally on the conservative side. If such information is lacking, reference is made to the discussion on engineering properties of rocks in Chapter 1. If rock samples are tested in unconfined compression machine, usually $\frac{1}{5}$ to $\frac{1}{8}$ of the test value is used as allowable bearing pressure.

When a socket is made in the bedrock and the concrete with or without a steel core is extended in it (Fig. 9-1), the bearing capacity of the caisson with steel shell may be considered as being consisted of three components: the bearing of the steel shell or pipe on the ledge of the rock socket, the bearing of the concrete with the core on the bottom of the socket, and the shear or friction between the concrete and the side of the rock socket. If the entire caisson (shell, core, and concrete) acts as one unit, as is always desired, the ultimate bearing capacity would be equal to the sum of these three components which are discussed below.

Because of irregular rock breakage, the end of the steel pipe is generally bearing partly on the edge of rock socket, and partly on the concrete plug which fills the socket. It is commonly assumed that the entire stress in the pipe is transferred to the concrete plug, not to the rock edge. This assumption is conservative, and leads to the common practice that bearing capacity of drilled caisson in rock socket consists of end bearing and side friction, or bond, regardless of the use of steel shell.

The bearing capacity of rock increases with increasing depth of the socket. Many building codes recognize this fact and permit a certain amount of additional allowable bearing pressure for each foot increase. Terzaghi (1946) quoted the following results of concrete tests

$$q_{\text{ult}} = q_u + 4.1p_c$$

where q_{ult} = ultimate bearing pressure of concrete under a confining pressure p_c,

q_u = unconfined compressive strength of concrete.

The true relationship between the depth of socket and the bearing capacity of rock is difficult to determine, because of the presence of joints, bedding planes, and fissures, etc., in rock formations. It is a common practice that when past experience in the locality is lacking the bearing capacity at the bottom of socket is assumed equal to that on the surface of rock.

In practice the problem is often further complicated because of presence of free water in the socket. Quite often there is a certain amount of seepage water flowing into the caisson either through the rock joints and fissures, along the rock surface, or from higher elevations. As a result, a concrete plug is usually poured under water, or else poured against an upward seepage pressure. The latter procedure is less desirable. In either case, the quality and strength of the concrete may control the bearing pressure.

Usually the walls of the rock socket is sufficiently rough to ensure large frictional resistance. In hard rocks, the resistance may be taken as the bond value between concrete and reinforcing bars. In soft rocks, the friction resistance is governed by the shear strength of the rock which must be determined by test of rock samples.

It should be re-emphasized that bearing values should be reduced if

1. Free water softens the soil or rock during construction.
2. Concrete is poured under water–pouring concrete against a seepage is not recommended.
3. Loose soil or mud accumulates on the bottom of caisson and is difficult to be removed completely.

9-5 *Skin Friction of Caisson Shaft*

A drilled caisson of relatively small diameter carried to a great depth in stiff or compact soil may derive a large portion of its bearing capacity from the skin friction between the shaft surface and the surrounding soil. In such cases, the computation of the ultimate bearing capacity should include the available skin friction which, unfortunately, is difficult to determine. In large projects where cost of load tests is justified, caisson load tests may be planned such that the capacity due to skin friction and that due to end bearing can be differentiated. This may be done in several ways. For example, one test caisson may be constructed with a small annular space between the caisson shaft and the soil, and another poured against the soil in the same way as the final construction. The difference in bearing capacity between these two types of test caissons is the load carried by skin friction. The value of skin friction may also be determined by making a test caisson with straight shaft with a layer of a compressible substance at the bottom (such as 3 in. of a pourous, compressible fiber board). The difficulty of these methods is that several sets of such tests must be made in order to reach some reasonable, sound conclusion. This is because large-size load tests seldom give consistent results even in an apparently uniform deposit, and a considerable judgement and interpretation of conditions is required.

For small and average sized projects, an approximate evaluation of the skin friction of caisson may be necessary. Such evaluation, as described below, must be used with conservative factor of safety.

The skin friction between cohesive soils and the caisson shaft cannot exceed the cohesion of the soil. Cohesion may be assumed to be equal to one-half of the unconfined compressive strength of the soil. In stiff and hard clays, however, the bond between the concrete and the soil may be smaller than the cohesion, particularly when the surface of the shaft excavation is relatively smooth. Therefore, the maximum design value for skin friction of all cohesive soils should be limited to about 1000 (or up to 1500) psf.

The skin friction between the caisson shaft and granular soils is equal to the total horizontal earth pressure against the shaft surface times the coefficient of friction between the concrete and the soil. Assuming the surface of shaft is perfectly rough, the coefficient of friction is equal to $\tan \varphi$ where φ is the angle of internal friction of soil. Since the caisson shaft is practically uniform in cross-section, the pressure against it is probably the earth pressure at rest. Hence the total skin friction f_s may be expressed as $(K_0 \gamma D) \times (D/2) \times \pi d \times \tan \varphi$. Assuming a conservative value of earth pressure at rest, and rearranging,

$$S_f = 0.7 \gamma d D^2 \tan \varphi \tag{9-3}$$

where S_f = total ultimate skin friction, lb;

γ = average unit weight of soil–use submerged weight for soil below water level, pcf;

d = diameter of caisson shaft, ft;

D = depth of caisson from ground surface to top of bell, ft;

φ = angle of internal friction of the soil.

9-6 *Stress on Lower Strata*

It is essential in all practical cases to check the bearing on lower strata due to the concentrated, large load from the caissons. For ordinary applications, the stress may be assumed to spread out on a slope of 2 vertical to 1 horizontal. Thus the average stress at a depth z below the bottom of a caisson due to a load Q is equal to $4Q/\pi(B + z)^2$, where B is the diameter of the bottom of the caisson.

The elastic equation (Boussinesq) and Newmark's influence chart commonly used to determine the vertical pressure on lower strata are derived for the condition of surface loading. A footing or a mat foundation is generally not too far below the ground surface and therefore the elastic theory is justified by ignoring the effect of the soil above the footing. The drilled caissons, however, are usually carried to great depths (large D/B ratio) and to neglect the effect of a large overburden of soil would probably give results considerably larger than actual values. On the other hand, a further refinement on the elastic theory by assuming that the soil above the footing is an elastic material is also erroneous because the soil, in the theory of elasticity, would be required to resist tensile stress.

Caissons carried to bedrock generally are safe as long as the design pressure is carefully determined. However, limestones often contain a complex system of cavities, fissures, and caverns resulted from slow dissolution of calcium carbonate in the long geologic past. These cavities have no general pattern and therefore their locations cannot be determined by economical procedure. In localities where cavities are predominent, some type of probing (such as drilling a hole at the bottom of the caisson to a depth of several feet) should be made at the important and heavily loaded caissons. Many local building codes recognize this problem and require a minimum number of rock cores in each job. Massive rocks, if free from cracks, etc., are capable of bridging over a large cavity.

9-7 *Settlement of Drilled Caissons*

Drilled caissons on sand or gravel will not be subjected to detrimental settlement, if the smaller value of q in Eqs. (9-2) is used, provided that the bottom of the caisson is cleaned off thoroughly to the original soil, that the bearing stratum has not been disturbed by upward seepage water, that the concrete is poured in the dry and not against seepage pressure, and that there is no weaker stratum at a greater depth.

The theoretical settlement of caissons on clays can be computed by the procedure discussed in Sec. 3-6. Settlement analysis is necessary for caissons resting on medium clays and soft clays. On hard clays where large bearing pressure is used, it is often necessary to ascertain the fact that the theoretical consolidation settlement due to the slow extrusion of moisture from the pore spaces is negligible. In practice, a small amount of settlement frequently takes place. The settlement may be a result of one or all of the following factors: softening of the upper layer of the clay due to moisture and disturbance, accumulation of loose material sloughed in, and swelling and upheaval of the soil. Even in a clay deposit, often a certain amount of seepage water accumulates at the bottom.

When caissons are carried to bedrock it is very likely to encounter water at the bottom of some or all of the caissons. Unless the water can be pumped out and the bottom thoroughly cleaned, a small amount of settlement is inevitable. Soft rocks may disintegrate upon soaking and cause large settlement. To avoid such settlement of uncertain magnitude, it is much desired to prevent free water from entering the caisson during the last stage of excavation or drilling. The methods commonly used for this purpose are discussed later.

9-8 *Design of Elements of Drilled Caissons*

A drilled caisson consists of a shaft with or without a bell and a cap. The design of each of these elements is discussed below.

1. Drilled caissons carried to materials other than hard rock are often enlarged at the bottom for the purpose of reducing the soil pressure to desired bearing value. The enlarged portion is generally known as a bell because of its shape. The ideal bell is in a shape of a frustum with a vertical side at the bottom. The vertical side may be 6 to 12 in. high, as required by local building codes, Fig. 9-2(a). The sloping side or roof of the bell should

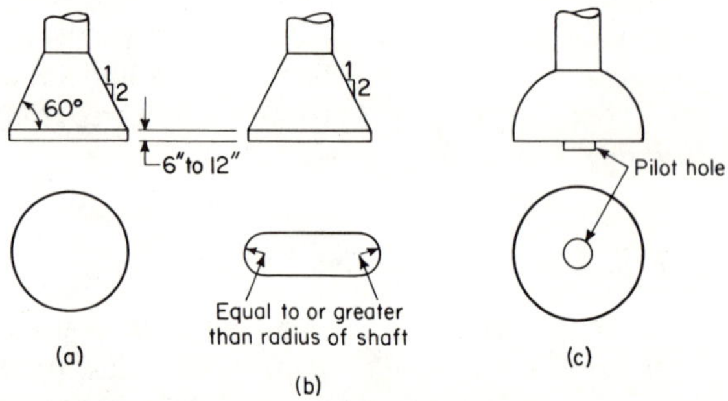

Fig. 9-2 Common shapes of caisson bells: (a) standard bell excavated by machine or hand; (b) oblong bell excavated by hand; (c) dome-shaped bell excavated by a certain reamer.

be at an angle at least 60 degrees with the horizontal (equivalent to a slope of 2 vertical on 1 horizontal) so that the soil would not cave in. Even with this slope, bells may not be formed in purely granular soils. Under an axial load the concrete in such a frustum is considered free from flexural stresses.

In special cases the bell may be excavated by hand in an oblong or elliptical shape, Fig. 9-2(b). This is necessary in the case of exterior caissons where a circular base would protrude beyond the property limit. Obstructions of existing constructions may also require such shape of bell.

The actual shape of the bells as constructed may be somewhat different from the ideal shape generally desired because of the irregular breakage of soil, the physical dimensions of the drilling or belling equipment, and other factors in the constructional procedure. In practice, it is generally required that:

(a) The furnished caisson should not be smaller than required and the overexcavation should be filled with concrete monolithically with the caisson proper.

(b) The center of the base area should not deviate from the required location by a certain prescribed amount, as discussed in Sec. 9-9.

When caisson bells are excavated by a bucket type equipment, the bell shape may differ considerably from the standard 60-degree frustum. Such bells are dome-shaped with a pilot hole of small diameter in the center. In order to fullfil the requirement (a) stated in the previous paragraph, it is necessary to modify the machine-excavated bell by hand. However, some engineers consider this type of bell satisfactory.

In the U.S.A. the excavating equipment is generally capable of making a bell with a diameter up to three times the diameter of the shaft. When bells with larger diameter are desired, they are generally enlarged by hand after the machine excavation. The hand labor would increase the cost of construction.

For the convenience of design engineers, the capacity of various sizes of bells is listed in Table 9-1.

2. A caisson shaft is principally a short column. The surrounding soil exerts somewhat confining pressure which tends to increase the compressive

Table 9-1 CAPACITY OF CAISSON BELLS (in kips)

Bell	*Area*	*Design bearing pressure (psf)*					
diam	*sq ft*	*6000*	*8000*	*10000*	*12000*	*15000*	*20000*
2′ 0″	3.14	18	25	31	37	47	62
2′ 6″	4.91	29	39	49	58	73	98
3′ 0″	7.07	42	56	70	84	106	141
3′ 6″	9.62	57	77	96	115	144	192
4′ 0″	12.57	75	100	125	150	188	251
4′ 6″	15.90	95	127	159	190	238	318
5′ 0″	19.64	117	157	196	235	294	392
5′ 6″	23.76	142	190	237	285	356	475
6′ 0″	28.27	169	226	282	339	424	565
6′ 6″	33.18	199	265	331	398	497	663
7′ 0″	38.48	230	307	384	461	577	769
7′ 6″	44.18	265	353	441	530	662	883
8′ 0″	50.26	301	402	502	603	754	1005
8′ 6″	56.74	340	454	567	681	851	1134
9′ 0″	63.62	381	508	636	763	954	1272
9′ 6″	70.88	425	567	708	850	1063	1417
10′ 0″	78.54	471	628	785	942	1178	1570
11′ 0″	95.03	571	760	950	1142	1425	1900
12′ 0″	113.10	678	905	1131	1356	1696	2262
13′ 0″	132.73	796	1062	1327	1592	1991	2654
14′ 0″	153.94	923	1232	1539	1846	2309	3078

NOTE : Bell diameter preferably not to exceed 3 × shaft diameter. Side slope should be 2 vertical: 1 horizontal. Check stresses if slope is less than 2:1.

strength of the concrete. Due to the relatively low modulus of elasticity of soil in comparison with that of concrete, the benefit of confining pressure is very small, particularly in soft soils near the ground surface. Consequently shafts of all drilled caissons are designed according to the rules for concrete columns.

Theory and experience have proven that a caisson (or a pile) receives sufficient lateral support even from weak surrounding soils to prevent it from buckling. Hence, the formula for short columns is applicable. For caissons projecting above ground surface and surrounded by extremely soft soils such as peat or loose river mud, the unsupported length is usually taken as the length of projection or the thickness of the extremely soft soil plus 5 ft (in stiff and compact soil) to 10 ft (in soft and loose soil).

Since concrete is more economical than steel in compression, plain concrete sections are commonly used. However, a small amount of reinforcing is highly desirable in the upper portion of the caisson because of several reasons. Generally a small amount of horizontal thrust and bending moment exists. The horizontal thrust may be resulted from wind pressure on the structure, and the bending moment resulted from bending or eccentric loading of the column. The caisson will be inevitably constructed with certain amount of eccentricity, Sec. 9-9. Furthermore, during the construction period, temporary bending moment may be introduced by construction equipment. Therefore a certain amount of vertical reinforcement equal to $\frac{1}{2}$ per cent of the cross-sectional area of the shaft is commonly provided in the upper several feet of shaft. Where the soil in the upper layer is very soft, this reinforcement should be extended below the soft layer.

A single drilled caisson supporting a column should not have too small a shaft. A 2 ft 0 in. diameter shaft is generally considered as the minimum size. Smaller sized shaft is not only considered too flexible to support a major column, but also prevents entrance of men for cleaning the bottom, for removal of boulders, and for inspection.

Except for large diameter drilled caissons, the top of caisson or the column base should be supported laterally by the structure of the basement or ground floor slabs.

The caisson shaft is designed in accordance with the rules for short columns. The allowable compressive strength for plain concrete shaft is generally equal to $0.18f_c'$,* f_c' being the ultimate strength of the concrete test cylinder. For the convenience of design engineers, the load capacity for commonly used shaft sizes is listed in Table 9-2.

The design stresses for caissons with steel shell and with or without a steel core may be taken from those for a composite column. In computation of the caisson capacity the thickness of the steel shell should be reduced to allow for

* A few building codes permit higher stresses.

corrosion. Generally $\frac{1}{16}$ in. is deducted from the shell, unless the conditions are adverse. According to ACI Building Code Requirements, a composite column may carry a load P.

$$P = f_c A_e + f_s A_s + f_r' A_r \tag{9-4}$$

where $f_c = 0.25 f_c'(1 - 0.000025\, h^2/K_c^2)$

f_c' = strength of the concrete test cylinder;

h = unsupported length of caisson, equal to zero when caisson is entirely surrounded by soil with moderate to high strength;

k_c = radius of gyration of concrete cross-section;

A_c = total cross-sectional area of concrete;

Table 9-2 CAPACITY OF CAISSON SHAFT (Plain Concrete)

Shaft diam	*Cross-sectional area (sq in)*	*Suggested minimum reinforcement*		*Design capacity (1000 lb)*		
		*Vertical**	*Round ties*	$f_c' = 3000$ *psi*	$f_c' = 4000$ *psi*	$f_c' = 5000$ *psi*†
1′ 6″‡	254.5	6-#5	#3 @ 12″	137	183	229
2′ 0″	452.4	6-#6	#3 @ 12″	244	326	407
2′ 3″	572.6	6-#6	#3 @ 12″	309	412	515
2′ 6″	706.9	6-#7	#4 @ 12″	382	509	636
2′ 9″	855.3	6-#7	#4 @ 12″	462	616	770
3′ 0″	1017.9	8-#7	#4 @ 15″	550	733	916
3′ 3″	1194.6	8-#8	#4 @ 15″	645	860	1075
3′ 6″	1385.4	8-#9	#4 @ 15″	748	997	1247
3′ 9″	1590.4	8-#9	#4 @ 15″	859	1145	1431
4′ 0″	1809.6	10-#9	#4 @ 18″	977	1303	1629
4′ 3″	2042.8			1103	1471	1839
4′ 6″	2290.2	10-#10	#4 @ 18″	1237	1649	2061
4′ 9″	2551.8	10-#10	#4 @ 18″	1378	1837	2297
5′ 0″	2827.4	10-#11	#4 @ 18″	1527	2036	2545
5′ 3″	3117.3	10-#11	#4 @ 18″	1683	2244	2806
5′ 6″	3421.2	12-#11	#4 @ 18″	1847	2463	3079
5′ 9″	3739.3	12-#11	#4 @ 18″	2019	2692	3365
6′ 0″	4071.5	12-#11	#4 @ 18″	2199	2931	2664

* Extend vertical reinforcing 4 ft to 8 ft below the very soft soil layer, mud, miscellaneous fill, or other weak stratum.

† Some building codes limit $f_c' = 4000$ psi for calculation of shaft capacity even if higher strength is provided.

‡ 18 in. shaft should not be used for supporting major columns.

f_s = 16,000 psi for steel core;

A_s = total cross-sectional area of steel core;

f_r' = 17,000 − 0.485 h^2/k_s^2 for Grade B pipe with minimum yield strength of 33,000 psi;

k_s = radius of gyration of the steel pipe (or shell);

A_r = cross-sectional area of steel pipe; a reduction of $\frac{1}{16}$ in. in pipe thickness should be used for calculation of k_s and A_r.

In localities where building codes govern, the stresses should be taken from the codes.

3. The bell and the shaft of a caisson are always constructed monolithically and the concrete is generally stopped at a level several feet below the top of the caisson. The remaining portion at the top that is poured later is often made larger than the shaft and is called the cap. To pour the cap at a later time permits the caisson to be concreted immediately after excavation and inspection. The cap is generally made at least six or eight inches larger than the shaft to allow exact placement of anchor bolts or column dowels while the center of the caisson might be out of line. Sometimes the caps are constructed with concrete of higher strength than that of the shaft, thus a higher bearing pressure can be used in the column base design. The cap may be square or circular in plan.

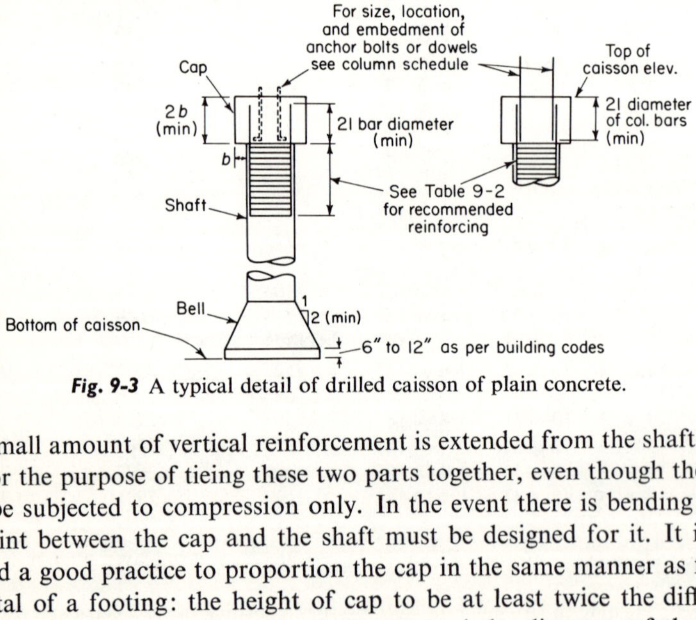

Fig. 9-3 A typical detail of drilled caisson of plain concrete.

A small amount of vertical reinforcement is extended from the shaft to the cap for the purpose of tieing these two parts together, even though the shaft may be subjected to compression only. In the event there is bending stress, the joint between the cap and the shaft must be designed for it. It is considered a good practice to proportion the cap in the same manner as for the pedestal of a footing: the height of cap to be at least twice the difference between the width, or diameter, of the cap and the diameter of the shaft.

The height of the cap should be also at least equal to the length of embedment of the anchor bolts or column dowels. A typical detail of the cap, shaft, and bell of drilled caisson of plain concrete is shown in Fig. 9-3.

In the case of a caisson with a steel shell and a steel core, the cap must be carefully designed to transfer the column load to these components of the caisson. Obviously we cannot cut off the steel shell and core at the bottom of the cap and expect the high stresses to be transferred from the cap to them by end bearing. To ensure a satisfactory stress transfer, various cap details have

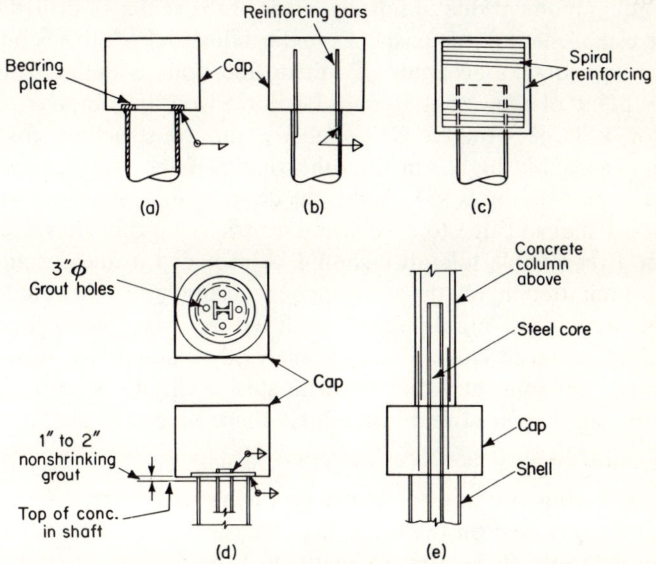

Fig. 9-4 Examples of cap details for drilled caissons with steel shell.

been used. Figure 9-4 shows a few of such details. The stress in the shell may be transferred to the cap by a bearing plate welded to the top of the shell, Fig. 9-4(a), by dowels welded to the shell and extended into the cap, or by embedment of the shell in a spirally reinforced cap, Fig. 9-4[(b) and (c)]. In the latter method, the outer surface of the shell may be provided with weldments thus to increase the bond stress between the shell and the concrete. The stress in the steel core may be also transferred by bearing plate or by dowels. The steel core itself may be extended into the cap and the concrete column above, Fig. 9-4(e). When the bearing plate occupies a large portion of the concrete section, and the concrete below the bearing plate may not come in direct contact with the plate, the concrete in the shaft is usually poured to an inch or two below the plate and the space is then filled with nonshrinking cement grout, Fig. 9-4(d).

9-9 Bending Moment and Eccentricity

Drilled caissons are invariably subjected to a certain amount of bending moment. The bending moment may be carried from the lower end of the column, caused by horizontal wind thrust, or may be due to the misalignment of the drilled caisson itself. In all cases, the stresses due to bending moment should be analysed. The amount of the wind thrust and the column bending should be computed when designing the superstructure.

The bending moment due to misalignment of the center of drilled caissons should be established on the basis of practicability and relative economy. A strict limitation on the accuracy of caisson locations is desirable from the designer's point of view, but it may be unrealistic and impose upon the constructor a large penalty. Consequently the construction cost would become unreasonably high. On the other hand, if the requirement on the accuracy of construction is too liberal, the designer must make provision for large bending moment due to excessive eccentricity he considers acceptable. Therefore, a reasonable tolerance should be provided in the design so that a realistic limitation upon the construction accuracy can be allowed. The permissible amount of misalignment or deviation from the ideal condition is called the tolerance. This problem is analysed separately for caisson bells, plain concrete caissons, and caissons with steel shells. Quite often the steel core section can be placed more accurately than the drilled shaft.

1. The center of a bell may be permitted to deviate 5 per cent from the diameter of the bell, if an overstress of 20 per cent is not considered objectionable. This statement is based on the following analysis.

In determination of bearing capacity of a foundation subjected to an eccentric loading, only the portion of the foundation which is symetrical about the load is assumed useful, see Sec. 6-9. With an eccentricity equal to 5 per cent of the bell diameter, the useful width is $d - 0.1d = 0.9\,d$ where d is the diameter of the bell. The ultimate bearing capacity becomes somewhat larger than $(0.9)^2 = 81$ per cent of the bearing capacity of a concentrically loaded caisson bell. The assumption above is applicable to cohesive soils, and is approximate for deep foundations in granular soils.

2. Caissons provided with $\frac{1}{2}$ per cent vertical reinforcing may have an eccentricity equal to 12 per cent of the diameter of the caisson shaft without overstressing the concrete. This can be proven by the following computation.

Let $A_g = \pi R^2$ = cross-sectional area of the shaft, R = radius;

p_g = vertical reinforcement as percentage of A_g;

A_s = area of vertical reinforcing $= \pi R^2 P_g$.

The vertical reinforcing may be considered to be equivalent to a steel ring with thickness $= t$

$$t = \frac{A_s}{2\pi(R-a)} = \frac{R^2 p_g}{2(R-a)}$$

where a = thickness of the concrete cover on the vertical reinforcing, usually about four inches to the center of the reinforcing.

The section modulus of the reinforced column is

$$S = \frac{\pi}{4}R^3 + (n-1)\pi(R-a)^2 t = \frac{1}{4}RA_g\left[1 + 2(n-1)\frac{R-a}{R}p_g\right]$$

The caisson shaft is designed for a load N with zero eccentricity

$$N = 0.18 f_c' \pi R^2$$

Based on the method of analysis in the ACI Codes 318

$$\frac{f_a}{F_a} + \frac{f_b}{F_b} = 1$$

where

$$f_a = \frac{N}{A} = 0.18 f_c'$$

$$F_a = 0.18 f_c' + f_s\frac{A_s}{A_g} = 0.18 f_c' + f_s p_g$$

$$f_b = \frac{Pe}{S} = \frac{0.18 f_c' A_g e}{\frac{1}{4}RA_g\left[1 + 2(n-1)\frac{R-a}{R}p_g\right]}$$

$$F_b = 0.45 f_c'$$

Rearranging the equation above

$$\frac{e}{R} = 3.47 p_g \frac{f_s}{f_c'} \frac{1 + 2(n-1)\frac{R-a}{R}p_g}{1 + 5.55 p_g fs/f_c'} \approx 1.74\, p_g \frac{f_s}{f_c'}$$

For $p_g = \frac{1}{2}\% = .005$, $f_s = 20{,}000$ psi, $f_c' = 3000$ psi, $e/R = 0.06$ or $e/D = 0.12$.

9-10 Horizontal Force on Drilled Caissons

A drilled caisson is capable of resisting a horizontal force in the same manner as a single pile. Usually no special provision is necessary for a horizontal force of less than 1500 lb acting on a drilled caisson in moderately compact or stiff soil. If the surrounding soil is very soft or if the horizontal force is large, it is generally more economical to rely on other means instead of the lateral strength of the caisson to resist the horizontal force, for example, by utilizing the passive earth pressure against basement walls. Drilled caissons may be installed on a small batter also.

A drilled caisson acted upon by a lateral force may be analysed by the theory of beams on elastic foundation (Hetenyi, 1946). If the elastic modulus of the soil (coefficient of horizontal subgrade reaction) is constant throughout the entire length, the following equation may be used.

Top of drilled caisson free to rotate:

$$\text{Maximum moment} = \frac{-P\,e^{-\pi/4}}{\lambda\sqrt{2}} = -0.322\frac{P}{\lambda} \tag{9-5a}$$

at $\pi/4\lambda$ from top of drilled caisson

$$\text{Maximum horizontal displacement} = \frac{2P\lambda}{Bk} \tag{9-5b}$$

at top of drilled caisson

$$\text{Maximum shear} = P \tag{9-5c}$$

at top of drilled caisson

$$\text{Moment at any point} = \frac{P}{2\lambda}(e^{-\lambda x}\sin\lambda x) \tag{9-5d}$$

at x from top of drilled caisson

Top of drilled caisson fixed:

$$\text{Maximum moment} = +\frac{P}{2\lambda} \tag{9-6a}$$

at top of drilled caisson

$$\text{Maximum horizontal displacement} = \frac{P\lambda}{Bk} \tag{9-6b}$$

at top of drilled caisson

$$\text{Maximum shear} = -P \tag{9-6c}$$

at top of drilled caisson

$$\text{Moment at any point} = \frac{P}{2\lambda}e^{-\lambda x}(\sin\lambda x - \cos\lambda x) \tag{9-6d}$$

at x from top of drilled caisson

where P = lateral force, lb;

$\lambda = 4\sqrt[4]{Bk/4EI}$, in.$^{-1}$; B = width (diam) of drilled caisson, in.;

k = coefficient of horizontal subgrade reaction, lb per cu in.;

E, I = modulus of elasticity and moment of inertia of drilled caisson, psi and in.^{4}, respectively.

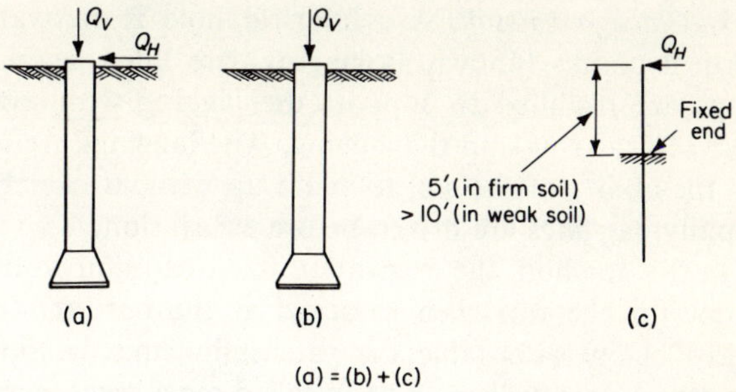

Fig. 9-5 Simplified procedures for estimating bending stress in drilled caisson due to lateral force.

The values of the coefficient of horizontal subgrade reaction are tabulated in Sec. 8-16.

In cases where the coefficient of subgrade reaction varies with depth, the differential equations derived by elastic theory may be solved by the method of finite differences (Gleser, 1953), but the procedure is more laborious. In practice, some simpler procedure is desired for the purpose of obtaining some very approximate idea as to the order of magnitude of flexural stress likely to exist in the drilled caissons. One very crude procedure assumes that the drilled caisson is fixed at an arbitrary depth below the ground surface, 5 to 10 ft depending upon the relative rigidity of the soil and diameter of caisson shaft, and that the drilled caisson behaves as a cantilever above the fixed point under the action of the lateral force.

9-11 Construction of Drilled Caissons

The construction of drilled caisson consists of the following procedures:

Excavation of the caisson

Placing of concrete

Inspection

Each of these procedures is discussed in the following. In addition, there are several problems encountered during the construction of drilled caissons which are discussed in Sec. 9-12. These problems are:

Safety

Underwater excavation

Other construction problems

A. Methods of excavation. Caissons may be excavated by hand or by machine. When excavated by hand, it is called a Chicago caisson or a Gow Caisson, depending on the procedure used. When excavated by machine, it is referred to as a drilled caisson.

1. *Chicago caisson.* A cylindrical hole is excavated by hand, and then vertical planks (known as lagging) are lined against the excavation. Steel rings are installed to support the lagging with wooden wedges driven in between the ring and the lagging. The laggings are 2 to 6 ft long depending on the ability of the soil to stand up without caving in. If the soil caves in rapidly, laggings are driven before excavation.

In this method, the excavation must be pumped free of excessive water in order for the workmen to stand at the bottom. In deposits of sand and gravel below water table, the water inflow may be too large for pumping, well points or deep wells may be installed outside the caisson excavation to lower the water pressure. The dewatering work then becomes costly, and furthermore, excessive removal of ground water will most likely cause settlement of the adjoining ground.

The concrete is placed against the laggings and steel rings which are left in place. The minimum size can be excavated by hand is about three feet in diameter.

2. *Gow caisson.* A Gow caisson is also excavated by hand where telescopic steel cylinders are driven as the excavation progresses. The steel cylinders are removed one section at a time by a drag line as the concrete is placed. The minimum diameter for Gow caisson is 4 ft.

3. *Machine drilled caissons.* The caisson shaft may be excavated by an auger type drill, a bucket, a chopping bit, or a chopping bucket. An auger type drill looks like a carpenter's auger and works in the same way. It has a cutting blade or a set of cutting teeth attached to the lower end, Fig. 9-6(a). As the auger is rotated and forced down it drills into the earth. Then the auger is lifted above the ground, and the earth retained on the blades is whirled out by rotating it with a large acceleration.

A bucket type drill is shown in Fig. 9-6(b). The bottom of the bucket has an opening with cutting edges or cutting teeth. As the bucket is rotated and forced down the cutting edges shave the earth and scoop it into the bucket. Then the bucket is lifted above the ground and emptied out by opening the bottom which hinges to the side.

The drill of either type is attached to a square shaft known as a *kelly* which is rotated by means of a yoke and ring gear assembly, Fig. 9-7. A motor mounted on the truck furnishes the power for driving the ring gear through a pinion or a chain drive. As the ring gear turns, it engages the yoke by the lugs on the inside of the ring. The downward force is furnished by the weight of the kelly and the drill. In addition, the rear end of the truck can be engaged to the kelly as a downward force.

When boulders or hard stratum is encountered, the caisson is usually excavated by a chopping bucket or a chopping bit. A chopping bucket has stout chopping teeth on the bottom. A chopping bit, Fig. 9-8, is made of

Fig. 9-6 (a) Bucket type caisson drill. Photograph courtesy of Calweld, Inc.

Fig. 9-6 (b) Auger type caisson drill. Photograph courtesy of Joslyn Manufacturing and Supply Co.

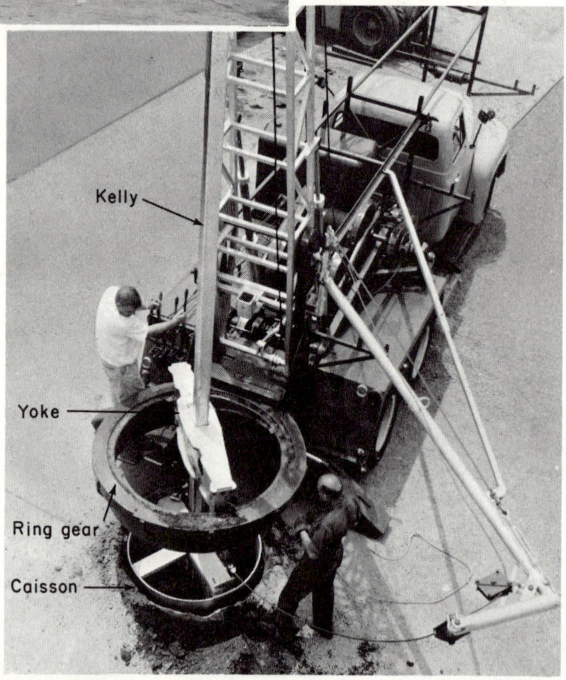

Fig. 9-7 (a) A common type of caisson drilling machine. Photograph courtesy of Calweld, Inc.

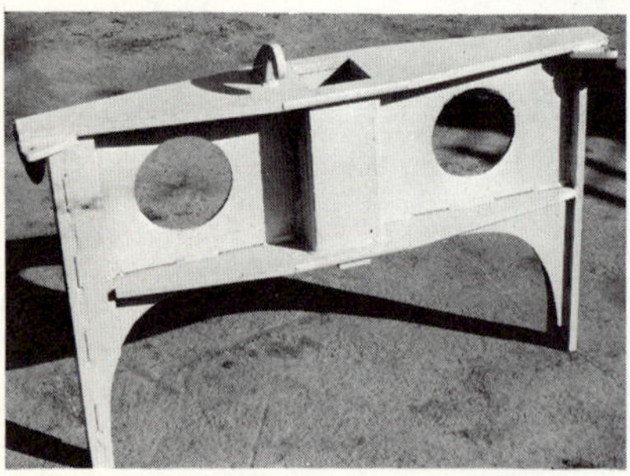

Fig. 9-7 (b) Yoke. Photograph courtesy of Calweld, Inc.

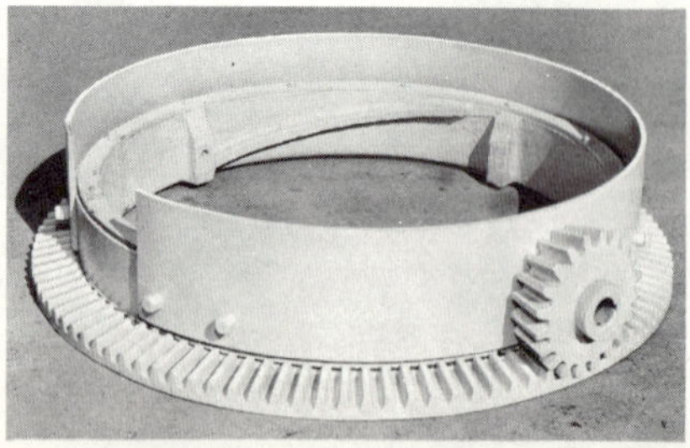

Fig. 9-7 (c) Ring gear. Photograph courtesy of Calweld, Inc.

solid metal with blunt cutting edges. Teeth or the chopping edges hit and break the rock by continued chopping action as the bucket or bit is dropped several feet by a drag line. During the chopping operation the hole is always filled with water which, being constantly stirred, keeps the rock chips in suspension and therefore enables the chopping action to be directed on a more or less fresh rock surface at the bottom of the hole.

Fig. 9-8 Rock bit. Photograph courtesy of Spencer, White and Prentis, Inc.

The caisson bell is excavated first by drilling with the equipment for the shaft. When the soil at the bottom of the drilled shaft is considered satisfactory for supporting the bell, the bell is then enlarged by means of a reamer. A reamer has two arms hinged on the top or bottom, Fig. 9-9. When the kelly is forced down, it tends to swing out the arms. At the same time the reamer is rotated thus the teeth or cutting edges on the arms ream the side of the shaft and enlarge it gradually into a bell. Plate 9 is a photograph of a bell made by the bucket type reamer.

The Benoto machine. The Benoto machine is a heavy duty caisson excavating equipment. The principal parts of this machine are a *hammer grab*, an oscillating boring tube, and hydraulic devices for moving and supporting the rig, as shown in Fig. 9-10.

The hammer grab is the drilling tool weighing about 3000 lb. Different cutting blades or jaws are fitted on the bottom for excavating different types of materials. The boring tube is a heavy casing with cutting edge at the lower end. It is held by a collar which transmits a reciprocating rotation and vertical motion to the boring tube. The reciprocating rotation is achieved by the operation of two opposite hydraulic jacks acting horizontally on a boom attached to the collar. Two other hydraulic rams operate vertically on the

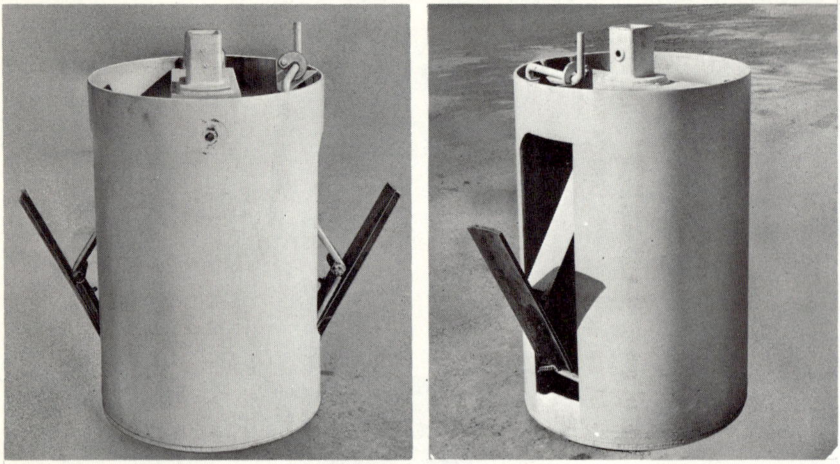

Fig. 9-9 Belling bucket. Photographs courtesy of Calweld, Inc.

same collar to raise and lower the tube, and to force the tube into the ground during the boring action. The combination of the two movements keeps the skin friction to a minimum thus facilitates the advancement of the tube. In soft soil the tube can be advanced ahead of the level of excavation in order to insure the stability of the ground. In hard soils or rock, the relative position is reversed and the excavation is advanced ahead of the tube.

When concreting, the rotationary and vertical movements tend to overcome the skin friction between the fresh concrete and the tube while it is being withdrawn; therefore, it reduces the possibility of uplifting the concrete filling.

The machine is mounted on a rig on wheels which are used for moving the machine on road. On arrival at the site, the jacks at the corners of the rig lift the entire rig to free the road wheels and the machine moves to the working position in any direction by the hydraulically operated *otary* device.

This machine may be used to drill caissons in a 12-degree batter, and to install overlapping caissons which form a continuous watertight wall.

B. Concrete placement. The major problems involved in the placement of concrete are segregation and underwater concreting. The problem of segregation of concrete aggregates becomes critical when concrete is dropped a large height such as from the ground surface to the bottom of a caisson.

Many engineers believe that the aggregate would segregate when fresh concrete hits the side of the hole during its fall. Therefore, a tremie or other type of sectional pipe is required to guide the fall of the fresh concrete. The concrete mix is usually made stiff (a slump of less than 3 in.). In any event, concrete hitting the side of the hole would cause sloughing in of the soil and

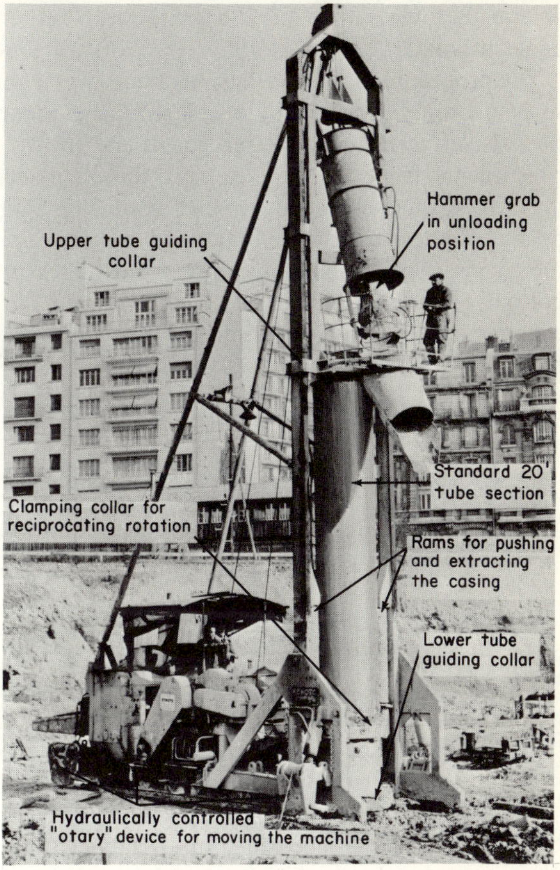

Fig. 9-10 Benoto caisson excavating machine. Photograph courtesy of Benoto, Inc.

that is objectional. In extreme cases, the concrete is deposited by means of a bucket with bottom dump doors. The concrete bucket is generally required when placing concrete under water, as is discussed in next paragraph.

When excavating in granular soils the hole is often cased with steel pipes (casings). This pipe is usually retracted as the concrete is deposited. Caution must be exercised not to pull the casing above the top of the fresh concrete, otherwise, soil may cave in and become sandwiched in the concrete.

Generally, only the concrete in the upper 10 ft or so is vibrated because it is believed that the weight of the fresh concrete will compact the concrete at greater depths. Vibration of concrete in the bell is not desirable because of the risk of introducing cave-in of the sloping side of the bell.

When possible, the caisson should be concreted in the dry. However, when there is a constant inflow of water in the caisson excavation, the seepage pressure may impair the strength of the concrete if the concrete is placed against it. In such cases, the seepage water should be cut off by other means or else the hole should be filled with clean water and the concrete placed under the water.

The procedure of placing concrete under water should be carefully planned so as to avoid the possibility of soil sediments accumulated on the bottom of the caisson before concrete placement and to avoid washing out of cement constituents from the concrete mix during placement. The latter can be avoided by using bottom dump bucket which is simply a pipe with doors that open up upon reaching the bottom of caisson. Another method is to deposit the concrete by a tremie pipe with a bottom-open flap. Upon reaching the bottom of the caisson the flap is opened and the concrete flows out. The concrete is continuously fed into the pipe while the lower end is kept always below the top of fresh concrete in the caisson.

C. Inspection of construction. A drilled caisson construction should be inspected for

1. Accuracy of alignment and dimensions,
2. The bearing capacity of the soil at bottom of the caisson,
3. The removal of loose material from the bottom of caisson,
4. The risk of loss of ground and settlement of the adjacent area,
5. The procedure of concrete placement.

The amount of permissible misalignment of a drilled caisson is generally two to three inches from the required exact location, and one per cent of the height or depth of caisson for plumbness or verticality. Such tolerances should be accounted for in the design procedure, Sec. 9-8.

The adequacy of the supporting soil at the bottom of the caisson should be inspected by spot checking. In a relatively uniform soil, one out of 5 to 10 caissons may be checked, whereas in erratic conditions it may be necessary to check everyone or every other one. The soil condition may be ascertained by taking hand auger samples or driving reinforcing rods at the bottom of the caisson before the bell is reamed out. The approximate strength of the clayey soil from the auger samples may be estimated by the thumb test as described in Sec. 1-8(a). To avoid human factors, a pocket size penetrometer may be used. It also has been proven expeditious and desirable to take tube samples from the caisson by attaching a Shelby tube to the drill rig.

In many instances the cleanness of the caisson bottom may be inspected by the aid of a mirror to reflect the sunlight into the hole. But the most reliable procedure is to examine the condition by descending down to the bottom of the caisson. The safety problem of caisson work is discussed later.

The inspector should always look out for possibility of causing ground settlement in an area adjacent to existing buildings. The settlement may be resulted from loss of ground or from excessive dewatering. By loss of ground it is meant that the soft soil squeezes in the hole and as a consequence the adjacent ground settles. Excessive pumping of water from silty, sandy, or gravelly soils tends to carry away the fine particles and causes the soil to settle. Lowering the water in clayey soil for an extended period of time may also cause settlement due to consolidation. When danger of settlement is confronted, alternative methods should be used, Sec. 9-12B.

9-12 *Problems Concerning Construction of Drilled Caissons*

A. Safety problem. In addition to the precautions required for all construction works the workmen should be guarded against the danger of cave-ins and the danger of gas explosion.

In unstable soils where cave-ins are possible, casing should be lowered in the excavation as the work progresses. Occasionally gas pockets are encountered by caisson workers in the hole. The gas may suffocate the workmen or may explode if ignited by sparks from power tools. The risk of suffocation is reduced if the compressed air is used for operating hand tools in the hole. It is always a good practice to have some emergency provisions on the site, particularly if the site is outside the city.

B. Underwater excavation. Since excessive dewatering from the site may cause ground settlement in adjacent area, it may become necessary to excavate the caissons with the hole filled with water. Underwater excavation may also be necessary when there is danger of a blowout of the caisson bottom. The blow-out is resulted from excessive hydraulic pressure at the bottom if the inside is kept dry. In either case, the concrete is deposited underwater unless the hole can be cased and sealed off.

In unstable soils where cave-in occurs, the use of drilling fluid has been proven successful. The hole is filled with a slurry of clay which exerts a fluid pressure against the side and tends to keep the hole from caving in.

9-13 *Design Example*

A simple example is given in DE 9 to illustrate the basic procedure of design of drilled caissons in clay. In this example, the load on the drilled caisson includes the effect due to negative skin friction.

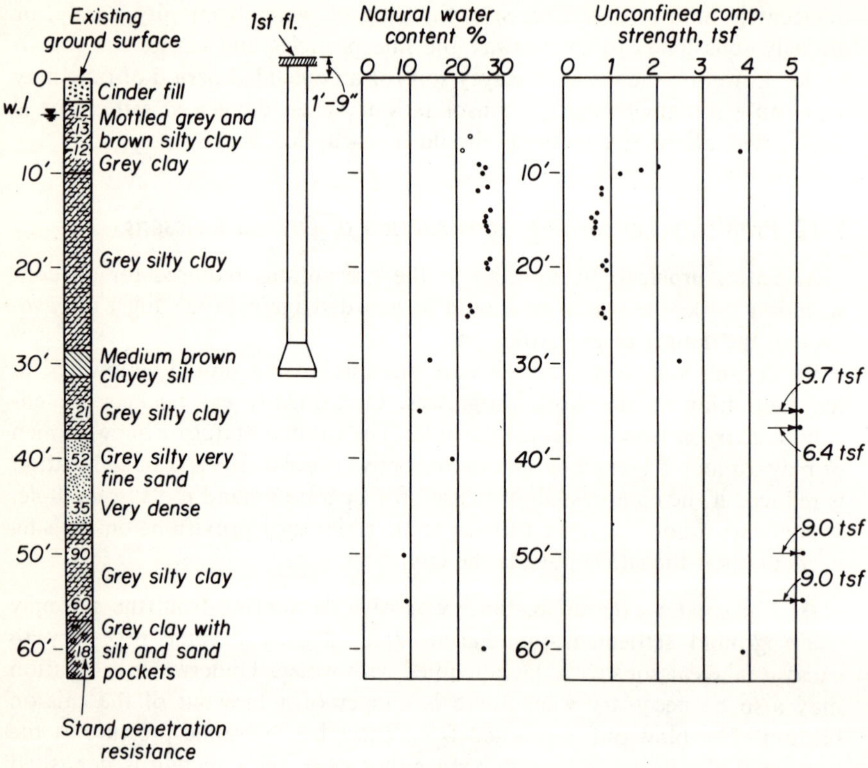
Existing ground surface
1st fl.
1'-9"
Natural water content %
Unconfined comp. strength, tsf
w.l.
Cinder fill
Mottled grey and brown silty clay
Grey clay
Grey silty clay
Medium brown clayey silt
Grey silty clay
Grey silty very fine sand
Very dense
Grey silty clay
Grey clay with silt and sand pockets
Stand penetration resistance
9.7 tsf
6.4 tsf
9.0 tsf
9.0 tsf

Settlement of 1st floor if it rests on fill

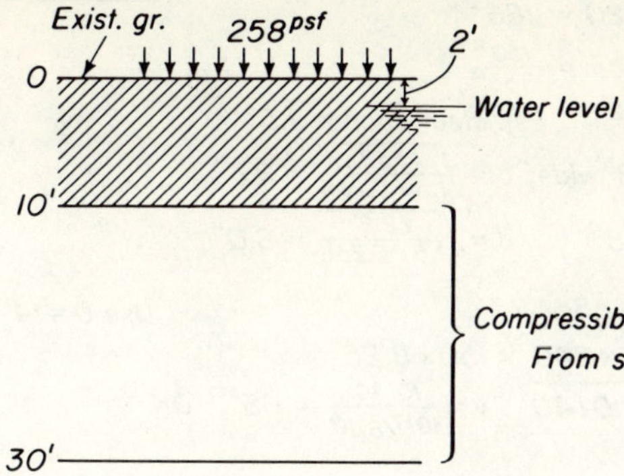

Moist unit wt $\gamma_b = 120\,pcf$
Buoyant unit wt $\gamma' = 65\,pcf$

From soil tests: $w = 27\%$ $\quad e_o = 0.74$

$$c = \frac{1}{2}q_u = 800\,psf$$

$$\frac{C_c}{1-e_o} = 0.14$$

3" floor finish + 6" slab	= 113 psf
1' fill	= 125
	DL = 238
	LL = 20 = $\frac{1}{2}$ design live load as permanent load
	258 psf

p_o at mid-height of comp. layer = 2' x 120 + 18 x 65 = 1410 psf
Δp = 258 psf in the interior
$\Delta p = 258 \times \frac{1}{4}$ = 65 psf at corner of building

$$\text{Settlement } S = \frac{C_c}{1-e_o} H \log \frac{p_o + \Delta p}{p_o}$$

$$= 0.14(240) \log(1 + \frac{258}{1410}) = 2.45''$$

$$= 0.14(240) \log(1 + \frac{65}{1410}) = 0.90''$$

Settlements are excessive, frame first and support it on drilled caissons

Design of 1st floor framing (interior 20' x 20' bays)

Assuming cost comparison indicates that two-way beam-and-slab system is most economical.

Slab:		
LL	= 40	$f_c' = 3{,}000\,psi$
Finish	= 38	$f_s = 20{,}000\,psi$
Slab	= 100	
w	= 178 psf	

$m = 1.0 \quad S = 20.0' \quad -M = 0.033\,wS^2 = 2.35^{1k}/_1 \quad d = \sqrt{\frac{2.350}{0.236}} = 3.2''$

$+M = 0.025\,wS^2 = 1.78^{1k}/_1 \quad d = \sqrt{\frac{1.780}{0.236}} = 2.75''$

Beams: $w' = \frac{1}{3}wS = \frac{1}{3}(178)20 = 1186^{\#}/_{1}$

wt of beam = $\underline{150^{\#}}$
$1336^{\#}/_{1}$

$-M = \frac{1}{11}(1.336)(16.5)^2$ Est. clear span = 3.5'

$= 33.1'^{k}$ Try 18" wide, $d = \sqrt{\frac{33.1}{1.5 \times 0.236}} = 9.7''$

$+M = \frac{11}{16} \times 33.1 = 22.8'^{k}$ $d = \sqrt{\frac{22.8}{1.5 \times 0.236}} = 8.0''$

$\frac{1}{8}(20)^2 \times 178 = 8900^{\#}$ Use D = 14"

wt of beam = $\underline{1240}$ = 150 x 8.25

$V = 10140$ $v = \frac{10140}{18(7/8)10} = 65^{psi}$ OK

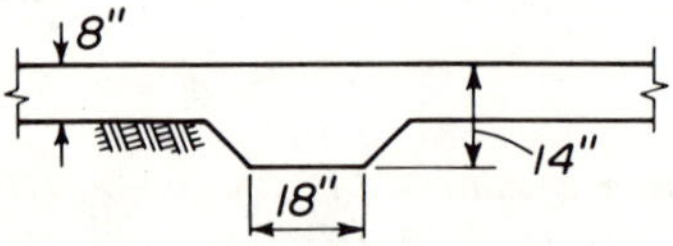

(Design of reinforcing is not shown here)

Load on Interior Drilled Caisson

Load including 1st floor:

$LL = 185^{k}$
$DL = \underline{220}$
405^{k}

From Table 9-2, use 2'-9" ϕ shaft

Negative skin friction:

$= 800 \times 2.75\pi \times 30' = 207^{k}$ adhesion on shaft
$= 125 \times 20 \times 20 = 50^{k}$ wt. of 1' fill in 20' x 20' bay
(Governs)

Total load on bell = 405 + 50 = 455^{k}

Note: 2'-9" ϕ shaft has a design capacity = $462^{k} > 455^{k}$, OK

Bearing Capacity on Silty Clay

Allowable $q = 1.4q_u = 1.4 \times 6.4 = 9.0^{tsf} = 18.0^{ksf}$

Assume some seepage water in excavation, use $q = 12^{ksf}$
From Table 9-1 requires a 7'-0" ϕ bell for 455^{k}

7'-0" < 3 x shaft ϕ, OK

An examination of soil samples from the layer of brown clayey silt indicates that this soil has sufficient cohesion to form bells.

Cap Detail

Column size = 20″ x 20″, 4 – #9 verticals

Min cap size = shaft diam + 2 x tolerance of caisson
= 2′-9″ + 2 x (3) = 3′-3″
Use 3′-6″ sq

Min cap depth = 21 bar diam + 3″
= 21 x 1.128 + 3″ = 27″ from top of slab
14
13 Use 2′-0″

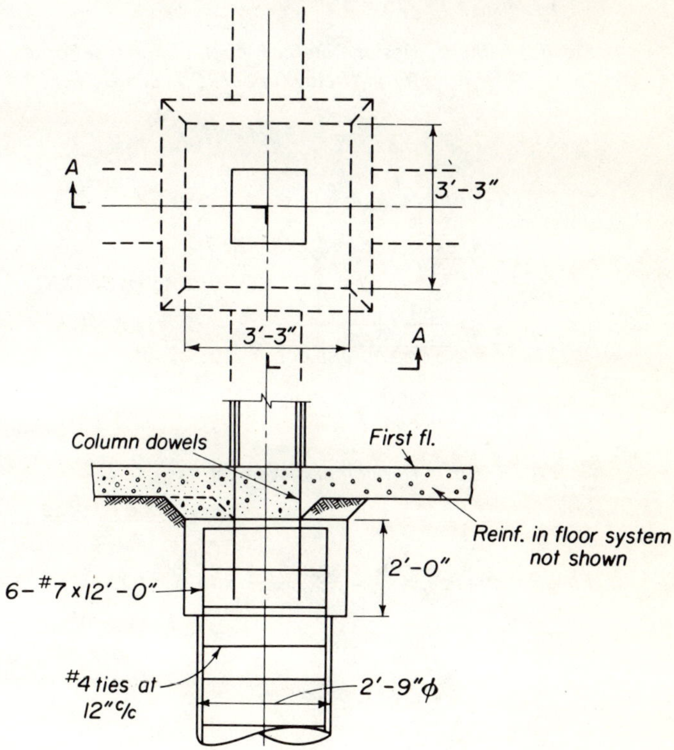

Section A-A

Settlement of Drilled Caisson

From the geological history of the area and tests made on the same soils elsewhere, it is known that the hard clays below the drilled caissons are preconsolidated and will settle very little under building foundations.

Horizontal Force

$$p = 2000^{\#}/\text{drilled caisson}$$

Simplified method: Assume fixity at 15′ below top of drilled caisson

$$M_{max} = 2^{k} \times 15 = 30^{1k}$$

Elastic method: Assume uniform coef. of horiz. subgrade reaction for stiff clay (upper 10′): $k_1 = 60$ pci (Art. 8-17)

$$k = \frac{k_1}{1.5B} = \frac{60}{1.5 \times 2.75} = 15\ \text{pcf}$$

Eq. (9-6a) $M_{max} = \frac{p}{\lambda}$

$$\lambda = \sqrt[4]{\frac{Bk}{4EI}}, \qquad E = 3{,}000{,}000$$

$$I = \frac{\pi}{4}R^4 = 0.78(16.5)^4 = 58{,}000$$

$$= \sqrt[4]{\frac{33 \times 15}{4(3{,}000{,}000)58{,}000}} = 0.0052\ \text{in.}$$

$$M_{max} = \frac{2000}{0.0052} = 385^{11k} = 32^{1k}$$

Eq. (9-6b) Horizontal movement $= \frac{p\lambda}{k}$

$$= \frac{2000 \times 0.0052}{15} = 0.69''$$

Flexural stress $f_b = \frac{M}{S} = \frac{32{,}000 \times 12}{58{,}000} \times 16.5 = 109$ psi

Less than comp. stress under dead load. OK. Use min reinforcing.

Plate Ten

A Floating Caisson being Towed to Site

The word caisson is derived from Latin which means a box or case. In civil engineering, it denotes a large substructure which is built in the dry and sunk into water and/or ground to a predetermined depth, thus becoming the base or foundation of a large structure. Piers and abutments under many renowned bridges have been constructed of caissons. Caissons are also frequently used in wharves and other waterfront structures.

Each job of caisson construction is a large endeavor, and each is a difficult job. A successful job relies upon the best understanding of both the design and construction procedures, and requires the best engineering know-how and ingenuity. This chapter deals with the fundamentals underlying the design and construction procedures.

10
CAISSONS

10-1 Use of Caissons

Generally the following structures are major areas where caissons are used:

1. Bridge piers and abutments in rivers, lakes, etc.;
2. Wharves, quay walls, docks;
3. Break waters and other structures for shore protection;
4. Large water front structures, such as pump houses, subjected to heavy vertical and horizontal loads.

Caisson construction is almost restricted to major foundation works because of the large construction cost. Usually a caisson is advantageous as opposed to other types of deep foundations when any or all of the following conditions exist.

1. The soil contains large boulders which obstruct penetration of piles or drilled caissons.
2. A massive substructure is required to extend to or below the river bed to provide resistance against destructive forces due to floating objects, sand scour, etc.
3. The foundation is subjected to large lateral forces.

If these conditions do not exist, caissons are generally more expensive than other types of deep foundations, such as pile foundations, drilled caissons, and piers constructed within cofferdams. For structures on land, piles, or drilled caissons are generally employed to carry the load to lower strata. In water, if the bearing stratum is not more than 40 to 50 ft below the water level, foundations constructed within cofferdams are generally less expensive.

Where the surface of bedrock is irregular and not suitable for seating caissons, pile foundation or cofferdam construction may be more suitable.

10-2 Types of Caissons

Caissons are classified into three types depending upon the method of construction, Fig. 10-1.

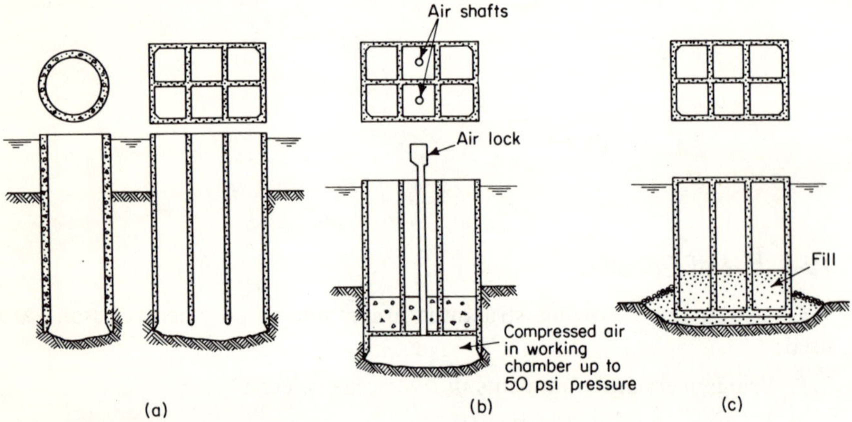

Fig. 10-1 Types of caissons: (a) open caissons; (b) pneumatic caisson; (c) box caisson (floating caisson).

A. Open caissons. The top and bottom are open during construction. The caisson is sunk in place as soil is removed by grab buckets through water. Upon reaching its final position a concrete seal, usually 5 to more than 15 ft thick, is deposited through water. After the concrete is matured, the caisson is pumped dry and filled with concrete.

Advantages of open caissons are:

1. Feasibility of extending to great depths,
2. Relatively low construction cost.

Disadvantages are:

1. Bottom of the caisson cannot be thoroughly cleaned and inspected.
2. Concrete seal placed in water is not as satisfactory.
3. Soil directly under the hauched portion near the cutting edges may require hand excavation by diver.
4. Construction is slowed down if obstruction of boulders or logs is encountered.

B. Pneumatic caissons. The top of a pneumatic caisson is closed, and compressed air is used to keep water from entering the working chamber, thus, excavation and concrete are done in the dry. The caisson is sunk as excavation proceeds. Upon reaching its final depth, the working chamber is filled with concrete.

Advantages of pneumatic caissons are:

1. All work is done in the dry; therefore, control over the work and foundation preparation are better.
2. Plumbness of the caisson is easier to control as compared with the open caissons.
3. Concrete placed in the dry obtains good and reliable quality.
4. Obstruction from boulders or logs can be readily removed. Excavation by blasting may be done if necessary.

Disadvantages are:

1. The construction cost is high due to the use of compressed air.
2. The depth of penetration below water is limited to about 120 ft (50 psi). Higher pressures are beyond the endurance of the human body.

C. **Box caissons** (floating caissons). The bottom of the box caisson is closed. This type of caisson is cast on land. After the concrete is matured, it is launched in water and towed to the site. It is sunk in position by filling the inside with sand, gravel, concrete, or water. Sometimes the caissons are provided with false bottoms (temporary bases made of timber) for floating to the site.

Advantages of box caissons are:

1. The construction cost is low.
2. It is used where construction of caissons at the site is costly or unfeasible.

Disadvantages are:

1. The ground must be level or excavated to a level surface.
2. This type is feasible if the stratum suitable for supporting the caisson is near the ground surface. Deep excavation is costly because the saturated soil tends to flow into the excavation.
3. Provisions must be made to protect against undermining by scour.
4. The bearing stratum is often not very compact.

Caissons may be constructed with any structural material: steel, reinforced concrete, or timber. However, timber caissons are used less frequently than before.

Reinforced concrete caissons utilize concrete to provide the structural strength as well as the weight for sinking. They are often more economical than the steel caissons. However, concrete caissons must be poured in sections (lifts), and the sinking operation must be interrupted while pouring each lift and while waiting for the concrete to mature. Every time the sinking is started from a stationary position, additional effort is required to overcome the static friction. Furthermore, the cyclic operation of stops and starts takes a long time to sink the caisson.

Steel caissons are made of steel skin plate, internal structural steel framing, and concrete ballast or fill. The concrete fill is used only to fill up the space in the steel caissons for the purpose of providing the necessary weight for sinking. As opposed to the concrete caisson, the steel caisson requires no form work and the sinking operation is continuous. Therefore, the construction time and the sinking effort are kept to the minimum. In locations where the water is deep, a large portion of the steel caisson can be floated to site. Figure 10-2 shows an example of a steel caisson.

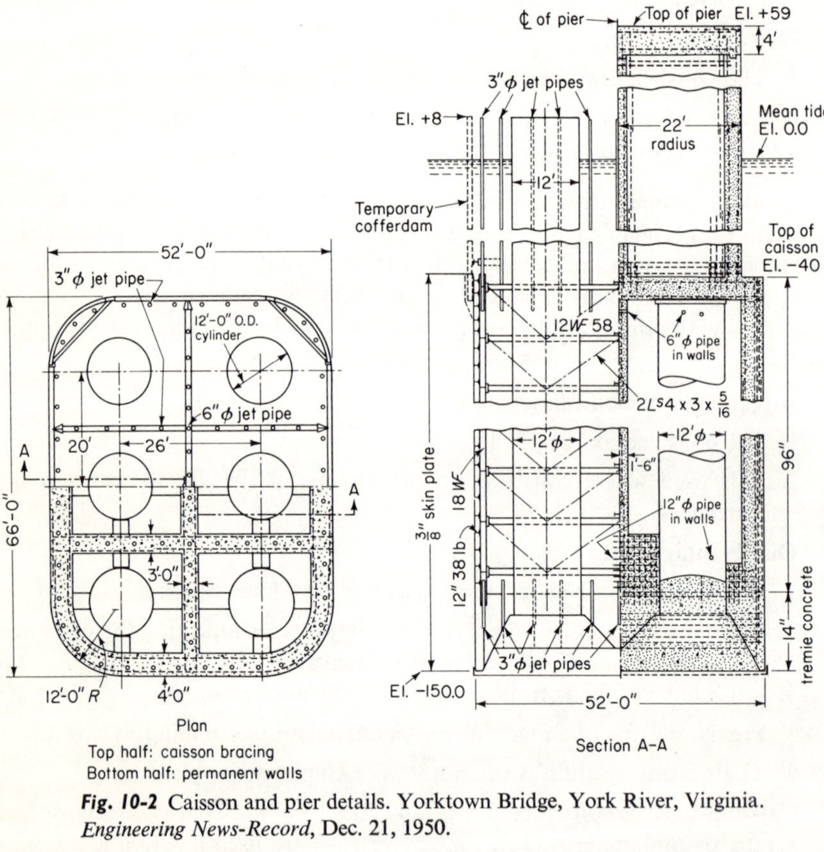

Fig. 10-2 Caisson and pier details. Yorktown Bridge, York River, Virginia. *Engineering News-Record*, Dec. 21, 1950.

10-3 Size and Shape of Caissons

All caissons are constructed with practically straight and vertical sides from top to bottom. The size of a caisson is generally governed by the following factors:

1. *Size of base or bearing plates.* The caisson should be proportioned at least 12 in. wider than the base of the superstructure on each side in order to allow for a reasonable amount of inevitable tilting and misalignment.
2. *Area required for bearing on supporting soil.* See Sec. 10-4 for determination of ultimate or allowable bearing pressure.
3. *Practical limit or minimum size.* Generally a cross-sectional area of about 8 ft × 8 ft is considered the smallest caisson economical and practicable for sinking. Small caissons are frequently more costly than other types of deep foundations.

The shape of caisson cross-section may be circular, square, rectangular, oblong, or any other shape. Generally the shape of a caisson is influenced by the following factors:

1. *Size of caisson.* The shape of caissons supporting large superstructures is governed by the outline of the base of the superstructure. For smaller caissons, however, circular sections are often economical. Sometimes a pair of circular caissons are used advantageously to support the legs of a bridge pier.
2. *Waterway restrictions.* For bridge piers where reduction of channel area may restrict the flow or the navigation, oblong shape may be desired.
3. *Ice floes or other floating objects.* Bridge piers should be made of circular or pointed sections on the upstream side extending to several feet above the high water level.

Caissons have been constructed in small to very large sizes and depths, and there is practically no limit. The caisson for the central anchor pier of the San Francisco-Oakland Bay Bridge was 197 ft × 97 ft in plan and 220 ft below the water. A caisson for the Tacoma (Washington) Narrow Bridge was 225 ft deep, and a caisson for the Mackinac Bridge, 195 ft deep.

Large caissons are always constructed with interior walls for the purpose of stiffening the caisson as a whole and reducing the flexural stresses in the exterior walls and the base or seal. The spaces in between the exterior and interior walls are called dredging wells which should be large enough to allow the operation of dredging buckets. The desirable size for dredging wells are from about 8 ft × 8 ft to 15 ft × 15 ft or larger. Dredging wells smaller than 8 ft are very difficult for excavation.

10-4 *Design of Caissons*

Foundations and substructures in general should be designed with sufficient knowledge of the field conditions and construction procedures. During the construction phase, it is often necessary to adjust the design to suit the field conditions actually encountered. Among all types of foundations, probably the design and the construction of caissons are most intimately related. It would be a gross mistake if the design is made without a thorough study of the contemplated steps of construction, and the difficulties which may occur during construction. For this reason it is not possible to set up a general design procedure which will encompass all the contingent conditions. The following are discussions on principles commonly encountered in the design of caissons.

A. Design loads. Any caisson must be designed to resist two types of loads which act at different times on the caisson, namely:

1. *Permanent loads.* Permanent loads are the maximum vertical and lateral forces acting on the caisson after it is constructed and sunk in place.

(a) Vertical load includes the load from superstructure plus the weight of caisson minus buoyancy force. The latter should be determined at low water level. The total vertical load is assumed to be carried to the bottom of the caisson if it penetrates a relatively shallow depth of soil. Actually a part of the vertical load is transmitted to the surrounding soil by skin friction along the surface of the caisson embedded in soil. For deep caissons, the design is usually made by assuming one-half of the skin friction is effective in supporting the vertical load. This skin friction should be computed for the most critical conditions when the soil is removed to the maximum depth of scour and when the water is at such a level as to produce the maximum net vertical pressure at the bottom of the caisson.

(b) Lateral loads include the forces due to wind pressure on the structure above water or ground surface and on the traffic over the bridge, the tractive force from traffic, ice pressure, and pressure due to current flow. In earthquake zones, the wind pressure should be replaced by the earthquake force.

(c) Lateral forces also include earth and water pressure. The earth pressure is generally taken at the active value. The combined earth and water pressure should be determined for the conditions producing the maximum lateral pressure. Under this lateral pressure, the flexual stresses in the caisson walls may be determined by the method of moment distribution.

2. *Temporary loads.* A caisson is likely to be subjected to large stresses during the construction period. Generally, it is considered desirable to provide the caisson with an ultimate strength to withstand any of the following contingencies.

(a) The caisson is hung up near the top by skin friction. The lower portion of the caisson is then subjected to tension. Sufficient strength should be provided in the caisson to carry the weight of the lower portion.

(b) The caisson is supported on one side only or on two opposite corners only.

(c) The caisson is subjected to unbalanced earth pressure. Under this condition the caisson may be analyzed as a vertical beam or a cantilever.

(d) The caisson is pulled to its correct position. Large racking force and earth pressure would be introduced by pulling.

(e) The caisson is dropped suddenly during sinking.

B. Skin friction. Skin friction is the shearing resistance between the soil and the exterior surface of the caisson encountered during the process of sinking. When possible, the caissons are so designed as to have sufficient weight in each lift to overcome the skin friction. Otherwise, additional ballast is necessary to sink the caisson. Therefore, a reasonable evaluation of skin friction is essential.

Values of skin friction vary within a wide range for each type of soil. Generally, lubrication by water jetting is employed to reduce this friction. Once the caisson sinking is started, the effort in maintaining the motion is smaller. Experience has indicated that the skin friction is not significantly reduced by enlarging the bottom of the caisson. For the purpose of design, the values of skin friction given in the accompanying table (Terzaghi and Peck, 1948) may be used.

Type of soil	*Skin friction (psf)*
Silt and soft clay	150–600
Very stiff clay	1000–4000
Loose sand	250–700
Dense sand	700–1400
Dense gravel	1000–2000

For example, if it is desired to proportion a circular caisson so that no ballast is necessary for sinking, the weight of concrete should be at least equal to the skin friction. Therefore,

$$\tfrac{1}{4}\pi(D^2 - d^2)H\gamma_c = f\pi DH$$

or

$$f = \frac{\gamma_c}{4D}(D^2 - d^2)$$

where f = skin friction, psf;

γ_c = unit weight of concrete (150 pcf above water and 87.5 pcf below water);

D = external diameter of caisson;

d = internal diameter of caisson;

H = depth of penetration.

C. Bearing capacity. Caissons are carried to compact sand, gravel, hard clay, or bedrock, but never to soft soil and decomposed rocks. The ultimate or allowable bearing pressure may be determined by the principles discussed in Secs. 3-3, 6-5, 6-9, 6-10, and 7-6. In reality, a caisson is a rigid mat foundation; therefore, the following equations are applicable.

$$q_{\text{ult}} = 4N^2BR_w + 12(100 + N^2)DR_w' \qquad \text{(granular soils)} \qquad (10\text{-}1)$$

$$q_{\text{ult}} = cN_c \qquad \text{(cohesive soils)} \qquad (10\text{-}2)$$

where q_{ult} = ultimate bearing capacity, psf; divide it by a factor of safety of 2 to 3 to obtain the allowable bearing pressure;

N = number of blows per foot in standard penetration test;

B = width of caisson, ft;

D = depth of caisson, ft;

R_w, R_w' = reduction factors for water level, see Fig. 6-4;

c = cohesion, psf;

N_c = bearing capacity factor for cohesive soils, see Fig. 6-5.

The allowable bearing pressure of caissons on bedrock should not exceed that for the concrete seal. Since the seal is invariably placed in water or in adversed working conditions, the bearing pressure is generally limited to about 500 psi.

D. Concrete seal. When an open pneumatic caisson is sunk to the bearing stratum at the planned depth, concrete is placed to plug the bottom of the caisson. Known as concrete seal or plug, it is actually the permanent base of the foundation. In open caissons the top of the seal is carried to a level several feet above the beveled portion of the cutting edge. In pneumatic caissons, the entire working chamber is filled with concrete.

During the period of construction, the concrete seal serves to seal off the inflow of water while placing concrete above it. The seal may be designed as a thick plate subjected to a unit bearing pressure under the maximum vertical load which is transmitted from the vertical walls of the caisson. Based on theory of elasticity,* the thickness of concrete seal is as follows:

$$t^2 = \frac{3W}{8\pi f_c}(3 + \mu) \qquad \text{(circular caisson)} \qquad (10\text{-}3)$$

---

* For example, Timoshenko, *Theory of Elasticity* (New York: McGraw-Hill Book Co., Inc., 1936). In reality the pressure distribution is subject to change with time due to plastic action and consequent pressure redistribution. However, caisson seals are usually constructed under adverse conditions and should have a conservative design; refinement on the design is seldom warranted.

$$= 1.18R^2\frac{q}{f_c}$$

$$t^2 = \frac{3qb^2}{4f_c(1 + 1.61\alpha)} \qquad \text{(rectangular caisson)} \qquad (10\text{-}4)$$

where t = thickness of concrete seal;
W = total bearing pressure on the base of caisson;
f_c = flexural strength of concrete seal, usually 200–300 psi, but not exceeding 500 psi;
μ = Poisson's ratio = 0.15 for concrete;
R = radius of caisson base;
q = unit bearing pressure against base of caisson;
b = width or short side of caisson;
α = width/length or short side/long side of caisson.

E. General design criteria for concrete caissons. The allowable stresses used for design of heavy foundations are generally lower than those normally used for ordinary concrete construction. The actual stresses may be very low where dead weight of concrete is required for sinking.

The exterior walls of concrete caissons are designed to withstand the flexual stresses caused by the combined maximum lateral pressure and the compressive stress caused by the vertical load including the weight of the caisson. The exterior walls and interior walls should be so designed that a large portion of the vertical load can be transferred from the exterior walls to interior walls, and vice versa. They should be checked for the temporary loads discussed in Sec. 10-4A.

In planning the horizontal construction joints, each lift should preferably be of a sufficient height so that the weight of the caisson will overcome the skin friction and hence no additional ballast is required to cause the caisson to sink.

F. General design criteria for steel caissons. A steel caisson consists of three major components:

Inner shell and outer shell
Steel frames
Concrete fill

The inner shell and the outer shell are made of skin plates $\frac{1}{8}$ in. to $\frac{1}{2}$ in. thick, braced together with steel frames which provide the structural strength and rigidity. The space between the inner and outer shells is filled with concrete in part or full height of the caisson. Thicker skin plate may be desirable in the part where no concrete fill is used. The skin plate is anchored to the concrete fill by means of welded anchor straps. The joints in the skin plate are made water tight by seal weld or by riveting (at 2 to 3 in. spacing) and calking.

The structural frames are made of angles or other structural shapes. The distance between the shells is determined by the required structural strength of the caisson and the required concrete fill to cause the caisson to sink. Greater distance results in deeper structural frames and consequently greater structural strength. Greater distance also provides large space for concrete fill which is necessary for caisson sinking in soils having large skin friction.

In planning the height of lifts, the steel shell should have sufficient height above the water after sinking of each lift.

10-5 Cutting Edge

Except for box caissons which are sunk through water only, the lower ends of caisson walls are made with an inside bevel which reduces the wall thickness to 4 in. to 18 in. at the bottom. The inside bevel is generally made 2 vertical to 1 horizontal. This beveled portion of the wall is called cutting edge, Fig. 10-3.

In concrete caissons, the lower portion of the cutting edge is wrapped with $\frac{1}{2}$ in. steel plates which are anchored to the concrete by means of steel straps.

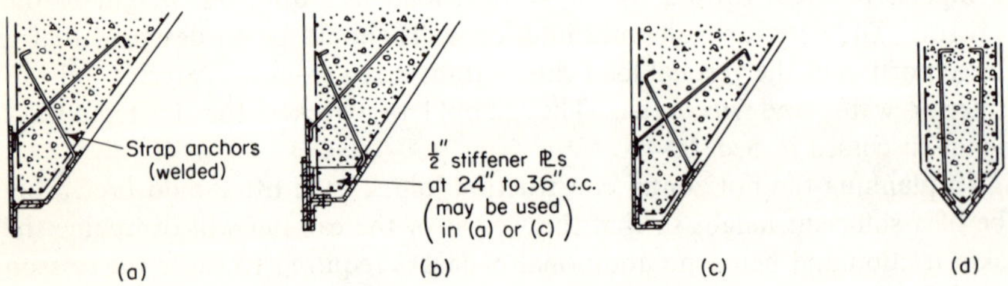

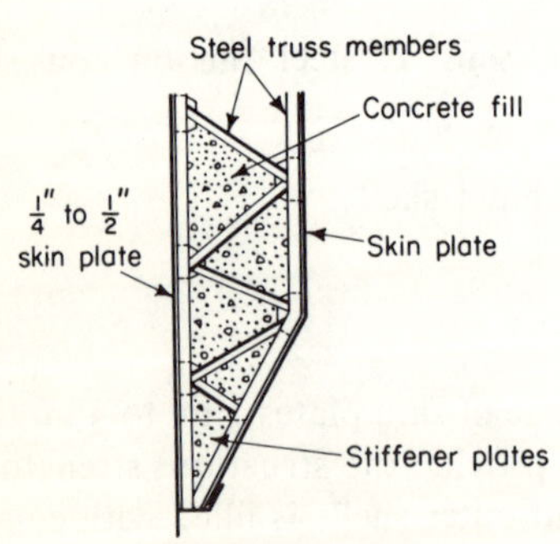

Fig. 10-3 Details of cutting edges: (a) for average condition; (b) for cutting in hard stratum; (c) for penetration in soft soil; (d) for interior walls; (e) cutting edge for steel caissons [may have arrangement similar to (a), (b), or (c)].

The skin plate near the lower end of steel caissons may be reinforced with additional plates, or a thicker skin plate may be used.

A sharp vertical edge is generally provided along the outside face of the caisson. Such edge facilitates the rate of sinking and prevents air leakage in the case of pneumatic caissons. However, it is prone to damage due to obstruction. Therefore, it may be advantageous to use a blunt cutting edge, Fig. 10-3(c), for caissons sunk through soft soil.

The beveled portion of the cutting edge is subjected to a bursting force due to the vertical reaction from the concrete seal bearing against the bevel face. With circular caissons, this bursting force may be resisted by ring tension if the caisson is designed with circumferential reinforcing. This bursting force may be eliminated by making the bevel surface into a series of vertical and horizontal surfaces (inverted steps) where no sloping surface is used to receive the vertical reaction.

10-6 Construction of Caissons

Construction of a caisson is a large engineering operation which requires heavy equipment. In each caisson construction, a variety of usual and unusual problems will be encountered which taps the ingenuity of the construction engineer. The general procedure of caisson construction is discussed before a brief description of some of the usual problems and difficulties.

A. General construction procedure. Caissons may be constructed in slipways, on barges, or on sand islands. Those constructed in slipways are launched and towed to their final location by floating. False bottoms made of wood are necessary for open or pneumatic caissons floated to position. Guide piles are commonly used for sinking the first few lifts of caissons.

Caissons are sunk by their own weight while the soil is being excavated from the dredging wells. As sinking operation progresses, additional sections (lifts) are successively installed. Upon reaching its final depth, the bottom of the caisson is plugged by a concrete seal. In open caissons, this plug is placed in water. The procedure of placing concrete in pneumatic caissons is discussed in Sec. 10-8.

When hard, cemented material is encountered, underwater blasting may be necessary. The charges of explosives are placed by divers. Divers may also be required to remove boulders, logs, or other obstructions.

The excavation of soil in open caissons is done by dredging with grab buckets or similar equipment. The volume of excavated soil always exceeds the volume of the embedded portion of the caisson, because the soil outside the cutting edge tends to flow into the excavation. In granular soils, the excess may be 100 per cent of the volume displaced. Grab buckets do not reach the area near and below the cutting edge; therefore, divers may be required to remove the soil under the cutting edge by hand if the soil does not flow into the excavation.

The sinking operation for concrete caissons must be stopped during the time of casting and curing.

After the concrete seal is matured, the water in the wells is pumped out (open caissons) or the air pressure is released and the equipment removed (pneumatic caissons). The top of the concrete seal should be cleaned and free of laitance or soils before placing concrete above it. A progress report should be kept to include the location, the elevation of the cutting edge, the amount of tilting or misalignment, and all other unusual events occurring during the construction period.

B. Jetting and lubrication. To facilitate the sinking, the exterior surface of caisson is sometimes applied with a film of grease, and/or the jetting is used. Jet pipes $1\frac{1}{2}$ in. to 2 in. diameter with nozzles are cast in the concrete, usually one series of jet pipes is provided on the sloping surface immediately above the cutting edge, and one or two series on the periphery of the caisson at several feet above the bottom of the cutting edge. All jets are arranged symmetrically to induce straight sinking. Since fixed jet pipes can readily become plugged, movable jets have been found more efficient. Eight inch diameter wells may be cast in the concrete for inserting movable jet pipes for inside jetting.

C. Tilting. Caissons are never sunk perfectly straight and true to position. A certain amount of deviation from the planned location should be permitted. For a deep caisson, the actual center may be 12 in. from the required location. It is important to keep the caisson in the vertical position during the entire process of sinking. As soon as it is 2 in. or 3 in. off center, corrective measures should be taken:

1. Excavating the high side ahead of the low side, but not stopping excavation on the low side;
2. Dredging on the outside of the high side;
3. Jetting on the outside and inside of the high side;
4. Pulling the caisson, attach cables to a deadman or dolphin and apply tension as the sinking proceeds.
5. Blocking under the cutting edge on the low side, this can be done readily in pneumatic caissons.

It should be noted that it is impossible to plumb a caisson without lowering it as a whole.

D. Placing concrete underwater. Concreting underwater should be carefully conducted to prevent honeycomb structure, and segregation between aggregates and cement paste. Underwater concrete is usually placed by the following methods.*

1. *Tremie pipes.* A tremie is a watertight pipe with a diameter of 6 to 10 in. The pipes are made in detachable lengths about 6 ft long. It extends from above the water level to slightly below the surface of the fresh concrete on the caisson bottom. The concrete is fed into the pipe by a hopper. A full column of fresh concrete is maintained in the pipe as the concrete is flowing out from the bottom. Sometimes a valve is equipped at the lower end of the tremie. In the absence of a valve, the lower end is plugged with cloth or burlap sacks. When the pipe is lowered into position and concrete fills the pipe, the plug is forced out by the weight of the concrete.

2. *Buckets.* A bucket is a large pipe with a bottom-dump door or doors. It is filled with concrete which is covered with a canvas or a lid and is lowered into the caisson. Upon reaching the bottom, the door opens and the concrete flows out.

3. *Grouted aggregates.* Coarse aggregate is first placed in the space where concrete is desired. Cement grout is then pumped into the voids through pipes extending to the bottom of caisson. The caissons for the Mackinac Bridge Piers are concreted by this method.

4. *Pumped concrete.* Pipe lines with plugged ends were lowered to the floor of the caisson bottom. Concrete pumped into the pipes forces out the plug. Pumping of concrete is continued until the pressure is too great. The pipes are then raised a foot or two and pumping continues again. The procedure is repeated until the concrete reaches the desired level.

5. *Concrete bags.* Canvas bags having one-half or two-thirds full of cement are lowered by divers. This method is expensive and not commonly used in caisson seals.

10-7 Sand Island Method

A sand island is simply an artificial island created at the site to provide a dry area for construction, Fig. 10-4. It is an additional construction expense but it offers many advantages:

1. The work can proceed without interruption as the top of the sand island is carried above the high water level.
2. The alignment of caisson during sinking is under better control.
3. The operation of construction is facilitated because of better access all around the caisson.

* *Concrete Information ST-12* (Chicago, Ill.: Portland Cement Assoc.).

Fig. 10-4 Cutting edge of a caisson being formed on sand island. Photograph courtesy of Cincinnati Gas & Electric Co.

4. The danger of blowup under the cutting edge is lessened during the sinking period.

The size of the sand island should be sufficient to provide working area around the caisson and to allow passage of construction equipment. The sand island method may be described in the following steps:

1. Sink a wooven willow mattress to the river bottom to cover an area considerably larger than the sand island.
2. Construct a timber staging (work platform) around the periphery of the intended sand island.
3. Install the steel shell. The shell may be sheet piles driven into ground, or $\frac{1}{2}$ in. thick steel plate fabricated in sections, about 10 ft long, joined together by flange angles, and lowered onto the mattress.
4. Cut the mattress along the inside face of the shell, and remove the inside mattress.

5. Fill the shell with sand.
6. Salvage the steel shell after the caisson and the construction is completed.

The purpose of placing a willow mattress around the steel shell is to provide protection against scour under the shell. A sand island restricts the flow in a river channel, consequently, it tends to aggravate the danger of scour. In some cases, the sand island was subjected to washout as a result of excessive scour. In such extreme cases, the river bed should be protected with large riprap in addition to the mattress.

10-8 Pneumatic Caissons

A. Essential parts of pneumatic caissons. The essential parts of a pneumatic caisson are shown in Fig. 10-5 and discussed below.

1. *Working chamber.* The working chamber is the space at the bottom of the pneumatic caisson surrounded by the beveled walls of the cutting edge and roofed by a concrete plug. Since it is the space for workmen to excavate the soil, it should be at least 8 ft high. The side walls and the roof are designed to withstand the maximum air pressure anticipated. This air pressure in

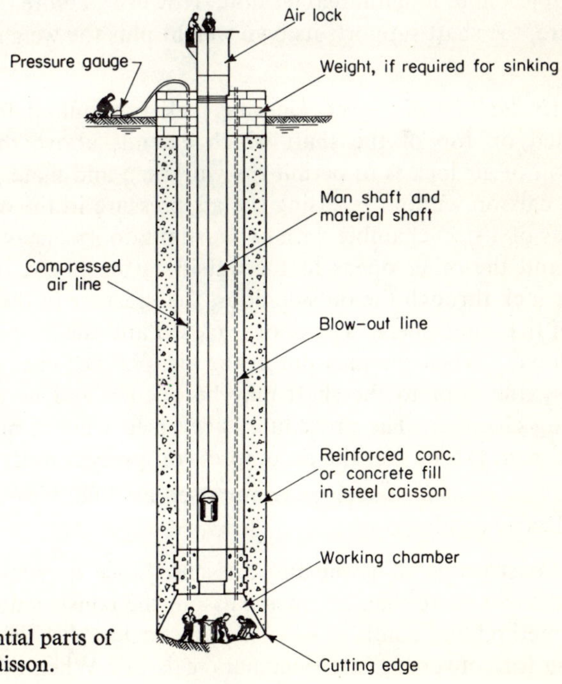

Fig. 10-5 Essential parts of a pneumatic caisson.

general is slightly greater than the pressure due to the head of water above the bottom of caisson. The chamber roof should also be strong enough to sustain the weight of concrete fill above it when no air pressure is used in the chamber. It should also be strong enough for the contingency when the roof may rest on the soil directly and takes a part of the vertical reaction.

2. *Air shaft or shafts.* Air shaft or shafts are the vertical passage ways for workmen and material. In smaller caissons, one shaft may be sufficient. With a ladder in the middle, one side of the shaft is for workmen and the other side for the material (removal of soil and placement of concrete). In larger caissons, however, two or three shafts are often provided, one for passage of workmen, the other or others for the material.

Shafts in concrete caisson may be made of steel or may be simply a cylindrical hole. In the latter case, the concrete must be reinforced to withstand the internal pressure due to the compressed air, and the air lock on top of the shaft must be anchored down to the concrete. Steel shafts are fabricated in short sections of 5 to 10 ft lengths. The joints between the sections are made airtight with rubber gaskets. As the sinking progresses, lengths of the shaft are added. During the time of adding shaft sections, the bottom of the shaft or the top of the working chamber is sealed with a steel plate.

The shaft must be designed to withstand the internal pressure, therefore the common cross-sections are circular, eliptical or figure eight. The shaft is also subjected to longitudinal tension. However, before the application of air pressure, the shaft supports its own weight plus the weight of air lock on top of it.

3. *Air lock or air locks.* One air lock is required for each shaft. It is mounted on top of the shaft which extends above the water level. The function of air lock is to permit the workmen and material to go in and out of the caisson without releasing the air pressure in the caisson. An air lock consists of a steel chamber with two airtight doors one of which opens to the shaft and the other opens to the outside atmosphere. When a man enters the air lock through the outside door, the pressure in the air lock is equal to that of the atmosphere. The door is closed and the air pressure is allowed to rise slowly. When the pressure in the air lock becomes equal to that in the caisson, the door to the shaft may be opened and he may descend to the working chamber. The procedure is reversed when a man comes out from the caisson, except that the decompression process must be done slowly.

4. *Other equipment* such as compressed air line, blow-out line, telephone, and electric conduits.

B. Construction of pneumatic caissons. Since a pneumatic caisson costs several times more than an open caisson, the construction should start with open method and continue as far as practicable. Provisions are made in the caisson for converting into pneumatic caisson. When the condition requires the pneumatic operation, the air shafts are lowered into the dredging wells, and the roof of the working chamber is formed by a concrete plug poured through water, or a steel diaphragm lowered into the dredging wells and bolted by divers to the bottom of the shaft. After the concrete is matured, the compressed air is applied and the water is forced out from the working

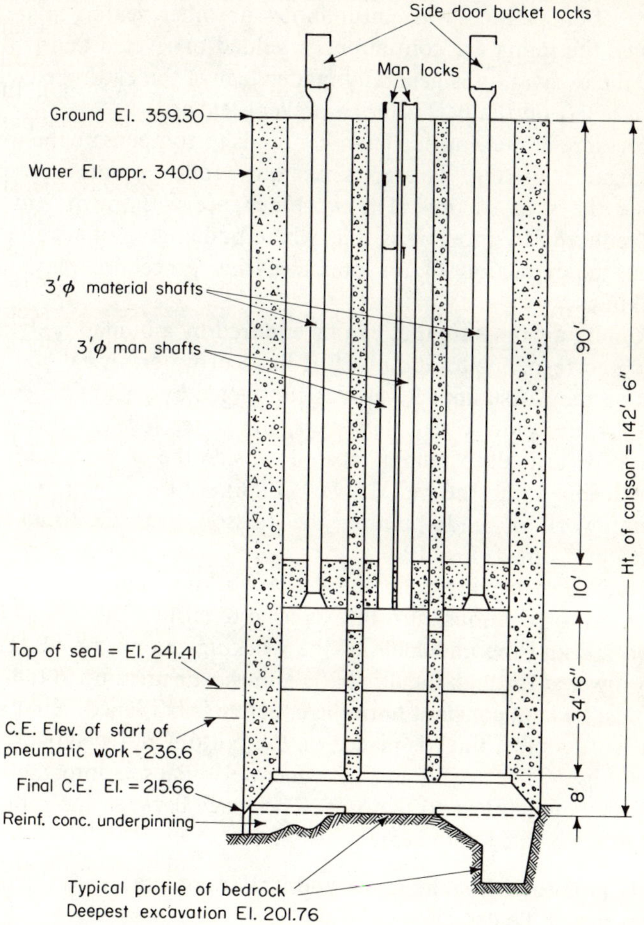

Fig. 10-6 Vertical section of a pneumatic caisson with unusual underpinning work. *Engineering News-Record,* May 12, 1955.

chamber. Then workmen can enter the air locks and descend to the working chamber.

The material is excavated by hand tools and is hoisted up in buckets through the air shafts. In granular soils the excavated material can be disposed by the blow-out method: A hose is connected to the blow-out pipe which rises from the working chamber to the atmosphere, Fig. 10-5. Upon opening the valve which is attached to the end of the hose, the granular material is blown out through the hose and pipe by the high air pressure inside the working chamber.

Pneumatic caissons must be made airtight. In concrete caissons, all construction joints must be carefully made as airtight as possible. The inside surfaces are often painted with bituminous or other sealing material. In steel caissons, the joints are continuously welded or riveted and calked. In either case, the excavation is generally made clear of the cutting edge, thus a berm of soil is left on the perimeter to prevent leakage.

The air pressure in the working chamber tends to compensate the weight of the caisson against sinking. To start sinking, the following procedure is used to overcome the skin friction. The workmen are withdrawn and the air pressure is reduced. In impervious soils where building up of a water head is slower than the reduction of air pressure, this procedure may start the caisson sinking.

The maximum air pressure that can be endured by a human body is about 50 psi which corresponds to about 120 ft head of water. A caisson under a bridge pier on the Mississippi River was conducted by pneumatic method to a depth of 145 ft below water. However, the water level in this case was lowered first by large deep pumps located outside the caisson, and later by air siphons dropped into the well (Newell, 1956). In the same job, a difficult and unusual work of underpinning the caisson over the cracks in the bedrock was performed, Fig. 10-6.

The procedure of placing concrete in the working chamber is of utmost importance. All precautions must be exercised to ensure full contact between the concrete fill and the underside of the working chamber roof. The fresh concrete is lowered through the air shafts and the air pressure in the working chamber must be kept constant until the concrete is hardened. First a slab of concrete about two feet thick is placed on the bottom and well packed under the cutting edge. Then a stiff mix of concrete is packed into the working chamber and brought close to the roof. The space between the roof and the concrete surface is filled with cement grout.

C. Safety problems. For the safety and welfare of workmen, the following precautions should be exercised.

1. *Accurate control of air pressure.* A gauge tender should watch the pressure gauge constantly, and the gauge should be accurate and in good working condition.
2. *Sufficient air circulation.* To avoid the air in the working chamber becoming stale, fresh air must be circulated into the working chamber constantly. This may be done by opening a valve in the air lock. In granular soils where certain amount of leakage takes place through the cutting edge and the soil, the air is automatically circulated.

3. *Slow decompression.* Men working under compressed air must be decompressed slowly. If coming out too fast, they are subjected to caisson disease. This disease is due to air bubbles formed in the blood and body tissues which are compressed while working under pressure. A period of about one-half hour is necessary for decompression from a pressure of 50 psi.
4. *Duplicate and spare equipment.* A spare or duplicate set of air compressors and other equipment for pneumatic operation should be provided in case of contingency.

10-9 Box Caissons (Floating Caissons)

Box caissons are generally made of reinforced concrete and floated to position after the concrete is matured.

A. Stability during floating. A box caisson should have at least a five feet free board above water and should be sufficiently safe against the danger of tipping or capsizing. The stability of a floating caisson can be analysed by the principles of hydrostatics. Figure 10-7(a) shows a caisson in equilibrium where point *c* is the center of gravity of the caisson and point *b* is the center of gravity of the displaced water, or center of buoyancy. If the caisson is tilted as shown in Fig. 10-7(b), point *c* remains at the same location with respect to the caisson itself, but the center of buoyancy is changed because the volume of displaced water is different. The point of intersection between the vertical line passing through point *b* and the center line of caisson itself is called the metacenter. The caisson is stable if the metacenter is located above point *c*. Otherwise the caisson is unstable and the proportion should be revised, or ballast should be used in order to prevent tipping of the caisson during floating. Caissons may also be floated in horizontal or inverted position and, upon reaching the final location, careened to the normal position.

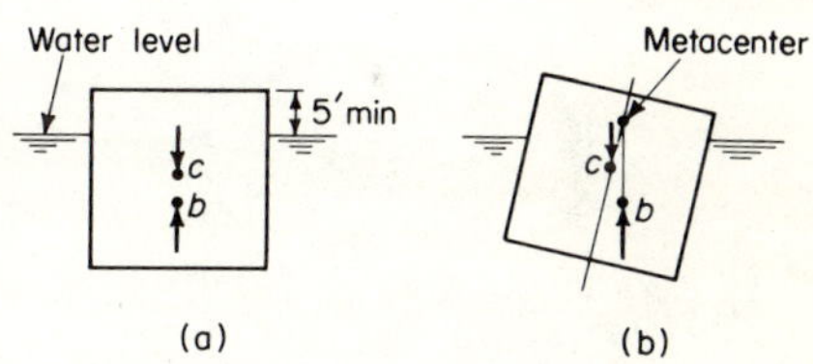

Fig. 10-7 Stability of floating caisson.

B. Design forces. Floating caissons must be designed to resist the following forces.

1. *Permanent forces.* These include loads from the superstructure, earth pressure, wave pressure, etc.
2. *External pressure.* This is the water pressure during floating. If the caisson is to be floated in rough water, it should be designed as a ship, and internal strutting may be desirable. In such cases, the caisson may be subjected to stresses due to hogging (the caisson is supported by a

wave in the middle), and sagging (two waves one at each end), torsion, and towing force.

3. *Internal pressure*. The caisson is subjected to maximum internal pressure exerted by sand and gravel fill at low water stage (box caissons are seldom filled with concrete).

C. Base preparation. The foundation bed upon which the caisson will rest should be excavated to a level surface. Unsuitable material should be dredged out before sinking the caisson. The toe of the caisson should be protected from scour action by depositing ripraps of sufficient sizes.

PART 3
RETAINING STRUCTURES

Plate Eleven

A Counterfort Retaining Wall Partly Backfilled

Generally, retaining walls are classified into five types: gravity walls, cantilever walls, counterfort walls, buttressed walls, and crib walls. Each type must be designed to satisfy the following criteria: (1) the structural strength of each component or section of the wall and (2) the stability of the wall as a whole including the foundation soil. Furthermore, it should include the selection and compaction of backfill material and provision for backfill drainage. All these criteria are presented in this chapter.

11 RETAINING WALLS

11-1 *Common Uses of Retaining Walls*

A retaining wall is a wall constructed for the purpose of supporting a vertical or nearly vertical earth bank which, in turn, may support vertical loads. It may also be used to retain water or other materials such as coal, ore, etc. It differs from other types of retaining structures because it does not require external bracing for stability. For this reason, retaining walls have been widely used in a variety of purposes, Fig. 11-1.

Occasionally retaining walls are constructed with external supports. For example, the toe of a retaining wall may be poured in direct contact with, or strutted to, a massive substructure from which the retaining wall derives a lateral support. When a retaining wall is constructed on top of sound bedrock, the base of the ball may be anchored to the bedrock by means of anchor rods grouted in drilled holes.

11-2 *Principal Types of Retaining Walls*

Retaining walls may be classified in six principal types shown in Fig. 11-2. These types are sometimes varied and modified to achieve the best economy. For example, the Lincoln Tunnel retaining wall designed by the Port of New York Authority consists of concrete piers poured in place with slots in the sides. Precast, prestressed panels 12 and 18 in. thick and 4 ft wide were placed into these slots to form the face of wall. For another example, the counterfort type retaining wall at the U.S. Air Force Academy at Colorado Springs, Colorado was modified to have two parallel continuous footings; one under the face of wall and another at the end of the counterfort (Teng, 1957).

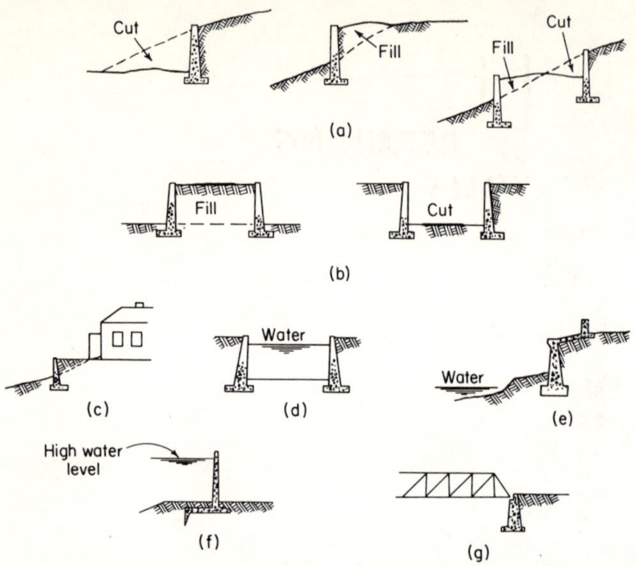

Fig. 11-1 Common uses of retaining walls: (a) hillside roads; (b) elevated and depressed roads; (c) landscaping; (d) canals and locks; (e) erosion protection; (f) flood walls; (g) bridge abutment.

When a retaining wall is used to support the end of a bridge span as well as retaining the earth backfill, it is called an abutment.

11-3 Design of Retaining Walls

The procedure of retaining wall design generally comprises of the following steps:

1. Assemble the general information: topographical and physical surveys, controlling dimensions.
2. Analyse the subsoil conditions: soil profile.
3. Establish surcharge loads: railway, highway, building and other structures, earthquake force.
4. Select type and tentative proportion of wall–Sec. 11-4.
5. Compute earth pressure and surcharge pressure–Sec. 11-5.
6. Analyse the structural stability–Sec. 11-6A.
7. Analyse the foundation stability–Sec. 11-6B.
8. Design structural elements–Sec. 11-7.
9. Select drainage in backfill–Sec. 11-8.
10. Predict settlement and movement of wall–Sec. 11-9.

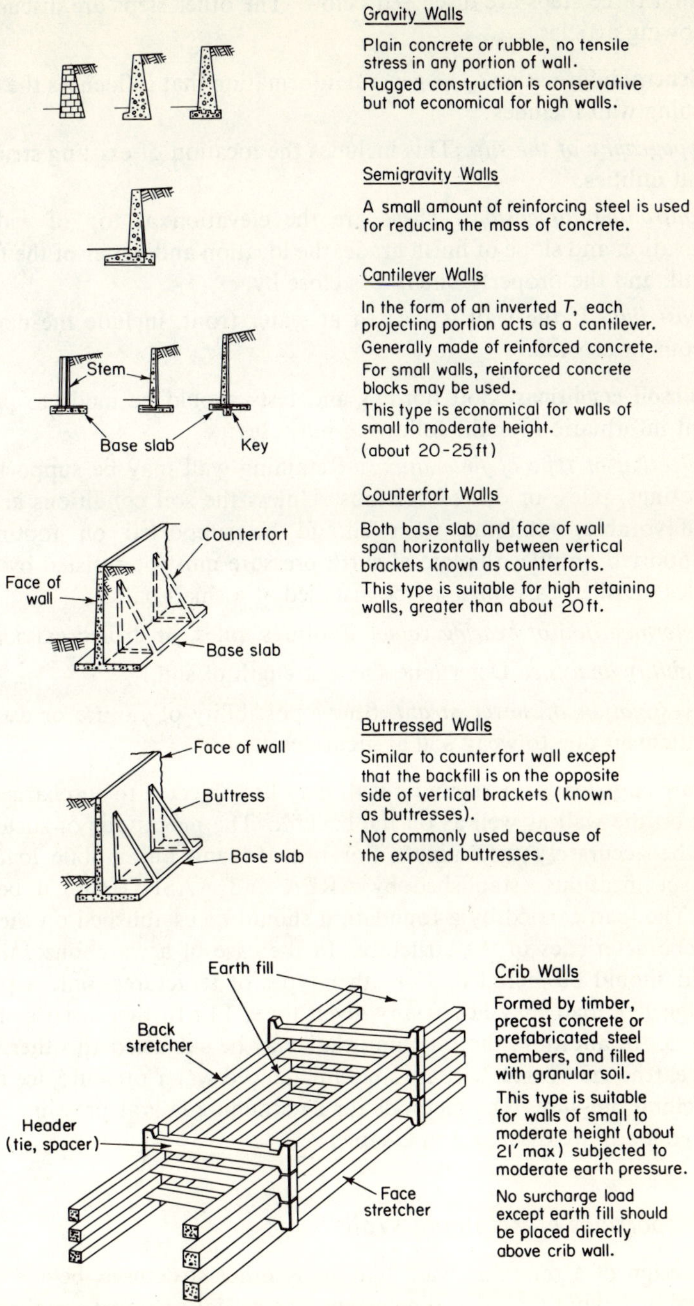

Fig. 11-2 Principal types of retaining walls.

The first three steps are discussed below. The other steps are discussed in the following articles.

A. General information. The general information that influences the design of retaining wall includes:

1. *Topography of the site.* This includes the location of existing structures and utilities.
2. *Controlling dimensions.* These are the elevation at top of wall, the elevation and slope of finish grade, the location and batter of the face of wall, and the property line if it is close by.
3. *Frost line.* If the wall is located at water front, include the depth of scour or erosion.

B. Subsoil conditions. Soil borings and tests should be made to provide sufficient information for the following purposes:

1. *Selection of type of foundations.* Retaining wall may be supported on footings, piles, or drilled caissons. Unless the soil conditions are very unfavorable, retaining walls should be supported on footings. If supported on piles the lateral earth pressure must be resisted by batter piles. Drilled caissons may be installed at a small batter.
2. *Determination of bearing value.* Footings, piles, or drilled caissons.
3. *Stability analysis.* Determine shear strength of soil.
4. *Investigation of lower strata.* Study possibility of failure or excessive settlement due to weak soil at great depth.

C. Surcharge loads. A retaining wall may be subjected to surcharge loads directly on the wall as well as on the backfill. The magnitude of such loads should be accurately determined. For railroad and automobile loads the design specifications established by AREA and AASHO should be consulted. The load carried by a foundation should be established on the basis of the characteristics of the structure. In the case of a warehouse, the full live load should be included. For other types of structures, only a part of the design live load may act at any given time. For further discussions, see Chapter 3. In addition, the retaining wall may be subjected to other forces such as earthquake force, seepage and unbalanced water pressure, ice thrust, and swelling pressure. The methods for determining lateral pressure due to surcharge loads are discussed in Chapter 4.

11-4 Proportions of Retaining Walls

The design of a retaining wall, like many other structures, begins with a trial section having tentative dimensions. This trial section is analysed for stability and structural adequacy. It is then revised to obtain the most satisfactory proportions under the given design conditions. The following general information may be of help for selecting the trial section.

Gravity walls. Gravity walls are primarily trapezoid in section, possibly with the base projecting beyond the face and back of wall. The projection may be only 4 to 6 in. beyond the back. The toe projection may extend further for the purpose of reducing the soil pressure. The top of wall should not be made less than 8 in. wide to allow proper placement of concrete. For ordinary cases the base width ranges between 30 per cent and 40 per cent of the height of wall.

Cantilever walls. The proportions of cantilever walls encountered under normal conditions are shown in Fig. 11-3. Since retaining walls are designed for active earth pressure which is associated with a small amount of tilting, it is advisable to provide the face of the wall with a small batter to compensate for the forward tilting. Otherwise a small amount of forward tilting may give the illusion of instability of the wall.

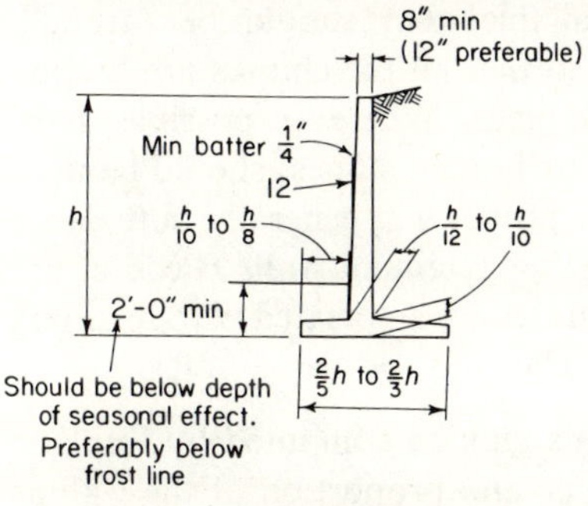

Fig. 11-3 Common proportions of cantilever retaining walls.

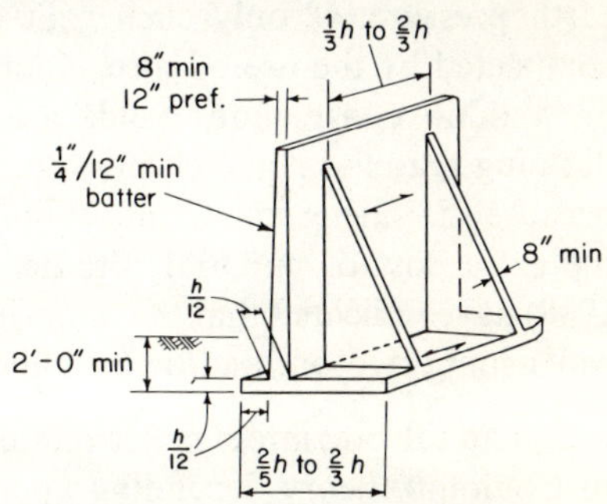

Fig. 11-4 Common proportions of counterfort retaining walls.

Counterfort walls. The proportions of counterfort walls, Fig. 11-4, vary to a greater extent than that of cantilever walls because the thickness of face and base slabs depends primarily on the spacing of counterforts. For walls of moderate height, the counterforts may be spaced as far as two-thirds of the height of wall. For walls higher than about 30 ft the spacing may be reduced to less than one-half of the height. From the construction point of view, counterforts should not be placed on a spacing less than about 8 ft. The toe projection is generally smaller than that for cantilever walls.

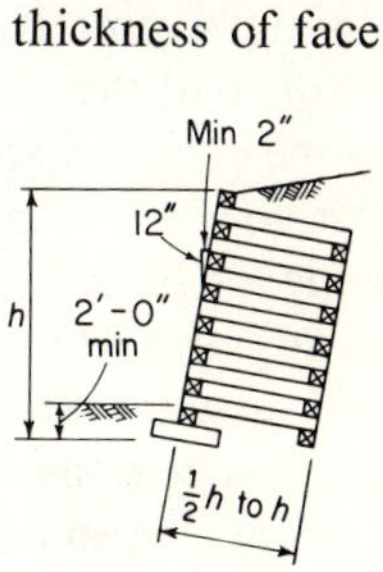

Fig. 11-5 Common proportions of crib walls.

Crib walls. Crib walls are usually constructed with a

minimum batter of 2 in./12 in. The base width or depth of wall ranges from 50 per cent to 100 per cent of the height of wall. If precast and metal crib walls are purchased from standard sizes, the manufacturers' catalogs should be consulted before making the design analysis. Figure 11-5 shows the common proportions of crib walls.

11-5 Earth Pressure Computation

A satisfactory earth pressure computation must be made on the basis of properly determined soil properties, and correct use of the earth pressure theory.

1. Since earth pressure varies considerably with the type of backfill, a careful examination of the contemplated backfill material is warranted. In Sec. 4-6, the procedure for determination of soil properties for earth pressure computation is discussed. Soils containing a large amount of clay, silt, or organic matter should be avoided because such soils exert excessively large earth pressure. If only such soils are available, they should be carefully compacted by the use of sheepsfoot roller so that all the chunks are broken up and no conspicuous voids are left in place. Wherever possible, free-draining granular soils (clean sand, gravel, or broken stones) should be used because the saving in cost of retaining wall proper will generally outbalance the extra cost of material. Studies of existing retaining walls (Peck et al., 1948) have indicated that the majority of failures are associated with retaining walls supported on, and/or backfilled with clay, silt, or clayey soils.

2. Lateral pressure against retaining walls may be computed by Rankine or Coulomb theory depending upon the type and proportion of the wall as well as the configuration of the ground surface. The Rankine theory deals with lateral pressure against a vertical plane *ab* inside the soil mass, Fig. 11-6(a), provided that the ground surface is a plane and that the shear zone, or sliding surface, *bc* is not obstructed by the stem or the back of the wall. Normally, this is the case of cantilever walls and counterfort walls. The position of the sliding surface *bc* in terms of the angle α may be determined by the equation shown in this figure.

In the case of gravity walls with small or no heel projection, the soil slides along the back of the wall. Therefore, the wedge theory using a plane surface of sliding (Coulomb theory) may be used to determine the pressure on the back of the wall. The angle of wall friction and the value of wall adhesion C_a must be estimated prior to the computation of earth pressure, Sec. 4-1(b).

Many retaining walls are so proportioned that the sliding surface is limited by a plane connecting the heel and the top of wall, as shown on the right side of Fig. 11-6(b). In this case, the wedge method can be used to determine the earth pressure on any section through the heel of the wall, commonly on the vertical section, *ab*. The angle φ' must be known in order to apply the wedge method. The φ' value ranges, in normal cases, from 0 to full value of φ, using $\frac{1}{3}$ to $\frac{2}{3}\varphi$ in the majority of cases. For plane ground surface the value of φ' may be assumed to be equal to the slope of ground β.

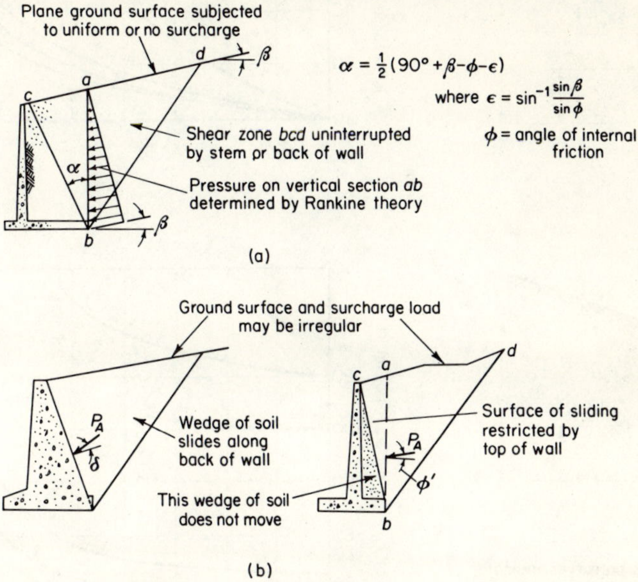

Fig. 11-6 Application of earth pressure theories in retaining wall design: (a) Rankine theory applies; (b) wedge theory applies.

For retaining walls less than 20 ft high, the empirical charts shown in Fig. 11-7 may be used.

11-6 Stability of Retaining Walls

A. Structural stability. The retaining wall as a unit should be proportioned to have the following minimum safety factors:

$$\text{Factor of safety against overturning} = 1.5 \quad \text{(for granular backfill)}$$
$$= 2.0 \quad \text{(for cohesive backfill)}$$
$$\text{Factor of safety against sliding} = 1.5$$

1. The lateral pressure due to the backfill and surcharge tends to tip the retaining wall over about its toe. This overturning moment is stabilized by the weight of the wall and the weight of the soil above the base of the wall, Fig. 11-8. The common practice is to proportion the wall so that the total

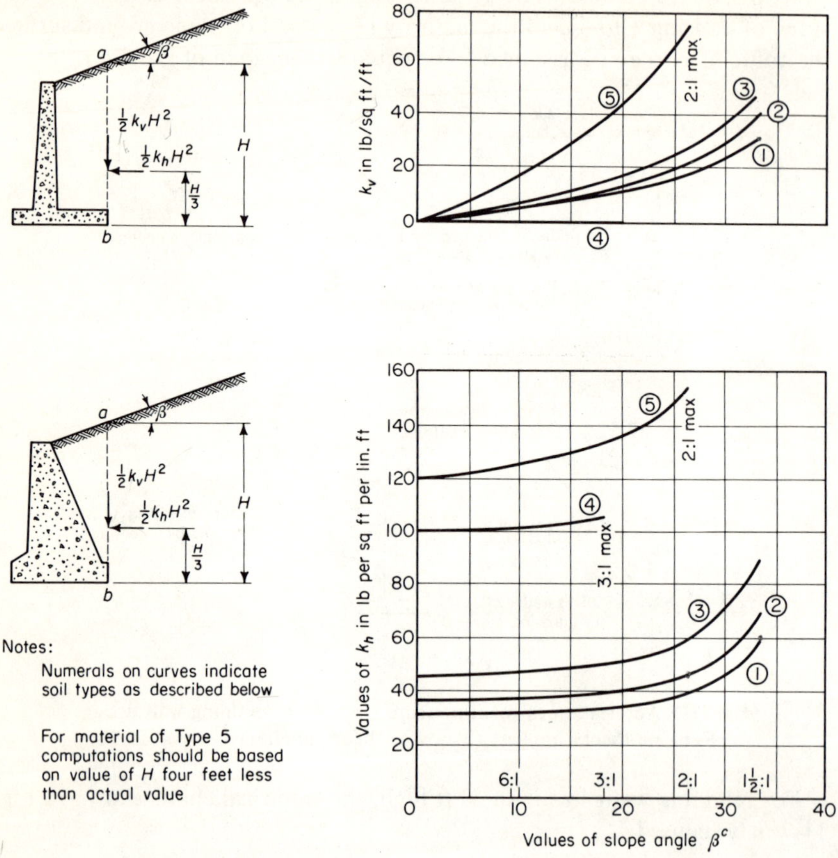

Types of backfill for retaining walls

① Coarse-grained soil without admixture of fine soil particles, very free-draining (clean sand, gravel or broken stone)

② Coarse-grained soil of low permeability due to admixture of particles of silt size

③ Fine silty sand; granular materials with conspicuous clay content; or residual soil with stones

④ Soft or very soft clay; organic silt; or soft silty clay

⑤ Medium or stiff clay that may be placed in such a way that a negligible amount of water will enter the spaces between the chunks during floods or heavy rains

Fig. 11-7 Earth pressure charts for retaining walls less than 20 feet high. From AREA Manual.

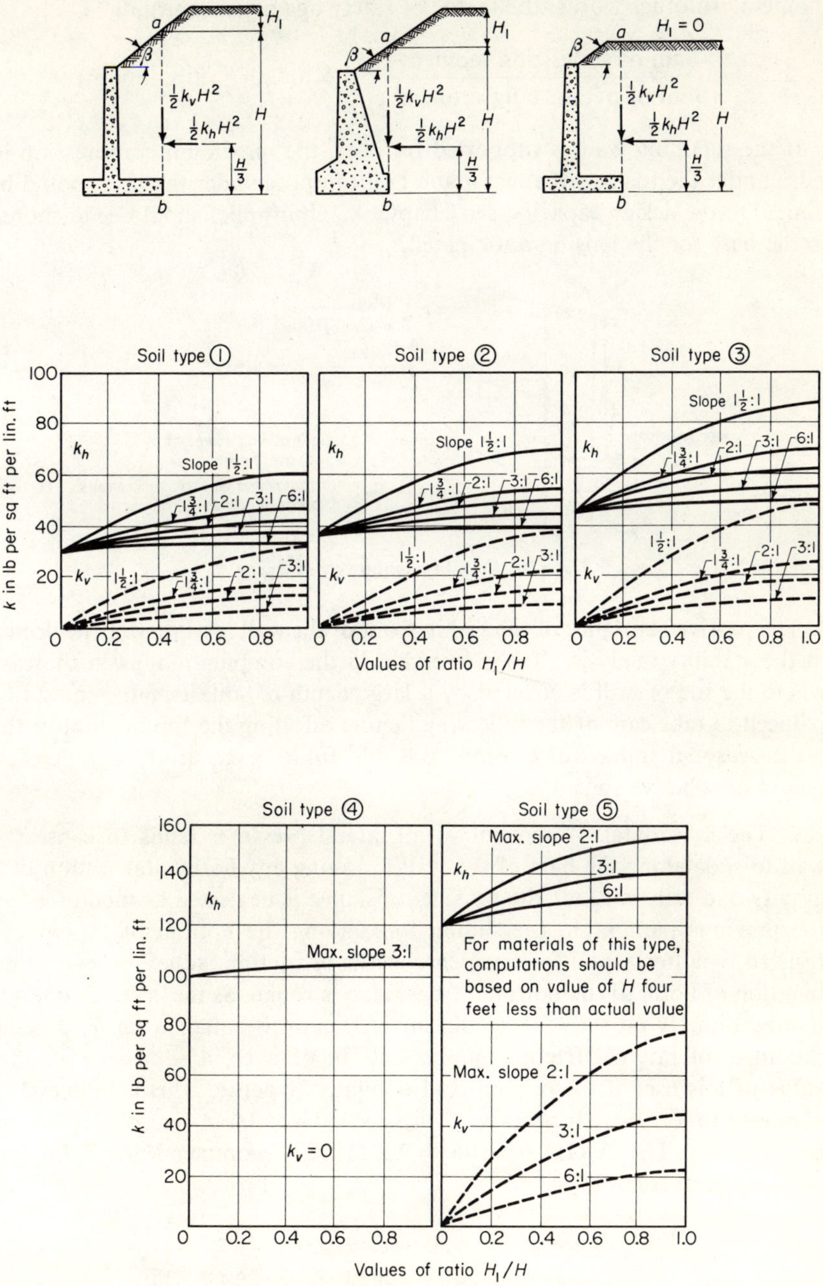

a
β
H_1
$H_1 = 0$
$\frac{1}{2}k_vH^2$
$\frac{1}{2}k_hH^2$
H
$\frac{H}{3}$
b
Soil type ①
Soil type ②
Soil type ③
Soil type ④
Soil type ⑤
k in lb per sq ft per lin. ft
k_h
k_v
Slope $1\frac{1}{2}$:1
$1\frac{3}{4}$:1
2:1
3:1
6:1
Max. slope 3:1
$k_v = 0$
Max. slope 2:1
For materials of this type, computations should be based on value of H four feet less than actual value
Values of ratio H_1/H

stabilizing moment is at least 50 per cent greater than the overturning moment. In other words, the factor of safety against overturning:

$$\text{F.S.} = \frac{\text{Sum of stabilizing moment}}{\text{Sum of overturning moment}} = 1.5 \text{ to } 2.0 \qquad \text{(minimum)}$$

If the retaining wall is supported on piles, the maximum compression in piles under the toe and the maximum uplift in piles under the heel should be limited to the design capacity, see Chapter 8. Uplift piles should be anchored in the base for the tension anticipated.

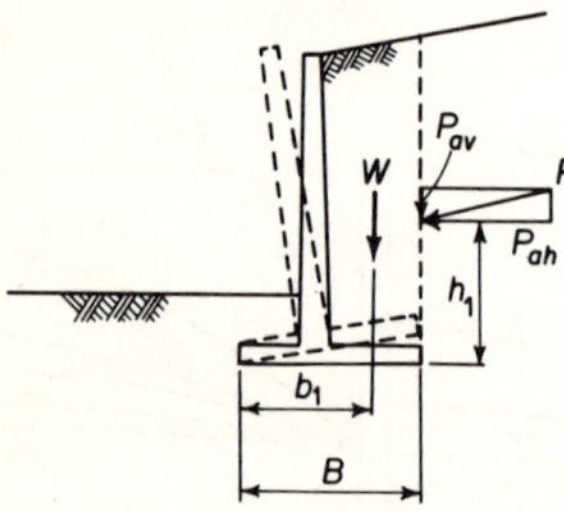

$$F.S. = \frac{\text{stabilizing moment}}{\text{overturning moment}}$$

$$= \frac{Wb_1}{P_{ah}h_1 - P_{av}B} = \text{at least } 1.5$$

where W = weight of wall + weight of soil above the base

P_{ah}, P_{av} = horizontal and vertical components of lateral pressure P_a, respectively

Fig. 11-8 Stability against overturning.

The passive resistance of the soil in front of the wall is commonly neglected in the stability analysis. If it is included in the computation, as in the case where the toe of wall is covered by a large depth of soil, its value should be reduced to take care of the following factors affecting the soil in front of the wall: seasonal influence, erosion, possible future excavation, and tension cracks in cohesive soils.

2. The horizontal component of all lateral pressures tends to cause the wall to slide along the base of the wall (or along any horizontal section of a gravity and crib wall). If the passive resistance is neglected as mentioned in the previous paragraph, the sliding force along the bottom of the wall is resisted by a horizontal force which consists of friction, adhesion, or a combination of both. If the bottom of base slab is rough, as the case of concrete poured directly on soil, the coefficient of friction is equal to tan φ, φ being the angle of internal friction of the soil. In practice, a somewhat smaller value of φ is used if the original soil is relatively dense. This is intended to compensate for the reduction in φ due to some inevitable disturbance during construction. The AREA specifications (1958) recommend the following coefficients of friction:

Course-grained soils (without silt)	0.55
Course-grained soils (with silt)	0.45
Silt	0.35
Sound rock (with rough surface)	0.60

In computing the sliding stability of retaining walls on cohesive soils the adhesion between the base slab and the soil is assumed to be equal to the cohesive strength of the clay and φ is assumed to be zero. If the clay is stiff or hard, the ground must be roughened before placing concrete to insure the full adhesion. In assessing the cohesive strength of the clay, the possibility of reduction in strength due to excavation, exposure to surface water, and construction disturbance must be taken into consideration.

If the retaining wall is supported on piles, the entire vertical and horizontal load should be assumed to be carried on piles. No frictional resistance and no adhesion should be assigned along the base slab.

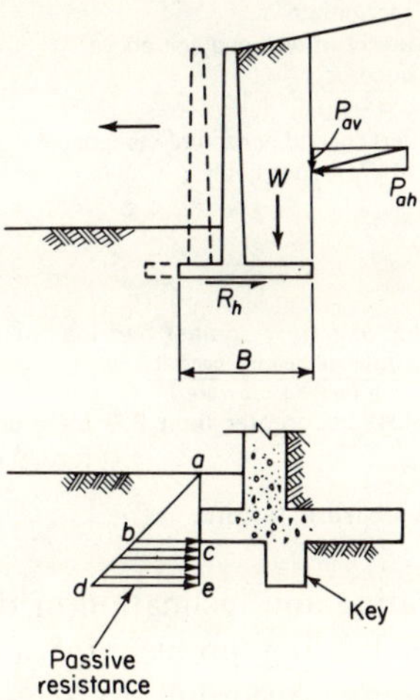

$$F.S. = \frac{\text{Horizontal resistance}}{\text{Horizontal force}}$$

$$= \frac{R_h}{P_{ah}} = \text{at least } 1.5$$

where $R_h = W \tan \phi'$ – for granular soils
$= cB$ for cohesive soils
W = weight of wall and soil above the base
$\phi' \leqq \phi$, usually taken at $\frac{1}{3}\phi$ to $\frac{2}{3}\phi$
c = cohesion, not exceeding 1000 psf

Fig. 11-9 Stability against sliding.

The common practice requires a minimum factor of safety against sliding of 1.5, Fig. 11-9. When this factor of safety is difficult to attain, a key may be constructed under the base slab. The key is generally located under the stem so that the vertical reinforcing bars may be extended into the key. The addition of a key, Fig. 11-9, in effect, increases the passive resistance from *abc* to *ade*. The benefit is generally small unless the key is embedded in rock or hard soil. In fact, the excavation of the narrow key with vertical sides as shown in the figure is likely to disturb the adjacent soil. If the soil is soft or purely granular, the sides of the key should have a slope of 1 vertical to at least $1\frac{1}{2}$ horizontal.

If layers or seams of fine sand or silt are encountered in the subsoil below the base slab, the danger of sliding along these layers should be investigated.

B. Foundation stability. A retaining wall must be also proportioned to have sufficient factor of safety against failure of the foundation soil. A satisfactory design should be one in which the shear strength of all foundation soils is sufficient to withstand the shear stresses introduced by the retaining wall and the backfill with an adequate margin of safety.

Figure 11-10 shows a retaining wall resting on one uniform soil layer extending to a large depth. At the level of the base, the supporting soil is subjected to a vertical force V equal to the sum of the weight of wall, the weight of soil above the base, plus the vertical component of the lateral pressure P. The soil is also subjected to a horizontal force equal to the horizontal component of the lateral pressure P. The resultant R of these forces acts at point b. The point of application and the magnitude of the force R can be determined graphically, as shown in Fig. 11-10, or by analytical method.

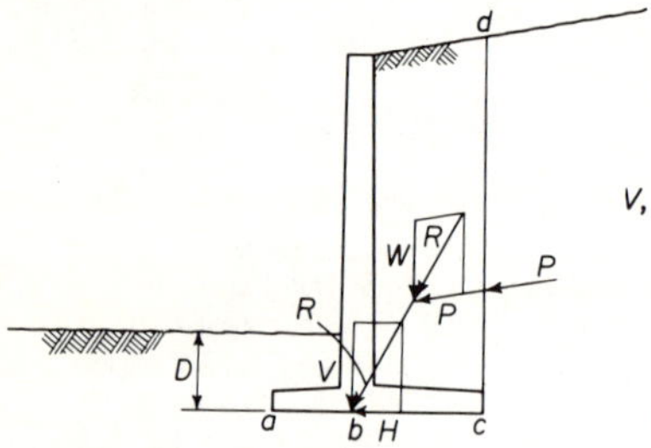

P = lateral pressure on vertical section cd
W = weight of wall and soil above base ac
$R = \sqrt{P^2 + W^2}$
V, H = vertical and horizontal component of R, respectively

Factor of safety against bearing failure

$$= \frac{\text{Ultimate bearing capacity}}{\text{Bearing pressure}}$$

Must be greater than 2.0 for granular soils
" " " " 3.0 for cohesive soils

Fig. 11-10 Stability against bearing failure.

With the magnitude, the point of application and inclination of the force R known, the stability problem becomes bearing problem of a footing subjected to an eccentric and inclined loading. According to Sec. 3-4, the ultimate bearing capacity of granular soil is

$$q_{\text{ult}} = \frac{1}{2}\gamma B N_\gamma\left(1 - 1.5\frac{H}{V}\right)^2 + \gamma D N_q\left(1 + 0.1\frac{D}{B}\right)\left(1 - 1.5\frac{H}{V}\right) \tag{3-2}$$

and the ultimate bearing capacity for cohesive soil is

$$q_{\text{ult}} = 5c\left(1 + 0.2\frac{D}{B}\right)\left(1 - 1.3\frac{H}{V}\right) + \gamma D \tag{3-3}$$

The actual bearing pressure is

$$q = \frac{V}{B'}$$

where B' is the useful width of footing, equal to $2ab$ or $2bc$ whichever is smaller, c is the cohesion of soil; γ is the unit weight of the soil; N_γ and N_q are bearing capacity factors shown in Fig. 3-1. Other symbols are self-explanatory in Fig. 11-10.

The bearing capacity of the foundation may be more accurately determined by first assuming the load V is applied at the center of the base, then multiplying by reduction factors shown in Figs. 6-16 and 6-19.

The factor of safety against bearing failure is defined as q_{ult}/q. In no case should the factor of safety be less than 2.0. When the backfill soil, the supporting soil, or both, contain a large amount of clay, a factor of safety of 3.0 is desirable.

Where there is a layer of weak cohesive soil, the bearing capacity of this soil must be reliably evaluated. A retaining wall should not be attempted if the weight of backfill will exceed the allowable bearing value of the underlying soil. In such cases, recourse may be made to the use of lightweight backfill, the use of structure instead of backfill, or replacing the soft underlying soil with compacted granular soil. If the layer of weak soil is located within a depth of about $1\frac{1}{2}$ times the height of the retaining wall, the stability of foundation soil should be investigated with the possibility of a sliding surface passing through this weak layer. The retaining wall should be proportioned so as to provide a factor of safety against deep foundation failure, at least equal to 2.0.

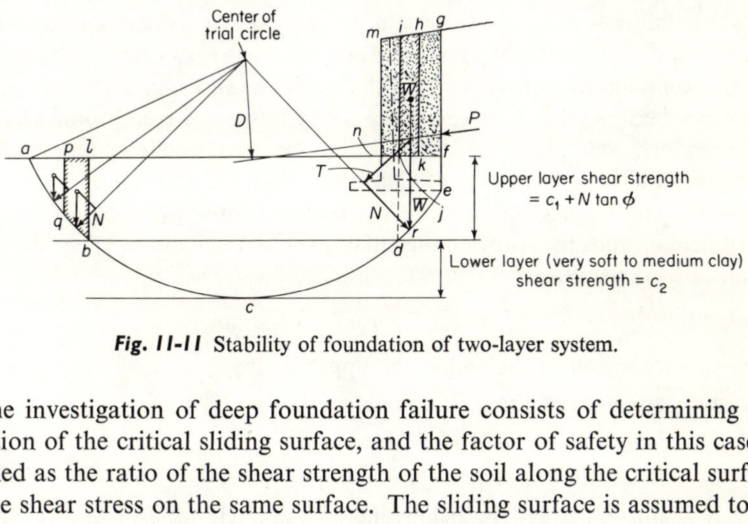

Fig. 11-11 Stability of foundation of two-layer system.

The investigation of deep foundation failure consists of determining the location of the critical sliding surface, and the factor of safety in this case is defined as the ratio of the shear strength of the soil along the critical surface to the shear stress on the same surface. The sliding surface is assumed to be circular or cylindrical. A number of trial circles must be made in order to determine the critical one. The critical surface of sliding, or surface of rupture, is one which gives the smallest factor of safety. This procedure is known as the Swedish circle method.

The Swedish circle method may be simplified for investigation of the most common case of retaining walls having horizontal soil stratifications and a horizontal ground surface in front of the retaining wall. A trial circle *abcde* is shown in Fig. 11-11. With a horizontal or a gently sloping surface of the backfill, the critical circle always intersects the heel of the base slab point *e*. If a vertical plane *eg* is drawn from the heel of the base slab, it can be readily seen that the forces that tend to mobilize the sliding failure are the pressure *P* acting on this plane, and the weight of the backfill in the area *fgmn* above the horizontal line *af*. The weight of soil in the area *abcdef*, for all practical purposes, is symmetrical about the vertical line passing through the center of trial circle; therefore, it may be omitted from the computation. The area *fgmn* is divided into three or four slices. The weight of each of these slices is determined by scaling the drawing, and shown by a weight vector *W* passing through the center of the slice. The force *W* intersects the sliding circle at a point *r*. This force is then resolved into a force *T* tangent to the circle and a force *N* perpendicular to the circle. *T* represents shearing force due to the weight of the slice *khij*, and the total shearing force due to the weight of the backfill *fgmn* may be denoted by ΣT. In addition, the lateral pressure *P* introduces a clockwise moment $P \times D$, *D* being the perpendicular distance of force *P* from center of the trial circle. Thus, it introduces a shearing force on the circle equal to PD/R, where *R* is the radius of the trial circle. Therefore, the total shearing force is $\Sigma T + PD/R$.

This shearing force is resisted by the shear strength or shear resistance along the sliding surface *abcde*. Again, the area *abcdef* is divided into slices, such as *apq* and *pqlb*. If the layer of weak soil is clay, as often is the critical case, no computation is necessary for the weight of slices lying directly above the arc *bcd* because the shear resistance along this arc is equal to one-half of the unconfined compression strength of the clay irrespective of the weight of soil above. In the upper layer of the foundation soil the weight vector for each of the slices *apg*, *pqlb*, and *gedi* is resolved into two components, one perpendicular and the other tangential to the trial circle. The shearing resistance of the soil is

$$S = c_1L_1 + c_2L_2 + \Sigma N \tan \varphi$$

where c_1 = cohesion of the soil in the upper layer;

c_2 = cohesion of the soil in the lower layer;

L_1 = length of arc *ab* and *de*;

L_2 = length of arc *bcd*;

N = normal stress (perpendicular to the circle) due to the weight of slices *apq*, *pqlb*, and *gedi* (for all practical purposes the normal stresses due to weight of area *efjd* may be assumed equal to that of *abl*);

φ = angle of internal friction of upper layer.

The factor of safety of the trial circle is equal to

$$\text{F.S.} = \frac{S}{\Sigma T + PD/R}$$

After a number of such trial circles are made, the minimum factor of safety can be determined. If the minimum safety factor is less than 2.0, the retaining wall proportions should be revised.

It is more logical to assign a larger safety factor for the cohesive resistance than the frictional resistance of soil; generally, F.S. = 3 and F.S. = 2 respectively. Then the shearing resistance may be written as

$$S' = \tfrac{1}{3}(C_1L_1 + C_2L_2) + \tfrac{1}{2}\Sigma N \tan\varphi$$

The critical circle is the one that gives the minimum value of $S'/\Sigma T + PD/R$. An ideal design would have the ratio of $S'/\Sigma T + PD/R \geqq 1.0$. If this ratio is smaller than 1.0, the retaining wall proportions should be revised.

11-7 Design of Structural Components

After the tentative proportions have been proved satisfactory with respect to stability, each element of the wall is designed to provide adequate structural strength. The principles for design of different types of walls are discussed below.

A. Gravity walls. Gravity walls are made of plain masonry, rubble stone, or concrete. In concrete walls a small amount of temperature reinforcement is commonly provided. The wall should be proportioned such that there is no

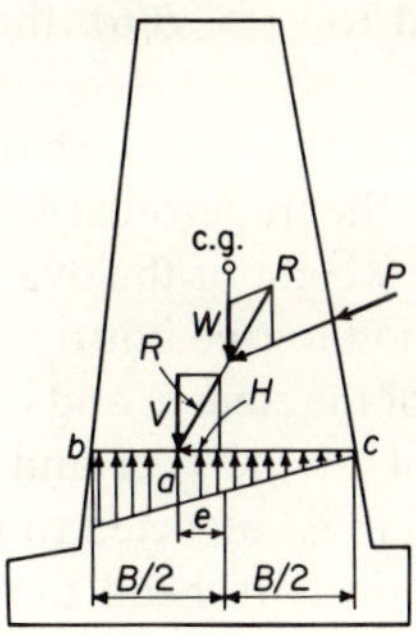

P = total lateral pressure acting on back of wall above pt. c

W = weight of wall above section bc

R = resultant of W and P

Proportion of wall must satisfy the following:

(1) Maximum vertical pressure at point b:

$$p = \frac{V}{B}\left(1 + 6\frac{e}{B}\right) \leqq \text{allowable compression}$$

(2) Minimum vertical pressure at point c:

$$p = \frac{V}{B}\left(1 - 6\frac{e}{B}\right) \geqq 0$$

(3) Horizontal shear along plane bc:

$$v = \frac{H}{B} \leqq \text{allowable shear}$$

Fig. 11-12 Stresses in gravity wall.

tensile stress at any point of the wall under any condition of loading. In favorable cases where the backfill consists of purely granular soil, a small amount of tension (not greater than 3 per cent concrete cylinder strength) may be permitted in the monolithic section of the wall.

A gravity wall may be analysed by the principle of simple statics. Any horizontal section of the wall is subjected to two forces: a lateral force due to earth pressure and surcharge, and a vertical force equal to the weight of the wall above. The magnitude, direction and point of application of the resultant R of these two forces can be readily determined, Fig. 11-12. Let the resultant force intercept the horizontal section at the point a, and let e be the distance from point a to the middle of the horizontal section, then this section is subjected to a vertical pressure q and a horizontal shear v

$$q = \frac{V}{B}\left(1 \pm 6\frac{e}{B}\right)$$

$$v = \frac{H}{B}$$

where V, H = vertical and horizontal component of the resultant force R,

B = width of the horizontal section under investigation.

B. Cantilever walls. A cantilever wall consists of three structural elements: the stem, the toe, and the heel. Each of these elements are designed as a cantilever, Fig. 11-13.

In order to design the base slab (toe and heel), the soil reaction (contact pressure) must be known. With the magnitude and the point of application of the vertical component V already determined in the stability analysis, the soil reaction is computed on the assumption of linear distribution. The soil reaction is trapezoidal if the force V is located within the middle third of the base. If the force V is outside the middle third (or $e > B/6$), the pressure distribution is triangular.

The toe is considered as a cantilever slab fixed at the front face of the stem *ce* and is acted by a large upward pressure (due to the trapezoidal soil pressure distribution) minus the weight of the toe and the weight of the overlying soil. The net pressure tends to bend the toe with tension on the bottom. Similarly the heel is a cantilever slab fixed at the back face of the stem *df* and is subjected to a smaller upward pressure minus the weight of the heel and the large weight of the soil above it. The end of the heel slab is subjected to the lateral earth pressure below point h. The net pressure tends to bend the heel with tension on top.

The stem is a vertical cantilever fixed at the base and is assumed to be subjected to a lateral pressure acting on the vertical section *gh*, Fig. 11-13 [(a) and (b)], where the line *dh* is drawn parallel to the ground surface. The stem is keyed to the base slab by means of a raised key, a depressed key, or a roughened surface. The AREA Manual (1958) permits a shear stress of $0.25 f_c'$ at the root of a key.

C. Counterfort walls. The forces acting on the various elements of a counterfort wall are similar to that on a cantilever wall, Fig. 11-13. The toe slab is a cantilever in both the cantilever wall and the counterfort wall. The difference lies on the method of structural function of the face slab and the heel slab.

The face slab spans horizontally between the counterforts. It is subjected to a negative bending moment with tension on the inside face and a positive moment with tension on the outside face. The horizontal reaction of the face slab is transmitted to the counterforts by reinforcing bars tieing these two elements together. In the simplest form, each horizontal strip of the face

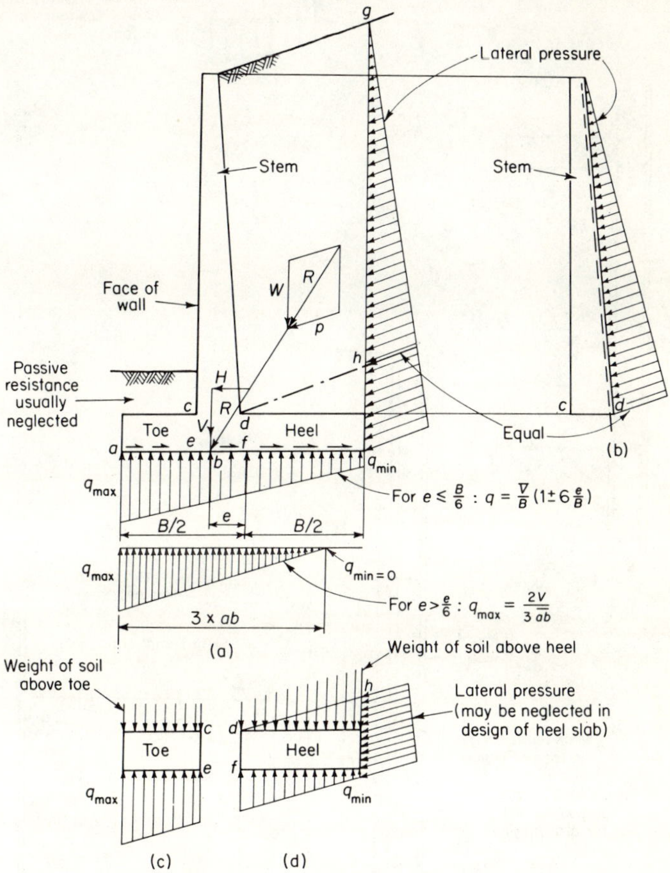

Fig. 11-13 Forces acting on cantilever and counterfort walls.

slab is designed as a continuous slab subjected to a uniform horizontal pressure.

The counterfort may be designed as a wedge-shaped cantilever fixed at the base and subjected to the horizontal reaction from the face slab, Fig. 11-14. When it is made an integral part with the face slab it is a T-beam with the face of wall as the compression flanges of the beam. The common arrange-

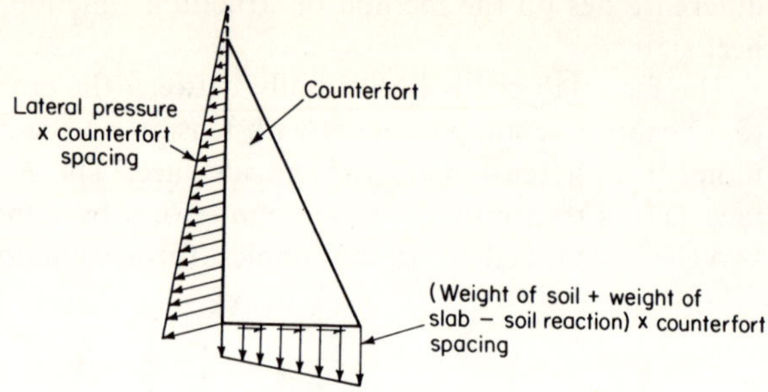

Fig. 11-14 Forces acting on a counterfort.

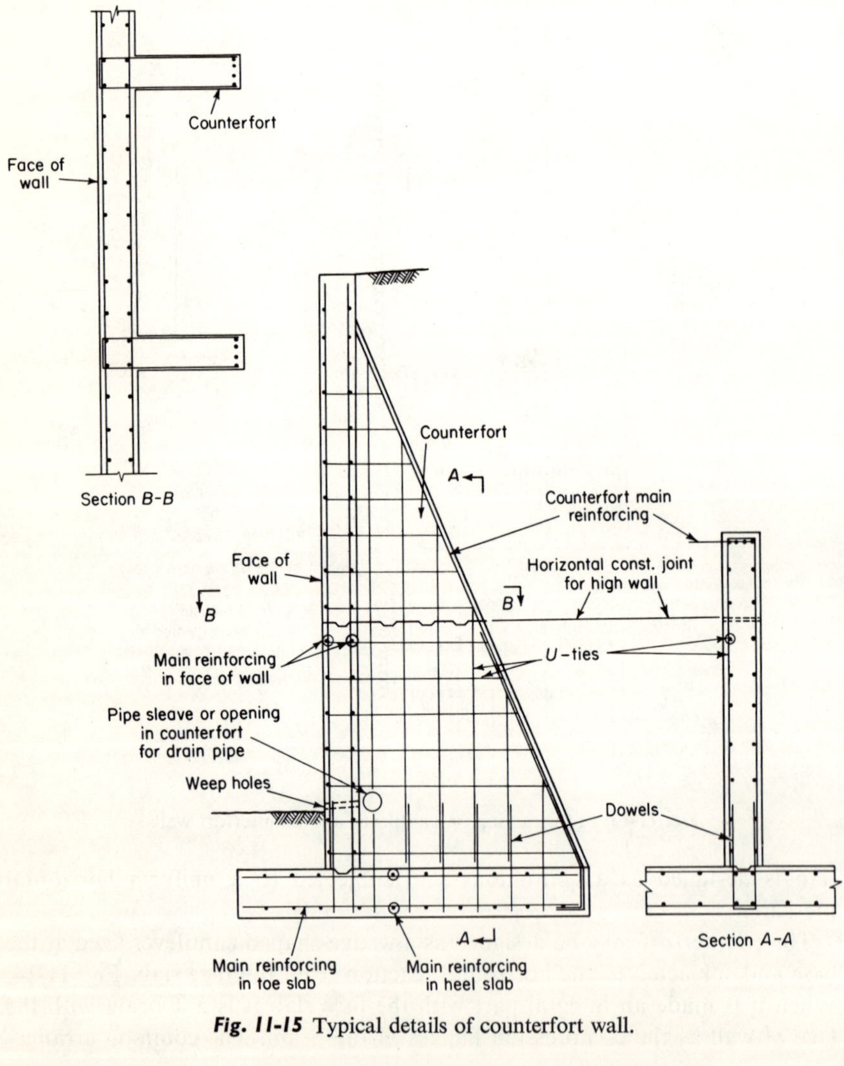

Fig. 11-15 Typical details of counterfort wall.

ment of reinforcement and other construction details in a counterfort wall is shown in Fig. 11-15.

The heel slab is subjected to net downward pressure due to the large weight of soil above it. This pressure causes a negative bending moment at the counterfort with tension on the top of slab and a positive bending moment at mid-span between counterforts. The base slab must be tied to the counterfort to transmit the vertical reaction. In addition to this bending parallel to the length of the wall, the heel slab is also subjected to a bending moment from the cantilevered toe slab.

This procedure of design of counterfort walls gives rather conservative design of the face slab because it ignores the benefit of the slab rigidity and

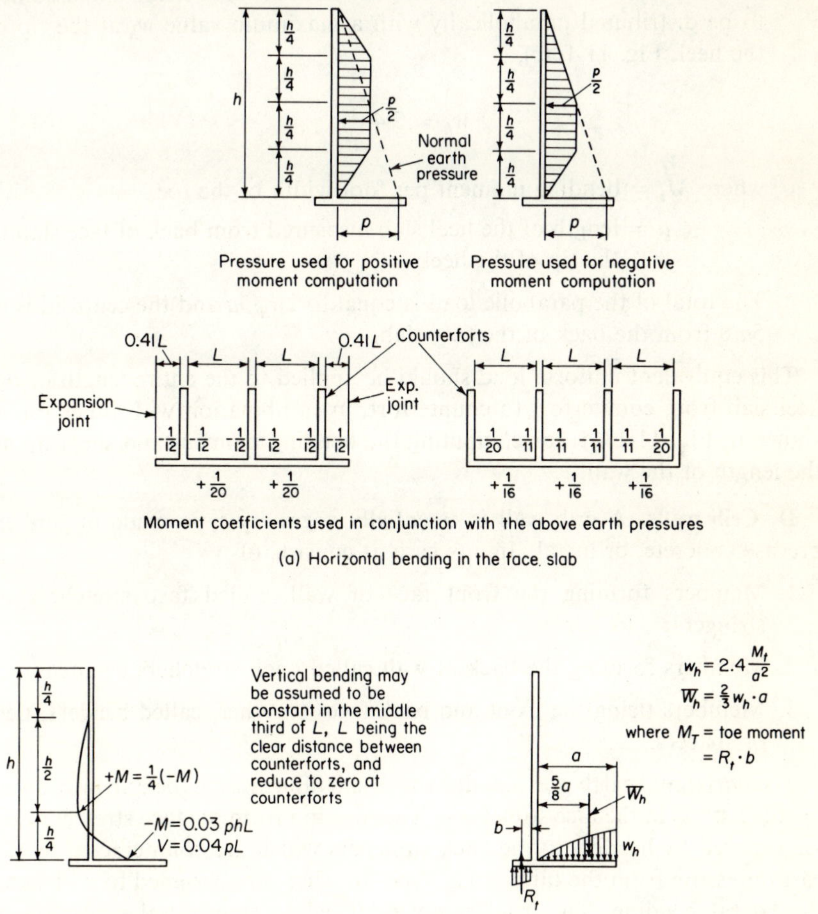

(b) Vertical bending in one foot strip of face slab

(c) Equivalent load (downward) on heel slab due to moment in the toe slab

Fig. 11-16 Huntington's design procedure for counterfort walls.

the fixity, or partial fixity, of the lower edge of the face of wall. Furthermore, no established rule of design takes into account the stresses in the heel slab as a result of bending moment of the toe slab. For these reasons, Professor Huntington (1957) proposed the following design procedures.

1. The face slab is designed as continuous horizontal slab to resist the lateral pressure shown in Fig. 11-16(a).
2. The face slab is also designed as a vertical slab subjected to the negative and positive bending moments shown in Fig. 11-16(b).
3. The heel slab is subjected to an equivalent downward force in addition to the forces shown in Fig. 11-13(d). This equivalent force is evaluated to include the effect of the bending moment of the toe slab and assumed to be distributed parabolically with a maximum value w_h at the tip of the heel, Fig. 11-16(c),

$$w_h = 2.4\frac{M_t}{a^2}$$

where M_t = bending moment per foot width of the toe,

a = length of the heel slab measured from back of face slab to the tip of the heel.

The total of the parabolic load is equal to $2w_h/3a$ and the centroid is at $5a/8$ from the back of the face slab.

This equivalent uniform load should be applied to the entire length of the heel slab from counterfort to counterfort, in combination with other forces shown in Fig. 11-13(d) for computing the bending moment and shear along the length of the wall.

D. Crib walls. A crib wall is essentially a gravity type made of timber, precase concrete, or metal. In any case, it consists of

1. Members forming the front face of wall, called face stretchers or stringers;
2. Members forming the back of wall called back stretchers or anchors;
3. Members tieing the front and back faces together, called headers, ties, or spacers.

The horizontal earth pressure due to the soil filling inside the crib introduces bending stress in the face stretchers. The reaction from the face stretchers are transmitted by headers to the back stretchers which are counteracted by the earth pressure from the filling. The face stretchers are designed to withstand the lateral bending. The headers are designed to transmit the reaction in tension from the face stretches and to support the weight of the soil superimposed on them.

Precast concrete and prefabricated steel crib walls may be purchased from manufacturers. Timber cribs may be made from whole logs (undressed lumbers), Fig. 11-17(a), and are used for temporary work only. The life of lumber may be increased by treatment with wood preservatives (generally creosote, see Sec. 8-3B). Logs should be adzed at points of contact between members to provide for good bearing.

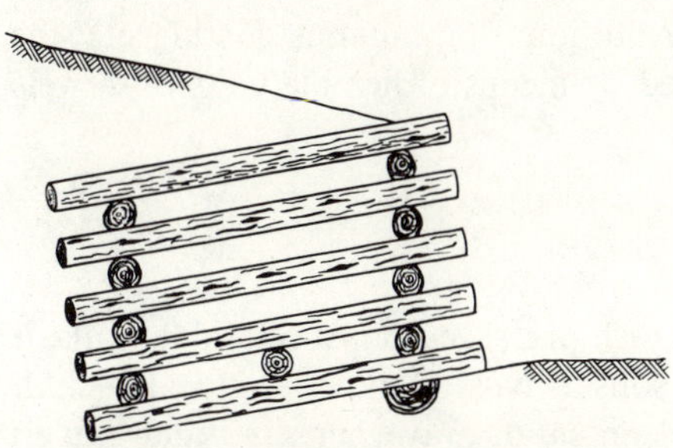

(a)

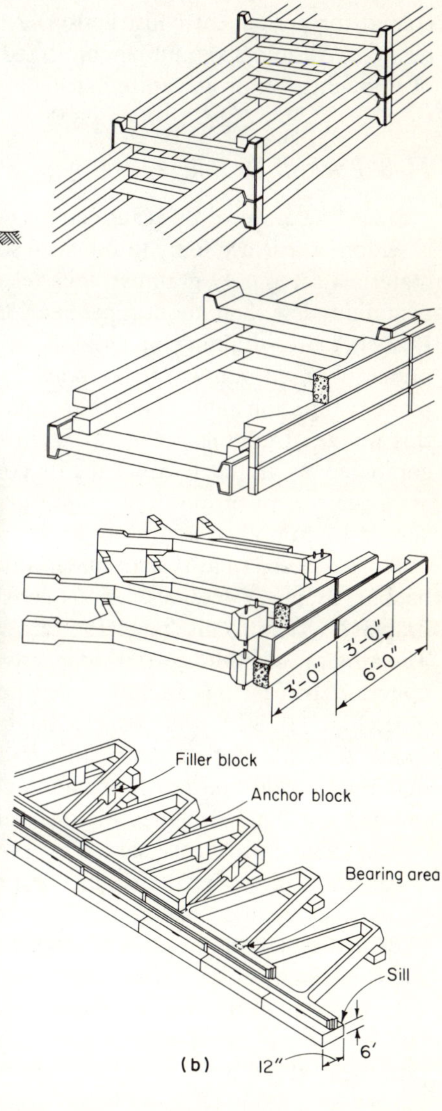

(b)

(c)

Fig. 11-17 Members of crib wall: (a) timber crib wall; (b) concrete crib walls (after Portland Cement Association); (c) metal crib walls (bin walls). Photograph courtesy of Armco Drainage & Metal Products Inc.

Concrete crib walls may have open-face or closed-face construction. The space between the face stretchers in the open-face construction should not exceed 8 in. so as to properly retain the filling. The back of concrete crib walls may have a variety of arrangement. Fig. 11-17(b) shows some of the commonly known types. In any arrangement, each of the precast members should bear at two points only, and should be reinforced with a minimum of 0.9 per cent of its cross-sectional area (AREA Manual, 1958).

Metal crib members should be not less than 16 gauge and should be zinc coated for permanent construction. Additional bituminous coating may be desired. All joints should be connected by means of flexible bolts or dowels of wrought iron or galvanized steel.

11-8 Backfill Drainage

Since fine-grained soils cause large earth pressure against retaining walls, it is seldom a true economy to use such soils as backfill material. Ideal backfill materials are purely granular soils (clean sand, gravel, or sand and gravel) containing less than about 5 per cent of very fine sand, silt, or clay particles. If such a free-draining material is expensive in the locality, it should be preferably used in a wedge bounded by the back of the retaining wall and a plane rising at an angle of not more than 60 degrees with the horizontal. With this wedge of granular material against the wall, the earth pressure may be computed as if the entire backfill consists of the granular soil. If other arrangements or proportions of granular backfill are used, the earth pressure should be computed according to Sec. 12-12.

Regardless of what type of material is used for the backfill, the problem of backfill drainage deserves serious consideration. There should be adequate drainage system to discharge the rain water that infiltrates the backfill soil. The amount of drainage work depends upon the permeability of the backfill material. In pervious backfill, weep holes, Fig. 11-18(a) or a line of drain pipe, Fig. 11-18(b), will suffice. Semipervious backfills (soils containing a small amount of fine sand, silt, or clay particles) require strips of filter material in addition to the drain pipes or weep holes, Fig. 11-18(c). To provide adequate drainage in fine-grained backfill, a drainage blanket or double blankets are necessary, Fig. 11-18[(d) and (e)].

All drain pipes or tiles should be provided with adequate clean-outs for periodical cleaning. The drain tiles or pipes should be embedded in selected filter material so as to prevent clogging the tile or pipes and carrying away soil particles by water. The criteria for filter material are discussed in Sec. 5-5.

In the case of flood walls constructed to protect useful land against river floods, a line of drain pipes is generally installed at the toe of the wall. The purpose of this drain is to reduce the possible damage due to the phenomenon known as piping or boil. When soil grains are subjected to an upward seepage pressure greater than their buoyant weight, the grains are uplifted in a manner similar to a boiling water. Adequate drainage at the toe will reduce the uplift pressure, therefore, eliminate the possibility of piping.

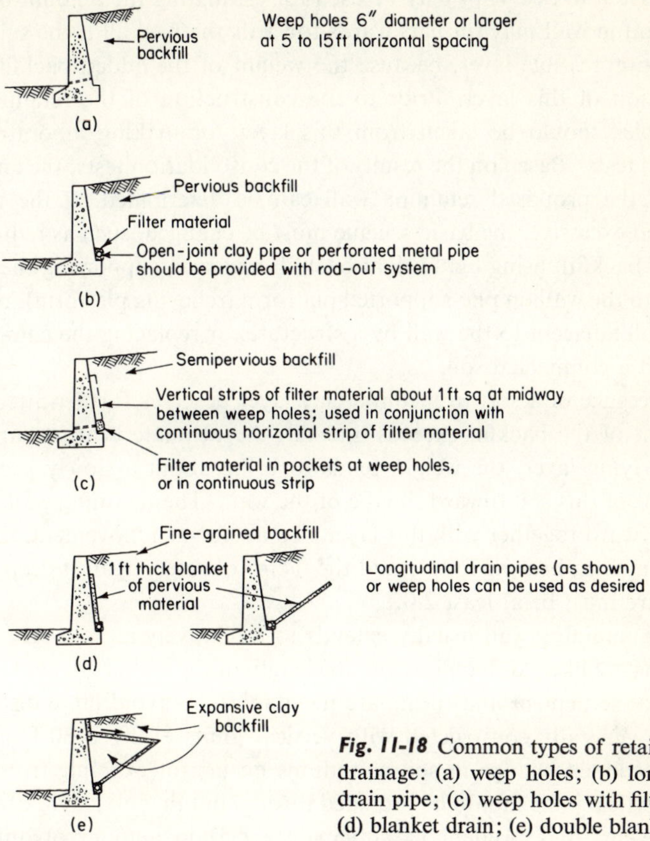

Fig. 11-18 Common types of retaining wall drainage: (a) weep holes; (b) longitudinal drain pipe; (c) weep holes with filter strips; (d) blanket drain; (e) double blanket drain.

11-9 Settlement and Tilting of Retaining Walls

Like any other structures on footings or piles, retaining walls are expected to have a certain amount of settlement. The basic principles of settlement analysis for footings and pile foundations are applicable. When the subsoil consists of soft and medium clays, a settlement analysis must be made.

In addition to normal settlement, a retaining wall is subjected to tilting as a result of eccentric pressure on the base of retaining wall. In the majority of cases, the wall tilts forward because the resultant force generally intercepts the base slab at a point between the center and the toe. The amount of tilting is difficult to determine and can only be roughly estimated. If the stability requirements are met in accordance with the common design procedure, the amount of tilting may be expected to be in the order of magnitude of less than one-tenth of one per cent of the height of wall. However, if the subsoils consist of compressive layer, this amount may be exceeded. In such cases, the discussion in Sec. 6-14 may be used for estimating the amount of tilting.

A retaining wall may tilt backwards towards the backfill if the subsoil has a thick compressible layer, because the weight of the added backfill causes compression of this layer. Prior to the construction of the retaining wall, soil samples should be taken from this layer for making laboratory consolidation tests. Based on the results of the consolidation tests, the amount of tilting of the proposed retaining wall can be determined. If the tilting is considered excessive, the basic scheme must be changed, such as reducing the height of backfill, using lightweight backfill material, supporting the backfill adjacent to the wall on pile-supported platform (relieving platform), replacing the backfill adjacent to the wall by a structure, or replacing the compressible layer with a compacted soil.

The presence of a soft to medium clay layer deserves further attention. If the weight of the backfill exceeds one-half the ultimate bearing capacity of the underlying layer, the weight tends to bring about a slowly progressing movement of this soil toward the toe of the wall. The retaining wall, in turn, moves forward together with this layer. To avoid such movement, the factor of safety against bearing failure and the factor of safety against deep foundation failure must be at least 2.0.

Since a retaining wall usually extends to a relatively large length, the soil conditions are likely different within the length of the wall, and, consequently, differential settlement and tilting are inevitable. To avoid large distress, the walls are generally constructed with vertical joints at 20 to 30 ft intervals. Where a sudden change in soil conditions occur, for example from soil to rock, a vertical joint should be provided throughout the full height of the wall. In any event, it is prudent to anticipate a certain amount of differential settlement and unequal tilting and to provide for them.

11-10 Construction of Retaining Walls

With the exception of masonry gravity walls and crib walls, all retaining walls are constructed of concrete. The construction of masonry walls should be made in accordance with the specifications* for the type of masonry used. The essential features of crib wall construction are discussed in Sec. 11-7D. The following discussion applies to the construction of concrete retaining wall.

* *Minimum Building Code Requirements for Masonry.*

A. Joints. A retaining wall may have some or all of the following joints:

Construction joints: Vertical or horizontal joints between successive placements of concrete.

Contraction joints: Vertical joints from top of base slab to top of wall, to permit the concrete to shrink during hardening without causing noticeable and harmful shrinkage cracks.

Expansion joints: Vertical joints from top of base slab to top of wall, to care for temperature changes.

Construction joints are necessary between the base slab and the upper portion of the wall. In high walls, one or two horizontal joints are needed at midpoint or third points of the height of the wall. Such a joint must be constructed to resist shear force resulting from the total lateral pressure above the joint. Figure 11-19(a) shows the common types of such a joint: a

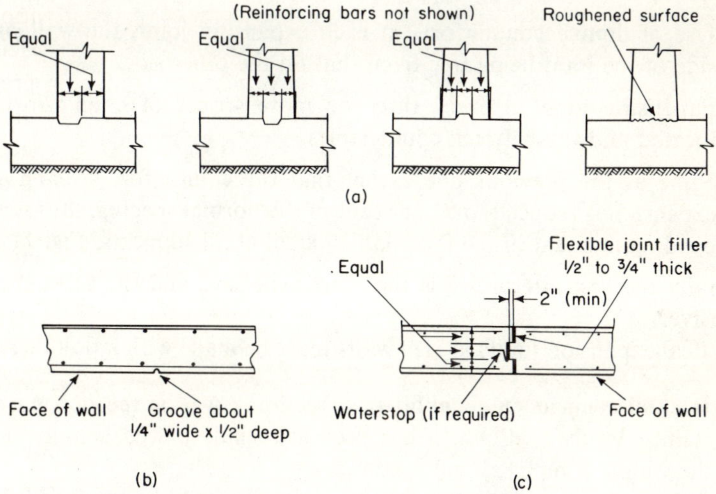

Fig. 11-19 Common types of joints in concrete retaining walls: (a) construction joints; (b) contraction joints; (c) expansion joint.

depressed shear key, a raised shear key, or a roughened joint. Before placing the fresh concrete, the laitance (a scum of very fine cement) should be cleaned off by a wire brush from the surface to be joined. If roughened surface is used, the most desirable is one with one-half of the coarse aggregates exposed.

A contraction joint is simply a shallow groove on the face of wall and running full height of the face slab. Reinforcing bars are running continuously through the joint as normal. This groove weakens the section and therefore, permits a tension crack to take place, Fig. 11-19(b).

An expansion joint permits relative movement of wall on each side of the joint. No reinforcing bars can run across this joint. Usually a key is provided to maintain the wall in alignment. A flexible type joint filler should be inserted, as shown in Fig. 11-19(c). If water is anticipated behind the wall, water stop should be used. Water stops may be made of rubber, thin steel ($\frac{1}{8}$ in.) or copper strip. To permit movement, metal water stops should be bent to a V shape at the joint, Sec. 5-8.

The location of construction joints, contraction and expansion joints should be kept to a minimum and should be predetermined and shown on the engineering drawing. The spacing for contraction joints is commonly 20 to 30 ft and the spacing for expansion joints is usually limited to about 90 ft. Where there is abrupt change in section a contraction or expansion joint is desirable.

Vertical joints in gravity and cantilever retaining walls may be located any place because each vertical strip of wall is designed as an independent element. In counterfort wall, there are three alternative designs:

1. Use of double counterforts at each expansion joint, the wall on one side of the joint being free from that on the other side.
2. Expansion joint in every three or more spaces of counterforts and located midway between counterforts.
3. Same as the previous one except that the counterfort spacing at the expansion is reduced to 81 per cent of the normal spacing, thus negative bending moment in the face slab is equal at all supports, Fig. 11-16(a).

Generally the first alternative is the most expensive and the third the least expensive.

Vertical expansion joints in crib walls must be made with a double header.

B. Backfill compaction. Backfill of a retaining wall is the portion of the soil retained by the wall which has been artificially placed behind the wall after the wall is completed and matured.

As discussed in Sec. 11-8, the backfill material should be carefully selected. In addition, it should be compacted to prevent large ground subsidence due to consolidation under its own weight. A select material compacted to a very dense state will suffer very little settlement. The amount of compaction required for each job depends on the material used, and on the nature of the job. More strict control of compaction is necessary where the backfill is cohesive soil and where the ground settlement is objectionable. Even when granular backfill is used and a small settlement is tolerable, the material should be placed in thin layers not thicker than 9 in. each. Each layer should be compacted before the next one is placed. In cases where ground settlement or bearing capacity is of importance, the compaction should be controlled in the field.

Caution should be exercised not to overcompact the backfill since overcompaction will cause excessive earth pressure. In general the soil between the back of the wall and a 45-degree line starting from the heel of the wall should be compacted by hand-operated tampers or other lightweight compactors. If compaction rollers are used in this area, the stresses in the wall must be investigated by the principles discussed in Sec. 4-9, and the stability of the wall should also be analysed. It should not be allowed to dump the material in sloping layers toward the wall, thus forming segregated layers of potential sliding surfaces. It is always a good practice to place an impervious soil in the upper layer of the backfill for the purpose of cutting down the amount of infiltration from the rain water.

C. Dampproofing and Waterproofing. To avoid mar of the face of wall due to seepage water through cracks and joints, the back of the wall is commonly dampproofed by application of hot bituminous material before placing the backfill. In the cases where permanent water level is above the ground elevation in front of the wall, the back of the wall should be provided with waterproofing membrane, and all joints should be provided with continuous water stops to a level about two feet above the water level.

11-11 Design Example

This example illustrates the complete design procedure for a cantilever retaining wall which is probably the most commonly used type of all retaining walls. A typical section of wall, one foot long, is analysed. All general information necessary for the design purpose are assembled in a sketch shown on sheet 1 of the Design Plates. It is assumed that soil information concerning the subsoil and the backfill material have been analysed. The typical soil profile representing the more critical condition is shown and is used for design.

Before earth pressure is computed, the applicable theory must be selected. Reference is made to Fig. 11-6(a). It is found in this example that the line *bc* intercepts the back of the wall. Hence Rankine theory is theoretically not applicable and the earth pressure should be computed by the wedge theory with an appropriate value of φ', Fig. 11-6(b). However, in general practice, φ' is assumed to be zero in design of cantilever retaining walls for the following reasons:

1. In most cases, the line *bc* intercepts near the top of the wall, and hence, the value of φ' is nearly zero.
2. The value of φ' is rather difficult to ascertain.
3. A refined evaluation of φ' is unwarranted since the value of φ and other basic index properties are only approximately assessed.

Therefore, in design of cantilever retaining walls of the ordinary proportions, the assumption of $\varphi' = 0$ is justified and Rankine theory is used. This gives a slightly conservative result.

The bearing capacity of the foundation soil is calculated by both methods discussed in Sec. 11-6(b). The method using the approximate equation (6-lb) results in a higher factor of safety than the alternate method using reduction factors. This is to be expected because the application of the useful width in Eq. (6-lb) is equivalent to the assumption that the reduction in bearing capacity due to eccentricity of footings on sand is in straight line proportion with e/B, whereas the theoretical reduction factor varies parabolically with e/B. The first method gives approximate result and is on the unsafe side.

Data

Surcharge	*2'-0" soil (AASHO specifications)*
Backfill	*Bank run sand and gravel compacted to 90% max density*
Foundation soil (see sketch)	
Concrete	$f_c' = 3{,}000$ psi
reinforcing	$f_s = 20{,}000$ psi

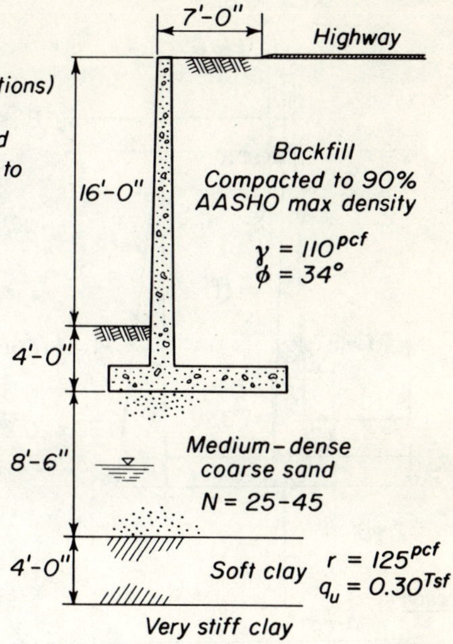

Tentative Proportions

$h = 20'\text{-}0''$

Toe projection	*1'-9"*
Base width	*10'-0"*
Base thickness	*1'-9"*
Stem	*0'-9" top*
	1'-9" bottom

Earth Pressure

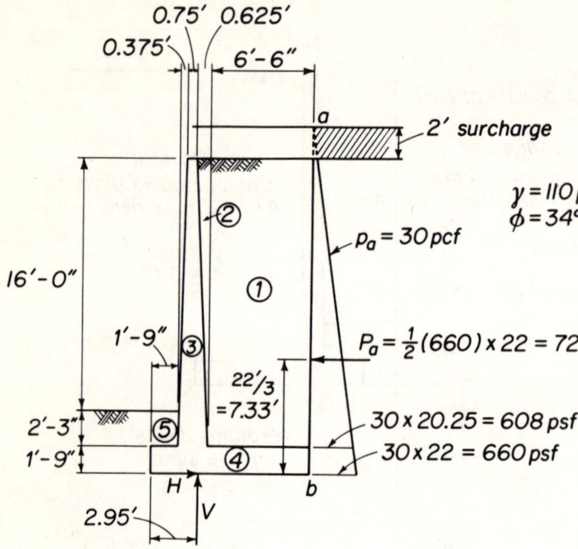

$\gamma = 110$ pcf
$\phi = 34°$ $\quad K_a = \tan^2(45 - \phi/2) = 0.28$
$p_a = \gamma K_a = 30$ pcf

Reaction

	Area		γ	W	Distance from toe	M
①	6.5′ x 18.25′ =	118.5	110	13.04	6.75′	88.0
②	$\frac{1}{2}(0.625') \times 18.25' =$	5.7	110	0.63	3.29′	2.1
③	$\frac{1}{2}(0.75 + 1.75) \times 18.25' =$	22.8	150	3.42	2.60′	8.9
④	1.75 x 10.0 =	17.5	150	2.62	5.00′	13.1
⑤	1.75 x 2.25 =	3.9	110	0.43	0.88′	0.4
				V = 20.14^k		112.5
		$P_a \times 7.33 =$				53.2
						59.3$^{ft\text{-}k}$

$H = P_a = 7.26^k$ $\qquad \times \frac{1}{20.14} = 2.95'$

Overturning

$$F.S. = \frac{112.5}{53.2} = 2.11 \text{ OK}$$

Sliding

$$\frac{20140 \times 0.55}{7260} = 1.52 > 1.50 \text{ OK}$$

(0.55: for clean sand)

(effect of passive resistance neglected)

Bearing Pressure

Using Eq. (3−1):

$$q_{ultimate} = \frac{1}{2}\gamma B N_r (1-1.5\tfrac{H}{V})^2 + \gamma D N_q (1+0.1\tfrac{D}{B})(1-1.5\tfrac{H}{V})$$

$\gamma = 110$

B = effective width $= 2 \times 2.95 = 5.90'$

$D/B = 4/5.9 = 0.66$

$H/V = 7.26/20.14 = 0.36$

$< \tan\phi$ OK

From Fig. 2−2

For $N = 25 \quad \phi = 34°$

From Fig. 3−1(b)

$N_r = 33,\ N_q = 32$

$$q_{ult} = \frac{1}{2}110(5.9)33(1-1.5\times 0.36)^2 + 110(4)32(1+0.1\times 0.66)(1-1.5\times 0.36)$$

$= 2250 + 6900 = 9150^{psf}$

$\times 0.75 = 6860^{psf}$

(Reduce for water level at $\frac{B}{2}$, Fig. 6−4)

$$q = \frac{V}{B} = \frac{20160}{5.90} = 3420^{psf}$$

$$F.S. = \frac{6860}{3420} = 2.01 \quad > 1.50 \quad OK$$

Using alternate method

For $e = 0,\ H = 0$, Eq. (6−1b):

$$q_{ultimate} = 3N^2 B R_w + 5(100 + N^2) D R_w'$$

$= 3(25)^2(10) + 5(100 + 25^2)4(0.75)$

$= 18{,}750 + 10{,}860 = 29{,}610^{psf}$

For $e = 5.0 - 2.95 = 2.05$, $\quad e/B = 0.205$

From Fig. 6−16 $\quad R_e = 0.36$

For $H/V = 7.26/20.14 = 0.36$, $D/B = 0.4$

From Fig. 6−19(a), $\alpha = \tan^{-1} 0.36 = 20°$

$R_i = 0.30$

$\therefore q_{ult} = 29{,}610 \times 0.36 \times 0.30 = 3200^{psf}$

$$q = \frac{V}{B} = \frac{20160}{10} = 2016$$

$$F.S. = \frac{3200}{2016} = 1.59 \quad > 1.50 \quad OK$$

Deep Foundation Stability

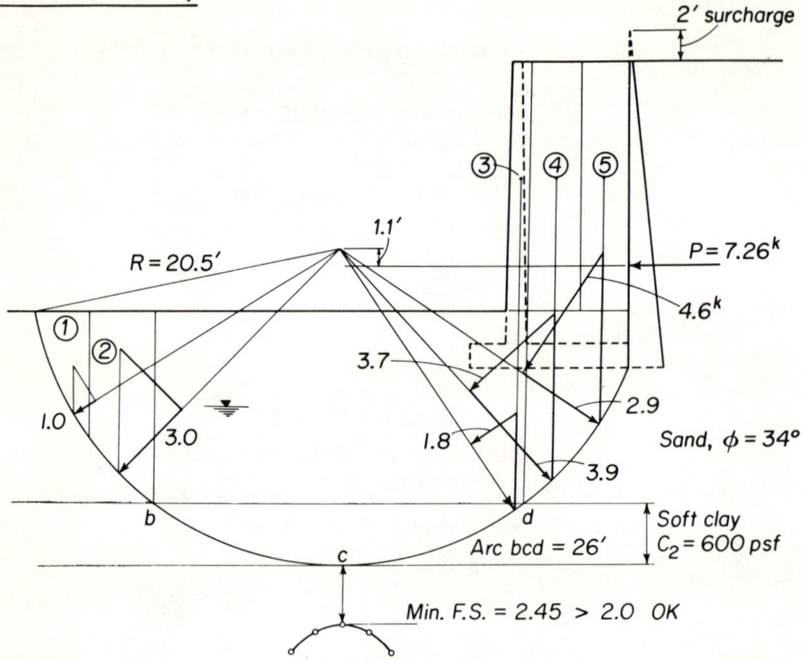

	Area		γ	W	N	T	Remarks
①	$\frac{1}{2}(4.0)(8.5)$	= 17.0	110	1.82^k	1.0 × 2 = 2.0	–	
②	(4.0)(10.5)	= 42.0	90	4.18	3.0 × 2 = 6.0	–	Av unit wt = 90 pcf
③	(1.5)(16.0)	= 24.0	130	3.12	–	1.8	Av unit wt = 130 pcf
④	(3.1)(16.0)	= 49.6	110	5.45	3.9	3.7	
⑤	DO		110	5.45	2.9	4.6	
						0.4	= 7.26 × 1.1/20.5
					$\Sigma N = 14.8^k$	$10.5^k = \Sigma T$	

$$S = C_2 L_2 + \Sigma N \tan\phi$$

$$= 26 \times 0.6 + 14.8 \times 0.67 = 25.6^k$$

$$F.S. = 25.6/10.5 = 2.45$$

(This happens to be the critical circle. Calculation of F.S. for other trial circles is not included here, but results are plotted above)

DE 11
Cantilever Retaining Wall
Sh. 5 of 7

<u>Base Slab</u>

$$e = \frac{10'}{2} - 2.95 = 2.05 > \frac{10}{6}$$

$$p_{max} = \frac{2 \times 20160}{3 \times 2.95} = 4550 \text{ toe pressure}$$

$$\times \frac{7.10}{8.85} = 3650$$

$$\times \frac{5.35}{8.85} = 2750$$

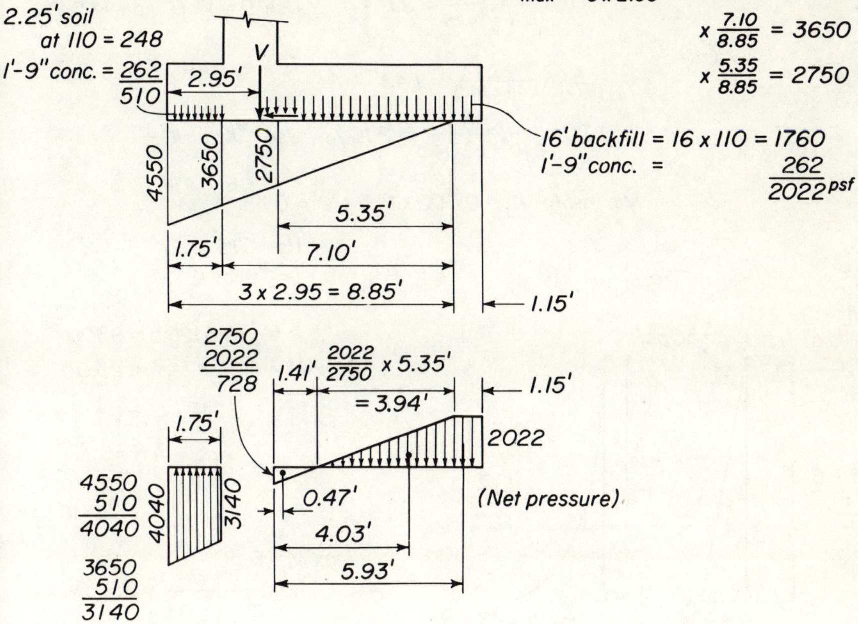

<u>Toe:</u> $V = \frac{1}{2}(4040 + 3140)1.75 = 6290^{\#}$

$$v = \frac{6290}{12 \times 7/8^{*} \times 17.5} = 34^{psi} \quad \text{OK}$$

$$M = 6290 \times \frac{1.75}{2}^{\pm} = 5.50^{\prime k}$$

$$As = \frac{5.50}{1.44 \times 17.5} = 0.22^{s.i.} \quad \text{Use min reinf.}$$

<u>Heel:</u>

$\frac{1}{2}(0.728)(1.41)$	$= -0.51$	$\times 0.47 =$	-0.2
$\frac{1}{2}(2.022)(3.94)$	$= 3.98$	$\times 4.03 =$	16.3
2022×1.15	$= 2.33$	$\times 5.93 =$	13.8
	$V = 5.80^{k}$		$M = 29.9^{\prime k}$ (max)

Max. $V = 3.98 + 2.33 = 6.31^{k}$

*ACI Building Code is used. Values in 1956 Code are marked with *

$$\left.\begin{aligned} \text{Req'd. } d &= \sqrt{\frac{29.9}{0.236}} = 11.3'' \\ d &= \frac{6310}{12 \times \frac{7}{8}^{*} \times 75} = 8.1'' \end{aligned}\right\} \quad \text{Use } D = 1'\text{-}6'' \quad d = 14.5$$

$$A_s = \frac{29.9}{1.44 \times 14.5} = 1.43\ \square''/'$$

$$\Sigma O = \frac{6310}{240^{*} \times \frac{7}{8}^{*} \times 14.5} = 2.10''/' \quad \text{Use } \#10 \text{ at } 10\tfrac{1}{2}''$$

Min reinf. $A_s = 0.002 \times 14.5 \times 12 = 0.35\ \square''/'$

Use #10 at 21"
or #6 at 12

<u>Stem</u>

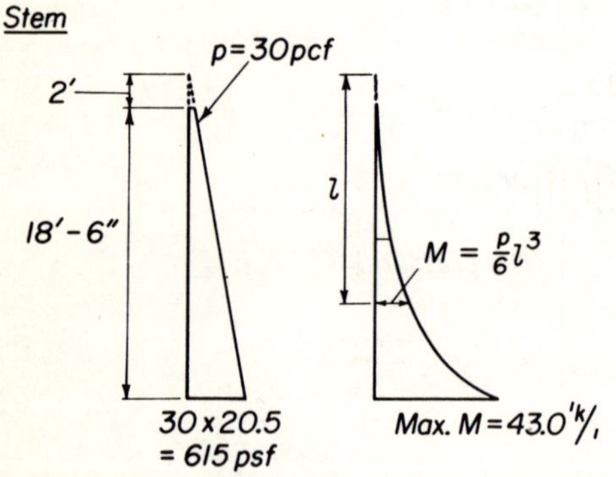

$$P \doteq \tfrac{1}{2}(615)20.5 = 6300^{\#}$$

$$\text{Max } M = 6.30 \times \frac{20.5}{3} = 43.0^{'k}$$

$$d = \sqrt{\frac{43.0}{0.236}} = 13.5''$$

$$\underline{2.5} \text{ cover}$$

$$16.0''$$

21" OK
(d = 18.5)

$$\text{Max } V = 6.30^{k}$$

$$v = \frac{6300}{12 \times \frac{7}{8}^{*} \times 18.5} = 33 \quad \text{OK}$$

$$\text{Max } A_s = \frac{43.0}{1.44 \times 18.5} = 1.61\ \square''/'$$

$$\Sigma O = \frac{6.300}{240^{*} \times \frac{7}{8}^{*} \times 18.5} = 1.64''/'$$

use #8 at 6
($A = 1.58\ \square''/'$)

<u>Cut off $\frac{1}{3} A_s$,</u> $M = \frac{p}{6} l^3$

$$= \frac{2}{3} \cdot \frac{p}{6} H^3$$

$$\left(\frac{l}{H}\right)^3 = \frac{2}{3}, \quad \frac{l}{H} = 0.87$$

$$\text{Cut-off} = (1 - 0.87)H + 21 \text{ bar dia.}$$

$$= 2.67' + 1.75 = 4.45' \quad \text{Use } 5'\text{-}0''$$

Cut off $\frac{2}{3} A_s$

$(\frac{l}{H})^3 = \frac{1}{3} \quad \frac{l}{H} = 0.67$

Cut-off $= 0.33H + 21$ bar dia.

$= 6.8' + 1.75' = 8.55'$ Use 9′-0″

Min. Reinf.

9″ thick : $A_s = 0.002 \times (9-2.5) \times 12 = 0.156$ □″/′

1′-9″ thick : $A_s = 0.002 \times (21-2.5) \times 12 = 0.444$ □″/′

Vert., exterior face $A_s = \frac{1}{3}(0.444) = 0.15$ □″/′ Use #4 at 15 Full height

Horiz. Top portion $A_s = \frac{1}{2}(0.156) = 0.08$ □″/′

Bot. portion $A_s = \frac{1}{2}(0.44) = 0.22$ □″/′ Use #4 at 12 ($A_s = 0.20$) / #4 at 10 ($A_s = 0.24$) E.F.

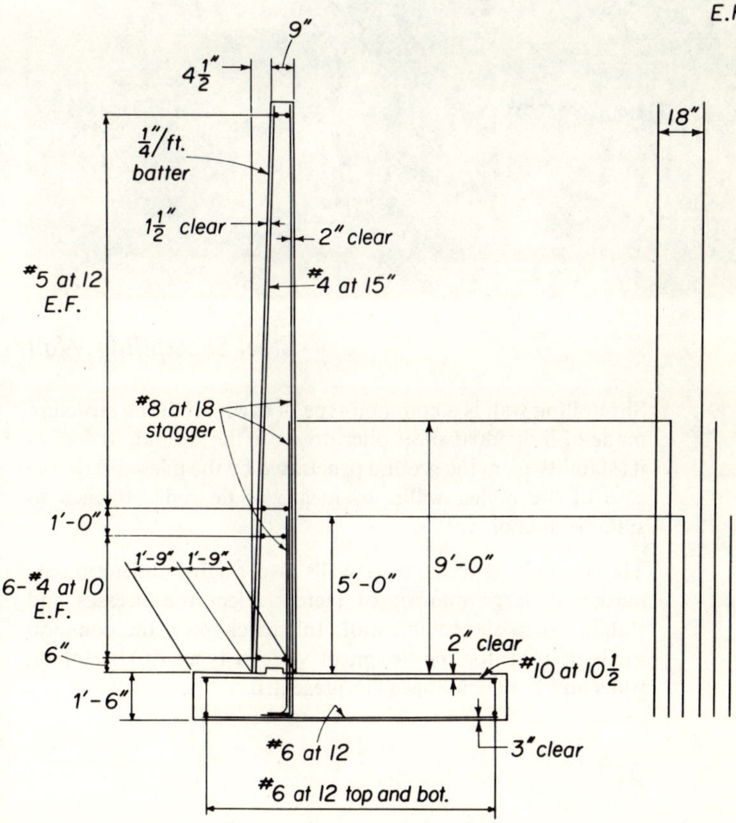

Plate Twelve

Steel Sheetpiling Wall

Sheetpiling wall is a common type of earth retaining structure made of individual sheet piles driven in the ground. It derives its stability from the ground penetrated by the piles and, in the case of the higher walls, by means of tie rods attached to suitable anchorages.

The analysis of sheetpiling walls is of highly indeterminate nature. A large number of factors affect the stresses and stability of a sheetpiling wall. In this chapter, the common methods used for the design of various types of sheetpiling walls and the anchorages are presented.

12 SHEETPILING WALLS

12-1 *Use of Sheetpiling Walls*

A sheetpiling wall consists of a series of sheet piles driven side by side into the ground thus forming a continuous vertical wall for the purpose of retaining an earth bank. It is also known as a single sheetpiling wall as opposed to the double walls and cellular cofferdams. Sheetpiling walls are commonly used for:

1. Water front construction where other types of retaining walls would require dewatering the site.
2. Temporary construction because of the high salvage value of sheet piles.
3. Lightweight construction at locations where the upper layer or layers of subsoil are inadequate for supporting retaining walls.

Because of these advantages, sheetpiling walls are employed as bulkheads in piers, docks, and wharves, and in sea walls, breakwaters, and other shore protection construction. They are also used for constructing jetties, groins and boat basins as well as for retaining earth on land areas. However, sheetpiling walls are not suitable for the following cases:

1. Very high walls which require disproportionately high flexural strength of the pile section.
2. Inadequate depth of penetration due to boulders in the subsoil or high bedrock which prevents pile penetration.

12-2 *Common Types of Sheet Piles*

Sheet piles are prefabricated or precast members to be driven vertically into the ground to form a continuous vertical wall. There are a variety of sheet piles, ranging from simple wood planks and light gauge sheet metals to heavy sections made of reinforced concrete and structural steel members. The characteristics of the various types of sheet piles are discussed below.

A. Wood sheet piles. Wood sheet piles are simply wood planks of 2 in. × 12 in. size or thicker, Fig. 12-1. In the simplest form, single planks are driven edge to edge. If double planks are used, they are nailed together in staggered

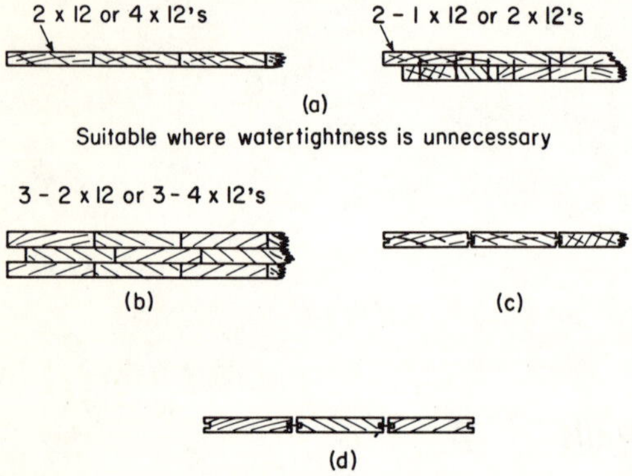

Fig. 12-1 Wood sheet piles: (a) planks; (b) Wakefield; (c) tongue and groove (tongue and groove cut in the mill); (d) splined (grooves cut in the mill, splines driven after piles are in place).

position, thus forming a lap at each joint. The Wakefield type is formed by nailing three planks together with the center plank offset 2 to 4 in. In tongue-and-groove piles and splined piles, the tongues and grooves are cut in the mill.

Wood sheet piles may be considered permanent if they are permanently underwater, or if they are impregnated with preservatives and mopped with preservatives on sawed or cut surfaces. This type of sheet pile is difficult to drive into compact soil even when water jetting is used. The lower end of the pile may be cut with a bevel and provided with a driving shoe made of $\frac{1}{16}$ to $\frac{1}{8}$ in. thick metal.

B. Concrete sheet piles. Concrete sheet piles are precast members designed to withstand the permanent stresses during service and the handling stresses during construction. The procedure for determination of permanent stresses is described in the discussion on design methods. The handling stresses are produced by the weight of the pile when it is picked up in more or less horizontal position. A short pile may be handled with a sling looped approximately at the third point from the top. For long piles, two or more pickup points may be necessary in order to reduce the handling stresses. The locations of pickup points must be clearly marked on the piles. To avoid extensive cracks due to shrinkage and handling, a certain minimum amount of reinforcing is desirable, as shown in Fig. 12-2, even if the calculated stresses are low. Also the reinforcing ties should be placed at closer spacing at the top and bottom of the pile to reduce the possibility of damage due to driving impact. The bottom of the pile is usually made with a bevel on the tongue side so that the pile will tend to be located tight against the one previously driven. Sheetpiling walls are generally provided with flexibility to allow for relative movement between sections of wall. Concrete piles if grouted for watertightness should have expansion joints above the dredge line (the ground surface in front of the sheet pile wall).

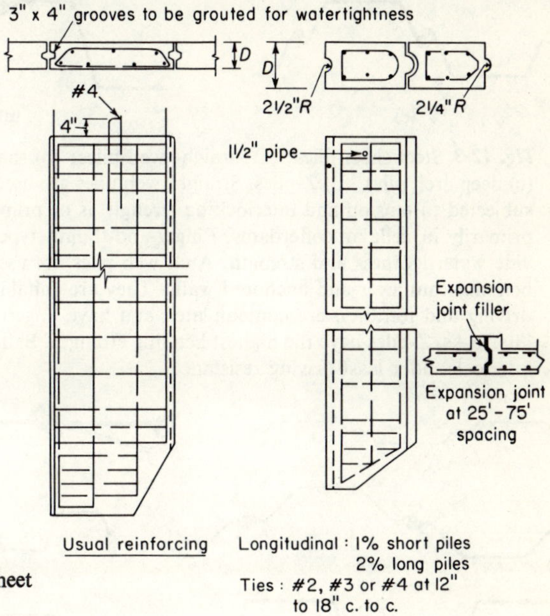

Fig. 12-2 Concrete sheet piles.

Concrete sheet piles are heavy and bulky, therefore, require heavier equipment to drive and handle. Because of the large volume displacement they also encounter greater driving resistance. Furthermore, concrete sheet piles, because of their greater rigidity, would be subjected to greater bending stresses than other types of piles, as will be discussed in Sec. 12-10.

C. Steel sheet piles. Steel sheet piles are rolled structural members with interlockings to engage with one another. There are a variety of steel sheet piles. In the American sections which are suitable for heavy construction, Figure 12-3, there are two types of interlocks:

Finger-and-thumb type

Ball-and-socket type

The interlocks have somewhat different forms in sheet piles produced by different manufacturers, and only a few sections manufactured by the same manufacturer will interlock with each other. Therefore, if different sections are contemplated for the same job, the manufacturers' catalogues should be

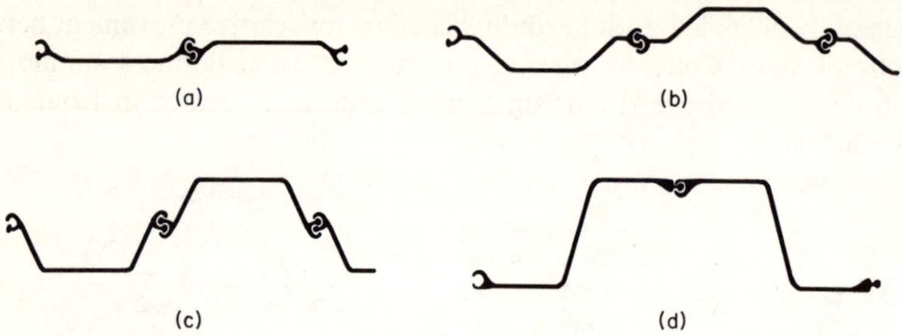

Fig. 12-3 Steel sheet piles: (a) straight web piles; (b) shallow arch piles; (c) deep arch piles; (d) Z-piles. Straight web piles are used where piles are subjected to tension and interlocking strength is of primary importance–primarily in cellular cofferdams. Finger-and-thumb type interlocks provide watertightness and strength. Arch web piles are used to resist large bending–cantilever and anchored walls. They are suitable where difficult driving and retration are contemplated and have finger-and-thumb type interlocks. Z-piles have the highest bending strength. Ball-and-socket type interlocks offer least driving resistance.

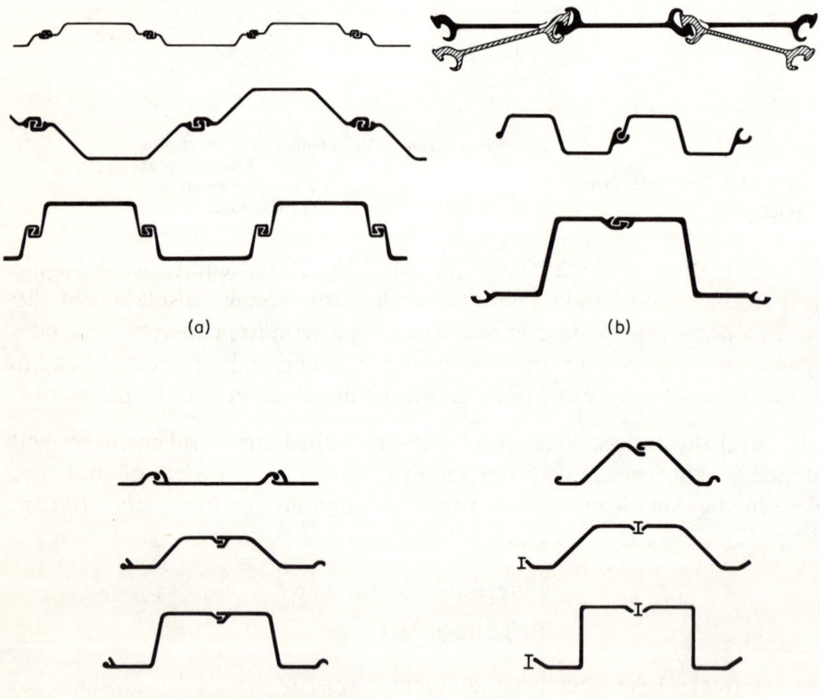

Fig. 12-4 Some European steel sheet piles: (a) Larsen piles (after Western Service Corp.); (b) Arbed-Belval piles (after Amerlux Steel Products Corp.); (c) Hoesch piles; (d) Krupp piles.

consulted. The American sheet piles are made with fairly loose interlocks and sometimes it may be desired to drive the piles in reversed position with arches all on one side of the wall. This may result a large saving in the quantity of concrete in the cases where sheet piles are used as forms for concrete walls. However, in this reversed position, the pile alignment is more difficult to maintain and the interlocks may easily become disengaged during driving. Therefore driving in reversed positions is ordinarily not recommended.

In American practice, the interlocks are assumed as offering no friction resistance and the section modulus is calculated about the neutral axis of each pile. There is undoubtedly a certain amount of friction between piles, and the actual flexural strength is therefore larger than the calculated value. All steel sheet piles are normally rolled in the mills. Pieces at corners and joints, in the forms of Y or T are fabricated, either by riveting or welding. Steel used for sheet piles has a higher carbon content than ordinary structural steel. Under permanent load, the allowable flexural stress is generally taken as 18,000 psi.

While the variety of sheet piles rolled by different mills in the U.S.A. are practically identical, the geometry of the pile sections manufactured in Europe is quite variegated. Figure 12-4 shows some of the commonly used European sections.

The sheet piles shown in Fig. 12-3 have web thickness of either $\frac{3}{8}$ in. or $\frac{1}{2}$ in. and they are suitable for retaining earth banks of moderate to large height. For retaining a shallow excavation or earth bank, sheet piles made of light gauge metal may be more economical. Examples of such light gauge sheetings are shown in Fig. 12-5.

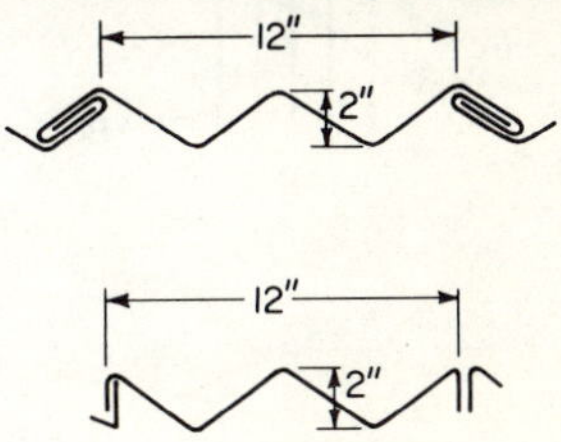

Fig. 12-5 Lightweight steel sheeting. Armco Steel Corp.

12-3 Common Types of Sheetpiling Walls

Sheetpiling walls may be classified into four types as shown in Fig. 12-6.

A. Cantilever sheetpiling. Sheet piles are driven to a sufficient depth in the ground to become fixed as a vertical cantilever in resisting the lateral earth pressure. This type is suitable for moderate height. It is subject to large lateral deflection and is readily affected by scour or erosion in front of the wall.

B. Anchored sheetpiling. An anchored sheetpiling derives its support against the lateral pressure by embedment in the ground (as the cantilever sheetpiling) and by the use of tie rods near the top of the piling. This type is suitable for moderate to high walls. For walls higher than about 35 ft, two or three tiers of tie rods may be necessary in order to reduce the required pile penetration and the flexural stresses.

C. Sheetpiling with relieving platform. A relieving platform is a pile-supported deck constructed for the purpose of supporting the weight of earth fill and other heavy load, thus relieving the lateral pressure from the sheet-piling. This type is especially advantageous in the cases where railway tracks or crane runways are close to the sheetpiling.

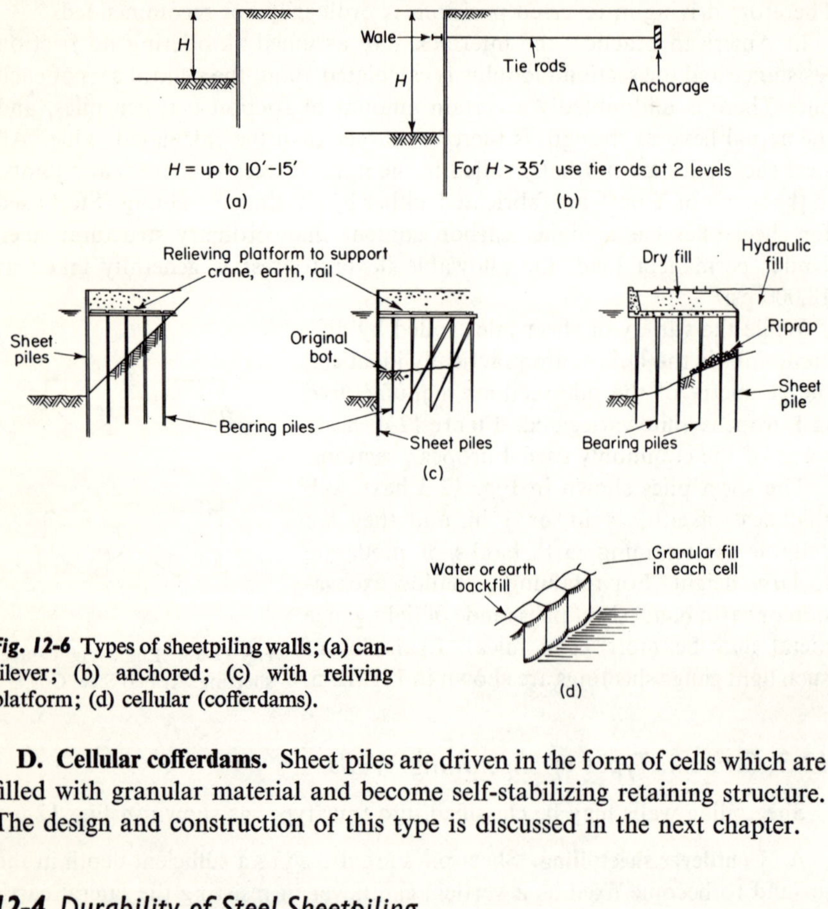

Fig. 12-6 Types of sheetpiling walls; (a) cantilever; (b) anchored; (c) with reliving platform; (d) cellular (cofferdams).

D. Cellular cofferdams. Sheet piles are driven in the form of cells which are filled with granular material and become self-stabilizing retaining structure. The design and construction of this type is discussed in the next chapter.

12-4 Durability of Steel Sheetpiling

Steel sheetpiling walls are constructed on dry land, in fresh water, and in sea water. The rate of corrosion differs considerably between these locations. On dry land, the usual practice of allowing an additional thickness of $\frac{1}{16}$ in. for corrosion is satisfactory in normal conditions where the subsoil does not contain corrosive constituents or a stray current. Only in occasional cases, other methods of protection discussed in Sec. 8-7D may be necessary.

The problem concerning the durability of steel sheetpiling was studied extensively by Beach Erosion Board (1952). A total of ninety-four structures located along the Gulf Coast of Florida and the Atlantic Coast of the United States were investigated. Most of the sheetpilings under this investigation were exposed to normal sea water or sea water only moderately diluted with fresh water, and a few were exposed to fresh water with occasional intrusions of sea water. This study indicated that the principal factors affecting the rate of deterioration are:

1. *Type of structure.* The pilings were classified into three types: harbor bulkheads, beach bulkheads, and groins. Harbor bulkheads were those in wharves, piers, slips, etc., all of which are protected to some extent from wave action. Beach bulkheads were located along the shore to protect the land from storm waves. Groins were generally built normal to the shore line to retard littoral currents; part of their surface is subjected to wave action. The average rate of loss of thickness for each type of structure were:

Harbor bulkheads	0.0033 in./yr
Beach bulkheads and groins	0.016 in./yr

It is noted that the durability of steel sheetpiling in harbor bulkheads was about five times that in other shore structures.

2. *Geographical location.* The average rates of loss of thickness are listed in Table 12-1.

3. *Zone relative to tidal planes.* The rate of deterioration varied with the height along the piling as shown in Table 12-1.

4. *Sand, earth, or other cover.* Sand, earth, or other covers on one or both sides protect the piling and greatly reduce the rate of loss. A comparison of the measured results indicates that the average rate of loss for sheetpiling covered on both sides ranged between 0.0016 to 0.0026 in. per year. But without cover, the rate might be increased 3 to 16 fold.

5. *Exposure to salt spray.* Salt spray from sea water increased the rate of corrosion as shown in Table 12-1.

6. *Paint protection.* The benefit of painting can be readily seen from the average rate of loss of thickness shown in Table 12-1.

Few of the structures studied were painted regularly. The fact that occasional painting reduced the rate of loss substantially indicates that regular painting would result in greater reduction. However, the justification of regular painting of sheetpiling depends upon the relative economy. For example, painting of groins or jetties in the tidal zone subject to abrasive action probably could not be justified.

In any case, it is desirable to paint the sheetpiling before driving with a coat of tar, bituminous paint, or other protective material. Wales should be provided with drain holes to prevent retention of water in the trough between flanges of the member. Tie rods should be coated with bituminous paint or tar, wrapped with fabric of tough and durable quality, and coated again with paint or tar. Protection of tie rods with concrete cover is not considered desirable due to the likelihood of cracks in the concrete thus permitting penetration of moisture.

Table 12-1 FACTORS AFFECTING THE RATE OF DETERIORATION OF STEEL SHEETPILING

	Harbor bulkheads (in./yr)	*Beach bulkheads (in./yr)*	*Groins and jetties (in./yr)*
Geographical location			
South (south of Welmington, N.C.)	0.0062	0.017	0.018
North (north of Pt. Pleasant, N.J.)	0.0023	0.0075	0.011
Zone relative to tidal planes			
8 ft above mean high water	0.0049 (8 ft to 2 ft above mean high water)	0.020	0.010 (8 ft to 2 ft above mean high water)
5 ft to 8 ft above mean high water		0.022	
2 ft to 5 ft above mean high water		0.0081	
Mean high water	0.0027	0.0074	0.0055
Mean tide level	0.0024	0.001	0.024
Mean low water	0.0035	0.002	0.028
Mean low water to ground line	Average of 4 values = 0.0036		
Below ground line	Average of 4 values = 0.0016		
Exposure to sait spray			
Heavy spray	0.0083	0.016	0.016
Moderate spray	0.0041	(beach bulkheads, groins, and jetties are considered to be subjected to heavy spray)	
Light spray or none	0.0024		
Paint protection			
None	0.0045	0.018	0.020
At least painted once	0.0027	0.011	0.010

12-5 Design of Sheetpiling Walls

The design procedure of sheetpiling walls generally comprises the following steps:

1. *Assemble the general information.* In addition to the topographical survey, the controlling dimensions must be included in the general information. The controlling dimensions are the elevation of top of wall, the dredge line (the elevation of ground surface in front of the wall), the maximum water level, the mean tide level (normal pool), and the low water level.

2. *Analyse the subsoil conditions.* The shear strength of each soil stratum should be determined by standard penetration tests (for granular soils) and unconfined compression strength (for cohesive soils).
   If considerable dredging is done, the effective pressure on the soil is reduced. This reduction in pressure tends to allow reduction in shear strength (Sec. 1-5). In such a case, unconfined compression tests may give unsafe results, and laboratory tests should be made to predict the anticipated conditions.
   A soil profile should be drawn for the most unfavorable conditions revealed by the soil borings. The borings should be carried to a very dense or hard layer or bedrock. The contemplated sheetpiling wall and the type of backfill material should also be shown in the profile.
3. *Select the type of wall.* Sections 12-2 and 12-3.
4. *Compute earth pressure and surcharge pressure.* Section 12-6.
5. *Determine the piling penetration.* Sections 12-7, 12-8, and 12-9.
6. *Determine the bending stress and design the piling.* Sections 12-7, 12-8, and 12-10.
7. *Design the tie rods.* Section 12-11.
8. *Design the anchorage.* Sections 12-3, 12-14, and 12-15.

12-6 Lateral Pressure Acting on Sheetpiling Walls

A sheetpiling wall may be subjected to some or all of the following types of lateral pressure.

Earth pressure: active and passive pressure

Lateral pressure due to surcharge load

Unbalanced water pressure and seepage pressure

Mooring pull, ship impact, etc.

Earthquake force, wave pressure, etc.

The procedure for calculation of earth pressure, unbalanced water pressure and seepage pressure are presented below. The lateral pressures due to other loads were discussed in Chapter 4.

A. Earth pressure acting on sheetpiling walls. The actual earth pressure acting on sheetpiling walls cannot be calculated by the classical theories (Rankine, Coulomb, and wedge theories). The classical theories are all based on the condition that the wall yields laterally, by sliding or by rotation about the bottom of the wall, to such an extent that the shear strength of the soil is fully mobilized. This condition is generally satisfied for ordinary retaining walls. The sheetpiling walls, however are supported differently, and they are more flexible, consequently they do not yield in the same manner as the ordinary retaining walls. An anchored sheetpiling wall, due to the elastic deflection of the wall, will bulge, or yield considerably more at a point between the tie rod level and the dredge line than other portions of the wall. This large yield tends to relieve the magnitude of pressure on that portion of the wall. The pressure distribution is further influenced by the amount of elongation of the tie rod and the depth of sheet pile penetration.

The earth pressure against sheetpiling walls can be determined by theories which take into account of the conditions of yield of the wall (Hansen, J. Brinch, 1953). Although the procedure is more laborious than the classical method it is recommended for larger projects. In practice, several empirical and semiempirical methods have been developed all of which use the classical earth pressure theories. The Coulomb theory has been employed to determine the active and passive earth pressures against the sheetpiling. Since this theory gives misleadingly large values for passive earth pressure, it should be used conservatively. For more accurate design, the wedge theory may be used for passive pressure calculations. The values of φ (angle of internal friction of soil) and δ (angle of wall friction) recommended for earth pressure calculations are shown in Table 12-2. The corresponding coefficients of earth pressure K_a and K_p for the condition of horizontal ground surface are also shown in the same table.

Table 12-2 UNIT WEIGHTS OF GRANULAR SOILS AND COEFFICIENTS OF EARTH PRESSURE

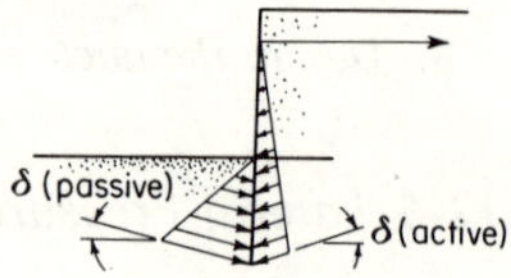

Type of soil	Unit wt of moist soil* γ		Unit wt of submerged soil* γ'		Coefficient of active earth pressure K_a				Coefficient of passive earth pressure K_p		
	Min	Max	Min	Max	For backfill	For soils in place	Friction angles† ϕ	Friction angles† δ	For soils in place	Friction angles† ϕ	Friction angles† δ
Clean sand:											
dense	110	140	65	78		0.20	38	20	9.0	38	25
medium	110	130	60	68		0.25	34	17	7.0	34	23
loose	90	125	56	63	0.35	0.30	30	15	5.0	30	20
Silty sand:											
dense	110	150	70	88		0.25	34	17	7.0	34	23
medium	95	130	60	68		0.30	30	15	5.0	30	20
loose	80	125	50	63	0.50	0.35	26	13	3.0	26	18

* In pounds per cubic foot.

† These angles, expressed in degrees, are ϕ, the angle of internal friction, and δ, the angle of wall friction, and are used in estimating the coefficients under which they are listed.

After Terzaghi, 1954.

B. Unbalanced water pressure and seepage pressure. Sheetpiling walls are widely used for water front construction. When the tide or river level is at the lowest stage, the sheetpiling is subjected to the maximum earth pressure. During a rain storm or a rapidly receding high water, the water level behind the wall may be several feet higher than that in front of the wall and the difference in water level introduces additional pressure on the piling. Furthermore, the receding water percolates downward through the soil behind the sheetpiling and then upward in front of the piling. The upward seepage reduces the effective weight of the soil and consequently reduces its passive resistance. Therefore, it is necessary to evaluate the unbalanced water pressure and the effect of seepage pressure in the cases where conditions of unequal water level may exist.

The height of water head lagging behind a rapidly receding tide or river stage depends primarily upon the type of backfill used. In coarse sand and gravel, the lag may be negligible. In fine or silty sand, the lag may be several feet. If the backfill is clay or silt, full hydrostatic pressure should be assumed below the highest possible position of water level.

If a sheetpiling is driven in granular soil with fairly uniform coefficient of permeability, the seepage water follows the paths indicated by the arrows shown in Fig. 12-7(a) This diagram is known as a flownet.* The unbalanced

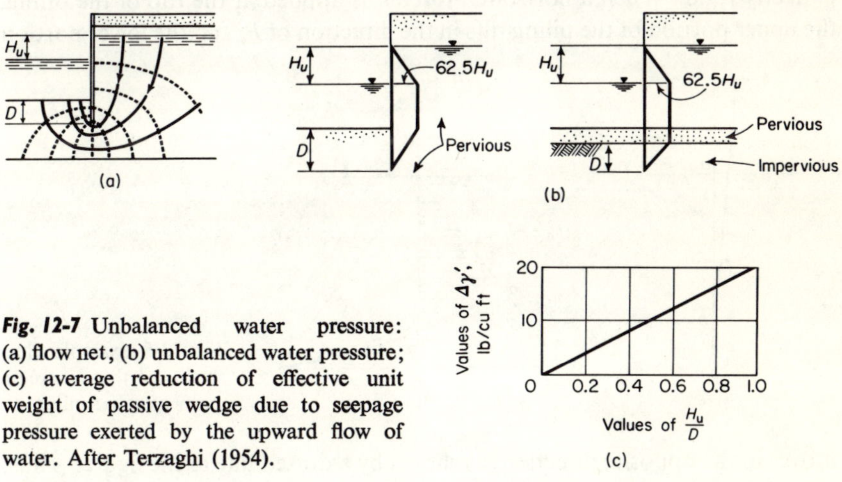

Fig. 12-7 Unbalanced water pressure: (a) flow net; (b) unbalanced water pressure; (c) average reduction of effective unit weight of passive wedge due to seepage pressure exerted by the upward flow of water. After Terzaghi (1954).

water pressure may be approximately established by the trapezoid in Fig. 12-7(b), where γ_w = the unit weight of water. If the permeability of soil varies widely in the vertical direction, the distribution of unbalanced water pressure must be determined by construction of a flow net.

The effective weight of soil below a static water table is the bouyant or

* See standard textbook of soil mechanics, e.g., Donald W. Taylor, *Fundamentals of Soil Mechanics*. (New York: John Wiley & Sons, Inc., 1948).

submerged weight. Under the action of upward seepage, the submerged unit weight is reduced approximately by the following amount,

$$\Delta'_\gamma = 20\frac{H_u}{D} \tag{12-1}$$

where Δ'_γ = reduction in submerged unit weight of soil, pcf; the effective unit weight to be used in the computation of passive pressure is $(\gamma' - \Delta'_\gamma)$;

H_u = unbalanced water head, ft;

D = as shown in Fig. 12-7(b).

The relationship between Δ'_γ and H_u/D is shown in Fig. 12-7(c).

The effect of downward seepage in the soil behind the piling is very small and may be neglected.

12-7 Design of Cantilever Sheetpiling Walls

A. General principle of design of cantilever sheetpiling. The action of earth pressure against a cantilever sheetpiling can be best illustrated by a simple case shown in Fig. 12-8(a). In this case, the sheetpiling is assumed to be perfectly rigid. When a horizontal force P is applied at the top of the piling, the upper portion of the piling tilts in the direction of P, and the lower portion

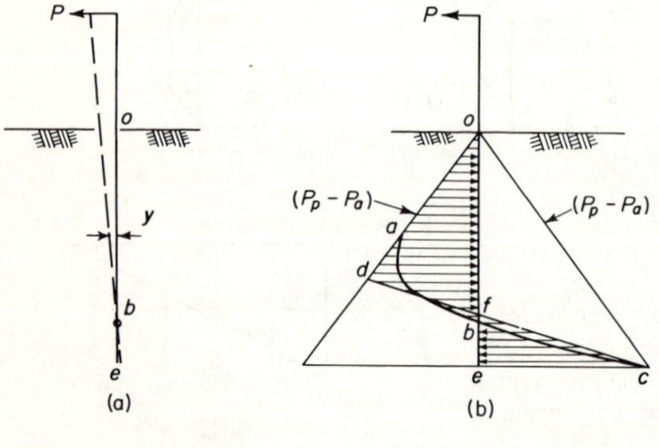

Fig. 12-8 Example illustrating earth pressure on cantilever sheetpiling.

moves in the opposite direction, as shown by a dotted line in the figure. Thus, the piling rotates about a stationary point b. The upper portion ob is subjected to a passive earth pressure from the soil on the left side of the piling, and the lower portion eb is subjected to a passive earth pressure from the soil on the right side of the piling. At point b, the piling does not move and therefore it is subjected to equal and opposite earth pressures (at-rest pressures from both sides) with a net pressure equal to zero. The earth pressure is represented by the diagram $oabc$ in Fig. 12-8(b). The lines oa and oc represent the net passive resistance which is equal to the passive earth pressure minus the active earth pressure acting in opposite directions. For the purpose of design, the curve of *abc* is replaced by a straight line *dc*. Point *d* is located at such a position on the line *oa* that the sheetpiling is in static equilibrium under the action of force *P* and the earth pressure represented by the areas *odf* and *fce*. The position of point *d* can be determined by a trial-and-error method.

This discussion leads to the conclusion that a cantilever sheetpiling derives its stability from passive earth pressure on both sides of the piling. However, the distribution of earth pressure is different between sheetpiling in granular soils and sheetpiling in cohesive soils. And, the pressure distribution is likely to change with time for sheetpilings in clay. Therefore, the design procedures for sheetpiling in both types of soils are discussed separately in the following.

B. Design of cantilever sheetpiling in granular soils. A cantilever sheetpiling to be driven to granular soils may be designed by the conventional method in accordance with the principles just discussed, or by an approximate method based on further simplifying assumptions. These methods are illustrated in Fig. 12-9 where the subsoil is assumed to consist of one layer of

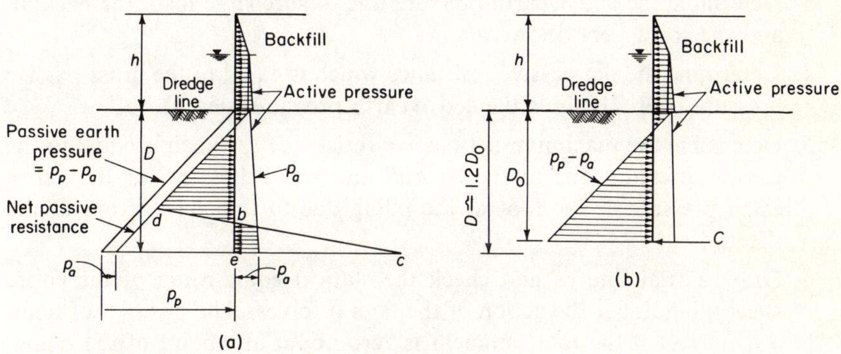

Fig. 12-9 Design of cantilever sheetpiling in granular soils: (a) conventional method; (b) simplified method.

soils throughout the piling penetration. For the cases where two or more layers are penetrated, the earth pressure distributions are different but the basic design concept remains the same.

In either design method, the earth pressure should be calculated by using appropriate values for γ (unit weight), φ (angle of internal friction), and δ (angle of wall friction). For ordinary projects, the values shown in Table 12-2 may be used. Note that the earth pressure coefficients listed in this table are for the case of horizontal ground surface. If the soil layers or the ground surface is sloping or irregular, the earth pressure should be determined by the wedge theory, Sec. 5-4, in which the surface of rupture is assumed to be a spiral surface. Since the application of the wedge theory is laborious, the Coulomb theory is often used instead. The passive pressure obtained by the Coulomb theory should be used conservatively because it is somewhat greater than the actual values.

The conventional method of design generally consists of the following steps:

1. Sketch a profile of the piling with a trial depth of penetration. Approximate depth of penetration may be taken as follows:

Soil	*Depth of penetration**
Dense	$0.75h$
Firm	$1.0h$
Loose	$1.5h$
Very loose	$2.0h$

* h = height of piling above the dredge line.

2. Determine the passive earth pressure in front of the piling. This is the gross passive resistance due to the weight of soil. Buoyant weight should be used for soil below water level.
3. Determine the active earth pressure due to surcharge load, the backfill, and the soil layers below.
4. Determine the net passive resistance which is equal to the gross passive pressure (step 2) minus the active earth pressure (step 3).
5. Determine the maximum net passive resistance *ce* which is equal to the passive pressure due to the backfill and soil below, minus the active earth pressure at the foot of the piling due to the soil in front of the sheetpiling.
6. Draw a trial line *cd* and check the statical equilibrium of the entire sheetpiling under the action of the lateral forces. The position of point *d* is correct if the total moment is zero about any point of the piling. When it is impossible to maintain equilibrium with any location of point *d*, the trial penetration is too small.
7. Add 20 to 40 per cent to the calculated depth of penetration. This will give a safety factor of 1.5 to 2.0 approximately. An alternate and more desirable method is the use of a reduced value of passive earth pressure for design. In this method, the maximum allowable earth pressure is limited to $\frac{1}{2}$ to $\frac{2}{3}$ the ultimate passive resistance.
8. Compute the maximum bending moment which occurs at the point of zero shear prior to increasing the depth by 20 to 40 per cent.

Fig. 12-10 Design charts for cantilever sheetpiling in granular soils. From Rombas, Paris.

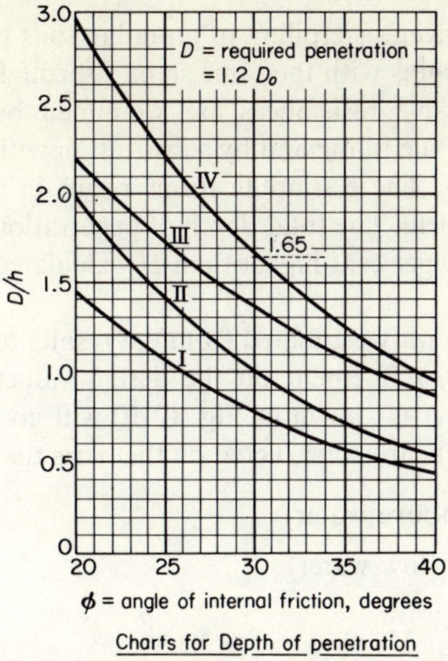

Charts for Depth of penetration

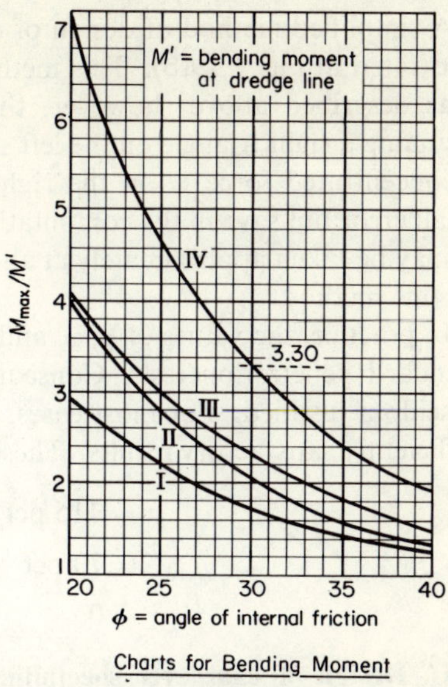

Charts for Bending Moment

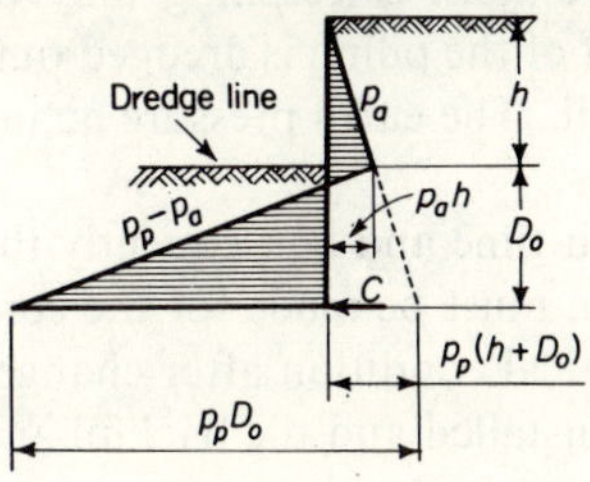

Case I Water below bottom of piling

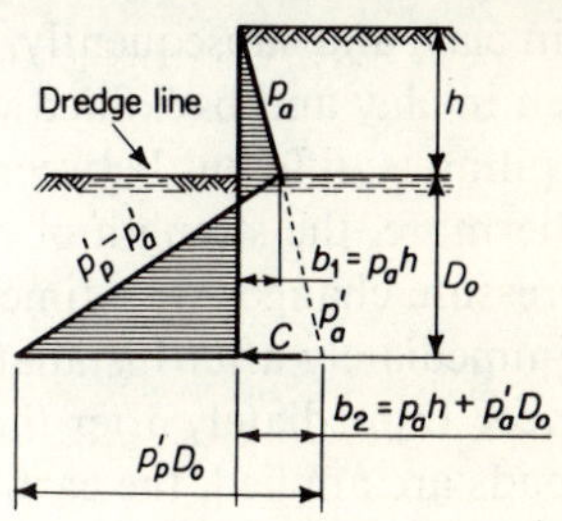

Case II Water level at dredge line

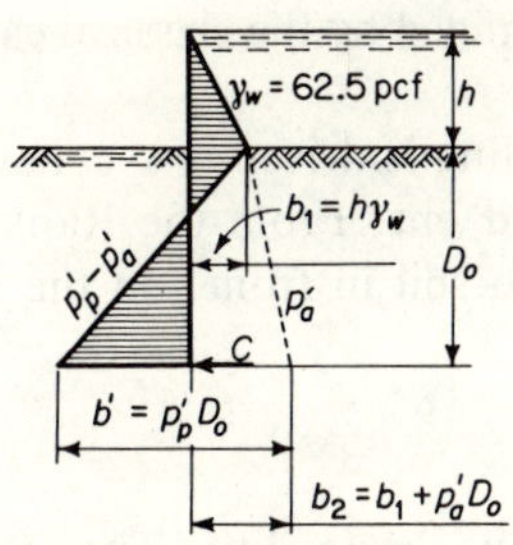

Case III Piling subjected to hydrostatic pressure only

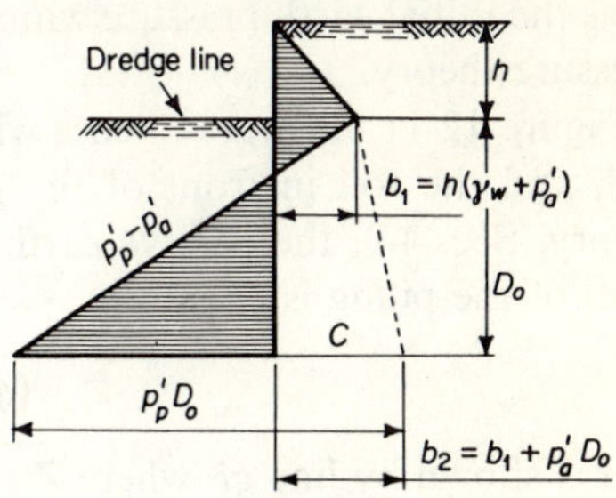

Case IV Water level at ground surface

Example

For case IV with h = 10 ft

$\phi = 30^\circ \quad \gamma' = 70\text{pcf} \quad p_a = 23.2\text{pcf}$

$\frac{D}{h} = 1.65 \quad D = 1.65 \times 10 = 16.5\text{ ft}$

$M = (62.5 + 23.2)\frac{10^3}{6} = 14.700\text{ ft lb}$

$\frac{M_{max}}{M'} = 3.30$

$M_{max} = 3.3 \times 14.700 = 48.500$ ft lb per ft of wall

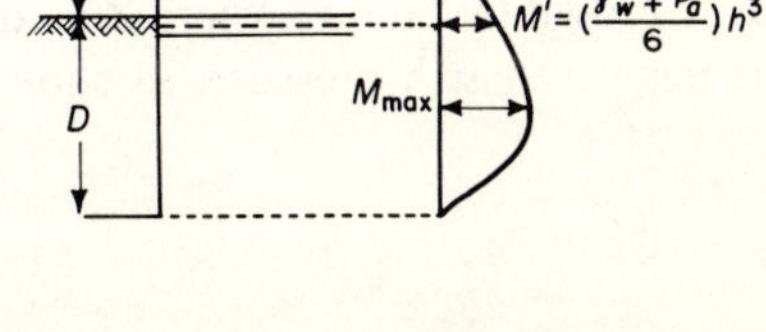

A simplified method of design of cantilever sheetpiling in granular soils is illustrated in Fig. 12-9(b). This method begins with the same steps 1 through 4 as described above; however, the passive resistances are simplified by assuming a right triangle on the left side of the piling and by substitution with a concentrated force C on the right side. These simplifications result in a small error but save in the computation work. The total depth of penetration D may be taken approximately at about 20 per cent higher than D_0 calculated by this method.

In practice, the values of γ, φ, and δ are only estimated from the results of standard penetration tests. Consequently, refinement on the design values is seldom justified. For most cases, the charts shown in Fig. 12-10 will give sufficiently satisfactory results. These charts are constructed on the basis that

$$\gamma = 115 \text{ pcf} \qquad \text{(above water)}$$
$$\gamma' = 70 \text{ pcf} \qquad \text{(below water)}$$
$$\delta = 0$$

C. Design of cantilever sheetpiling in cohesive soils. Sheetpiling may be driven in clay, and subsequently, the clay in front of the piling is dredged out; or driven in clay and backfilled with granular soil. The earth pressure acting on the piling is different between these two types.

Furthermore, the strength of clay changes with time and consequently the earth pressure changes with time also. The design must be made for the condition immediately after installation or for a critical condition after changes take place. Immediately after the sheetpiling is installed and the backfill and other loads are applied, the earth pressure may be calculated on the assumption that the angle of friction of clay φ is zero; and the cohesive strength c is equal to half the unconfined compression strength q_u. This case is referred to as the initial earth pressure which may be determined by the classical earth pressure theory.

Figure 12-11 illustrates a case where the sheetpiling is driven into cohesive soil, and the soil in front of the piling is dredged out. From the Rankine theory, Sec. 4-3, the passive earth pressure due to soil in front (on the left side) of the piling is equal to

$$(q_u + \gamma_e Z)$$

and is shown by line gi, where Z = depth below the dredge line. The active earth pressure due to the soil behind (or on the right side) of the piling is equal to

$$(\gamma_e Z - q_u)$$

and is shown by line jk, where Z = depth below the original ground surface. The negative earth pressure, as shown by the dotted line, is usually ignored because the soil may develop cracks in the upper portion. The net passive resistance is constant throughout the pile penetration D, and is equal to

$$(2q_u - \gamma_e h)$$

For the lower end of the piling where it moves toward the right side, the net maximum passive resistance ce is equal to

$$(2q_u + \gamma_e h)$$

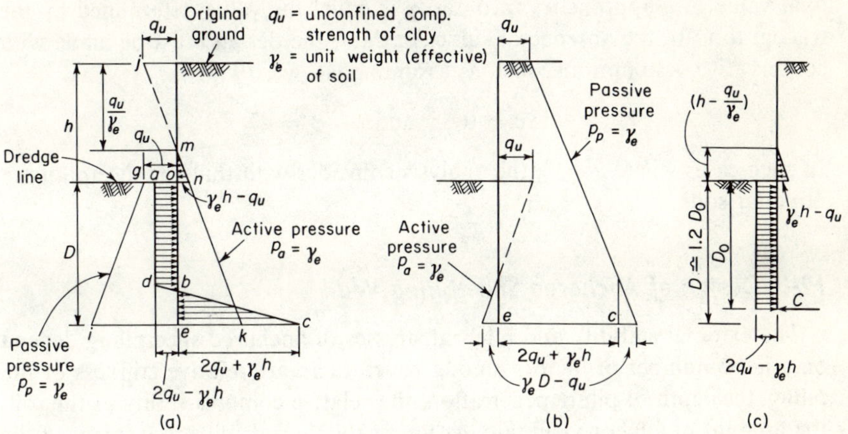

Fig. 12-11 Initial earth pressure for design of cantilever sheetpiling entirely in cohesive soil.

which is derived from Fig. 12-11(b). Based on the discussion of the general principle for cantilever piling, the earth pressure diagram is *mnoadbce*, where point d and the depth of penetration D are so chosen to satisfy the static equilibrium of the lateral forces. Similar to the simplified method for granular soil, the design may be made by the use of the pressure diagram shown in Fig. 12-11(c).

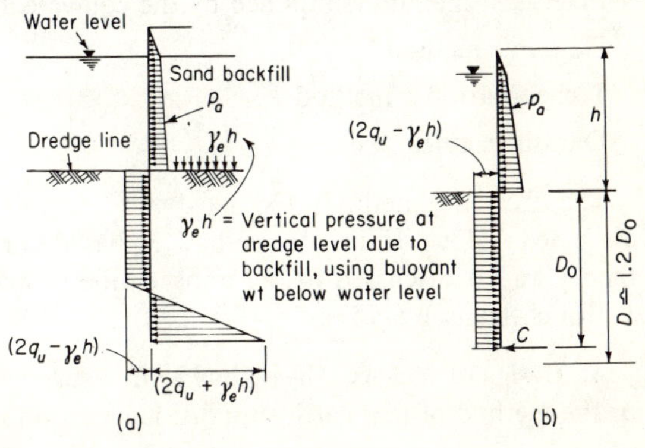

Fig. 12-12 Initial earth pressure for design of cantilever sheetpiling: (a) in cohesive soil; (b) backfilled with granular soil.

This reasoning may be readily extended to the case where the sheetpiling is driven in clay but backfilled with granular soil, Fig. 12-12(a). Again, the simplified method shown in Fig. 12-12(b) is expeditious.

The evaluation of the initial earth pressure presents no problem in design. In fact, the procedure is quite simple. However, the longtime stability of the sheetpiling, after the piling is in service for a period of years, is a difficult problem, because as time goes on, the values of c and φ tend to change. The final values are difficult to predict. Limited information indicates that the final value of c approaches zero and φ is equal the value determined by the triaxial test. In the absence of sufficient data, the design must be made with conservative assumptions, such as assuming (Rowe, 1957)

$$c = 0 \qquad \text{and} \qquad \varphi = 27$$

In such cases where $c = 0$, the analysis coincides with that for sheetpiling in granular soils.

12-8 Design of Anchored Sheetpiling Wall

The external stability and internal stresses of anchored sheetpiling depend on a large number of factors among which there are relative stiffness of the piling, the depth of piling penetration, the relative compressibility of the soil, the amount of anchor yield (elongation of the tie rods plus movement of the anchor block), etc. Every one of these factors affects the others in different degrees. For example, a sheetpiling driven to a great depth in granular soils would have a smaller flexural stress than one driven to a smaller depth, because in the former case the soil surrounding the lower portion of the piling tends to prevent it from rotation. Because of the large number of factors affecting each other, anchored sheetpiling has been designed by a number of methods which may be grouped as follows:

Free-end method (may be modified by Rowe's theory)

Fixed-end method (simplified by the equivalent beam method)

Hansen's method

Tschebotarioff's method

Danishish rules

The free-end method, the fixed-end method and Rowe's method are discussed below. Hansen's method is based on a rational earth pressure theory and is discussed in the book entitled *Earth Pressure Calculations* by J. Brinch Hansen (1953).

A. Design of anchored sheetpiling by free-end method. The free-end method, or the method of free earth support, is based on the following assumptions:

1. The sheetpiling is perfectly rigid as compared to the surrounding soils.
2. The soil pressure acting on the piling may be computed by the Rankine or Coulomb theory.
3. The sheetpiling is free to rotate but is not permitted lateral movement at the tie rod level. At its ultimate capacity the piling fails by outward movement rotating about the level of tie rods.

With these assumptions, the design becomes a problem in simple statics. Procedures for the design of anchored sheetpiling, Fig. 12-13, in granular soil and in cohesive soil are discussed below.

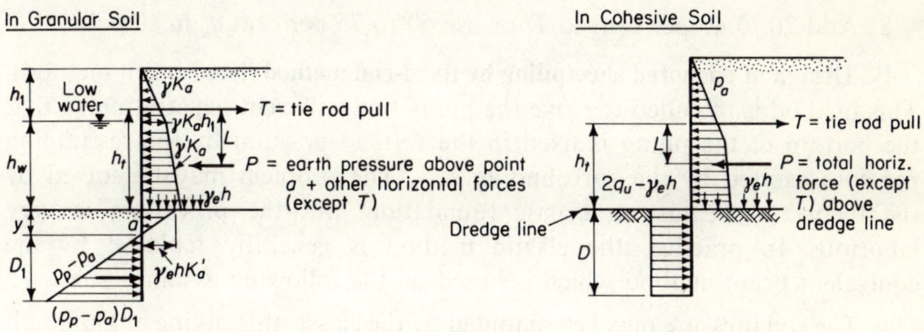

Fig. 12-13 **Design of anchored sheetpiling by free-end method.**

In granular soil:

1. Select appropriate values for active and passive earth pressure.
2. Calculate weight of overburden and surcharge load at the dredge level, $\gamma_e h$.
3. Locate point of zero pressure $y = \gamma_e h k'_a/(p_p - p_a)$
4. Take moment about tie rod level:

$$L_P - \tfrac{1}{2}(p_p - p_a)D_1^2(h_t + y + \tfrac{2}{3}D_1) = 0$$

   Solve for D_1.

5. Compute tie rod tension $T = P - \frac{1}{2}(p_p - p_a)D_1^2$
6. Determine maximum bending moment at point of zero shear, as in the case of ordinary flexural member.
7. Select pile section for the maximum moment or use the moment reduction theory, Sec. 12-10.
8. Add 20 to 40 per cent to D_1 to provide for safety margin, or divide p_p by a factor of safety of 1.5 to 2.0 in steps 3 and 4.

In cohesive soil:

1. Select appropriate values for active earth pressure p_a.

2. Calculate weight of overburden and surcharge load at the dredge level, $\gamma_e h$.
3. Evaluate the unconfined compression strength q_u of the cohesive soil.
4. Take moment about tie rod level

$$LP - (2q_u - \gamma_e h)D(h_t + \tfrac{1}{2}D)$$

Solve for D.

5. Compute tie rod tension $T = P - (2q_u - \gamma_e h)D$
6. Determine maximum moment at point of zero shear.
7. Select pile section for the maximum moment, or use the moment reduction theory, Sec. 12-10.
8. Add 20 to 40 per cent to D or use 50 to 75 per cent q_u in step 4 and 5.

B. Design of anchored sheetpiling by fixed-end method (fixed beam method). This method is intended to drive the piling to a sufficient penetration so that the bottom of the piling is fixed in the vertical position by the restraining pressure exerted by the surrounding soil. The problem may be solved by the theory of beams on elastic foundation, but the procedure is very laborious. In practice, the elastic method is generally replaced by the equivalent beam method which is based on the following assumptions:

1. The soil pressure may be computed by the classical Rankine or Coulomb theory.
2. The sheetpiling is free to rotate but not permitted to lateral movement at the level of the tie rod.
3. The point of inflection b is known, Fig. 12-14(b). From rigorous elastic analysis, the location of this point is found to be a function of φ, angle of internal friction of the backfill material, see Fig. 12-14(e).
4. The passive resistance *oac*, Fig. 12-14(c) is simplified by substitution with a triangular pressure diagram *ode* and a concentrated reaction C.

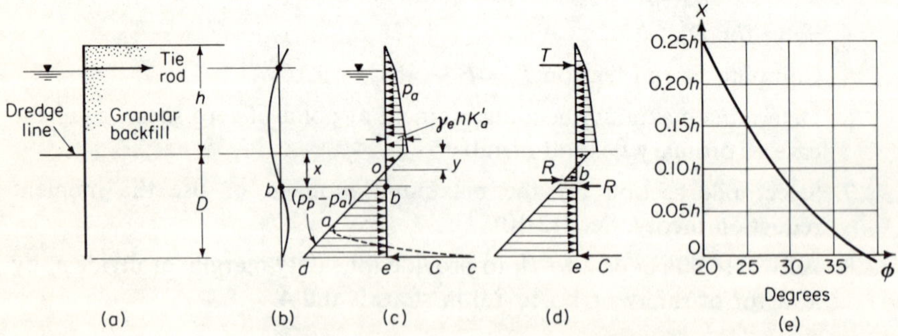

Fig. 12-14 Equivalent beam method.

Based on these assumptions, the sheetpiling may be analysed by the following steps, referring to Fig. 12-14(c).

1. Select appropriate values for earth pressures p_a, p_p', and p_a' where p_p' and p_a' are passive and active earth pressure for the natural soil deposit below the dredge line. Values shown in Table 12-2 may be used.
2. Determine depth of point *o* by

$$y = \frac{\gamma_e h k_a'}{p_p' - p_a'} \tag{12-2}$$

where $\gamma_e h$ = weight of backfill and surcharge load above the dredge line, using buoyant weight for soil below water level,

k_a' = coefficient of active earth pressure for the soil below the dredge line.

3. Locate the point of inflection by the use of chart shown in Fig. 12-14(e).
4. Determine the horizontal shear R at the point of inflection b. R is the horizontal reaction at point b by treating the piling above b as a simple beam supported at b and at the tie rod level, Fig. 12-14(d).
5. Treat the lower portion of the piling *be* as a simple beam, Fig. 12-14(d), and determine the dimension *be* by equating the moment about point e equal to zero.
6. The depth of penetration D is equal to the sum of the dimension *be* and x. To provide a margin of safety, either one of the following two methods may be used:
   (a) Add 20 to 40 per cent to the calculated depth of penetration D.
   (b) Use a reduced value of p_p' by dividing it by a safety factor of 1.5 to 2.0.

It should be noted that the equivalent beam method is valid only for sheetpiling driven in granular soil and backfilled with granular soil. In ordinary cases, the point of inflection b and the point of zero earth pressure o are located in close vicinity, and the value of x may be taken as equal to y. Therefore, the depth of penetration may be expressed in a simple equation:

$$D = y + \sqrt{\frac{6R}{p_p' - p_a'}} \tag{12-3}$$

where y = location of point *o*, point of zero earth pressure; use Eq. (12-2);

R = horizontal reaction at point *o* by assuming the piling is simply supported at point *o* and at the tie rod level;

p_p', p_a' = passive and active earth pressure in the subsoil below the dredge line.

12-9 Stability of Sheetpiling in Cohesive Soils

For sheetpiling driven in cohesive soils, whether it is a cantilever or an anchored sheetpiling and whether the backfill is granular soil or the original cohesive soil, the height of the sheetpiling is limited by the initial (or immediate) unconfined compression strength of the cohesive soil below the level of dredge line. The sheetpiling will fail if this limiting height is exceeded. Therefore, in the design of sheetpiling in cohesive soils, the first step should be the investigation of the limiting height. This investigation may be made in terms of the stability number S which is defined as (Rowe, 1957):

$$S = \frac{c}{\gamma_e h} \tag{12-4}$$

in this equation and in the following discussion, the following symbols are used:

c = the cohesion of the soil below the dredge line;

γ_e = the effective unit weight of the soil above the dredge line = moist unit weight above water level, and buoyant unit weight below water level;

h = the height of the soil above the dredge line, including the height of equivalent surcharge load, if any;

q_u = unconfined compression strength of the soil below the dredge line = $2c$;

γ = unit weight of the soil below the dredge line.

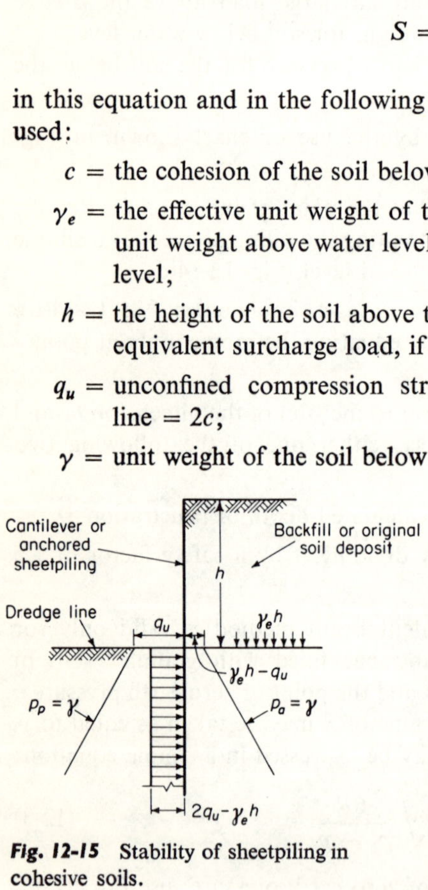

Fig. 12-15 Stability of sheetpiling in cohesive soils.

Figure 12-15 shows a sheetpiling in cohesive soil and the earth pressure acting on the portion of the piling below the dredge line. The initial passive earth pressure p_p acting on the left side of the piling is equal to $(q_u + \gamma z)$ and the initial active earth pressure acting on the right side is equal to $(\gamma_e h - q_u + \gamma z)$, where z is any depth measured from the dredge line. The net passive resistance is equal to

$$(q_u + \gamma z) - (\gamma_e h - q_u + \gamma z) = 2q_u - \gamma_e h = 4C - \gamma_e h$$

When the height of the piling h is such that the net passive resistance becomes zero, the sheetpiling will fail. At this state,

$$4C - \gamma_e h = 0$$

$$S = \frac{c}{\gamma_e h} = 0.25$$

To include the effect of adhesion c_a between the cohesive soil and the sheet-piling, this equation is modified as

$$S = \frac{c}{\gamma_e h}\sqrt{1 + \frac{c_a}{c}} \qquad (12\text{-}5)$$

where c_a = the wall adhesion. The value of $\sqrt{1 + c_a/c}$ may be taken as 1.25 for the design purposes and therefore $S = 0.3$. Any sheetpiling driven to cohesive soils should have a minimum stability number equal to 0.3 times a desired factor of safety.

12-10 Rowe's Theory of Moment Reduction

The design of anchored sheetpiling by the free-end method is based on the assumption that the piling is perfectly rigid and that the earth pressure distribution is hydrostatic, obeying the classical earth pressure theory. In reality, the sheetpiling is rather flexible and the earth pressure differs considerably from the hydrostatic distribution. There are a large number of factors that influence the behavior of an anchored sheetpiling. For anchored sheetpiling in granular soils, the most significant factors are (Rowe, 1952; Terzaghi, 1954):

1. The relative density of the soil.
2. The relative flexibility of the piling which is expressed in terms of the flexibility number

$$\rho = \frac{H^4}{EI}$$

where H = the total height of the piling,

EI = the modulus of elasticity and the moment of inertia of the piling.

For anchored sheetpiling in cohesive soils, the most significant factors are (Rowe, 1957 and 1958):

1. The stability number

$$S = \frac{c}{\gamma_e h}\sqrt{1 + \frac{c_a}{c}} = 1.25\frac{c}{\gamma_e h}$$

2. The flexibility number

$$\rho = \frac{H^4}{EI}$$

3. The relative height of piling α.

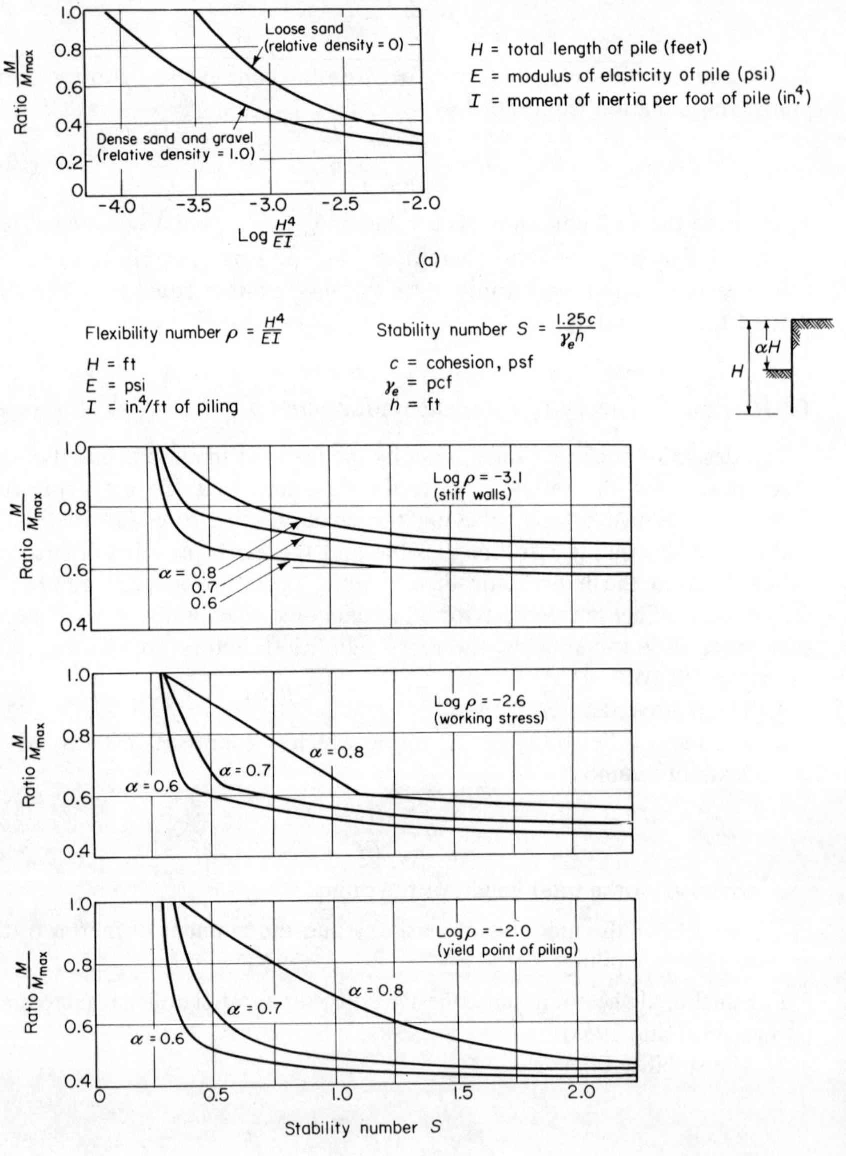

Fig. 12-16 Bending moment in anchored sheetpiling designed by free-end method: (a) in granular soils; (b) in cohesive soils. After P. W. Rowe.

The notations used in the equation for the stability number are described in connection with Eq. (12-5). The flexibility number ρ is calculated in the same manner as for the case of piling in granular soils. The notation α denotes the ratio between the height of the piling above the dredge level to the total height of the piling.

The relationship between the bending moment in the anchored sheetpiling and these factors is shown in the form of charts in Fig. 12-16. In these charts, the vertical axes indicate the moment reduction factor which is the ratio of (1) the moment to be used for design to (2) the maximum moment calculated by the free-end method. The use of these charts can be best illustrated by numerical examples shown in DE 12-1 and DE 12-2.

12-11 Wales and Tie Rods

In the system of anchored sheetpiling a wale (or waler) is a flexural member whose function is to receive the horizontal reaction from the sheetpiling and to transfer it to the tie rods. Usually it consists of a pair of structural channels placed with the web in the horizontal position, Fig. 12-17. For high sheetpiling walls, the channels may be reinforced with cover plates. The channels are spaced with a sufficient distance between the backs of the channels to clear the upset end of the tie rods. Pipe separators or other forms should be used for connecting the pair of channels together.

Tie rods are round or square bars made of structural steel, and usually have upset ends to avoid reduction in net area. In order to take up the slack, turnbuckles should be provided in every tie rod. Wire ropes are not recommended for tie rods in permanent construction.

In the majority of cases, the wale is a pure flexural member and the tie rod is a pure tension member. If the tie rods are constructed at an angle other than 90 degrees with the sheetpiling, the wales are subjected to bending and axial stress. The axial stress is caused by the component of the tie rod tension parallel to the face of the sheetpiling. In this case, there should be a positive connection between the wales and the sheetpiling to transfer this axial stress. In any event, the structural design of wales and tie rods is a relatively simple procedure and requires no further discussion. The arrangement of wales and tie rods and the nature of tie rod tension are discussed below.

A. Location of wales. Wales may be constructed either on the inside or outside face of the sheetpiling. If on the inside, positive provision is necessary to transfer the reaction of piling to the wales, usually every piece of piling is bolted to the wale. Unless a flush face of the piling is desired, the most economical location of wales is generally above low water level and on the outside face of the piling. When above low water level, wales should be provided with drain holes, and tie rods should be painted and covered for corrosion protection as discussed in Sec. 12-4.

B. Tie rod support. If there is any soft soil below the tie rods, even at a great depth, it will consolidate under the weight of the backfill and consequently cause the ground to settle. A small settlement will cause the tie rods to sag under the weight of the soil above them. In order to eliminate the stress due to such sagging, one of the following methods may be used:

1. Support the tie rods with vertical piles at 20 to 30 ft intervals. The piles must be driven to firm soil below the compressible layer.
2. Install a large pipe and lay the tie rod on the invert of the pipe. The inside diameter of the pipe should be larger than the total contemplated settlement so that the tie rod will be always free inside the pipe as the ground and the pipe settle.

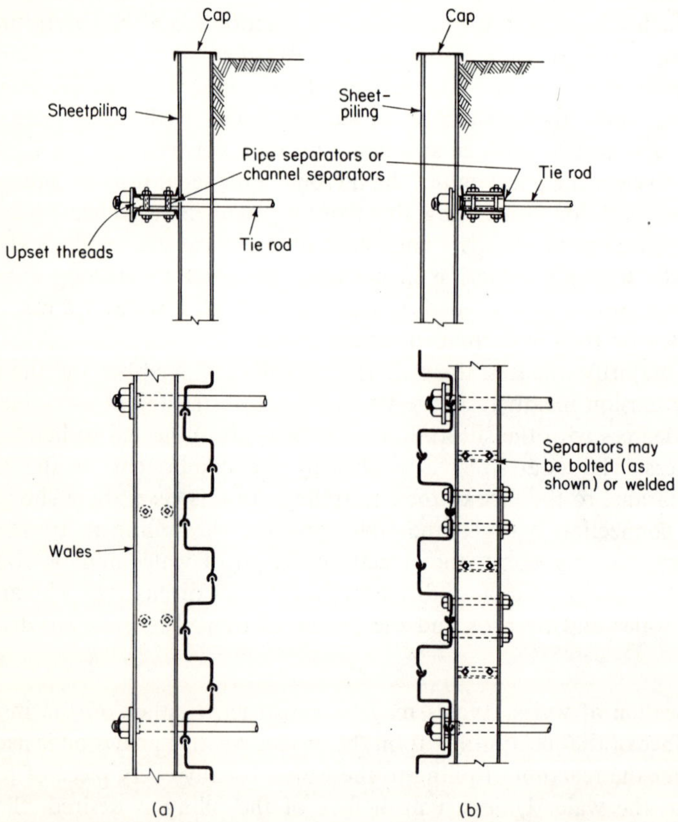

Fig. 12-17 Common arrangement of wales and tie rods: (a) wales on the outside face of piling; (b) wales on the inside face of piling.

C. Tie rod tension. The tie rod is frequently subjected to a tension considerably greater than the value calculated by the free-end method.* For design purposes the computed tie rod tension should be increased by at least 30 per cent for the tie rod proper, and 50 to 100 per cent at splices and connections where abrupt change of cross section introduces stress concentration. Tie rod tension is also higher for sheetpiling in cohesive soil where the stability number, Eq. (12-5), is less than 0.5.

12-12 Methods for Reducing Lateral Pressure

When a sheetpiling is subjected to large lateral pressure, the possibility of achieving an economic design by reducing the magnitude of the lateral pressure should be considered. Some of the causes of large lateral pressure and the methods of reducing it are:

1. If the upper layer of the original ground is soft, it exerts large pressure on the sheetpiling due to the weight of backfill. This layer may be dredged out and replaced with a layer of granular soil, Fig. 12-18(a).

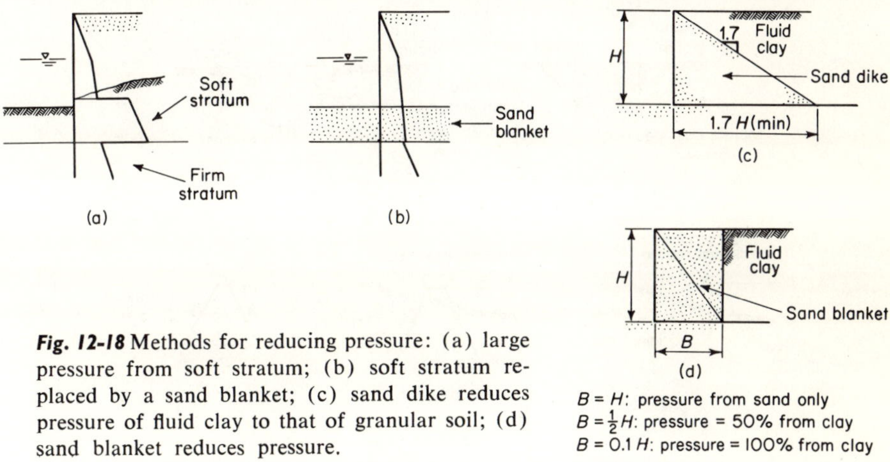

Fig. 12-18 Methods for reducing pressure: (a) large pressure from soft stratum; (b) soft stratum replaced by a sand blanket; (c) sand dike reduces pressure of fluid clay to that of granular soil; (d) sand blanket reduces pressure.

2. If granular soils are expensive in the vicinity of the construction site and cohesive soil must be used as the backfill material, the magnitude of lateral pressure may be reduced by the use of a sand dike or a sand blanket. The relationship between the magnitude of the lateral pressure and the size of the sand dike or blanket is shown in Fig. 12-8[(c) and (d)].

---

* This is due to the use of empirical or semi-empirical methods which do not take into consideration of realistic wall movements.

3. If heavy surcharge loads (railroads, crane rail, etc.) are close to the sheetpiling, the lateral pressure due to such surcharge loads may be eliminated by supporting the loads directly on bearing piles, Fig. 12-6(c).

12-13 Types of Anchorage

The reaction of tie rods may be carried to one of the following types of anchorages (Fig. 12-19):

1. *Deadmen* (anchor beams, etc.). Deadmen are short concrete blocks or continuous concrete beams deriving their resistance from passive earth pressure. This type is suitable when it can be installed below the level of the original ground.
2. *Brace piles.* Timber or concrete piles are driven in the form of A-frame. This type is employed where firm soil is at great depth. The analysis of brace piles was discussed in Sec. 8-17.
3. *Sheet piles.* Short sheet piles are driven to form a continuous wall which derives its resistance from passive earth pressure in the same

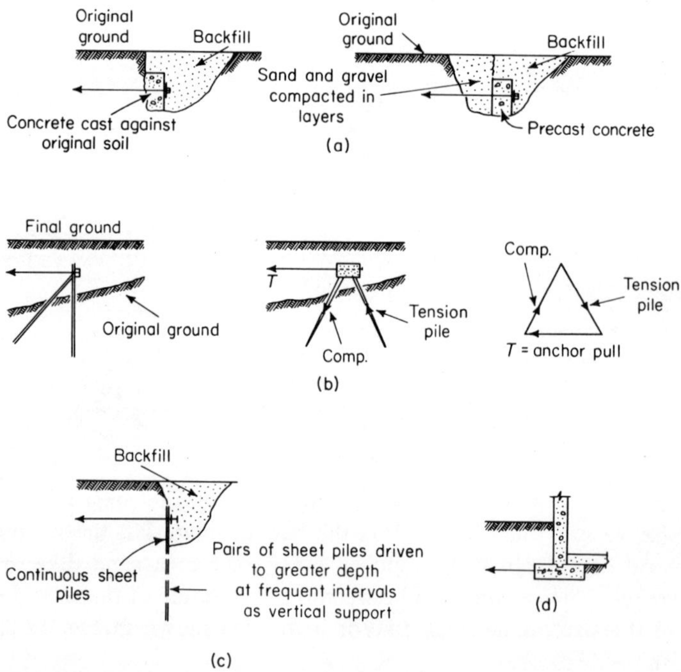

Fig. 12-19 Types of anchorage: (a) deadmen; (b) brace piles; (c) sheet piles; (d) large structure

manner as the deadmen. To provide some stability during installation of the piling and the wales, pairs of the piling should be driven to a greater depth at frequent intervals.

4. *Existing structures.* Tie rods can be connected to heavy foundations, such as buildings, retaining walls, crane foundations, etc.

12-14 Capacity of Deadmen

Deadmen (anchor beams, anchor blocks, or anchor plates) may be constructed near the ground surface or at a great depth, and, in short lengths or in one continuous beam. The capacity of each case in resisting horizontal anchor pull is discussed below.

A. Continuous deadmen near ground surface. A continuous deadman is one whose length is considerably greater than its depth. If the depth of the top of the deadman h is less than about one-third to one-half of the depth of the bottom of the deadman H, Fig. 12-20(a), the capacity may be calculated by assuming that the top of the deadman extends up to the ground surface. Therefore, the following equation holds true.

$$T_{ult} = P_p - P_a \tag{12-6}$$

where T_{ult} = ultimate capacity of the deadman, lb per lin. ft;

P_p = total passive earth pressure, lb per lin. ft;

P_a = total active earth pressure, lb per lin. ft.

The magnitude of P_p and P_a can be readily determined by the earth pressure theory, as shown in Fig. 12-20(a), on the basis that the friction and adhesion between the deadman and the soil is zero.

For deadman in cohesive soil, the distribution of P_p and P_a immediately after the application of anchor pull is referred as the initial pressure, and is shown in the right-hand side of Fig. 12-20(a). Notice that the active earth pressure is assumed zero in a depth $= 2c/\gamma$ which is the depth of tension cracks. As time elapses, the magnitude and distribution of earth pressure are likely to change slowly. In the absence of definite information, the design of deadman in cohesive soils should be made with a conservative factor of safety, see discussion at the end of Sec. 12-7.

B. Short deadman near ground surface. Figure 12-20(b) shows a deadman with a length L subjected to an anchor pull T. Experiments have shown that at the time of failure, the heave of the ground surface takes place in an area somewhat like *adbb'da'* which is longer than the length of the deadman. The surfaces of sliding at both ends is a curved surface *abde* and *a'b'd'e'*. The resistance to sliding along these wide and curved surfaces is undoubtedly smaller than the resistance along the vertical surfaces abe and $a'b'e'$. The total earth pressure (normal to abe) in granular soil is

$$\int_0^H \frac{H-X}{H}(H\sqrt{K_p}+H\sqrt{K_a})\,(dx)\,\gamma K_0=\frac{1}{6}K_0\gamma(\sqrt{K_p}+\sqrt{K_a})H^3$$

Therefore, the total ultimate capacity of a short deadman in granular soil near ground surface is

$$T_{\text{ult}} \leqq L(P_p-P_a)+\tfrac{1}{3}K_0\gamma(\sqrt{K_p}+\sqrt{K_a})H^3\tan\varphi \qquad \text{(12-7a)}$$

where L = length of the deadman, ft;

P_p, P_a = total passive and active earth pressure, lb per lin. ft;

K_0 = coefficient of earth pressure at rest. It may be taken as 0.4 for design of deadman;

γ = unit weight of soil, pcf;

K_p, K_a = coefficients of passive and active earth pressure;

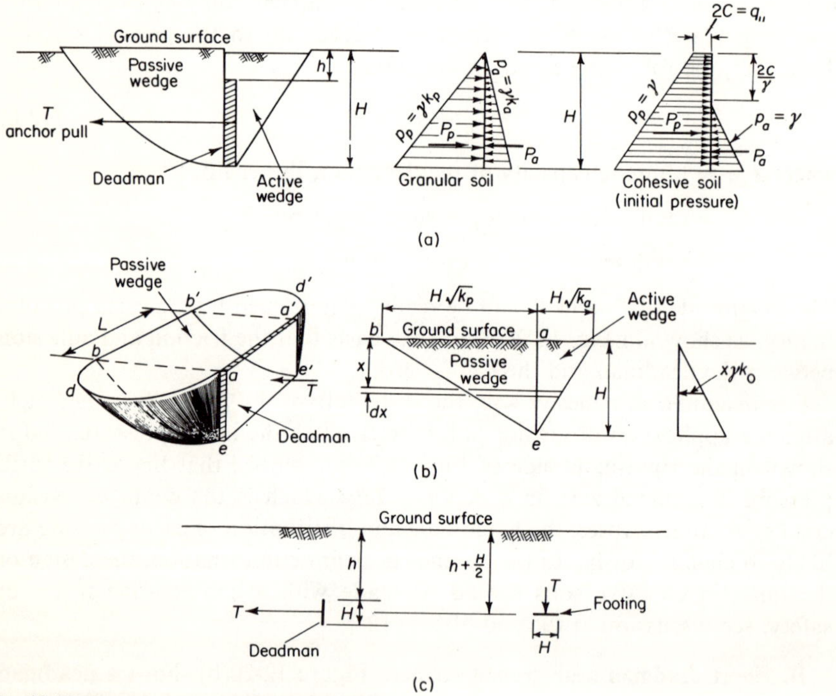

Fig. 12-20 Capacity of deadmen: (a) continuous deadmen near ground surface ($h/H<1/3\sim1/2$); (b) short deadmen near ground surface; (c) deadmen at great depth below ground surface.

H = height of deadman;

φ = angle of internal friction.

In cohesive soils, the second term of Eq. (12-7a) should be replaced by the cohesive resistance; therefore:

$$T_{\text{ult}} \leqq L(P_p - P_a) + q_u H^2 \tag{12-7b}$$

where q_u = unconfined compression strength of the soil, psf. Other notations are same as above.

C. Deadman at great depth. The ultimate capacity of a deadman at great depth below ground surface ($h > H$, Fig. 12-20(c)) is approximately equal to the bearing capacity of a footing whose base is located at a depth corresponding to the mid-height of the deadman (Terzaghi, 1943).

12-15 Location of Anchorage

The anchorage is useless if it is located within the sliding wedge of the backfill, Fig. 12-21(a). The capacity of the deadman is impaired if it is located in unstable ground or if the active wedge of the backfill interferes with the passive wedge in front of the deadman, Fig. 12-21(b). In the latter case, the reduction in capacity must be determined (see Terzaghi, 1943, p. 232ff).

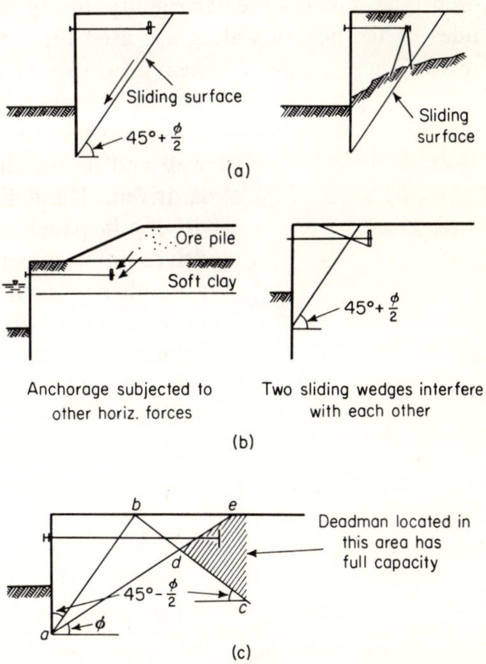

Fig. 12-21 Location of deadman: (a) offers no resistance; (b) efficiency greatly impaired; (c) full capacity.

Anchorage should not be constructed in an unstable ground.

Full capacity of a deadman is available when

1. The active sliding wedge of the backfill does not interfere with the passive sliding wedge in front of the deadman.
2. The deadman is located below the slope line starting from the bottom of the sheetpiling and making an angle φ with the horizontal, φ being the angle of internal friction of the soil.

To satisfy these two requirements, the deadman should be located in the shaded area bounded by the line *ae* and *bc*, Fig. 12-21(c).

12-16 Construction of Sheetpiling Walls

Sheet piles are driven in place by the same equipment for driving bearing piles, Chapter 8. The procedure and problems peculiar to the driving of sheetpiling are discussed here.

Sheet piles should be driven with the tongue (or ball) leading and are usually driven in pairs. The advantages of driving in pairs instead of driving one pile at a time are better economy and faster operation. Driving a pair of piles eliminates the energy required to overcome the interlock friction between the pair if they were driven separately. Two piles, particularly the arch-web type, possess greater rigidity during driving.

Guides or temporary waling are used for maintaining the sheet piles in alignment. These guides are heavy horizontal timbers or steel beams supported by stakes. Sheet piles tend to lean outward and also to creep along the plane of the wall and in the direction in which the piling is driven. These difficulties may be avoided by driving in panels. A panel of a few piles are driven to part penetration at a time. After a number of such panels are driven, the piling in the first few panels may be driven to the final penetration, Fig. 12-22.

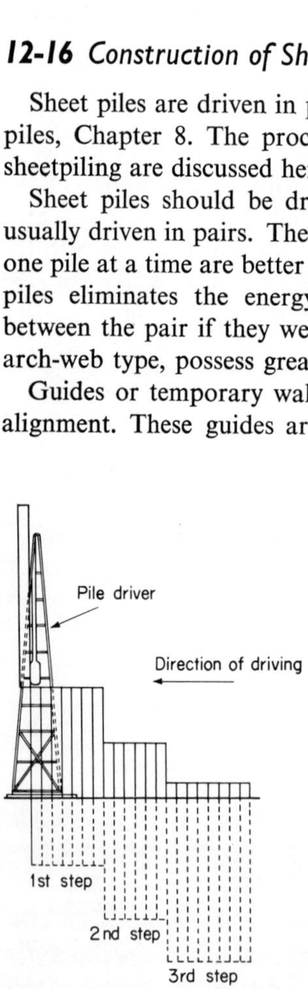

Fig. 12-22 Driving sheet piles in panels.

Leaning of sheetpiling may be the result of a number of factors:

Defective guides

Deformation of the pile web or interlocks

Water jetting

Improper driving.

When leaning is noticed, it should be corrected before the condition becomes worse. A number of corrective measures are effective (Fig. 12-23):

1. Pulling the heads of the successive piles during driving
2. Driving in panels with guide piles
3. Applying the driving energy at an angle
4. Use of piles with chamfers at the foot
5. Use of specially fabricated wedge-shaped pile or piles.

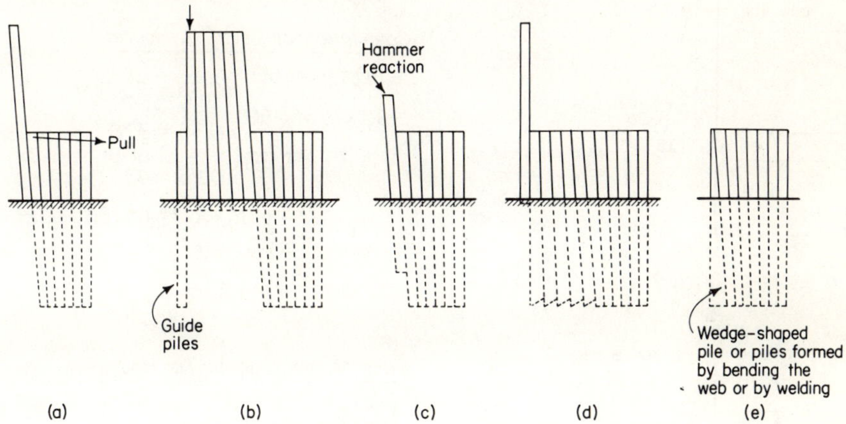

Fig. 12-23 Some methods for correcting the pile leaning.

12-17 *Design Examples*

Three examples are given to illustrate the design of an anchored sheet-piling: one in granular soil by the free-end method (DE 12-1), one in cohesive soil by the free-end method (DE 12-2) and one in granular soil by the equivalent beam method (DE 12-3). In DE 12-2 the shear strength for the clay is given in two sets of values: the initial strength ($C = 800$ and $\varphi = 0$) and the final strength ($C = 0$ and $\varphi = 27$ degrees). As can be seen, the final pressure condition governs the design.

Plate DE 12-1 shows a complete design analysis, whereas in DE 12-2 and DE 12-3, only the depth of penetration, tie rod tension, and maximum bending moment are computed, the design of tie rods, wales, and anchorage are similar to the previous example.

Design of Anchored Sheetpiling in Granular Soil by Free-end Method

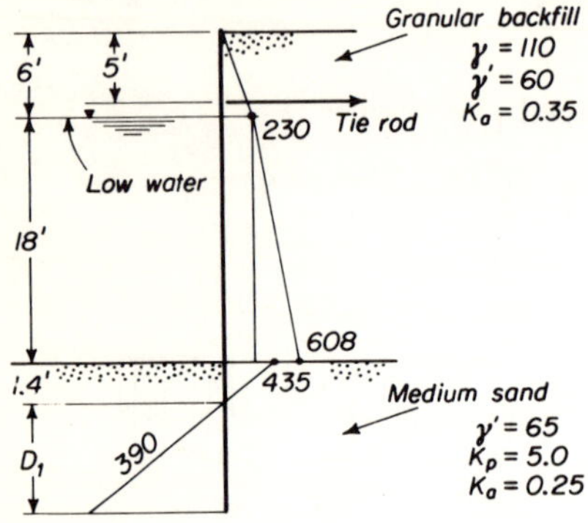

Calculation of Earth Pressure:

$$110 \times 6 = 660 \quad \times 0.35 = 230\ \text{psf}$$
$$60 \times 18 = 1080 \quad \times 0.35 = \underline{378}$$
$$\overline{1740} \qquad\qquad 608$$
$$\times 0.25 = 435$$

$$p_p' - p_a' = \gamma'(K_p - K_a) = 65(5.0 - 0.25)$$
$$= 309\ \text{psf}$$

$$y = {}^{435}/_{309} = 1.4'$$

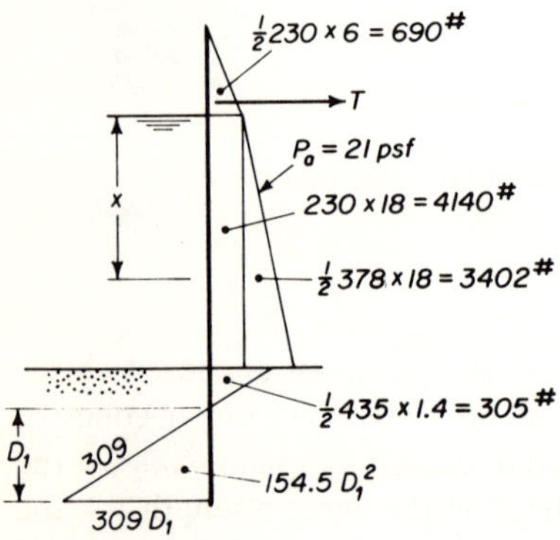

Take Moment about Tie Rod Level:

$$690 \times (-1.0) = \qquad -690$$
$$4140 \times 10.0' = 41{,}400$$
$$3402 \times 13.0' = 44{,}200$$
$$\underline{305 \times 19.5'} = \underline{5\,940}$$
$$8537^{\#} \qquad 91{,}540 - 690 = 90{,}850'^{\#}$$

$$154.5 D_1^2 \left(20.4 + \tfrac{2}{3} D_1\right) = 90{,}850$$

$$D_1^3 + 30.6 D_1^2 = 882$$

Solve D_1 by trial $\quad D_1 = 5.0'$

Total penetration required $= (1.4 + 5.0) \times 1.2 = 7.7'$ use $8'$

Total height of piling $= 6 + 18 + 8 = 32'$

Tie Rod Tension

$$T = 8537 - 154.5 D_1^2 = 4677 \times 1.3 = 6100^{\#}/'$$ use for design of tie rod and wale

(30% increase)

Max Moment

Let point of zero shear be at X below low water level:

$$4677 = 690 + 230X + \tfrac{1}{2}(21)X^2$$
$$X^2 + 21.9X - 380 = 0$$
$$X = 11.4'$$

$$690 \times 13.4' = 9250$$
$$11.4 \times 230 = 2620 \times 11.4/2 = 14{,}950$$
$$\tfrac{1}{2}(21)(11.4)^2 = \underline{1367} \times 11.4/3 = \underline{5200}$$
$$T = 4677^{\#} \qquad -29{,}400$$
$$\times\ 12.4 = \underline{57{,}900}$$
$$M_{max} = 28{,}500'^{\#}$$
$$\times \frac{12}{18{,}000} = 19.0''^3$$

Req'd. section modulus if no moment reduction (below) is desired

Selection of Piling Section by Rowes Moment Reduction Chart:

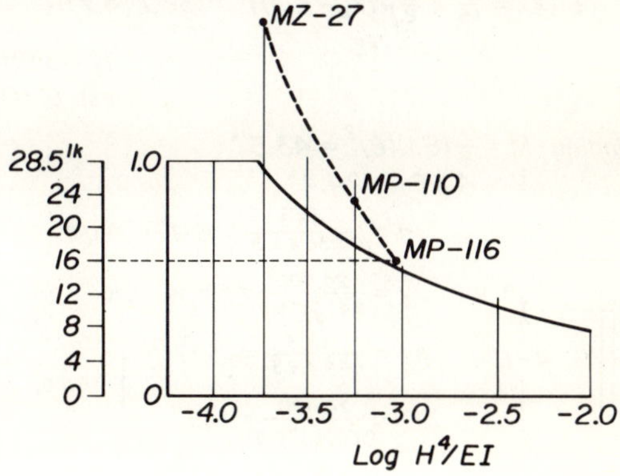

1. Select moment reduction curve corresponding to the relative density of the soil. In this case, the curve for medium sand is constructed by interpolation.
2. Draw the scale for moment on the vertical axis such that the maximum moment (28.5^{lk}) is at the top of chart.
3. Compute the values of log H^4/EI for several pile sections: $H^4 = (32)^4 = 1{,}049{,}000$; $E = 29(10)^6$

Pile Section	MZ-27	MP-110	MP-116	
I (per ft)	184	65.4	40.0	
$\rho = \frac{H^4}{EI}$	0.000196	0.000552	0.000904	
log ρ	$\bar{4}.29 = -3.71$	$\bar{4}.74 = -3.26$	$\bar{4}.96 = -3.04$	
Moment capacity (lk)	45.3	23.0	16.0	= Sect. mod. $\times \frac{18}{12}$

4. Plot log ρ and moment capacity for each pile on the chart. MP-110 is adequate for the bending moment. MP-116 has reserve strength. MZ-27 is wasteful. Any section falling below the curve would be inadequate.

Tie Rods

Try 8'–0" spacing

$T = 8 \times 6.1 = 48.8^k$

at $20^{ksi} = 2.44^{\square''}$ net use $1\frac{3}{4}''\phi$ upset to $2\frac{1}{4}''\phi$

Wales

$M \doteq \frac{1}{9}(4.677)8^2 = 33.3^{'k}$

$\times \frac{12}{20} = 20.0''^3$ req'd. use 2–9 ⊔ 13.4

Deadman $f'_c = 3000$ psi

Medium sand above water $\gamma = 110$ pcf $K_p = 5.0$ $K_a = 0.30$

$6100 = \frac{1}{1.5} \times \frac{1}{2}\gamma(K_p - K_a)H^2 = 36.7(4.7)H^2$

F.S. = 1.5

$H = 5.95'$

use 6'–0" deep cont. deadman

Long. bending $M = \frac{1}{9}(6.1)(8)^2 = 43.3^{'k}$

$V = 4 \times 6.1 = 24.4^k$

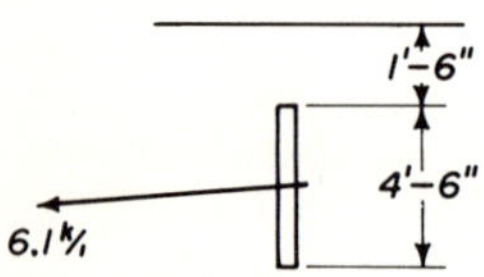

$d = \sqrt{\frac{43.3}{0.236 \times 4.5}} = 6.4''$ req'd. use 12" thick, $t = 9.5''$

$v = \frac{24{,}400}{54(\frac{7}{8})^* 9.5} = 55$ psi OK

$As = \frac{43.3}{1.44 \times 9.5} = 3.17^{\square''}$ } req'd.

$\Sigma O = \frac{24{,}400}{210^*(\frac{7}{8})^* 9.5} = 14.1''$ } req'd.

use 6–#7 longit.

Transverse bending $M \doteq \frac{1}{2}(6.1)(\frac{4.5}{4}) = 3.43^{'k/'}$

$V \doteq 3.05^{k/'}$

$As = \frac{3.43}{1.44 \times 9.0} = 0.27^{\square''/'}$ } req'd.

$\Sigma O = \frac{3050}{210^*(\frac{7}{8})^* 9.0} = 1.86''/'$ } req'd.

Min. $As = 0.002 \times 9.5 \times 12 = 0.23^{\square''/'}$ use #5 at 12

* indicates value permitted in 1956 ACI Code

Bearing Plates

$A = \frac{48.8}{1.125} = 43.4^{\square''}$

7×7 ℞ $= 49^{\square''}$

$3''\phi$ hole $= \underline{7}$

$42^{\square''}$ slightly under,

$M \doteq \frac{1}{2}(1.125)(3.5)^2 = 6.9^{''k}$

$\times \frac{1}{20} = 0.35''^3$ req'd use $1\frac{1}{2}''$ thick plate

DE 12-1
Anchored Sheetpiling
Sh. 4 of 4

Location of Deadman

49′ from sheetpiling as determined by graphical (or analytical) procedure.

$$\tan 32^\circ = \frac{32.0'}{X}$$

$$X = \frac{32.0'}{0.625} = 49'$$

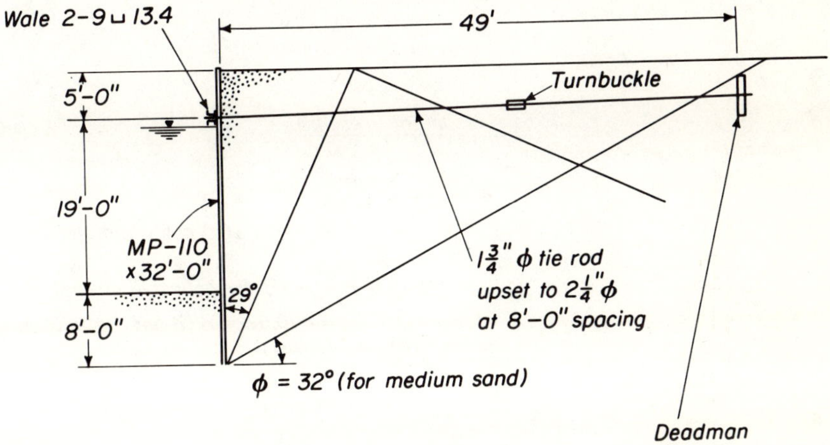

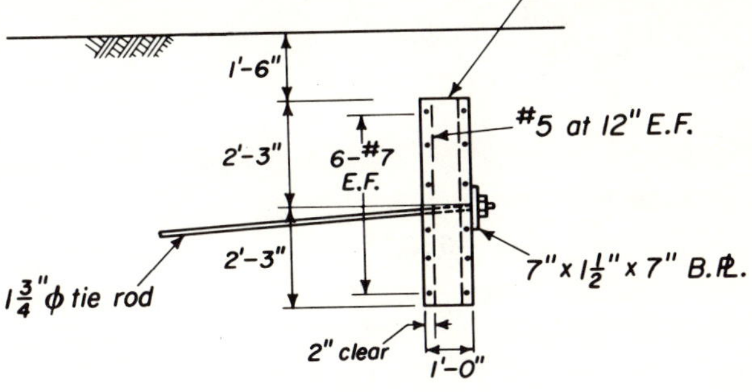

Design of Anchored Sheetpiling in Cohesive Soil by Free-end Method

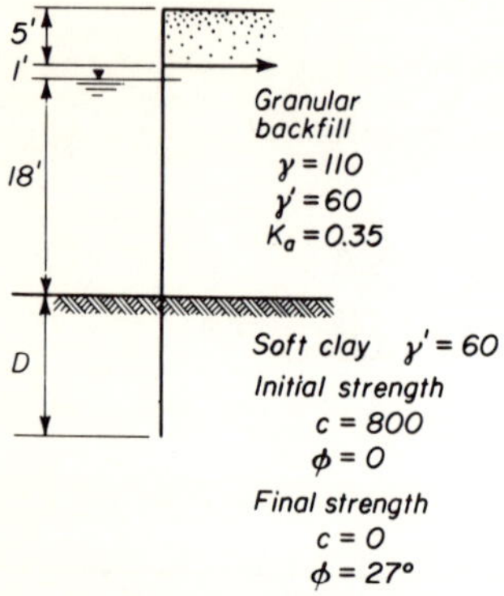

Stability

$$110 \times 6 = 660$$
$$60 \times 18 = 1080$$
$$\gamma_e h = 1740 \text{ psf}$$

$$S = \frac{1.25c}{\gamma_e h} = \frac{1.25 \times 800}{1740} = 0.575$$

$$F.S. = \frac{0.575}{0.30} = 1.92 \quad \text{ok}$$

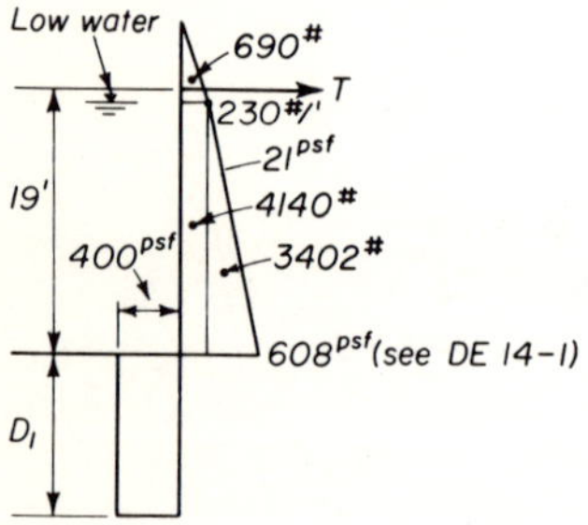

Initial Pressure

$$2q_u = 4c = 3200$$
$$\gamma_e h = 1740$$
$$1460 \text{ psf (ultimate)}$$

Use F.S. = 1.5 for cohesive strength

$$c' = 800/1.5 = 533 \text{ psf} \quad \text{allowable}$$
$$4c' = 2140$$
$$\gamma_e h = 1740$$
$$400 \text{ psf for design}$$

Take moment about tie level:

$$M = 41{,}400 + 44{,}200 - 690 \quad \text{(see DE 12-1)}$$
$$= 84{,}990 \text{ '\#}$$
$$= 400D_1(19' + \frac{D_1}{2})$$

$$D_1 = 9.0'$$

(final pressure governs) see next sheet

Take moment about $D_1/2$:

$$T = \frac{108{,}400}{23.5} = 4630^{\#}$$

Pt. of zero shear at X^{ft} below low water:

$$4630 = 690 + 230X + \frac{1}{2}(21)X^2 \qquad X = 11.3'$$

$$\text{Max } M = 4630 \times 12.3 - 690 \times 13.3 - \frac{1}{2}(230)11.3^2 - \frac{1}{6}(21)(11.3)^3$$
$$= 28{,}000 \text{ '\#}$$

DE 12-2
Anchored Sheetpiling
Sh. 2 of 2

<u>Final Pressure</u>

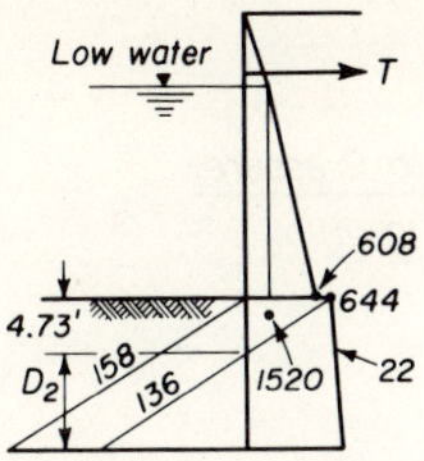

$\phi = 27°$ $K_a = \tan^2(45-13.5) = 0.37$ $\times 1740 = 644$

$K_p = \tan^2(45+13.5) = 2.65$

$60 \times 0.37 = 22$ psf/ft

$\times 2.65 = \underline{158}$ psf/ft

136

$\frac{644}{136} = 4.73'$ $\times \frac{1}{2}(644) = 1520^{\#}$

Take moment about tie level:

$84{,}990 + 1520 \times (19 + 1.58) = 116{,}290^{\#}$

$= \frac{1}{2} \times 136 D_2^{\,2}(23.73 + \frac{2}{3}D_2)$

$D_2 = 7.7'$

$\underline{4.73}$

Total penetration $= 12.43'$ Use 12'-6" (governs)

Take moment at $\frac{2}{3}D_2$:

$T = 5760^{\#}/'$ (governs)

Pt. of zero shear at X ft below low water:

$5760 = 690 + 230X + \frac{1}{2}(21)X^2$ $X = 13.9'$

$\text{Max } M = 5760 \times 13.9 - 690 \times 15.9 - \frac{1}{2}(230)(13.9)^2 - \frac{1}{6}(21)(13.9)^3$

$= 37{,}400'^{\#}$ (governs)

<u>Moment Reduction</u>

From Fig. 12-16(b)

$S = 0.575$ $\alpha = \frac{24}{36.5} = 0.66$

Reduction ratio = 0.56 for working stress

$M = 37.4 \times 0.56 = 21.0'^{K}$

$\times \frac{12}{18} = 14.0''^3$ req'd.

MP-110 = 15.3

Check MP-110

$\rho = \frac{H^4}{EI} = \frac{(36.5)^4}{29(10)^6 65.4} = 0.000936$

$\log \rho = 4.97 = -3.07$

Interpolated between $\log \rho = -3.1$ and -2.6

Reduction ratio = 0.62

$M = 0.62 \times 37.4 = 23.1'^{K}$

$\times \frac{12}{18} = 15.0''^3$

MP-110 = 15.3 ok

Design of Anchored Sheetpiling by Equivalent Beam Method

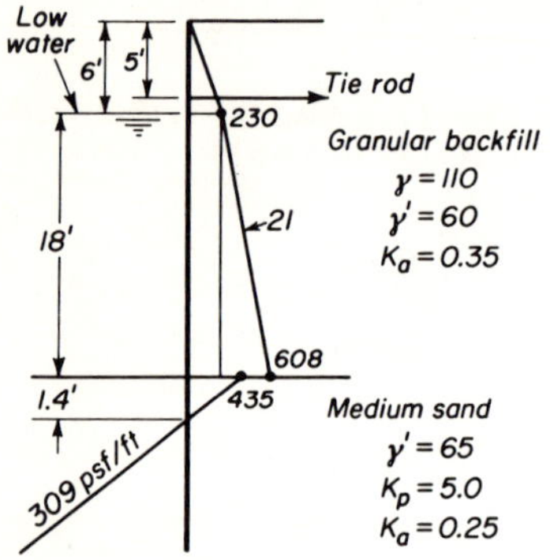

Calculation of Earth Pressure:

Same as in DE 12-1

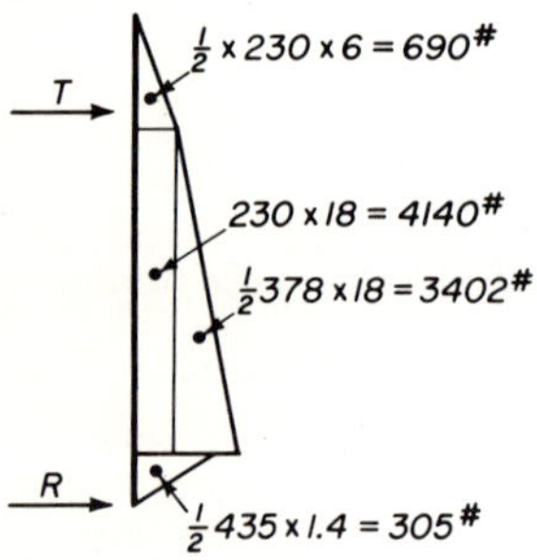

Take Moment About T:

690 × 1.0' =		690
4140 × 10.0' =	41,400	
3402 × 13.0' =	44,200	
305 × 19.5' =	5940	
8537	91,540 − 690 = 90,850	

$$R = \frac{90{,}850}{20.4'} = 4{,}450^{\#}$$

$$T = 8537 - R = 4087^{\#}$$

Depth of Penetration

$$D = y + \sqrt{\frac{6R}{p_p' - p_a'}} = 1.4 + \sqrt{\frac{6 \times 4{,}450}{309}} = 10.7'$$

$$\times 1.2 = 12.8'$$

Use 13'

Max Moment

Let pt. of zero shear be at X below low water level:

$$4087 = 690 + 230X + \tfrac{1}{2}(21)X^2 \qquad X = 10.5'$$

$$\text{Max } M \text{ at pt. of zero shear} = 4087 \times 11.5 - 690 \times 12.5 - \tfrac{1}{2}(230)(10.5)^2 - \tfrac{1}{6}(21)(10.5)^3 = 21.62^{\text{lk}}$$

$$\times \frac{12}{18} = 14.5''^3$$

Req'd. section modulus

Plate Thirteen

A Braced Cofferdam Using Three Tiers of Struts and Wales for Excavation in Soft Clay

Braced cofferdams are widely used for retaining earth and/or water because of their relative economy. While the procedures of design of braced cofferdams in water are relatively simple, the design of such cofferdams in open cut require a thorough understanding of the principles of earth pressure theories and their limitations. Furthermore, the stability of the bottom of excavation is of utmost importance. This chapter presents the procedure of analysis and details of construction.

13 BRACED COFFERDAMS

13-1 Common Uses of Braced Cofferdams

A braced or strutted cofferdam is a temporary enclosure consisting of vertical or horizontal sheeting with internal struts. It is probably the most common type of all cofferdams because it is relatively economical when utilized for excluding shallow water, earth, or both. Its use ranges from supports for shallow trench excavations to bracing systems for multi-level basement construction where adjacent ground must be protected from collapse or settlement damages. This is almost invariably the case when excavation is required in urban areas. In localities where the adjacent ground need not be protected, the excavation is generally made with sloping sides without the use of such cofferdams.

Because of their relative economy, braced cofferdams are frequently used in the construction of bridge piers and abutments. Generally speaking, the braced cofferdam is more economical than cellular cofferdams and caissons up to a depth about 40 ft of water. Even when caissons are sunk to greater depths, the upper portion of the bridge piers may still be constructed inside a braced cofferdam. In such a case, the upper portion of the caisson is provided with connections to braced cofferdams.

The construction of braced cofferdams for large and deep excavations, particularly in soft soils, is difficult and expensive. Therefore, the engineer should investigate other possible schemes of excavation before it is adopted. For instance, the site may be excavated in narrow strips and the portion of the substructure within each strip is constructed before adjacent strips are excavated. Sometimes it may be feasible to leave an earth berm all around the site, and the central portion of the substructure is constructed first. Then, portions of the berm are excavated and the construction completed in predetermined steps. Such schemes sometimes result in reducing the construction cost and the possibility of damage to the adjacent ground.

13-2 Common Types of Braced Cofferdams

A braced cofferdam consists of three or four principal components:

Lagging: Horizontal timber planks placed by hand as the excavation proceeds.

Sheetpiling: May be timber, concrete, or steel and is installed vertically by driving.

Soldier beams: Vertical beams (timber or steel) for taking reactions from the lagging. Soldier beams are not required in the case where sheetpiling is used.

Wales: Horizontal beams for transfering the reaction from the lagging or the sheeting to the struts.

Struts: Compression member for carrying the reactions from one side to the other side of the excavation.

According to the arrangement of the components, braced cofferdams may be classified into three general types, Fig. 13-1. The procedures of construction of each type are discussed in the following.

The type of bracing shown in Fig. 13-1(a) is most useful in small scale excavation because it does not require heavy construction equipment. When the excavation is advanced to a depth equal to one length of the soldier beam, lagging, soldier beams, and struts are placed. Wedges are driven at one end of the strut to tighten up the system. Additional sets of lagging, soldier beams, and struts are installed as excavation progresses.

Another type of bracing, Fig. 13-1(b), consists of driving soldier beams prior to any excavation. Lagging is placed between soldier beams as excavation progresses. Wales and struts are installed as soon as the excavation reaches the level of each tier. This type of bracing is suitable for deep excavation. The soldier beams driven in the soil afford some stability of the system and also offer some resistance against heave of the bottom of excavation. Soldier beams are sometimes left in place and used as support of the permanent wall as shown in the alternate section *B-B*.

If condition permits driving of sheetpiling, the type shown in Fig. 13-1(c) may be used advantageously. The sheetpiling is driven to the desired depth prior to excavation. This type has the following advantages:

1. In soft soil where the danger of bottom heave is critical, sheetpiling driven to a great depth offers considerable resistance against such a danger.

2. In purely granular soils where loss of ground would occur due to excavation prior to placement of lagging, the sheetpiling eliminates such a danger.

3. In excavation of granular soil below water level, the sheetpiling serves as a cut-off wall and therefore reduces the amount of dewatering cost as well as reduces the danger of piping, boil, or blow-out, Chapter 5.

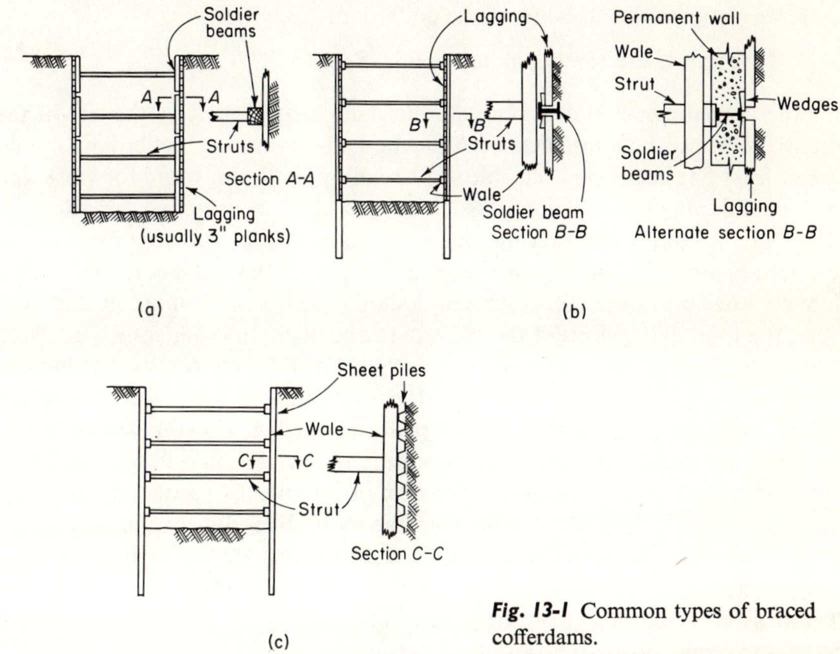

Fig. 13-1 Common types of braced cofferdams.

Cofferdams are generally made just large enough to permit the operation of the construction within. Therefore, the shape in plan is governed by the physical dimensions of the substructure. However, when conditions permit, circular cofferdams may be advantageous. In a circular cofferdam, the wales act as compression rings requiring no internal strutting, thus eliminating the cost of struts and also providing large clear working space within the cofferdam.

13-3 Design of Braced Cofferdam

The design of a braced cofferdam generally consists of the following steps:

1. Assemble the controlling dimensions–see discussion below.

2. Study the subsoil conditions and sketch a soil profile extending to bedrock or to a depth about 0.707 times the width of the excavation.

The maximum water level should be established.

3. Compute the lateral pressure–Sec. 13-4.
4. Select the type and material for the components of the cofferdam–Sec. 13-2.
5. Design the various components–Sec. 13-5.
6. Investigate the stability of the cofferdam–Sec. 13-6.
7. Investigate the safety against piping–Sec. 13-7.

The plan dimension of a braced cofferdam is controlled by the size of the substructure to be constructed within the cofferdam. The cofferdam should be at least 5 ft larger on each side so as to allow sufficient room for work and the installation of formwork. Space should also be provided for sumps if they are used for dewatering the cofferdam. If the substructure will be poured against the lagging or sheeting, the plan size of the cofferdam is, of course, the size of the substructure. It should be borne in mind, in this case, that the final dimensions of the cofferdam are likely to differ somewhat from the design dimensions. Therefore, it is essential to allow for such variations in the design. The amount of variation between the design and the actual dimensions depends on the characteristics of the soil and the workmanship of the constructor. In unfavorable conditions, the variation may exceed 6 in. Sheetpiling tends to creep in the direction of driving. Consequently, the total length of wall is generally longer than the theoretical dimension.

In water, the height of a cofferdam is controlled by the high water level expected during the useful life of the cofferdam. The high water level also controls the necessary depth of penetration of the piling in order to prevent boil and heave of the bottom of excavation.

13-4 Lateral Pressure on Braced Cofferdams

A braced cofferdam may be subjected to earth pressure, surcharge load, hydrostatic pressure, wave pressure, and earthquake forces. With the exception of earth pressure, all these forces may be evaluated by the principles discussed in Chapter 4. The earth pressure, however, cannot be calculated by the classical earth pressure theories (Rankine, Coulomb and wedge theories) because the yield (movement) of cofferdam walls does not satisfy the basic assumptions of these theories. For cofferdams in sand and soft to medium clays, the trapezoidal rules proposed by Terzaghi and Peck (1948) may be used for design purposes. In stiff clays, the earth pressure is very small in the beginning but frequently increases with time and may approach a value corresponding to the earth pressure assuming a zero cohesion (Kirkdam, 1958). These cases are discussed below.

A. Earth pressure on braced cofferdam in sand. The actual earth pressure at a given vertical profile of the cofferdam varies considerably from place to place. The greatest load that a strut may be required to take can be calculated by the trapezoidal pressure diagrams shown in Fig. 13-2(a). These trapezoids represent the maximum pressure which may exist on some of the profiles in a given braced cofferdam in sand. However, the actual loads on many struts are likely to be smaller.

For ordinary design, the value of φ to be used for calculation of P_a may be

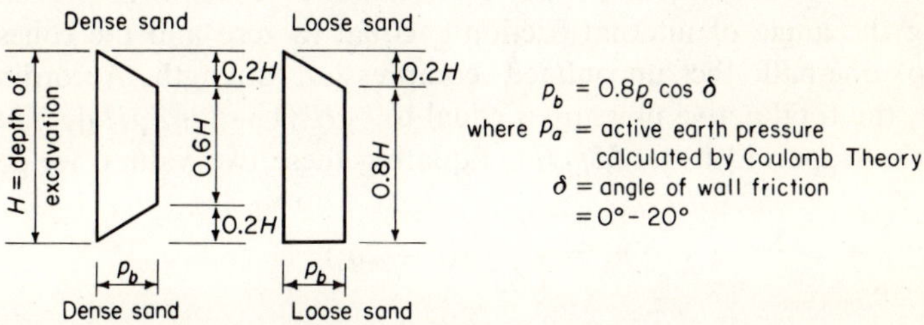

(a)

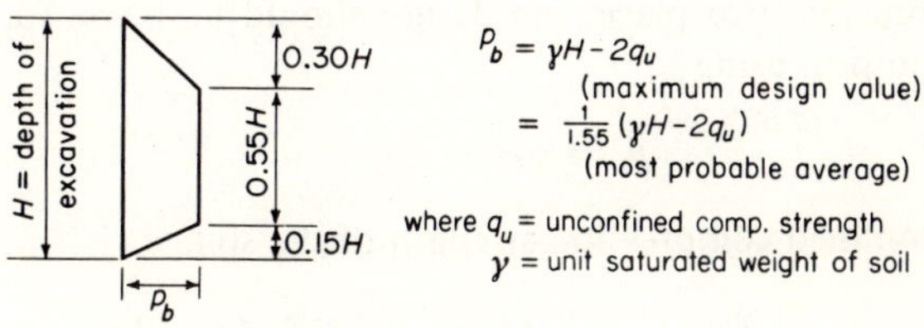

(b)

	Theoretical rupture surface	Av measured pressure t/m^2	Values used for computing earth pressure, in percentage of triaxial values	
			ϕ	C
Shortly after excavation (September)	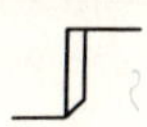	2.27	100	80
September 20th		2.27	100	60
October 21st	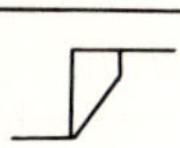	4.29	100	15
November 15th	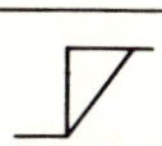	7.13	100	0

(c)

Fig. 13-2 Earth pressure on braced cofferdams: (a) in sand; (b) in soft and medium clays (after Terzaghi and Peck); (c) on an experimental cofferdam in stiff fissured clay (after Kirkdam).

estimated on the basis of the standard penetration resistance using the relationship between φ and N shown in Table 1-1. The angle of wall friction δ should be assumed zero unless the sheeting or soldier beams are driven to dense soil. In the latter case, the value of δ may be taken up to 20 degrees.

B. Earth pressure on braced cofferdams in soft and medium clays. Similar to the case of braced cofferdams in sand, the earth pressure varies from place to place and may be approximated, for the design purpose, by a trapezoid shown in Fig. 13-2(b). The total earth pressure represented by the area of the trapezoid is probably equal to the total Rankine active earth pressure, by assuming the angle of internal friction φ equal to zero and the cohesion c equal to one-half the unconfined compression strength. According to Rankine, the total active pressure is equal to $\frac{1}{2}\gamma H^2[1 - (2q_u/\gamma H)]$. The total area of the trapezoid is $\frac{1}{2}(1.55p_bH)$. Equating these two values, we get

$$p_b = \frac{1}{1.55}(\gamma H - 2q_u)$$

This value is the probable average pressure. Since the actual pressure varies considerably from place to place, the design should be made for the most unfavorable condition using

$$p_b = \gamma H - 2q_u \qquad (13\text{-}1)$$

where q_u = unconfined compression strength of the soil.

C. Earth pressure on braced cofferdams in stiff clays. Immediately after the installation of a cofferdam, the earth pressure due to stiff clays is relatively small. However, as time goes on, the fissures in the stiff clay become softened and the earth pressure increases considerably. This phenomenon has been studied in a test trench, the results of which are shown in Fig. 13-2(c). At a time shortly after excavation, the average earth pressure was 2.27 tons per square meter, corresponding to φ = 100 per cent, and c = 80 per cent of the values determined by triaxial test. At the end of about two months, the earth pressure increased to 7.13 tons per square meter, corresponding to φ = 100 per cent and c = 0 per cent of the test values. The results of this experiment leads to the conclusion that the final maximum earth pressure may be calculated on the basis of $c = 0$ and φ = the value determined by tri-axial test.

D. Earth pressure on braced cofferdams in stratified soils. When layers of sand and soft clay are encountered, the trapezoidal distribution shown in Fig. 13-2(b) may be used if the unconfined compression strength q_u is substituted by the average value q_a and the unit weight of the soil by the average value γ_a (Peck, 1943).

$$q_a = \frac{1}{H}[\gamma_s K_s H_s^2 \tan \varphi_s (H - H_s) n q_u] \tag{13-2}$$

$$\gamma_a = \frac{1}{H}[\gamma_s H_s + (H - H_s)\gamma_c] \tag{13-3}$$

where γ_s = unit saturated weight of sand;

K_s = hydrostatic pressure ratio for the sand layer, may be taken as 1.0 for design purposes;

H_s = thickness of the sand layer;

φ = angle of internal friction of the sand;

H = total depth of excavation;

q_u = unconfined compression strength of the clay;

γ_c = unit saturated weight of the clay;

n = coefficient of progressive failure, the value ranges usually from 0.5 to 1.0. This value varies with the creep characteristics of the clay, the length of time during which the excavation remains open, and the care excercised in construction. In Chicago clay, the value ranges between 0.75 to 1.0.

Fig. 13-3 Interior of a braced cofferdam. Courtesy of Spencer, White & Prentis.

13-5 Components of Braced Cofferdams

A. Lagging. Usually made of 2 in. or 3 in. planks, the lagging is installed by hand after a depth of several feet is excavated. The maximum depth made each time before a series of lagging is placed depends on the soil characteristics. Soft clay and loose sand must be planked in short depths to reduce the amount of soil moving or running into the excavation. Immediately after placement of lagging, wedges should be driven to force it tightly against the soil. Voids behind the lagging should be packed by hand to reduce the amount of loss of ground. Figure 13-3 shows an interior view of a braced cofferdam using lagging.

Due to flexibility of wood, lagging usually deflects considerably. This deflection tends to reduce the magnitude of the earth pressure at the midspan of the lagging. Consequently, lagging is seldom subjected to high bending stress even if the calculated value is high.

B. Sheetpiling. Generally sheetpiling is designed as a continuous member supported at the strut levels. In addition, a fictitious support is introduced at the bottom of excavation and the sheetpiling is assumed to be hinged at this level, Fig. 13-4(a). The struts may be arranged at such levels that either

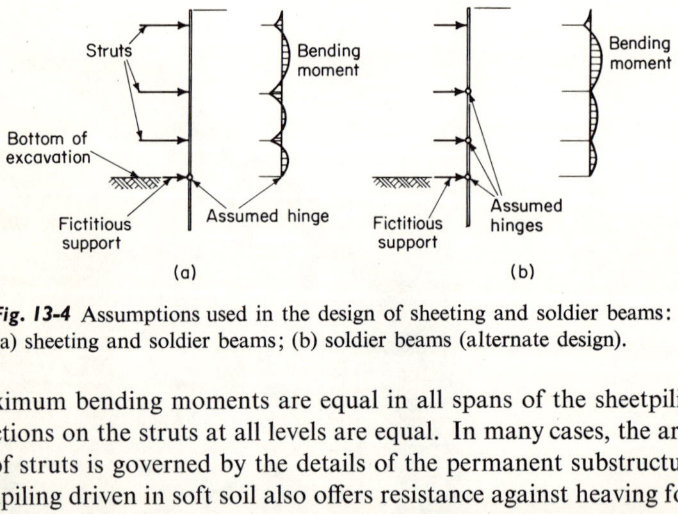

Fig. 13-4 Assumptions used in the design of sheeting and soldier beams: (a) sheeting and soldier beams; (b) soldier beams (alternate design).

the maximum bending moments are equal in all spans of the sheetpiling, or the reactions on the struts at all levels are equal. In many cases, the arrangements of struts is governed by the details of the permanent substructure.

Sheetpiling driven in soft soil also offers resistance against heaving force of the soil at the bottom of excavation. This is discussed in Sec. 13-7.

C. Soldier beams. Soldier beams may be designed as fully continuous over-all supports as discussed above, or as simple spans as shown in Fig. 13-4(b). The latter assumption gives a more conservative design.

D. Wales. Wales are designed to resist horizontal reactions from soldier beams or sheetpiling. In addition they are also subjected to axial load due to reaction from the perpendicular wales at the corner of the cofferdam. Usually moment splices are provided in wales and in such cases, they are designed as continuous beams subjected to lateral and axial loads.

E. Struts. Struts are pure compression members. They must be tightened against the wales for several reasons. Excessive yielding of struts means horizontal movement of the sides of excavation and therefore, causing settlement of the adjacent ground. In fissured clays, such lateral movement opens up the fissures which readily absorb moisture and become soft. Once the shear strength is reduced on the fissures, large earth pressure will be inevitable. Furthermore, unyielding supports are essential to maintain minimum bending moment in the wales.

Long struts required in large excavations are also subjected to considerable expansion and contraction due to temperature variations. At high ambient temperature, additional stress exists in the struts.

It is always advantageous to provide posts at some intermediate points for the purpose of reducing l/r value of the horizontal struts. It also reduces undue stresses due to excessive sagging of the struts.

Although close spacing of struts may be desirable from the designer's point of view, a spacing of 8 ft is considered minimum for construction operations. The struts and wales in each tier may be located at the same elevation in both directions. At the point of intersection, the strut in one direction is continuous and the perpendicular struts are framed into it. Each point of intersection is a field connection which must be designed and constructed to carry the axial stress of the strut. This field connections may be eliminated by running continuous struts in both directions, one strut being immediately above the other. Since the struts meet the wales at the same elevation, this arrangement will have wales meeting at different elevations at each corner of the cofferdam. The horizontal reaction transferred from one wale to the other introduces axial stress as well as bending moment resulting from the eccentricity of the reaction. In such cases, the wale must be designed for the combined stress due to the axial force, the bending moment from the lateral earth pressure, and the bending moment from the eccentrically applied longitudinal force.

13-6 Stability of Braced Cofferdams

A braced cofferdam may fail as a unit due to unbalanced external forces or by heaving of the bottom of excavation.

A. Unbalanced external forces. If the external forces acting on the opposite sides of cofferdams are unequal, the stability of the entire cofferdam as a unit should be analyzed. For example, a cofferdam driven to shallow depth in a river may be damaged by forces from swift current during a flash flood. Soil adjacent to the cofferdam may also be removed by scour and thus endanger the safety of the cofferdam. If the ground surface is sloping or at different levels on opposite sides of the cofferdam, the stability of the cofferdam and the surrounding mass of soil must be studied. In this case, removal of soil from inside of the cofferdam may cause instability of a large mass of soil which is originally stable. Furthermore, the reaction of the upper tiers of struts must be carried to lower tiers where adequate soil reaction is available. This is usually done by providing diagonal members in the bracing system as illustrated in Fig. 13-5.

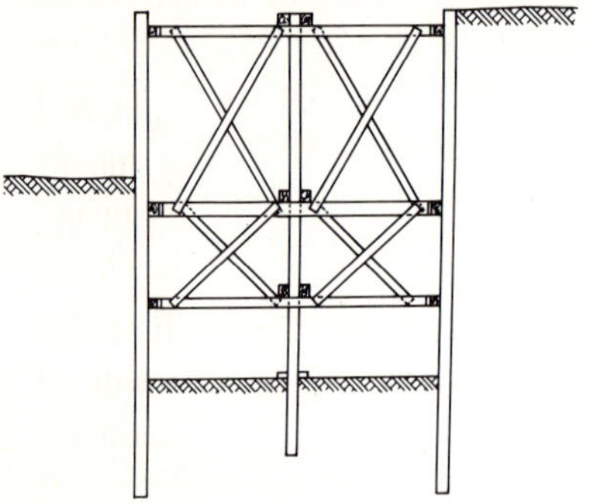

Fig. 13-5 Diagonal bracing for transferring unequal external forces.

The instability of a cofferdam installed near a sloping ground can be best examplified by the case shown in Fig. 13-6. The cofferdam was installed on a river bank for the construction of a bridge pier. When the excavation inside the cofferdam was almost completed, the entire river bank slid downwards and the cofferdam was destroyed. The failure took place at a time when the river suddenly subsided from a high water to a low stage, leaving a saturated mass of soil above the river stage. Another cofferdam had to be built and the bank was stabilized by removing a section of the bank from the high side and

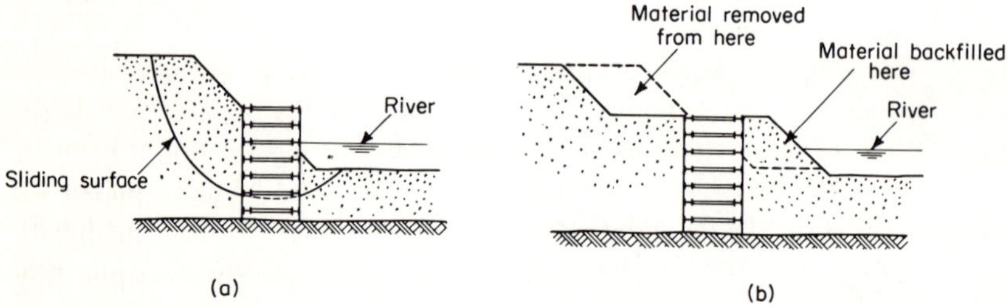

Fig. 13-6 Example of cofferdam failure due to instability of surrounding soil: (a) original cofferdam before failure; (b) rebuilt stable cofferdam. After White and Prentis.

placing the material on the low side of the cofferdam, as shown in Fig. 13-6(b). Whenever such conditions are encountered, the stability of the surrounding area should be analyzed by the Swedish circle method.

B. Heaving. If the soil below the bottom of excavation is soft, the danger of heaving should be investigated. The weight of the column of soil *abcd*, Fig. 13-7(a), acts as a surcharge load on this layer of soft soil. If the surcharge load is equal to or greater than the bearing capacity of the soil, the bottom of the excavation will fail by heaving. The stability of the bottom against heaving may be determined by the $\varphi = 0$ analysis whereby the shear strength is assumed equal to one-half of the unconfined compression strength and the angle of internal friction φ is assumed to be zero.

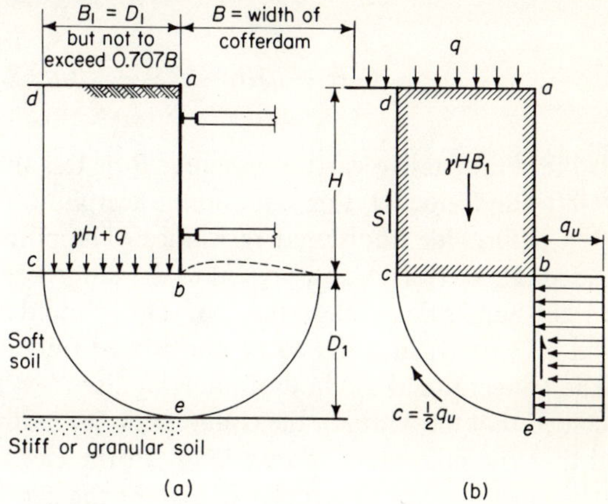

Fig. 13-7 Heaving of bottom of excavation in clay.

In the conventional method of analysis (Terzaghi, 1943; Tschebotarioff, 1951), the surface of rupture is assumed circular and the width B_1 of the soil column *abcd* is equal to the depth D_1 of the soft soil. It can be shown by geometry that the width B_1 cannot exceed 0.707B, B being the width of the cofferdam.

As the column of soil *abcd* tends to move downwards, a shear resistance S is developed along the surface *cd*. If the soil above the line *bc* is also cohesive, the shear resistance S is

$$S = \frac{1}{2}q_u\left(H - \frac{q_u}{\gamma}\right)$$

where q_u = unconfined compression strength,

γ = unit weight of the soil.

If the soil above the line *bc* is granular, the shear resistance S is

$$S = \tfrac{1}{2}K_a\gamma H^2 \tan\varphi$$

where K_a = coefficient of active earth pressure,

φ = angle of internal friction of the soil.

A free body diagram of the sliding block *abecd* is shown in Fig. 13-7(b). The weight of this block including the surcharge load q is resisted by the shear resistance S on surface *cd*, by the cohesive resistance c on surface *ce*, and by the passive earth pressure on surface *be*. Ignoring the weight of soil below the line *bc* on both sides of *be*, the net passive earth pressure is equal to the unconfined compression strength q_u, as shown in the figure. By taking moment about point *b*, we get the condition of equilibrium

$$\frac{1}{2}(\gamma H + q)B_1^2 - SB_1 - \frac{\pi}{4}B_1^2 q_u \leq \frac{1}{2} q_u B_1^2$$

or

$$(\gamma H + q)B_1 - 2S - \frac{\pi}{2}B_1 q_u \leq q_u B_1 \tag{13-4}$$

If the total passive earth pressure $q_u B_1$ is less than the sum of the terms in the left-hand side of this equation, sheetpiling or soldier beams must be driven to provide additional resistance. The piling or soldier beams should be adequate to resist a force equal to the difference between the left-hand and the right-hand sides of the equation. They should be driven to a depth at least equal to $\frac{2}{3}$ of D_1 in order to be considered fully effective.

The conventional method of analysis discussed above gives reliable results provided that the width of the cofferdam is larger than the depth of excavation and that the cofferdam is very long. In the cases where the cofferdams are square, rectangular, or circular in plan or the depth of excavation exceeds the

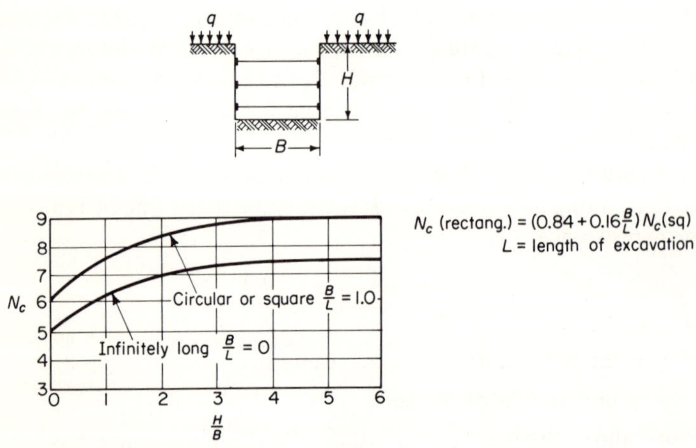

Fig. 13-8 Stability of bottom of excavation. After Bjerrum and Eide.

width of the cofferdam, the following analysis (Bjerrum and Eide, 1956) should be used. In this analysis, the cofferdam is visualized as a deep footing whose depth and horizontal dimensions are identical to those at the bottom of the cofferdam. This deep footing would fail in an identical manner to the bottom of cofferdam failed by heave, with the exception that the shear stresses in these two cases are in opposite directions. Using the theory of

Skempton for footings in clay, the factor of safety against bottom heave is

$$\text{F.S.} = N_c \frac{c}{\gamma H + q} \tag{13-5}$$

where N_c = the bearing capacity factor.

c = the cohesion of the clay = $\frac{1}{2}$ unconfined compression strength.

γ = the unit weight of soil above the bottom of excavation.

H = the depth of excavation.

q = the uniform surcharge load on the area surrounding the cofferdam.

The values of N_c are shown in Fig. 13-8 and are identical to that shown in Fig. 6-5, except that the symbols are different. Note that Eq. (13-5) does not include the additional resistance of sheetpiling or soldier beams if they are used.

13-7 Piping

Sheetpiling is generally used in the cases where the soil is pervious and where the excavation must be dewatered by pumping from the bottom of the excavation. Sufficient penetration of sheetpiling must be provided to reduce the amount of seepage water and to avoid danger of piping.

Piping is a phenomenon of water rushing up through pipe-shaped channels due to large upward seepage pressure. When piping takes place, the weight of the soil grains is counteracted by the upward hydraulic pressure. Virtually the soil at the bottom of excavation has no contact pressure between the grains. Therefore, it offers no lateral support to the sheetpiling, and as a result, the sheetpiling may collapse. Furthermore, the soil, after being subjected to piping or boil, loses its original compactness and becomes very loose. If it is intended to support a heavy foundation, it may not have the bearing power, as required by the original design. Therefore, it is essential to avoid the danger of piping.

The investigation of piping, as any other hydraulic problems in soils, is complicated by the nonhomogeneity and anisotropy of the natural soil deposits. Natural soils frequently include pockets and lenses of different material or similar material with different coefficient of permeability. The aggregate property of the soil deposits as a whole, including the effect of the lenses and pockets, is of practical significance in actual problems. Theories are developed on the basis of uniform and isotropic material. In stratified soils, however, the hydraulic properties are anisotropic because the permeability in the horizontal direction is considerably higher than that in the vertical direction. To investigate the case of stratified soils, the soil profile may be transformed by using a reduced horizontal scale equal to $\sqrt{k_v/k_h}$ and

Table 13-1 MINIMUM PENETRATION (D_1) OF SHEETPILING* (F.S. = 1.5)

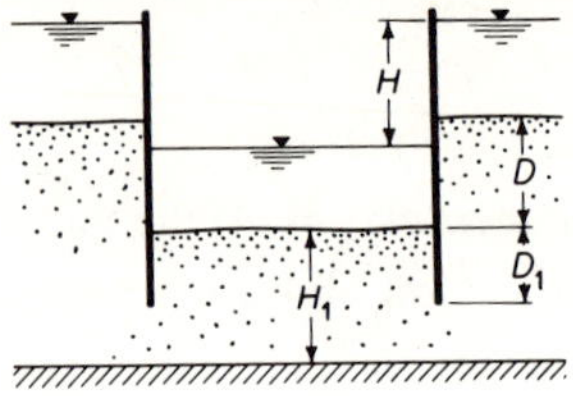

Site conditions	Width of excavation					
	$8D$	$4D$	$2D$	D	$\frac{1}{2}D$	$\frac{1}{4}D$
1. Homogeneous with infinite depth						
Loose sand	$0.7D$	$0.8D$	$0.9D$	$1.0D$	$1.2D$	$1.4D$
Dense sand	$0.4D$	$0.5D$	$0.6D$	$0.8D$	$1.0D$	$1.3D$
2. Homogeneous, impervious layer below (dense)						
$H_1/D = 1$		$0.4D$	$0.4D$	$0.6D$	$0.8D$	$0.9D$
$H_1/D = 2$		$0.4D$	$0.5D$	$0.8D$	$1.1D$	$1.3D$
3. Coarse layer under fine layer						
At a depth > width below pile tip	As in 1.					
Between excavation level and a depth < width below pile tip	Use flow nets					
Coarse layer reaches above excavation level	Homogeneous values are safe					
4. Coarse layer above fine layer						
	Generally safer than homogeneous					
5. Fine layer on a homogeneous soil mass						
At a depth ≧ width below pile tip	Use homogeneous values with impervious base at upper level of layer					
Lying below excavation base but above a depth ≧ width below pile tip	Very dangerous					
Layer above excavation base	Complete excavation safer than homogeneous case but dangerous conditions may arise during construction					

* After Marsland.

the soil may be considered as homogeneous and isotropic with a coefficient of permeability equal to $\sqrt{k_v k_h}$, where k_v and k_h are coefficients of permeability in the vertical and horizontal directions, respectively.

In the more common conditions, the penetration of sheetpiling required to prevent piping may be found from Table 13-1. Values in this table are based on a factor of safety of 1.5 against the danger of piping.

13-8 Design Example

This example, Plate DE 13, illustrates the essential steps of a design for the bracing of a deep open cut in a subsoil consisted of sand and clay. Since the maximum earth pressure is the envelope for the most unfavorable condition, large overstress is allowed.

As the wales are stressed by the lateral pressure, the lateral deflection of the member tends to reduce the magnitude of pressure at the point of large deflection. Consequently, the wales are seldom stressed to the value as calculated by the trapezoid rule. Therefore, a larger overstress factor of the wales is justified.

An interior view of the braced cofferdam exemplified in this design example is shown in Fig. 13-8.

General Arrangement

Arrangement largely governed by the details of permanent substructure which is to be built within the cofferdam

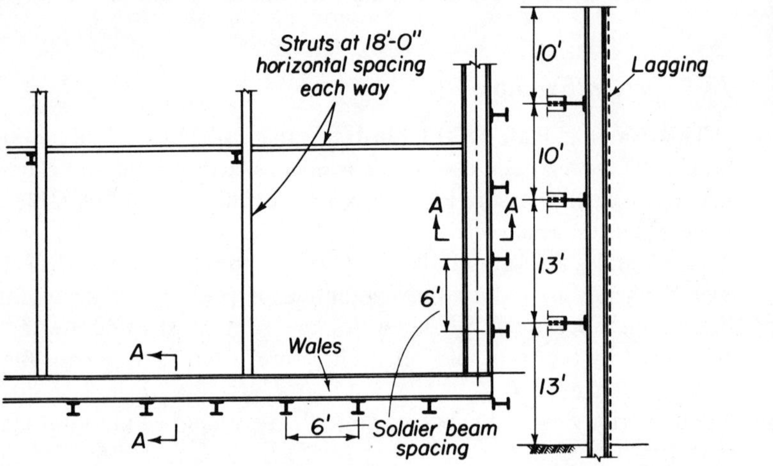

Plan at corner of cofferdam

Section A–A

Boring Data

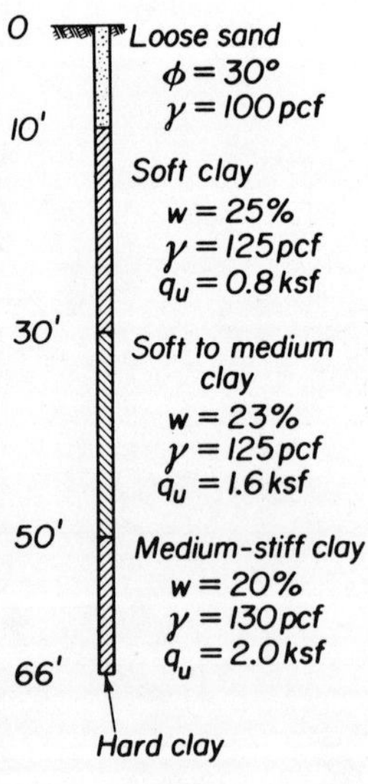

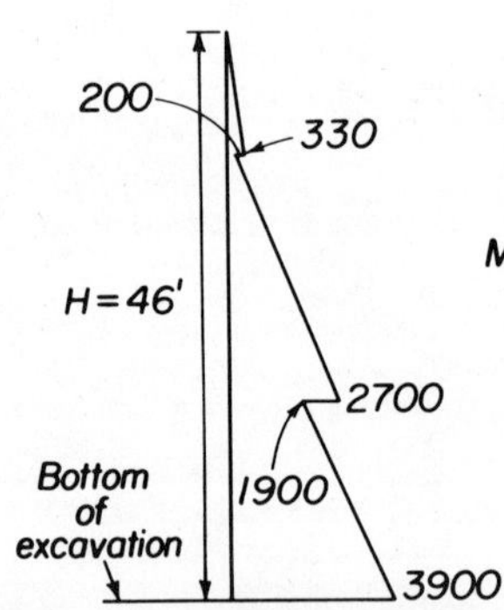

Rankine

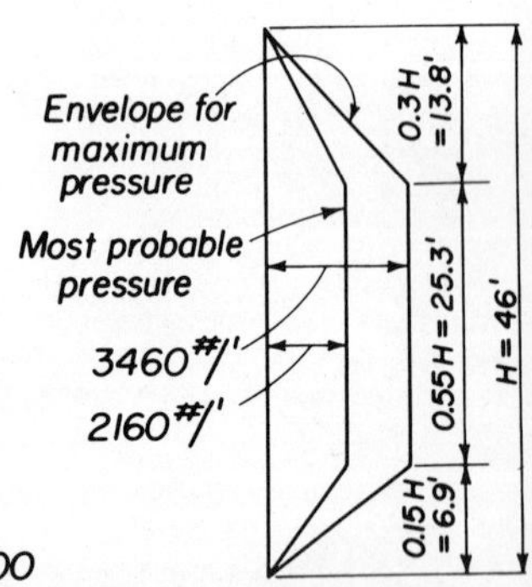

Peck

Earth Pressure Calculation (Rankine)

Sand layer: $k_a = \tan^2(45 - \frac{30}{2}) = 0.33$

$p_a = \gamma k_a = 33$ psf/ft

$\times 10 = 330$ #/′

$\times \frac{1}{2}(10) = 1650$ #/′

Clay layers:

γH		q_u	$\gamma H - q_u$	Total P
100 x 10 =	1000	800	200	1,650 #/′
125 x 20 =	2500			$\frac{1}{2}(200+2700)20 = 29{,}000$
	3500	800	2700	
		1600	1900	$\frac{1}{2}(1900+3900)16 = 46{,}400$
125 x 16 =	2000			$\Sigma P = 77{,}050$ #/′
	5500	1600	3900	

Redistribution

$\frac{1}{2} p_b(46 + 25.3) = 77{,}050$ $p_b = 2160$ #/′ (most probable)

Maximum Pressure for Design (Peck's Rule)

Eq. (13-2), $q_a = \frac{1}{H}[\gamma_s k_s H_s^2 \tan\phi_s + (H - H_s) n q_u]$

Eq. (13-3), $\gamma_a = \frac{1}{H}[\gamma_s H_s + (H - H_s)\gamma_c]$

$H = 46'$
γ_s = unit wt of sand = 100 pcf
$k_s = 1.0$
$H_s = 10$
$\phi_s = 30°$
$q_u = 800$ and 1600
$n = 1.0$

q_a = aver. unconfined comp. strength

$= \frac{1}{46}[100 \times 1.0\,(10)^2 \tan 30 + 20 \times 8.00 + 16 \times 1600] = 1030$ psf

$\gamma = \frac{1}{46}[100 \times 10 + (46 - 10)125] = 120$ pcf

Max probable pressure $= p_b = \gamma_a H - 2q_a = 120 \times 46 - 2 \times 1030 = 3460$ psf

DE 13
Braced cofferdam
Sh. 3 of 4

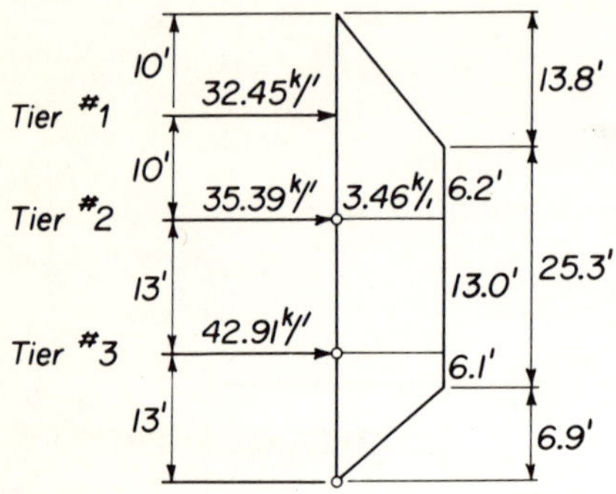

Load on Wales:

Take moment about Tier #2:

$\frac{1}{2}(3.46)(6.2)^2 = 66.5$

$\frac{1}{2}(13.8)(3.46) \times 10.8 = 258.0$

$10 \,\overline{)\,324.5}$

Tier #1, $P = 32.45^{k/'}$

Take moment about Tier #1:

$(3.46)6.2(6.9) = 148.0$

$\frac{1}{2}(13.8)(3.46)\times(-0.8) = -19.1$

$10 \,\overline{)\,128.9}$

12.89

$\frac{1}{2}(3.46)(13) \quad 22.50$

Tier #2, $P = 35.39^{k/'}$

Take moment about bottom of excavation:

$6.1(3.46)(9.95) = 210.0$

$\frac{1}{2}(3.0)6.9(4.6) = 54.9$

$13 \,\overline{)\,264.9}$

20.41

22.50

Tier #3, $P = 42.91^{k/'}$

Struts

Tier #1: Axial load $= 32.45^k \times 18 = 584^k$

$\times \frac{1}{1.5} = 390^k$ equivalent load allowing 50% overstress*

Try 14WF87 $L/r = \frac{18 \times 12}{3.70} = 59$ $F_a = 15.31^{ksi}$

$\times 25.56 = 390^k$ OK use 14WF87

Wales (Upper Tier)

Assume all wales and struts in each tier are at same elevation

Axial load $= 32.45 \times 9 = 292^k$

$\times \frac{1}{1.65} = 177^k$

$Q = 32.46 \times 6 = 195^k$

$\times \frac{1}{1.65} = 118^k$

(allowing 65% overstress*)

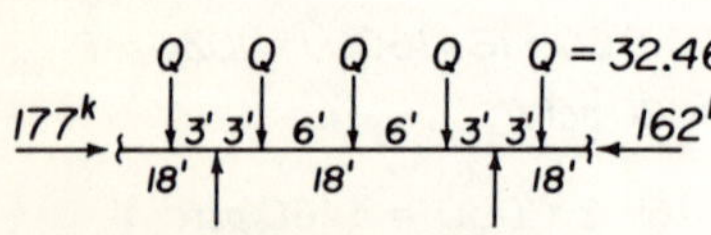

Interior spans

$\text{Max } M = \frac{Q(3)(15)}{18} + \frac{Q(18)}{8} = 561^k$

*Such large overstress may be used only when (1) soil borings and tests are adequate and carefully made, (2) cofferdam is installed by experienced contractor, and preferably (3) stresses in the structs are checked at frequent time intervals by means of strain gages.

Try 27 WF 160 $L/r = 18 \times 12/3.12 = 69 \quad F_a = 14.69^{ksi}$

$$\frac{f_a}{F_a} + \frac{f_b}{F_a} = \frac{177/47.04}{14.69} + \frac{561 \times 12/444.5}{20} = 0.26 + 0.76 = 1.02 \approx 1.0$$

Use 27 WF 160

Soldier Beams

$$\text{Max. } M = \frac{1}{12}(3.46 \times 6)(13)^2 = 292^{1k}$$

Allowing 65% overstress, $S = \frac{292 \times 12}{20 \times 1.65} = 106''^3$ req'd

Use 14 BP73 at 6'-0" spacing

Stability at Bottom of Excavation:

If the cofferdam is very long and the width is greater than 46', the conventional analysis applies:

$$\text{Eq. (13-4)} \quad \gamma H B_1 - 2S - \frac{\pi}{2} B_1 q_u \leq q_u B_1$$

Consider the possibility of failure in 4' of soft-medium clay below bottom of excavation:

$B_1 = 4'$

$S = 20' \times 0.8 + 16' \times 1.6 = 41.6^k$ (ignoring frictional resistance in sand)

$\gamma H = 5500$

$5.500 \times 4 - 2 \times 41.6 - \frac{\pi}{2}(4)(1.6) = \text{negative}$ Net pressure = 0

$\therefore$ safe

Consider the possibility of failure in 20' of soft-medium and medium-stiff clays below bottom of excavation (4 ft soft-medium clay and 16 ft medium-stiff clay):

Weighted average $q_u = \frac{1}{20}(1.6 \times 4 + 2.0 \times 16) = 1.92^{ksf}$

$B_1 = 20'$

$5.500 \times 20 - 2 \times 41.6 - \frac{\pi}{2}(20)1.92 = \text{negative}$ OK

If the cofferdam is only 20' wide and 25' long, use eq. (13-5):

From Fig. 13-7, for $B/L = 20/25 = 0.8$

$H/B = 46/20 = 2.3$

N_c square = 8.3

N_c rectangular = $(0.84 + 0.16 \times 0.8)0.83 = 8.02$

$$F.S. = 8.02 \, \frac{1920}{5500} = 2.8$$

Plate Fourteen

A Cellular Cofferdam

A cellular cofferdam is constructed by installing steel sheet piles in a form of series of cells which are subsequently filled with suitable soil. They are used primarily in water front construction.

The procedure of design of cellular cofferdams includes the analysis of stability and internal stresses. The procedure of construction consisting of template driving, pile driving, and cell filling, are discussed in this chapter.

14 CELLULAR COFFERDAMS

14-1 Common Types of Cofferdams

A cofferdam is a temporary enclosure in a river, lake, etc., to keep water from the enclosed area in order to construct a permanent structure or structures in the dry. A cofferdam is constructed in the water, and when it is completed, the enclosed area is dewatered by pumping. During the entire life of a cofferdam, a certain amount of pumping is constantly needed because some water will leak through the cofferdam and the foundation.

A variety of cofferdams have been used, Fig. 14-1, and the most common types and their relative merits are discussed as follows:

1. *Cantilever sheet piles:* suitable for small cofferdams; susceptible to large leakage and flood damage.
2. *Braced cofferdams:* economical for small to moderate height; susceptible to flood damage.
3. *Earth embankments:* no height limitations.
4. *Double-wall cofferdams:* suitable for moderate height.
5. *Cellular cofferdams:* suitable for moderate and large height.

The double-wall cofferdam and cellular cofferdam are fundamentally the same in design principles which are discussed in this chapter. The other types of cofferdams, except earth embankment, are discussed elsewhere in this book.

14-2 Common Uses of Cellular Cofferdams

A cellular cofferdam is made by driving straight-web steel sheet piles to form a series of cells which are subsequently filled with a suitable soil. The

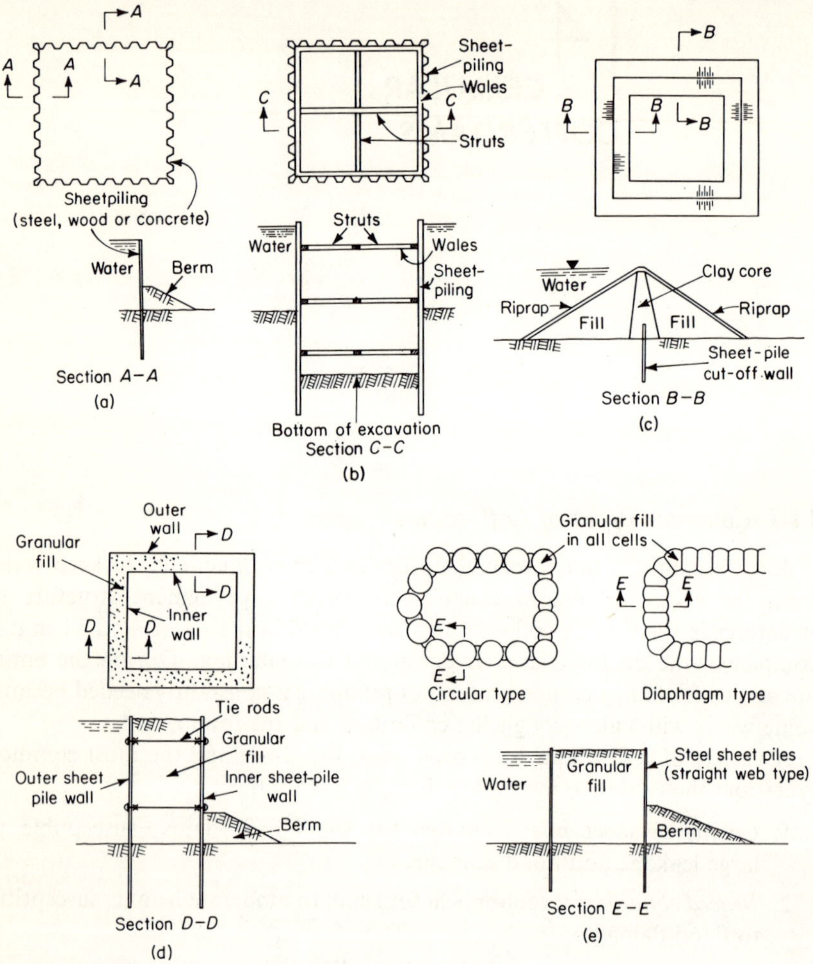

Fig. 14-1 Common types of cofferdams: (a) cantilever sheet pile; (b) braced cofferdams; (c) earth embankment; (d) double sheet pile wall; (e) cellular cofferdam.

cells are interconnected for watertightness and are self-stabilizing against the lateral pressure of water and earth.

The advantages of the cellular type cofferdams, as opposed to other types, are:

1. *Construction economy*. The cellular type is more economical than the braced cofferdam where the enclosed area and/or the water head is large.
2. *Watertightness*. The cellular type is more watertight than the braced cofferdam.

3. *Channel restriction.* The cellular cofferdam is smaller than the embankment type which occupies a large area and restricts channel flow or navigation.

4. *Salvage value.* Steel sheet piles are pulled out for reuse after completion of the job.

5. *Relatively simple construction.*

Therefore cellular cofferdams are often used for large waterfront constructions. In addition, singular cells may be constructed in the waterfront for scour protection, as dolphins, as supports for heavy equipment and industrial structures, as coal handling stations, as mooring stations, etc., Fig. 14-2. The condition favorable for cellular cofferdams is a bedrock

Fig. 14-2 Single cells used as supports for water front structures and mooring stations. Courtesy of Public Service Company of Indiana, Inc.

covered with sand and gravel but free from boulders that may prevent pile penetration. If there is soft material, it may have to be removed and replaced by sand and gravel, as will be discussed later.

Cellular cofferdams may also be used to retain earth backfill in the same manner as retaining walls.

14-3 Common Types of Cellular Cofferdams

Cellular cofferdams are generally classified according to the shapes or arrangement of the cells, Figs. 14-3 and 14-4.

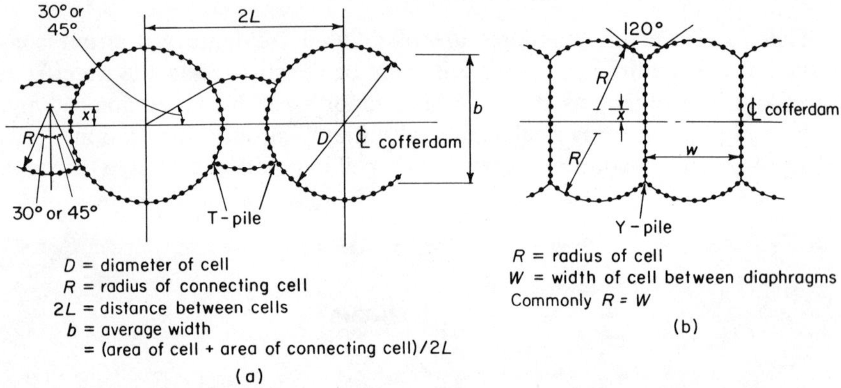

Fig. 14-3 Basic types and common arrangement of cellular cofferdams: (a) circular types; (b) diaphragm type.

Circular type consists of individual large diameter circles connected together by arcs of smaller diameter. These arcs usually intercept the circles at a point at 30 degrees or 45 degrees with the longitudinal axis of the cofferdam. They are often perpendicular to the circle, but occasionally different angles may be used.

Diaphragm type consists of two series of arcs connected together by diaphragms perpendicular to the axis of the cofferdam. Generally the radii of these arcs are made equal to the distances between the diaphragms. At the intersection point the two arcs and the diaphragm make an angle of 120 degrees between each other.

Cloverleaf type may be considered as a modification of the circular type in cases where a large diameter (required by stability) would result in excessive high interlock stresses if no diaphragms are added.

Modified types have been used (TVA, 1957) to retain earth fill. In the circular type, one side of the small arcs may be omitted if stability is not a problem. The front face of the diaphragm type cells may be anchored by extending the diaphragms a sufficient distance into the earth bank and thus eliminating the necessity of sheet piles in the back face.

The relative advantages and disadvantages of each of the two basic types (circular and diaphragm) are as follows (TVA, 1957). In normal cases, the circular type is far more favorable than the diaphragm type.

Advantages of circular type: The major feature of the circular type is that each cell is self-supporting and independent of the next cell, therefore:

1. It can be used singularly, in a group, or at end.
2. It will not collapse in the event of failure of adjoining cells (due to interlock damage, flash flood, or any other causes).

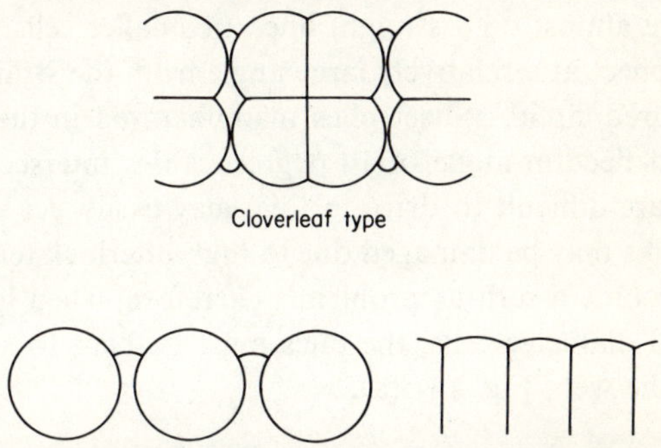

Fig. 14-4 Modified cellular cofferdams.

3. It can be filled with no regard to the progress of the other cells. This permits the fill to be placed by hydraulic method.* The cell may be filled immediately after completion of pile driving for each cell, and the construction equipment may be operated on top if it.
4. It requires fewer number of piles per lineal foot of cofferdam as compared with the diaphragm type of an equal design.

Advantages of diaphragm type:

1. It has uniform interlock stress throughout the section at any given level. The stress is smaller than that at the joint of a circular cell of a comparable design.
2. It can be widened readily by increasing the length of the diaphragms if it is required by stability. This will not increase the interlock stress which is a function of the radius of the arc.

From the design point of view, cellular cofferdams are classified with respect to the type of supporting stratum:

1. Cellular cofferdams on rock foundation
2. Cellular cofferdams in earth foundation.

14-4 *Components of Cellular Cofferdams*

The major components of cellular cofferdams are the steel cells and the cell fill. Sometimes earth berms are also constructed on the dry side of the

---

* A hydraulic fill is one in which the material is transported by water.

cofferdam to increase its stability. In the circular type the main cells are circular, or cylindrical, whereas the connecting cells are arcs of smaller radii. In the diaphragm type, the front and back faces of the cells are circular but the diaphragms are straight walls.

The circular portions of the cells are actually polygons, with one sheet pile forming one side of the polygon. In large diameter cells any two adjacent piles are almost on a straight line. In smaller cells, however, each sheet pile must deflect at a relatively large angle from the straight line in order to form the desired circle. Sheet piles manufactured in the U.S.A. permit a maximum deflection angle of 10 degrees. Piles intersecting at larger deflection angles are difficult to drive in line, may easily get out of interlock, and the interlocks may be damaged due to high interlock tension, and watertightness may become a serious problem. Therefore, when larger angles are required in small diameter cells, the piles must be bent to such an angle in the middle of the web, Fig. 14-5(a).

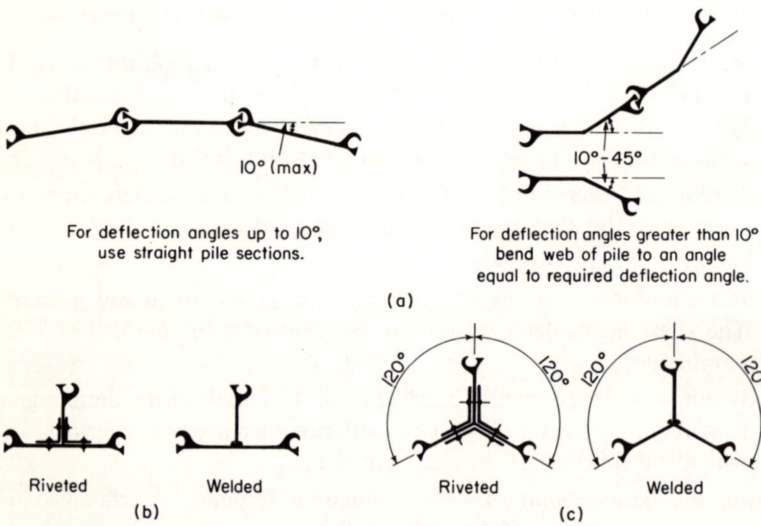

Fig. 14-5 Components of cellular cofferdams: (a) deflection angle; (b) T-piles; (c) Y-piles.

The pieces at the junction points in the circular cofferdams are commonly 90 degree Tees, and those in the diaphragm cofferdams are commonly 120 degree Wyes. These pieces are prefabricated by riveting or welding whole and half sections together, Fig. 14-5[(b) and (c)].

The cell fill should be coarse-grained soil which may be deposited through water. The criteria for selecting cell fill is discussed in Sec. 14-6.

14-5 Design of Cellular Cofferdams

Cellular cofferdams may be designed by the following steps:

1. Assemble controlling dimensions: height of cofferdam, maximum and minimum water level–see discussion below.
2. Make site and soils survey–see discussion below.
3. Select cell fill–Sec. 14-6.
4. Select tentative proportion and average width–Sec. 14-7.
5. Analyse stability–Secs. 14-8 and 14-9.
6. Determine interlock stress–Sec. 14-10.

In order to establish the height of the cofferdam, the flood records must be carefully studied. The top of cofferdam should be at least at the level of the anticipated high water during the life of the cofferdam. The total height of cofferdams measured from top to bottom often exceeds 50 ft. The Kentucky Dam designed by TVA had a total height of 98 ft with the top of dam 48 ft above the earth berm and overburden. For high cofferdams, earth berms are often used to reduce the relative height of the cofferdam above the ground.

The site survey should include information of the elevation of the existing ground surface and the depth of scour. Usually the depth of scour is difficult to establish. Geological reports published by public agencies may contain some description of damages in the locality due to floods in the past. Local bridge engineers may have some experience concerning the nature of flood and the possible depth of scour.

Soil borings should be made down to bedrock so that a sufficient knowledge is obtained concerning the soil strata and the general configuration of the bedrock surface. Several borings should be cored into the bedrock in order to eliminate the too frequent error of stopping the borings on top of boulders and believing mistakingly that the bedrock is encountered.

The depth of soft soils (soft clay, silt, organic soils, etc., if encountered) should be carefully determined, since such soils must be removed and replaced by granular soils for high cofferdams.

14-6 Material Survey and Tests

Since cellular cofferdams acquire their stability largely from the shear resistance of the cell fill, it is of utmost importance to use the most suitable material available in the locality, and the physical properties of the cell fill must be accurately determined before any design analysis is attempted. A desirable fill material should have the following properties (TVA, 1957):

1. Free-draining, granular soils with little fine particles.

2. High shear strength; high coefficient of friction.

3. High unit weight.

4. Large resistance to scour and leakage; well-graded soils are most suitable.

As a rule, natural deposits of mixed sand and gravel possess all of these desirable properties. If granular soils are not available in sufficient quantities in the locality, the better material should be placed on the lower portion of the cells.

With the exception of small cells, the shear strength (the angle of internal friction) of the fill material should be determined by laboratory tests on a sufficient number of samples. The lowest test values should be used in the design.

Even with free-draining materials as cell fill, it is a common practice to provide weep holes in the cells (see later discussion).

When a berm is constructed on the dry side of the cofferdam for the purpose of acquiring additional stability of cell, the materials to be used for the berm should have similar properties as the cell fill itself. If the cofferdam is constructed in swift current or is subjected to scour, a layer of riprap is necessary.

14-7 Average Width of Cofferdams

After the height of the cofferdam is established on the basis of flood records and the pertinent physical properties of the underlying soils as well as of the cell fill are determined, the design of cofferdam begins with a tentative proportion which is subsequently analyzed for stability and other safety requirements. The final design is one that satisfies all the design requirements set forth in the next article.

The design is usually made on the basis of a section one foot long with a uniform average width b, instead of working with the actual cells having varying widths. The average theoretical width is that of a rectangular section having a section modulus equal to that of the actual cofferdam from center to center of cells. For design purposes a simple procedure may be used whereby the average width is determined such that the area in the rectangular section and that in the actual cofferdam are equal. Let b = average width and $2L$ = distance center to center of cells, then:

$$b = (\text{area of main cell} + \text{area of one connecting cell})/2L$$

TVA (1957) engineers have found that these two methods give almost identical results, and that the average widths by the method of equal area vary up to 6 per cent higher than that by the method of equal section modulii. The method of equal area, therefore, may be slightly on the unsafe side. The relationship between the average width b and the diameter D used by TVA engineers is as follows:

$$\left.\begin{aligned} b &= 0.785D \qquad \text{with 90-degree T's} \\ b &= 0.875D \qquad \text{with 60-degree T's} \end{aligned}\right\} \qquad (14\text{-}1)$$

After the required average width of cofferdam is proved satisfactory by the stability analysis, the final arrangement can be worked out.

14-8 Stability Analysis–Cofferdams on Rock

Since a large portion of the cell fill is saturated, the location of the line of saturation becomes the first step in the stability analysis. Based on observations of several cellular cofferdams, TVA has arrived at the conclusion shown in Fig. 14-6. If the cell is perfectly free-draining, the lower one-half of the cell

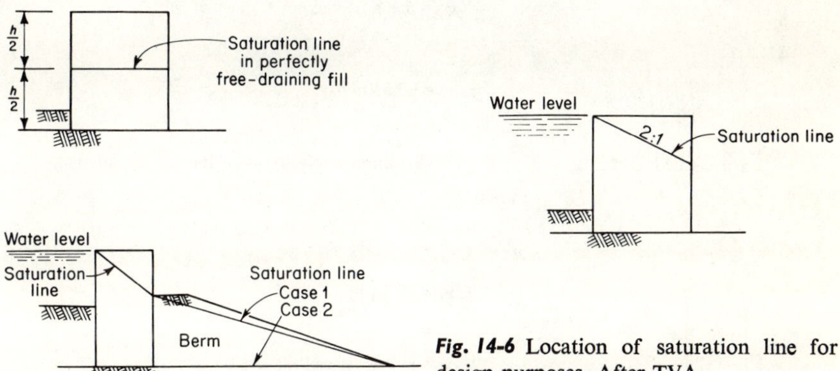

Fig. 14-6 Location of saturation line for design purposes. After TVA.

may be assumed saturated. With other type of cell fill, the saturation line is on a 2:1 slope. In the cases where earth berm is used, the saturation line slopes down to the top of the berm. In the earth berm, two locations of saturation line should be used so as to make provision for the more critical condition.

For cellular cofferdam carried down to bedrock, the following possibilities of failure should be analysed.

A. Sliding. The cofferdam is subjected to a horizontal force due to the maximum water head and the soil pressure, Fig. 14-7(a). This horizontal force is stabilized by the frictional resistance along the bottom of the cofferdam and the passive resistance of the berm, if used. The frictional resistance is equal to the total weight of fill (use buoyant weight for fill below the saturation line) in the cell times the coefficient of friction which is usually taken as equal to tan φ, φ being the angle of internal friction of the granular fill. The passive resistance of the berm is generally determined by the Coulomb theory, which gives values on the unsafe side.

B. Slipping. As the cell is subjected to an excessive lateral thrust, the sheet piles along the river side (pressure side) creep up with respect to the cell fill and cell assumes the shape shown in Fig. 14-7(b). As a result, the fill material runs out at the heel of the cell. While the piles on the river side creep upward as one whole unit, the piles on the dry side slip with each other along the interlocks.

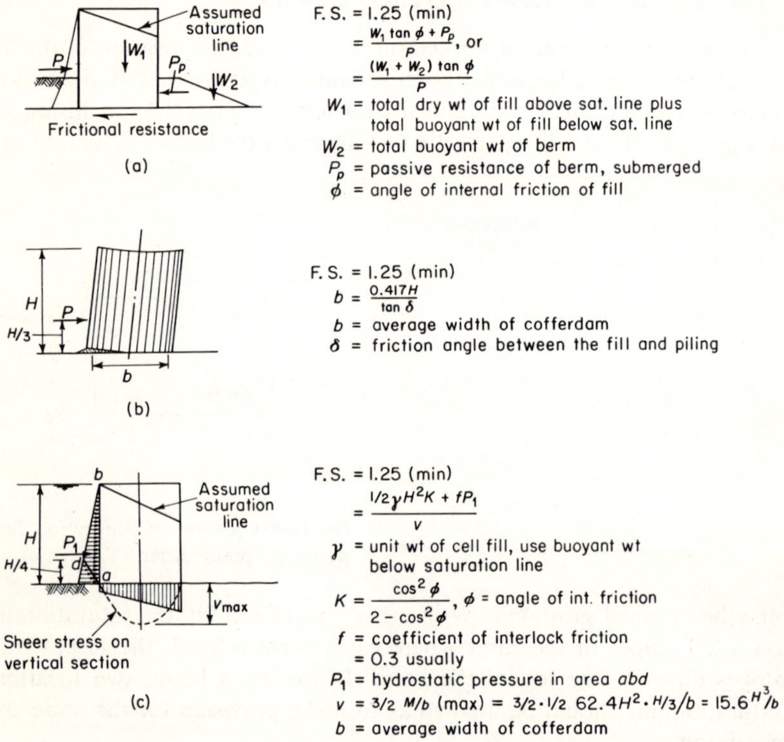

Fig. 14-7 Stability of cellular cofferdam on rock: (a) sliding; (b) slipping; (c) vertical shear. After TVA.

This phenomenon indicates that the frictional resistance between the fill and the piling on the river side of the cell is smaller than the frictional resistance along interlocks. In order to prevent such upward slippage, the average width b of the cell should satisfy the following equation:

Frictional resistance against upward movement of piling = F.S. $\times$ upward force

where F.S. = factor of safety = 1.25 minimum.

If the lateral pressure between the fill and piling on the river side is assumed equal to the external pressure P, this equation may be written as

$$Pb \tan \delta = 1.25 \frac{H}{3} P$$

Solving for b,

$$b = \frac{0.417H}{\tan \delta} \tag{14-2}$$

where b = average width of the cofferdam;
H = height of the cofferdam;
δ = friction angle between the fill and piling.

In the case where the piling is driven through an overburden, there is an additional resistance against upward movement. This resistance may be assumed equal to the total active earth pressure times $\tan \delta$.

C. Vertical shear. As any structural member acted upon by lateral forces, the cofferdam is subjected to shearing stresses. The magnitude of this shear stress is maximum along the center of the cell, and may be determined by

$$V_{max} = \frac{3M}{2b} \tag{14-3}$$

where M = bending moment due to external lateral force above the base (or above any horizontal section under consideration), b = average width of the cofferdam. The shear resistance s' at any given point on a vertical section in the cell fill may be determined by the following equation:*

$$s' = \gamma z K$$

$$s' = \gamma z \frac{\cos^2 \varphi}{2 - \cos^2 \varphi} \tag{14-4}$$

where γ = unit weight of cell fill, use buoyant weight for portion of fill below saturation line;
z = depth of the point under consideration;
φ = angle of internal friction of cell fill.

Note that the value K is greater than coefficient of active earth pressure K_a.

In addition, the friction on the interlocks also offer resistance against the vertical shear. This friction is equal to the interlock tension times coefficient of friction. Let the horizontal internal pressure due to the cell fill and water = P_1, then the frictional resistance along interlocks = fp_1.

* TVA, 1957 and Krynine, 1944, *Disc. on Terzaghi* paper.

TVA engineers use the pressure shown in the triangle *adb*, Fig. 14-7(c), because of the fact that the lower end of piling bites into the rock and therefore the later pressure is resisted by vertical bending not by ring tension.

The total maximum ring tension along the center line of cofferdam is LP_1 and the frictional resistance is

$$\frac{LP_1}{L}f = fP_1$$

The total shear resistance is then

$$S = \frac{1}{2}\gamma H^2 \frac{\cos^2 \varphi}{2 - \cos^2 \varphi} + fP_1$$

14-9 Stability Analysis–Cofferdams in Deep Soil

The procedure of stability analysis of cofferdam on rock foundation must be applied to the cofferdams on deep strata of sand or clay. In addition, the following requirements must be satisfied.

1. The sheetpiling must be driven to a sufficient depth so that its bearing capacity is equal to at least 1.5 times the vertical force which acts on the piling.

As a lateral force is applied to a cofferdam, it increases the reaction at the toe (or dry side) of the cell and the sheetpiling is dragged down by the cell fill. Since the maximum vertical pressure introduced to the sheetpiling by the cell fill never exceeds the friction between the fill and the piling, therefore,

$$F_1 = \tfrac{1}{2}\gamma H^2 K_a \tan \delta \tag{14-5}$$

where F_1 = downward force on a unit length of sheetpiling;
γ = unit weight of cell fill;
H = height of cell above top of sand stratum;
K_a = coefficient of active earth pressure;
δ = angle of friction between cell fill and piling.

2. Cellular cofferdams on a deep sand bed should be designed to prevent boiling at the toe due to seepage water.

Due to high permeability of sand, water retained behind the cofferdam percolates at a relatively large speed under the base of the cofferdam and rises up in front of the toe. In Fig. 14-8, the arrows represent the paths of seepage flow. It is seen that the seepage flow in front of the toe of the cofferdam is nearly vertical, thus, exerting an uplift pressure against the weight (or the buoyant weight) of the soil particles. If this seepage pressure becomes equal to the buoyant weight of the soil, shear strength of the soil is reduced to that of a pure fluid which is equal to zero. Under such circumstances the toe of cofferdam is, in effect, acted by water pressure only. Therefore, it is imperative to prevent the possibility of boil or any reduction in the effective weight of soil particles.

The danger of boil can be readily eliminated by the use of loaded filter (Terzaghi, 1945). The purpose of a filter is to provide a counter weight which is at least equal to the upward seepage pressure. The magnitude of this pressure can be determined by the flow net (see e.g. Terzaghi, 1943) such as the one shown in Fig. 14-8, and the filter material must be selected in accordance with the principle discussed in Sec. 5-5. The sheetpiling in sand should be driven to a depth equal to two-thirds of the height of the cofferdam above the ground surface or until it bears on a hard stratum.

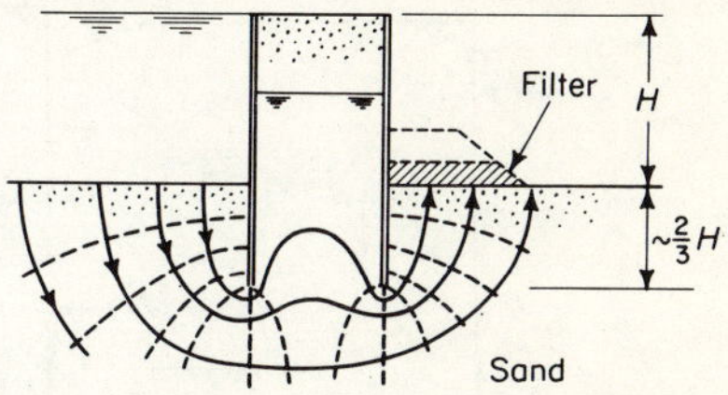

Fig. 14-8 Seepage in cellular cofferdam in sand. After Terzaghi.

3. Cellular cofferdams on clays should be investigated for the bearing capacity of the clay and the tilting due to large compression at the toe of the cofferdam.

The ultimate bearing capacity of a clay stratum can be determined by

$$H = 5.7\frac{c}{\gamma}$$

If a minimum factor of safety of 1.5 is used, the maximum height of cofferdam on deep clay is

$$H = 3.8\frac{c}{\gamma} \qquad (14\text{-}6)$$

where H = maximum height of cofferdam above ground surface;

c = cohesion of the clay;

γ = unit weight of cell fill, use buoyant weight for fill below saturation line.

If the cofferdam rests on soft to medium clay, or other compressible material, the unequal pressure may introduce a large amount of tilting. The tilting can be estimated if the compressibility of the soil is determined by laboratory tests.

14-10 Hansen's Theory of Stability of Cellular Cofferdams

In the masterly treatise on earth pressure, J. Brinch Hansen (1953 and 1958) proposes an analysis of cofferdam stability by the extreme method (or the maximum-minimum method). The cofferdam is analysed as an entire unit with a convex and concave surface of rupture, Fig. 14-9. For cofferdams on rock, the rupture takes place on a convex surface located above the foot of the sheetpiling. For cofferdams in soil, the rupture surface may be convex or concave depending on the penetration of the piling. With considerable depth of sheetpiling penetration, the rupture surface may develop above a point at

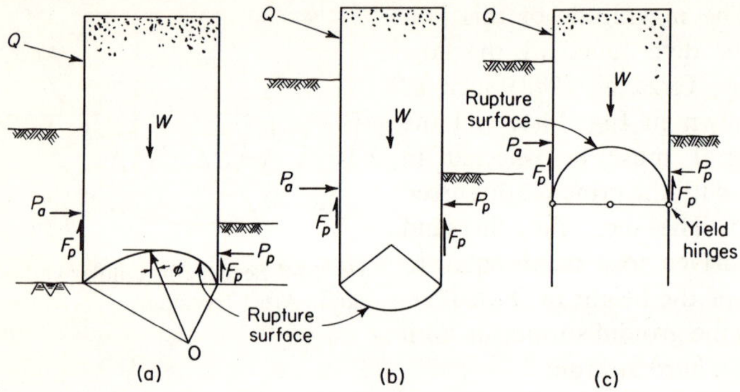

Fig. 14-9 Stability analysis of cellular cofferdams by Hansen's theory: (a) on rock; (b) in soil; (c) with yield hinges.

which the sheetpiling yields plastically. In any case, the location of rupture surface is determined by a number of trial calculations. The critical rupture surface is one which offers the least resistance against the external forces.

The rupture surface is approximated by a logarithmic spiral represented by

$$r = r_0\, e^{\theta \tan \varphi}$$

where r and θ = variables in the polar coordinate system;

φ = the angle of internal friction of the soil which is a known value.

A trial position of the spiral is chosen as shown in Fig. 14-9(a). The weight of soil above the surface of rupture W and the external forces Q, P, and F can be calculated. All the frictional forces acting on the spiral surface are directed towards the pole o and therefore, give no moment about the pole. Consequently, the stability of the cofferdam can be readily analysed by taking moment about the pole. If the stability is expressed as the ratio between the stabilizing moment due to W, P_p, F_a and F_p and the overturning moment due to Q and P_a, the critical rupture is one that has the smallest ratio between these two moments. To facilitate the solution by this method, engineers may utilize the graphs and tables prepared by Brinch Hansen (1953).

14-11 Interlock Stress

The interlock stresses at the junction between the main cells and connecting arcs are not easy to determine due to the difficulty arising in the evaluation of the lateral earth pressure in confined areas. However, on plane *oa* along the center line of the cofferdam, Fig. 14-10, the ring tension t is obviously equal to pL. Similarly, cutting a plane bc, the ring tension t_{max} is

$$t_{max} = pL \sec \theta \qquad (14\text{-}7)$$

where p is the horizontal pressure due to cell fill, L and θ are as shown in the figure. The difference of ring tension, t_{max} and t, is taken up by friction between the piling and the cell fill.

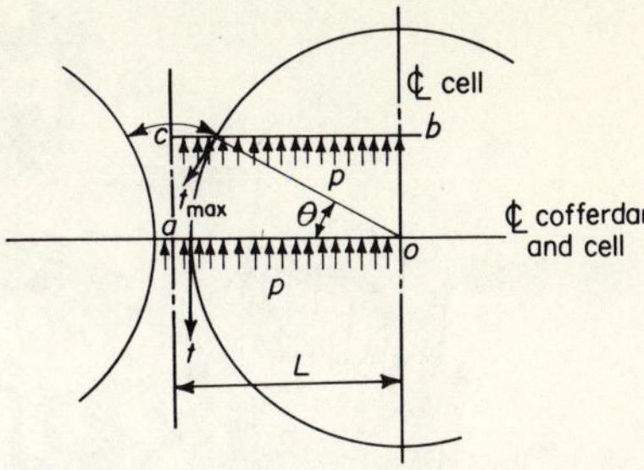

Fig. 14-10 Interlock stress in circular cofferdams. From TVA.

In computing the maximum ring tension, the maximum unit horizontal pressure p should be taken at a point above the bottom of cell, because the piling being well driven into the ground or to rock is restricted by the overburden, or rock, from developing the full ring tension. As already discussed in the preceding article, TVA engineers use the active earth pressure plus the hydraulic pressure at the lower quarter point above the rock or above the ground surface as the maximum design pressure. If the cells are filled by a hydraulic method, the fill immediately after placement is practically a fluid, and the full fluid pressure equal to the saturated weight should be used in connection with the low water stage of the river. This unit pressure may exceed the value calculated for the dry side of the cell.

The computed maximum interlock stress should not exceed the maximum stress specified by the manufacturers of the steel sheetpiles. A minimum of 12,000 pounds per inch is commonly assumed for the straight web type sheet piles manufactured in the U.S.A. If a factor of safety of 1.5 is desired, the design stress should be 8000 lb per in. At special request, a maximum guaranteed tension of 16,000 lb per in. may be obtained. For an interlock tension of 12,000 lb per in., the tensile stress in a $\frac{1}{2}$ in. thick web may reach 24,000 psi and $\frac{3}{8}$ in. thick web, 32,000 psi. Therefore, it can readily be seen that the interlock stress, not the web tension, is usually the governing factor.

14-12 Construction of Cellular Cofferdams

A. Template setting. The construction of a cofferdam begins with the setting of a template supported on four or more spuds. At locations where the current is swift the templates may be anchored by guey wires. The templates may be made of pipes or timber and may be on the outside or inside, or on both inside and outside of the cell. Metal templates can be reused many times, and therefore are often more economical. Special sizes, such as those for the closure cell, and the tie-in cells, are often made of timber.

Figure 14-11 shows an example of sectional pipe templates.

Fig. 14-11 Template for construction of cellular cofferdams. Courtesy of TVA.

B. Pile driving. Sheet piles and template supports may be driven by a pile driver supported on barges, on timber trestles, or on cells being just filled. All piles must be set around the cell and the junction piles (T-piles or Y-piles) must be located at accurate position before any pile is driven.

Sheet piles can be purchased in relatively long pieces. However, splicing may be necessary for the following reasons:

1. For excessively long piles, the unit cost is higher than those below the maximum available length. The engineer should check with the manufacturer in order to find out the non-premium length of piling. The cost of splicing may be below that of the extra premium for the long piles.
2. Shipping cost of long pieces may also be high.

Splicing may be made before driving, or when the first length is driven. The splice may be welded or riveted. If butt-welded, side plates are desirable.

When piles are driven to rock foundation, it is imperative to drive them home but not to overdrive them.

The lower portion or the full height of piling on the dry side of cofferdam should be provided with weep holes to prevent large pressure inside the cells. Weep holes may be burned after driving, or made beforehand. The diameter of weep holes commonly ranges from $1\frac{1}{2}$ in. to 4 in., and the spacing is generally 3 to 6 ft centers. The weep hole diameters should be determined in accordance with the principle of filter materials, Sec. 5-5.

The American sheet piles are made with either $\frac{3}{8}$ in. or $\frac{1}{2}$ in. thick webs. For relatively easy driving condition, a $\frac{3}{8}$ in. web should be sufficient and results in a slightly better economy. Where the piles must be driven through dense material, particularly with boulders, however, the $\frac{1}{2}$ in. web may be preferred to avoid damage due to hard driving.

C. Cell filling. The cell fill material may be placed by hydraulic or mechanical methods.

In the hydraulic method the material is dredged from the source, usually in the river bottom, and pumped directly to the cell. This method is economical, provided that suitable fill material is available within a radius about 2500 ft of the cell. Otherwise, the fill material must be trucked or shipped by other means to the site and placed by clam shell buckets.

The circular cells may be filled individually and independent of the fill depth in the adjacent cells. The connecting cells which are in between the main circular cells, however, should not be filled until the two adjacent main cells are completely filled. The diaphragm type cofferdam must be filled with a relatively small difference in fill height, about 5 ft maximum, because the diaphragm walls are not capable of resisting lateral load due to large difference in fill height on both sides.

D. Closure. The last cell in a cofferdam in a river presents one of the most difficult steps in the construction procedure. Usually the opening left for the last cell varies considerably from the planned cell size, and a special cell must be constructed to fit. Furthermore, the swift current may have to be diverted in order to maintain the empty cell until the fill is placed.

E. Foundation treatment. Cofferdams on rock foundation should be sealed against leakage by grouting before and/or after the construction of the cofferdam.

14-13 Design Example

A complete design of a cellular cofferdam on rock is given in DE 14. The material in this example is taken largely from TVA (1957).

DE 14
Circular Cofferdam
Sh. 1 of 4

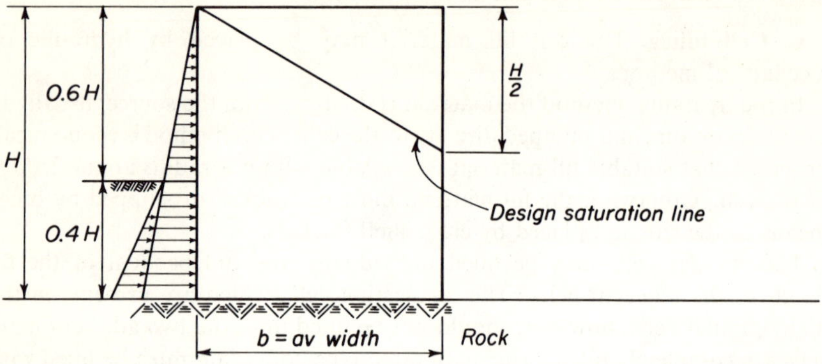

Given Data

Weight of dry material	$= 110^{pcf}$	from laboratory tests
Weight of submerged material	$= 65^{pcf}$	
Coef. of friction – gravel on rock	$= 0.5$	
– gravel on steel ($\delta = 21°-50'$)	$= 0.4$	
ϕ	$= 28°-50'$	
$\tan\phi$	$= 0.55$	
Coef. of friction – steel to steel (interlock)	$= 0.3$	
Slope of berm 3:1	$= 18°-26'$	

Allowable stresses in piling:

Tension in web $= 32^{ksi}$ (for $70{,}000^{psi}$ ultimate)

Tension in interlocks $= 8^{k}/^{in.}$ (for $12{,}000^{psi}$ min guaranteed

Solution

$$\text{Weight of cell fill} = 65\,bH + (110-65)\tfrac{1}{2}\,b\left(\tfrac{H}{2}\right) \qquad = 76.25\,bH$$

Lateral pressure:

$$\text{Hydrostatic} = \tfrac{1}{2}\,62.5\,H^2 \qquad = 31.25\,H^2$$

$$\text{Active earth pr.} = \tfrac{1}{2}\,\gamma' h^2 \cdot \frac{\cos^2\phi}{\left(1+\sqrt{\frac{\sin(\phi+\delta)\sin\phi}{\cos\delta}}\right)^2}$$

$$= \tfrac{1}{2}\,65(0.4H)^2(0.29) \qquad = \underline{1.51\,H^2}$$

$$P = 32.76\,H^2$$

Sliding

Frictional resistance = $W \tan\phi = 76.25bH(0.55) = 42.0bH$

For factor of safety = 1.25, $1.25P = W \tan\phi$

$1.25 \times 32.76H^2 = 42.0bH$, $b = 0.975H$

Slipping of Piles

Take moment about toe of cofferdam:

Hydrostatic $= 31.25H^2 \times \frac{H}{3} = 10.42H^3$

Active pr. $= 1.51H^2 \times \frac{1}{3}(0.4H) = 0.20H^3$

$M_t = 10.62H^3$

Force tends to uplift back face of cell $= 10.62\frac{H^2}{b}$

Frictional resistance between piling and fill $= P\tan\delta$

$= (32.76H^2 + 1.51H^2)0.4 = 13.70H^2$ (32.76 inside, 1.51 outside)

For factor of safety = 1.25

$1.25 \times 10.62\frac{H^3}{b} = 13.70H^2$ $b = 0.969H$

Vertical Shear

Try $b = 1.0H$ for vertical shear computation:

External moment about center of width b:

Hydrostatic, Active earth $= 10.62$

Vertical frictional resistance between piling and earth $= 1.51H^2(0.4)(\frac{H}{2}) = -0.32H^3$

$M_c = 10.30H^3$

Shear force on ℄ of cell $= \frac{3}{2}\frac{M_c}{b} = \frac{3 \times 10.30H^3}{2H} = 15.50H^2$

Shear resistance offered by cell fill, $s' = P\tan\phi$

s' (on river side) $= 32.76(0.55) = 18.0H^2$

s' (on dry side) $= \frac{1}{2}k\gamma H^2 - \frac{1}{2}k(\gamma - \gamma')(\frac{H}{2})^2$

$= \frac{1}{2}(0.623)(110)H^2 - \frac{1}{2}(0.623)(45)(\frac{H}{2})^2$

$= 16.9H^2$

Where $k = \frac{\cos^2\phi}{2 - \cos^2\phi}$

s' (at C) = average $= 17.45H^2$

Shear resistance offered by interlock friction = s'':

(Assume cell to be filled with water at top of overburden, and the inside overburden not to be removed until cells have been filled)

$(k_a = 0.29)$

At top of overburden $p = 0.29 \times 110 \times 0.6H \qquad = 19.13H$

At bot. of cell (pressure balanced below top of overburden) $19.13H$

Total pressure abcd:

$19.13H \times 0.6H \times \frac{1}{2} = 5.74H^2$

$19.13H \times 0.15H = 2.87H^2$

$19.13H \times 0.25H \times \frac{1}{2} = 2.39H^2$

$P = 11.00H^2$

$s'' = fP = 0.3 \times 11.00H^2 = 3.3H^2$

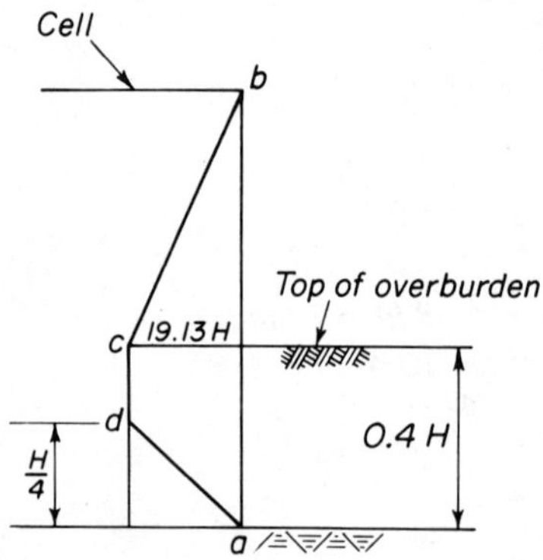

Total shear resistance $= s' + s'' = 20.75H^2$

$$F.S. = \frac{20.75H^2}{15.50} = 1.34$$

<u>Interlock Stress</u>

$t_{max} = pL \sec\theta$

For a cell filled hydraulically, $p = (62.5 - \overset{k_a}{0.29} \times 65)h = 81.4h$

for $h = 0.6H$ (at top of overburden)

$p = 81.4 \times 0.6H = 48.8H$

For a cell with 58.9^{ft} diameter (148 piles) and 10-connecting piles

$L = 30.83' \quad \theta = 34^\circ\text{-}03'$, and

$t_{max} = 48.8H \times 30.83 \times 1.207 = 1815H$ (#/')

If $b = H$, $H = 0.785D = 46.2'$

then $t_{max} = \frac{1}{12} 1815 \times 46.2 = 7000$#/''

8000#/'' OK

Cell Arrangement

$$D = 1.27B, \; b = H$$
$$\therefore D = 1.27H$$

With a given value H, select the cell diameter to suit

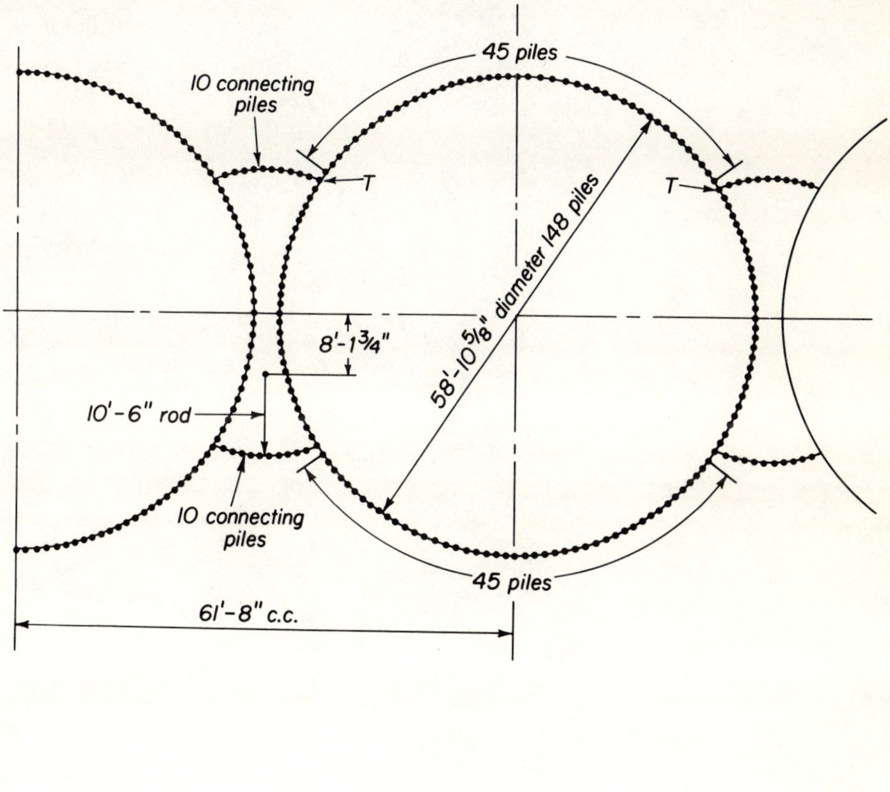

APPENDIX

A-1 Soil Classification

Soils are classified differently in various fields, namely:

Engineering: Unified Classification
Highway Research Board (or AASHO) Classification
Federal Aviation Agency (FAA) or Civil Aeronautical Administration (CAA) Classification

Agriculture: Textural Classification
Pedological Classification

Geology: Geological Classification

Foundation engineers commonly use the Unified Classification and the Geological Classification. For preliminary and reconnaissance study, information published by pedologists are useful. These three classification systems are presented as a reference source. AASHO and FAA systems are widely used by highway and airfield engineers. For information concerning the latter two systems, reference should be made to the publications of the respective agencies.

A-2 Unified Soil Classification System

The system was adopted in 1952 by the Bureau of Reclamation and the Corps of Engineers. Soils are categorized in 15 groups each of which has distinct engineering properties (Bureau of Reclamation, 1953). The criteria for classification are:

1. Percentage of gravel, sand, and fines–in accordance with the grain sizes.
2. Shape of grain-size curve–for coarse-grained soils.

3. Plasticity and compressibility characteristics–for fine-grained soils and organic soils.

The soil groups are denoted by two letters. The first letter indicates the component of the soil:

G = gravel and gravelly soil
S = sand and sandy soil
M = *mo* (Swedish word for silt) = fines
C = clay
O = organic soil
Pt = peat

The second letter describes the outstanding characteristics of the soil:

W = well-graded (having particle sizes ranging from coarse to fine)
P = poorly graded
H = high plasticity
L = low plasticity

In borderline cases a combination of symbols are used. Table A-1 shows the 15 soil groups, their identification and description. Table A-2 is a chart showing the engineering uses of each of these 15 soil groups. In this chart, the numerical ratings indicate the relative desirability for use as construction or foundation soil, in descending order; number 1 is most desirable.

A-3 *Highway Research Board (or AASHO) Soil Classification System*

The American Association of State Highway Officials classification (AASHO Designation M 145-49) divides soils into two main groups: the granular materials and the silt-clay materials. They are further divided into seven groups A-1 through A-7. This classification system is based on the suitability of soils as highway construction materials. For detailed information, reference is made to the Standard Specification for Highway Materials and Methods of Sampling and Testing, AASHO.

A-4 *Pedological Classification System*

Pedology (*pedo* = soil) is the branch of science dealing with the earth's surface material which is adapted to the support of plant life. Soil survey maps and soil reports concerning the pedological characteristics are available for a large portion of the U.S.A. This information has been utilized by civil engineers, particularly those engaged in the highway field.

Pedologists use soil profiles and the process involved in producing the soil as the bases for classification. A hypothetical, complete soil profile consists

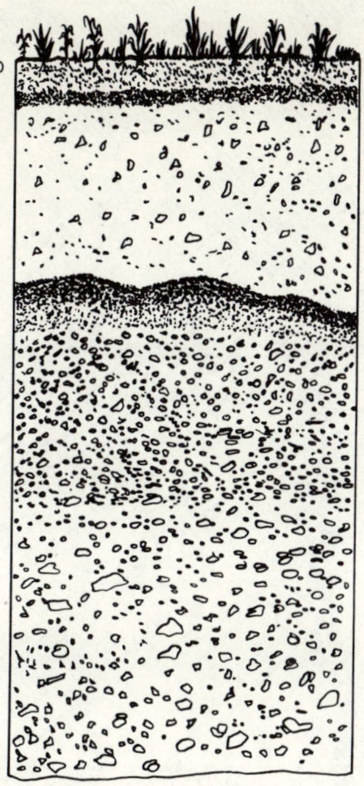

Fig. A-1 Diagrammatic representation of a pedological soil profile. After G. W. Robinson.

of four major layers known as *A*, *B*, *C*, and *D* horizons. Each horizon may be subdivided as shown in Fig. A-1. Profiles of most of the natural soils possess only part of these horizons.

Soils are classified by pedologists into three main divisions: zonal, intrazonal, and azonal soils, depending on the amount of profile development. They are divided into suborders and further divided into great soil groups on the basis of the combined effect of climate, vegetation, and topography, Table A-3. For example, the great soil group *chernozems* is developed under grass vegetation in temperate subhumid areas, and the group *laterite* is formed in areas of abundant rainfall and high temperature. Soils in one series have the same profile except for the texture of the top soil or *A* horizon. The soil series were originally named after a town, county, stream, etc., such as "Barnes" where the soil series was first identified. The final classification unit is the soil type which is made up of the soil series name plus the textural classification of the *A* horizon. (ASTM, Suggested Method of Test For Pedological Classification of Soils, 1958.)

The Soil Survey Division of the Department of Agriculture has made surveys and reports over vast areas of the U.S.A. The soil survey maps, usually covering a county, show the distribution of soil types, and the soil reports describe the characteristics of the successive horizons, and, of

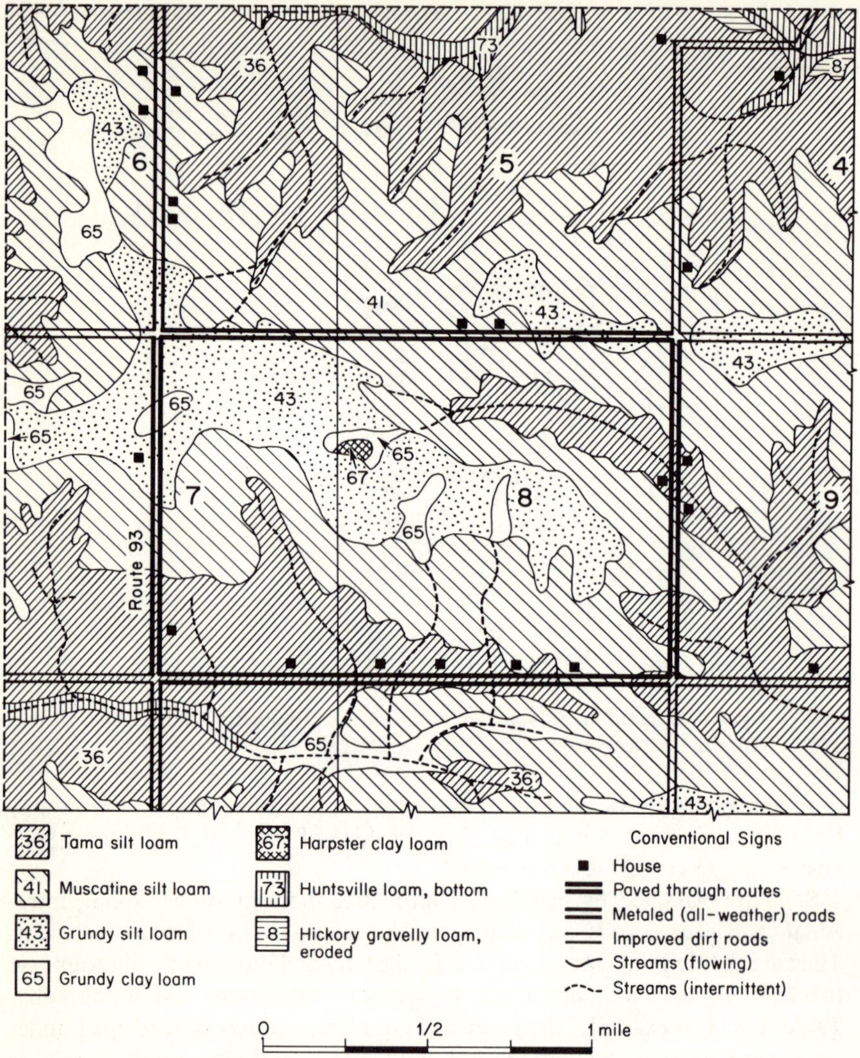

Fig. A-2 Example of a detailed soil survey map.

particular interest to the engineers, the nature of the parent material including geological origin, texture, and chemical constituents. An example of such a soil survey map is shown in Fig. A-2. A typical description of a soil type is taken from *Soil Primer*, Portland Cement Association:

> Grundy silt loam (Type 43). A dark soil developed on nearly level topography. It occurs, for the most part, in association with the larger areas of Muscatine silt loam (Type 41). The surface varies from a dark-brown, faintly granular silt loam to silty clay loam, 8 to 10 in. thick. (This is the *A* horizons.) The subsurface extends to a depth of 16 or 18 in. and is a little heavier and usually

darker than the surface. (This is the *B* horizon.) The subsoil is a brownish or yellowish-drab clay loam having dark-coated and angular structural particles. (This is the C_1 horizon.) The lower part of the subsoil becomes more friable. (This is the C_2 horizon.) Surface drainage of this soil type is slow because of its smooth topography, and underdrainage is good where a satisfactory outlet is available. The dark color of the *A* and *B* horizons is indicative of high organic matter.

A-5 *Classification of Soils by Geological Origins*

All soils can be classified into two general groups; namely, residual soils and transported soils. Residual soils are those formed by rock weathering in place. Transported soils are those carried away from the location of rock weathering and deposited elsewhere by ice, water, wind, or gravity. Due to the fundamental difference in the process of soil formation and transportation, each type is markedly different from the other. The general characteristics of these types bear considerable significance in their engineering properties.

The profile of a residual soil consists of three principal layers (Vargas, 1953):

1. A surface layer of matured residual soil,
2. A middle layer of young residual soil, whose main characteristics exhibit the original structure of the parent rock,
3. A lower layer of disintegrated rock which often requires explosives to remove.

The properties of residual soils are largely influenced by the characteristics of the parent rocks. This influence is summarized as follows:

Parent rock	*Common characteristics of residual soils*
Sandstone:	Silty sand, sandy clay or silty clay.
Limestone:	Composed chiefly of insoluble residue of limestone. Highly plastic soil due to large amount of clay particles. Iron oxide causes red color.
Shale:	High in clay constituents.
Igneous and metamorphic rocks:	In the mountain areas where rocks are exposed on steep slopes, the soils are often shallow and consequently unimportant in engineering problems. In gentle and level areas, three distinct layers in residual soils are developed. Such soils are often plastic and expansive.

Transported soils cover most of the land areas. Many of them have special geologic names. Their common characteristics are described in Table A-4.

Table A-1 UNIFIED SOIL CLASSIFICATION CHART*

UNIFIED SOIL CLASSIFICATION
INCLUDING IDENTIFICATION AND DESCRIPTION

			FIELD IDENTIFICATION PROCEDURES (Excluding particles larger than 3 inches and basing fractions on estimated weights)	GROUP SYMBOLS	TYPICAL NAMES
COARSE GRAINED SOILS — More than half of material is larger than No. 200 sieve size	GRAVELS — More than half of coarse fraction is larger than No. 4 sieve size.	CLEAN GRAVELS (Little or no fines)	Wide range in grain size and substantial amounts of all intermediate particle sizes.	GW	Well graded gravels, gravel-sand mixtures, little or no fines.
			Predominantly one size or a range of sizes with some intermediate sizes missing.	GP	Poorly graded gravels, gravel-sand mixtures, little or no fines.
		GRAVELS WITH FINES (Appreciable amount of fines)	Non-plastic fines (for identification procedures see ML below).	GM	Silty gravels, poorly graded gravel-sand-silt mixtures.
			Plastic fines (for identification procedures see CL below).	GC	Clayey gravels, poorly graded gravel-sand-clay mixtures.
	(For visual classifications, the $\frac{1}{4}$" size may be used as equivalent to the No. 4 sieve size.) SANDS — More than half of coarse fraction is smaller than No. 4 sieve size.	CLEAN SANDS (Little or no fines)	Wide range in grain sizes and substantial amounts of all intermediate particle sizes.	SW	Well graded sands, gravelly sands, little or no fines.
			Predominantly one size or a range of sizes with some intermediate sizes missing.	SP	Poorly graded sands, gravelly sands, little or no fines.
		SANDS WITH FINES (Appreciable amount of fines)	Non-plastic fines (for identification procedures see ML below).	SM	Silty sands, poorly graded sand-silt mixtures.
			Plastic fines (for identification procedures see CL below).	SC	Clayey sands, poorly graded sand-clay mixtures.

(The No. 200 sieve size is about the smallest particle visible to the naked eye)

		IDENTIFICATION PROCEDURES ON FRACTION SMALLER THAN No. 40 SIEVE SIZE: DRY STRENGTH (CRUSHING CHARACTERISTICS)	DILATANCY (REACTION TO SHAKING)	TOUGHNESS (CONSISTENCY NEAR PLASTIC LIMIT)	GROUP SYMBOLS	TYPICAL NAMES
FINE GRAINED SOILS — More than half of material is smaller than No. 200 sieve size.	SILTS AND CLAYS — Liquid limit less than 50	None to slight	Quick to slow	None	ML	Inorganic silts and very fine sands, rock flour, silty or clayey fine sands with slight plasticity.
		Medium to high	None to very slow	Medium	CL	Inorganic clays of low to medium plasticity, gravelly clays, sandy clays, silty clays, lean clays.
		Slight to medium	Slow	Slight	OL	Organic silts and organic silt-clays of low plasticity.
	SILTS AND CLAYS — Liquid limit greater than 50	Slight to medium	Slow to none	Slight to medium	MH	Inorganic silts, micaceous or diatomaceous fine sandy or silty soils, elastic silts.
		High to very high	None	High	CH	Inorganic clays of high plasticity, fat clays.
		Medium to high	None to very slow	Slight to medium	OH	Organic clays of medium to high plasticity.
HIGHLY ORGANIC SOILS		Readily identified by color, odor, spongy feel and frequently by fibrous texture.			Pt	Peat and other highly organic soils.

FIELD IDENTIFICATION PROCEDURES FOR FINE GRAINED SOILS OR FRACTIONS

These procedures are to be performed on the minus No. 40 sieve size particles, approximately $\frac{1}{64}$ in. For field classification purposes, screening is not intended; simply remove by hand the coarse particles that interfere with the tests.

DILATANCY (Reaction to shaking)

After removing particles larger than No. 40 sieve size, prepare a pat of moist soil with a volume of about one-half cubic inch. Add enough water if necessary to make the soil soft but not sticky.

Place the pat in the open palm of one hand and shake horizontally, striking vigorously against the other hand several times. A positive reaction consists of the appearance of water on the surface of the pat which changes to a livery consistency and becomes glossy. When the sample is squeezed between the fingers, the water and gloss disappear from the surface, the pat stiffens, and finally it cracks or crumbles. The rapidity of appearance of water during shaking and of its disappearance during squeezing assist in identifying the character of the fines in a soil.

Very fine clean sands give the quickest and most distinct reaction whereas a plastic clay has no reaction. Inorganic silts, such as a typical rock flour, show a moderately quick reaction.

DRY STRENGTH (Crushing characteristics)

After removing particles larger than No. 40 sieve size, mold a pat of soil to the consistency of putty, adding water if necessary. Allow the pat to dry completely by oven, sun, or air drying, and then test its strength by breaking and crumbling between the fingers. This strength is a measure of the character and quantity of the colloidal fraction contained in the soil. The dry strength increases with increasing plasticity.

High dry strength is characteristic for clays of the CH group. A typical inorganic silt possesses only very slight dry strength. Silty fine sands and silts have about the same slight dry strength, but can be distinguished by the feel when powdering the dried specimen. Fine sand feels gritty whereas a typical silt has the smooth feel of flour.

*Bureau of Reclamation, 1952.

INFORMATION REQUIRED FOR DESCRIBING SOILS		LABORATORY CLASSIFICATION CRITERIA		
Give typical name; indicate approximate percentages of sand and gravel; max. size; angularity, surface condition, and hardness of the coarse grains; local or geologic name and other pertinent descriptive information; and symbol in parentheses. For undisturbed soils add information on stratification, degree of compactness, cementation, moisture conditions and drainage characteristics. **EXAMPLE:-** Silty sand, gravelly; about 20% hard, angular gravel particles ½-in. maximum size; rounded and subangular sand grains coarse to fine; about 15% non-plastic fines with low dry strength; well compacted and moist in place; alluvial sand; (SM)	Use grain size curve in identifying the fractions as given under field identification	Determine percentages of gravel and sand from grain size curve. Depending on percentage of fines (fraction smaller than No. 200 sieve size) coarse grained soils are classified as follows:- Less than 5% GW, GP, SW, SP. More than 12% GM, GC, SM, SC. 5% to 12% Borderline cases requiring use of dual symbols.	$C_U = \frac{D_{60}}{D_{10}}$ Greater than 4 $C_C = \frac{(D_{30})^2}{D_{10} \times D_{60}}$ Between one and 3	
			Not meeting all gradation requirements for GW	
			Atterberg limits below "A" line, or PI less than 4	Above "A" line with PI between 4 and 7 are borderline cases requiring use of dual symbols.
			Atterberg limits above "A" line with PI greater than 7	
			$C_U = \frac{D_{60}}{D_{10}}$ Greater than 6 $C_C = \frac{(D_{30})^2}{D_{10} \times D_{60}}$ Between one and 3	
			Not meeting all gradation requirements for SW	
			Atterberg limits below "A" line or PI less than 4	Above "A" line with PI between 4 and 7 are borderline cases requiring use of dual symbols.
			Atterberg limits above "A" line with PI greater than 7	
Give typical name; indicate degree and character of plasticity, amount and maximum size of coarse grains; color in wet condition, odor if any, local or geologic name, and other pertinent descriptive information; and symbol in parentheses. For undisturbed soils add information on structure, stratification, consistency in undisturbed and remolded states, moisture and drainage conditions. **EXAMPLE:-** Clayey silt, brown; slightly plastic; small percentage of fine sand; numerous vertical root holes; firm and dry in place; loess; (ML)		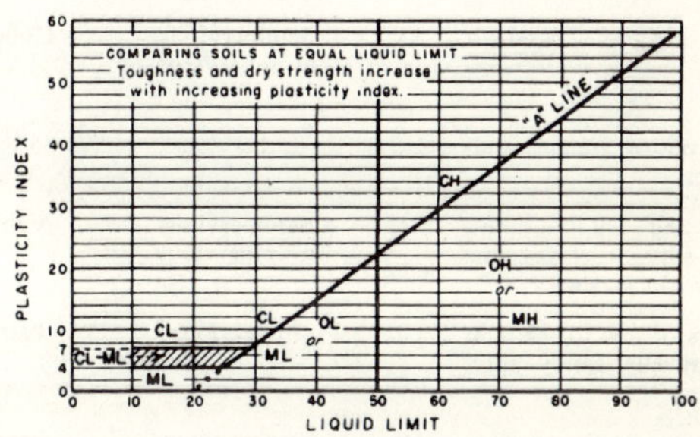 PLASTICITY CHART FOR LABORATORY CLASSIFICATION OF FINE GRAINED SOILS		

PLASTICITY CHART

FOR LABORATORY CLASSIFICATION OF FINE GRAINED SOILS

[1] Boundary classifications:- Soils possessing characteristics of two groups are designated by combinations of group symbols. For example GW-GC, well graded gravel-sand mixture with clay binder.

[2] All sieve sizes on this chart are U.S. standard.

TOUGHNESS (Consistency near plastic limit)

After removing particles larger than the No. 40 sieve size, a specimen of soil about one-half inch cube in size is molded to the consistency of putty. If too dry, water must be added and if sticky, the specimen should be spread out in a thin layer and allowed to lose some moisture by evaporation. Then the specimen is rolled out by hand on a smooth surface or between the palms into a thread about one-eighth inch in diameter. The thread is then folded and rerolled repeatedly. During this manipulation the moisture content is gradually reduced and the specimen stiffens, finally loses its plasticity, and crumbles when the plastic limit is reached.

After the thread crumbles, the pieces should be lumped together and a slight kneading action continued until the lump crumbles.

The tougher the thread near the plastic limit and the stiffer the lump when it finally crumbles, the more potent is the colloidal clay fraction in the soil. Weakness of the thread at the plastic limit and quick loss of coherence of the lump below the plastic limit indicate either inorganic clay of low plasticity, or materials such as kaolin-type clays and organic clays which occur below the A-line.

Highly organic clays have a very weak and spongy feel at the plastic limit.

Table A-2 ENGINEERING USE CHART FOR SOILS CLASSIFIED BY UNIFIED SOIL CLASSIFICATION SYSTEM*

		Important properties			
Typical names of soil groups	*Group symbols*	*Permeability when compacted*	*Shearing strength when compacted and saturated*	*Compressibility when compacted and saturated*	*Workability as a construction material*
Well-graded gravels, gravel-sand mixtures, little or no fines	GW	Pervious	Excellent	Negligible	Excellent
Poorly graded gravels, gravel-sand mixtures, little or no fines	GP	Very pervious	Good	Negligible	Good
Silty gravels, poorly graded gravel-sand-silt mixtures	GM	Semipervious to impervious	Good	Negligible	Good
Clayey gravels, poorly graded gravel-sand-clay mixtures	GC	Impervious	Good to fair	Very low	Good
Well-graded sands, gravelly sands, little or no fines	SW	Pervious	Excellent	Negligible	Excellent
Poorly graded sands, gravelly sands, little or no fines	SP	Pervious	Good	Very low	Fair
Silty sands, poorly graded sand-silt mixtures	SM	Semipervious to impervious	Good	Low	Fair
Clayey sands, poorly graded sand-clay mixtures	SC	Impervious	Good to fair	Low	Good
Inorganic silts and very fine sands, rock flour, silty or clayey fine sands with slight plasticity	ML	Semipervious to impervious	Fair	Medium	Fair
Inorganic clays of low to medium plasticity: gravelly, sandy, silty, and lean clays	CL	Impervious	Fair	Medium	Good to fair
Organic silts and organic silt-clays of low plasticity	OL	Semipervious to impervious	Poor	Medium	Fair
Inorganic silts, micaceous or diatomaceous fine sandy or silty soils, elastic silts	MH	Semipervious to impervious	Fair to poor	High	Poor
Inorganic clays of high plasticity, fat clays	CH	Impervious	Poor	High	Poor
Organic clays of medium to high plasticity	OH	Impervious	Poor	High	Poor
Peat and other highly organic soils	Pt				

* Bureau of Reclamation, 1953.

Relative desirability for various uses

Rolled earth dams			*Canal sections*		*Foundations*		*Roadways*		
							Fills		
Homogeneous embankment	*Core*	*Shell*	*Erosion resistance*	*Compacted earth lining*	*Seepage important*	*Seepage not important*	*Frost heave not possible*	*Frost heave possible*	*Surfacing*
		1	1			1	1	1	3
		2	2			3	3	3	
2	4		4	4	1	4	4	9	5
1	1		3	1	2	6	5	5	1
		3 If gravelly	6			2	2	2	4
		4 If gravelly	7 If gravelly			5	6	4	
4	5		8 If gravelly	5 Erosion critical	3	7	8	10	6
3	2		5	2	4	8	7	6	2
6	6			6 Erosion critical	6	9	10	11	
5	3		9	3	5	10	9	7	7
8	8			7 Erosion critical	7	11	11	12	
9	9				8	12	12	13	
7	7		10	8 Volume change critical	9	13	13	8	
10	10				10	14	14	14	

Table A-3 PEDOLOGICAL SOIL CLASSIFICATION

Order		*Suborder*	*Great soil groups**	*Soil series†*	*Soil type‡*
Zonal soils Soils whose characteristics are primarily determined by the climate and associated vegetation	*six climatic zones*	1. Soils of the cold zone	Tundra soils		
		2. Light-colored soils of arid regions	Desert soils Red desert soils Sierozem Brown soils Reddish-brown soils		
		3. Dark-colored soils of semi-arid, subhumid, and humid grasslands	Chestnut soils Reddish chestnut soils Chernozem soils Prairie soils Reddish prairie soils	EXAMPLE Richfield Moody Barnes Vernon etc.	EXAMPLE Barnes silty clay, Barnes clay etc.
		4. Soils of the forest-grassland transition	Degraded chernozem Noncalcic brown or Shantung brown soils		
		5. Light-colored podzolized soils of the timbered regions	Podzol soils Gray wooded or Gray podzolic soils** Brown podzolic soils Gray-brown podzolic soils Red-yellow podzolic soils**		
		6. Lateritic soils of forested warm-temperate and tropical regions	Reddish-brown lateritic soils** Yellowish-brown lateritic soils Laterite soils**		

Order	*Suborder*	*Great soil groups**	*Soil series†*	*Soil type‡*
Intrazonal soils Soils of well-developed characteristics resulting from local effects such as topography, drainage or parent rock	1. Halomorphic (saline and alkali) soils of imperfectly drained arid regions and littoral deposits	Solonchak or Saline soils Solonetz soils Soloth soils		
	2. Hydromorphic soils of marshes, swamps, seep areas, and flats	Humic-glei soils** (includes wiesenboden) Alpine meadow soils Bog soils Half-bog soils Low-humic-glei** soils Planosols Groundwater podzol soils Groundwater laterite soils		
	3. Calcimorphic soils	Brown forest soils (braunerde) Rendzina soils		
Azonal soils Soils recently developed and not yet modified by climate or vegetation		Lithosols Regosols (includes dry sands) Alluvial soils		

* Soils having similar soil profile.

† Soils having same *B*, *C*, and *D* horizons and same characteristics and drainage.

‡ Soils having same profile and same characteristics throughout.

** New or recently modified great soil groups.

Modified from James Thorp and Guy D. Smith, "Higher categories of Soil Classification: Order, Suborder, and Great Soil Groups," *Soil Science*, LXVII (1949), and *PCA Soil Primer*, Portland Cement Association, Chicago, Illinois.

Table A-4 TYPES OF TRANSPORTED SOILS AND THEIR USUAL CHARACTERISTICS

Transportation agents	*Deposits*	*Usual characteristics*
1. Ice	Glacial deposits	Extremely variable
Glaciers	Glacial till (moraines, till plains, drumlins, . . .)	Unstratified, heterogeneous mixture of clay, sand, silt, gravel, and boulders
Glacial water	Fluvio-glacial deposits (outwash plains, terraces, eskers, kames, . . .)	Stratified, often gravelly and granular
Glacial lake	Glacial lake deposits	Silt, clay thinly stratified; may be associated with peat and muck
2. Water		
River	Alluviums (alluvial deposits)	Deposited along the present and ancient rivers; usually stratified; particles of different sizes are usually assorted
Lake	Lacustrine deposits	Fine-grained soils
Ocean	Marine deposits	Fine-grained soils, lime mud
Beach	Beach deposits (bars, etc.)	Coarse-grained soils (sands and gravels), tidal marshes
3. Wind	Aeolian deposits	Loose soils of uniform particle sizes
	Dunes	Sand deposited by rolling along the ground
	Loess	Wind-borne silt with root holes and cementing material
	Volcanic ash	Porous and light in weight
4. Gravity	Colluviums (collovial deposits)	
	Talus, detritus	Disintegrated rocks at mountain slopes

GEOPHYSICAL MAPS
OF CONTINENTAL U.S.A.

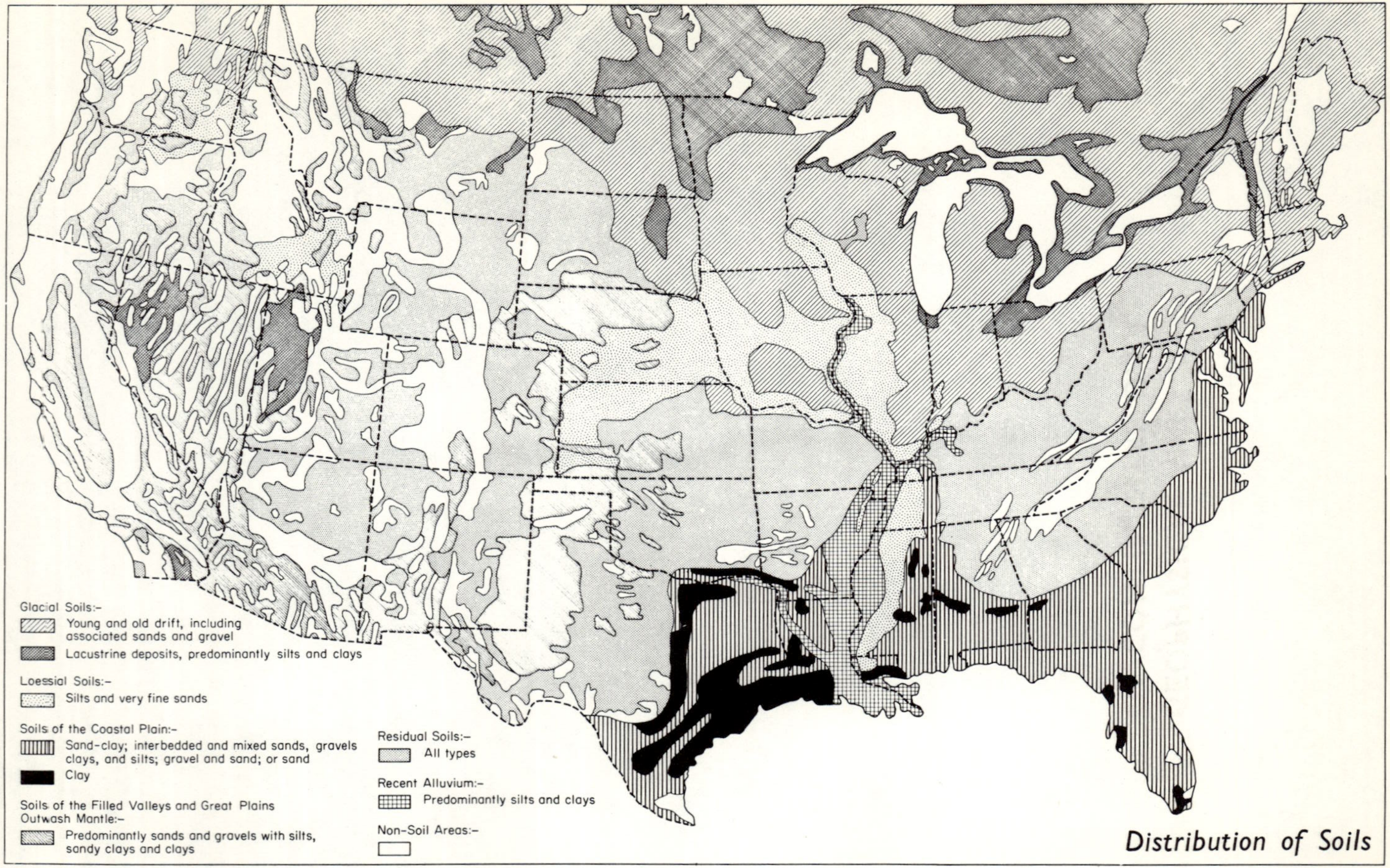

After Bletcher *et al.*

Earthquake Zones

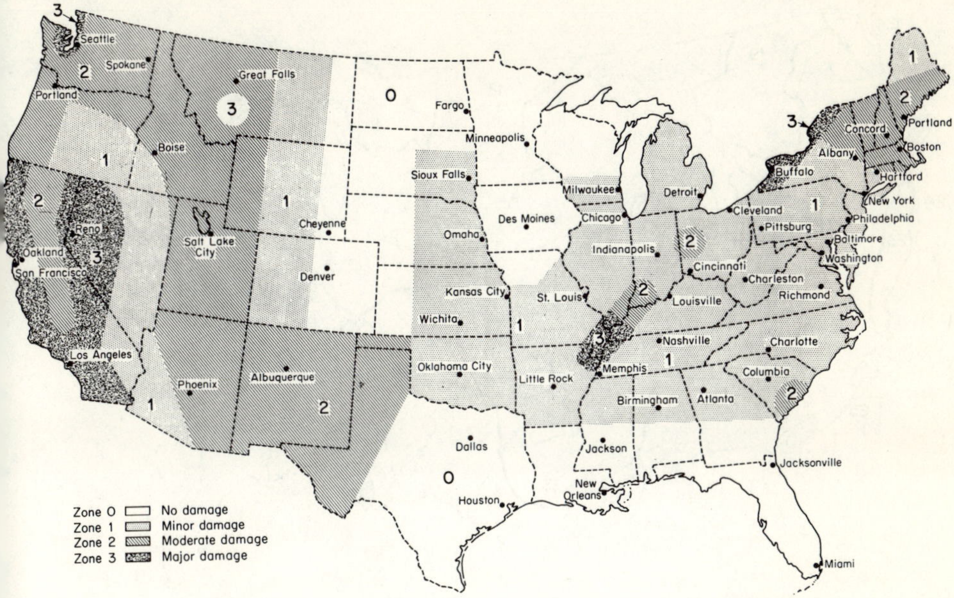

HORIZONTAL FORCE FACTOR K FOR BUILDINGS OR OTHER STRUCTURES†

Type or arrangement of resisting elements	*Value of K*
All building framing systems except as hereinafter classified	1.00
Buildings with a box system	1.33
Buildings with a complete horizontal bracing system capable of resisting all lateral forces, which system includes a moment resisting space frame, which when assumed to act independently is capable of resisting a minimum of 25 per cent of the total required lateral force	0.80
Buildings with a moment resisting space frame which when assumed to act independently of any other more rigid elements is capable of resisting 100 per cent of the total required lateral forces in the frame alone	0.67
Structures other than buildings	1.50

* From Uniform Building Code.

† Where wind load would produce higher stresses, this load shall be used in lieu of the loads resulting from earthquake forces.

*Physical Divisions and Their Characteristics**

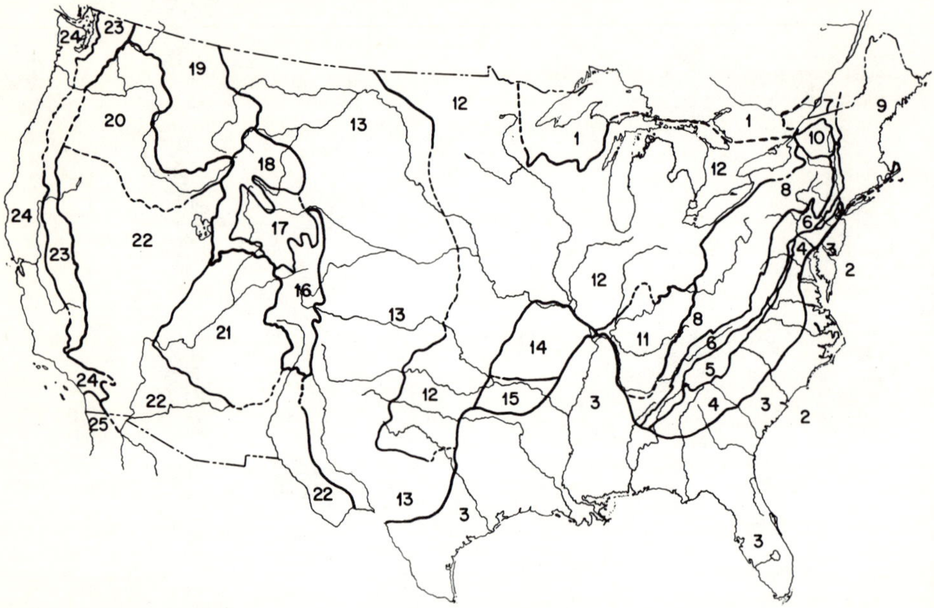

Laurentian Upland

1. SUPERIOR UPLAND. Submaturely dissected, recently glaciated peneplain or crystalline rocks of complex structure.

Atlantic Plain

2. CONTINENTAL SHELF. Sloping submarine plain of sedimentation.

3. COASTAL PLAIN. *Embayed section:* Submaturely dissected and partly submerged, terraced coastal plain. *Sea Island section:* Young to mature terraced coastal plain with submerged border. *Floridian section:* Young marine plain, with sand hills, swamps, sinks, and lakes. *East Gulf Coastal Plain:* Young to mature belted coastal plain. *Mississippi Alluvial Plain:* Flood plain and delta. *West Gulf Coastal Plain:* Young grading inland to mature coastal plain.

Appalachian Highlands

4. PIEDMONT PROVINCE. *Piedmont Upland:* Submaturely dissected peneplain on disordered resistant rocks; moderate relief.† *Piedmont Lowlands:* Less uplifted peneplain on weak strata; residual ridges on strong rocks.

5. BLUE RIDGE PROVINCE. *Northern section:* Maturely dissected mountains of crystalline rocks; accordant altitudes. *Southern section:* Subdued mountains of disordered crystalline rocks.

*Prepared by Nevin M. Fenneman and Douglas W. Johnson.

† Degrees of relief are herein spoken of as low, moderate, strong, and high. As used here *high* relief is measured in thousands of feet; *moderate* relief in hundreds of feet. *Strong* relief may be anything approaching 1,000 feet with a wide latitude on both sides.

6. VALLEY AND RIDGE PROVINCE. *Tennessee section:* Second-cycle mountains of folded strong and weak strata; valley belts predominate over even-crested ridges. *Middle section:* The same, but even-crested ridges predominate over valleys except on east side. *Hudson Valley:* Glaciated peneplain on weak folded strata.

7. ST. LAWRENCE VALLEY. *Champlain section:* Rolling lowland, glaciated; in part covered by young marine plain. *Northern section:* Young marine plain with local rock hills.

8. APPALACHIAN PLATEAUS. *Mohawk section:* Maturely dissected glaciated plateau; varied relief and diverse altitudes. *Catskill section:* Maturely dissected plateau of mountainous relief and coarse texture (glaciated). *Southern New York section:* Mature glaciated plateau of moderate relief. *Allegheny Mountain section:* Mature plateau of strong relief; some mountains due to erosion of open folds. *Kanawha section:* Mature plateau of fine texture; moderate to strong relief. *Cumberland Plateau section:* Submaturely dissected plateau of moderate to strong relief. *Cumberland Mountain section:* Higher mature plateau and mountain ridges on eroded open folds.

9. NEW ENGLAND. *Seaboard Lowland section:* Peneplains below 500 feet postmaturely eroded and glaciated; few monadnocks. *New England Upland section:* Dissected and glaciated peneplains on complex structural features; monadnocks. *White Mountain section:* Subdued glaciated mountain masses of crystalline rocks. *Green Mountain section:* Linear ranges of subdued and glaciated mountains and residual plateaus. *Taconic section:* Maturely dissected and glaciated mountains and peneplain on resistant folded strata.

10. ADIRONDACK PROVINCE. Subdued mountains and dissected peneplain, glaciated.

Interior Plains

11. INTERIOR LOW PLATEAUS. *Highland Rim section:* Young to mature plateau of moderate relief. *Lexington Plain:* Mature to old plain on weak rocks; trenched by main rivers. *Nashville Basin:* Mature to old plain on weak rocks; slightly uplifted and moderately dissected. *Possible western section:* Low, maturely dissected plateau with silt-filled valleys.

12. CENTRAL LOWLANDS. *Eastern lake section:* Maturely dissected and glaciated cuestas and lowlands; moraines, lakes, and lacustrine plains. *Western lake section:* Young glaciated plain; moraines, lakes, and lacustrine plains. *Wisconsin Driftless section:* Maturely dissected plateau and lowland invaded by glacial outwash. (Margin of old eroded drift included.) *Till Plains:* Young till plains; morainic topography rare; no lakes. *Dissected Till Plains:* Submaturely to maturely dissected till plains. *Osage Plains:* Old scarped plains beveling faintly inclined strata; main streams intrenched.

13. GREAT PLAINS PROVINCE. *Missouri Plateau, glaciated:* Glaciated old plateaus; isolated mountains. *Missouri Plateau, unglaciated:* Old plateau; terrace lands; local badlands; isolated mountains. *Black Hills:* Maturely dissected domed mountains. *High Plains:* Broad intervalley remnants of smooth fluviatile plains. *Plains Border:* Submaturely to maturely dissected plateau. *Colorado Piedmont:* Late mature to old elevated plain. *Raton section:* Trenched peneplain surmounted by dissected, lava-capped plateaus and buttes. *Pecos Valley:* Late mature to old plain. *Edwards Plateau:* Young plateau with mature margin of moderate to strong relief. *Central Texas section:* Plateau in maturity and later stages of erosion.

Interior Highlands

14. OZARK PLATEAUS. *Springfield-Salem plateaus:* Submature to mature plateaus. *Boston "Mountains":* Submature to mature plateau of strong relief.

15. OUACHITA PROVINCE. *Arkansas Valley:* Gently folded strong and weak strata; peneplain with residual ridges. *Ouachita Mountains:* Second-cycle mountains of folded strong and weak strata.

Rocky Mountain System

16. SOUTHERN ROCKY MOUNTAINS. Complex mountains of various types; intermont basins.

17. WYOMING BASIN. Elevated plains in various stages of erosion; isolated low mountains.

18. MIDDLE ROCKY MOUNTAINS. Complex mountains, mainly anticlinal ranges; intermont basins.

19. NORTHERN ROCKY MOUNTAINS. Deeply dissected mountain uplands, not anticlinal ranges; intermont basins.

Intermontane Plateaus

20. COLUMBIA PLATEAUS. *Walla Walla Plateau:* Rolling plateau with young incised valleys. *Blue Mountain section:* Complex mountains and dissected volcanic plateaus. *Payette section:* Young plateaus of prevailingly weak rocks; broad alluvial terraces. (Applies to northern part only.) *Snake River Plain:* Young lava plateau. *Harney section:* Young lava plateau; features of recent volcanism; ineffective drainage.

21. COLORADO PLATEAU. *High Plateaus of Utah:* High block plateaus, in part lava-capped; terraced plateaus on south side. *Uinta Basin:* Dissected plateau; strong relief. *Canyon Lands:* Young to mature canyoned plateaus; high relief. *Navajo section:* Young plateaus; smaller relief than Canyon Lands, into which it grades. *Grand Canyon section:* High block plateaus, trenched by Grand Canyon. *Datil section:* Lava flows entire or in remnants; volcanic necks.

22. BASIN AND RANGE PROVINCE. *Great Basin:* Isolated ranges (largely dissected block mountains) separated by aggraded desert plains. *Sonoran Desert:* Widely separated short ranges in desert plains. *Salton Trough:* Desert alluvial slopes and delta plain; Gulf of California. *Mexican Highland:* Isolated ranges (largely dissected block mountains) separated by aggraded desert plains. *Sacramento section:* Mature block mountains of gently tilted strata; block plateaus; bolsons.

Pacific Mountain System

23. CASCADE-SIERRA MOUNTAINS. *Northern Cascade Mountains:* Sharp alpine summits of accordant height; higher volcanic cones. *Middle Cascade Mountains:* Generally accordant summits; higher volcanic cones. *Southern Cascade Mountains:* Volcanic mountains variously eroded; no very distinct range. *Sierra Nevada:* Block mountain range tilted west; accordant crests; alpine peaks near east side.

24. PACIFIC BORDER PROVINCE. *Puget Trough:* Lowlands of diverse character; in part submerged. *Olympic Mountains:* Generally accordant crests; local alpine peaks. *Oregon Coast Range:* Uplifted peneplain on weak rocks, dissected; monadnocks of igneous rock. *Klamath Mountains:* Uplifted and dissected peneplain on strong rocks; extensive monadnock ranges. *California Trough:* Low fluviatile plain. *California Coast Ranges:* Parallel ranges and valleys on folded, faulted, and metamorphosed strata; rounded crests of subequal height. *Los Angeles Ranges:* Narrow ranges and broad fault blocks; alluviated lowlands.

25. LOWER CALIFORNIA PROVINCE. Dissected westward-sloping granite upland (in northern part).

REFERENCES

Akai, K., *The Physical Properties of Silt*, Report No. 22. Aachen, Germany: Soil Mechanics Institute, Technical University, 1960.

Anderson, Paul, *Substructure Analysis and Design*, 2nd ed. New York: Ronald Press Co., 1956.

AASHO, *Standard Specifications for Highway Materials and Methods of Sampling and Testing*. Washington D.C., I and II, 1955; III, 1958.

———, *Manual of Highway Construction Practices and Methods*. Washington D.C., 1958.

AREA, *Manual of Recommended Practice*. Chicago: Construction and Maintenance Section, Engineering Division, Association of American Railroads, 1958.

ASTM, *Procedures for Testing Soils*. Philadelphia, 1958.

Baracos, A., and M. Bozozuk, "Seasonal Movements in Some Canadian Clays," *Proc. 4th Intern. Conf. on Soil Mech. and Foundation Engineering*, I, 264–68. London, 1957.

Beach Erosion Board, *Durability of Steel Sheet Piling in Shore Structures*, Technical Memorandum No. 12. Washington, D.C.: Corps of Engineers, U.S. Department of the Army, 1952.

ABBREVIATIONS

AASHO	American Association of State Highway Officials, Washington D.C.
ACI	American Concrete Institute, Detroit, Michigan.
ASCE	American Society of Civil Engineers, New York.
ASTM	American Society for Testing and Materials, Philadelphia.
AREA	American Railway Engineering Association, Chicago, Illinois.

Belcher, D. J., L. E. Gregg, D. S. Jenkins, and K. B. Woods, *The Origin, Distribution, and Airphoto Identification of United States Soils.* Washington, D.C.: Civil Aeronautics Administration, U.S. Department of Commerce, 1946.

Beyer, Kurt, *Die Statik im Stahlbetonbau.* Berlin: Springer–Verlag, 1956.

Bjerrum, L., "Norwegian Experiences with Steel Piles to Rock," *Geotechnique*, VII, 2, 73. London, 1957.

Bjerrum, L., and O. Eide, "Stability of Strutted Excavation in Clay," *Geotechnique*, VI, 1, 31. London, 1956.

Bjerrum, L., and R. Kirkedam, "Some Notes on Earth Pressure in Stiff Fissured Clay," *Brussels Conference on Earth Pressure Problems*, I, 15–27. Brussels: Belgian Group of the International Society of Soil Mechanics and Foundation Engineering, 1958.

Bolognesi, A. J. L., and O. Moretto, "Properties and Behavior of Silty Soils Originated from Loess Formations," *Proc. 4th Intern. Conf. on Soil Mech. and Foundation Engineering*, I, 9–12. London, 1957.

Bretschneider, C. L., "Revised Wave Forecasting Relationships," *Proc. 2nd Conf. on Coastal Engineering.* 1952.

Bureau of Reclamation, *Unified Soil Classification System*, Supplement to *Earth Manual.* Denver, Colorado, 1953.

Burwell, E. B., Jr. and R. H. Nesbitt, "The NX Bore-hole Camera," *Mining Engineering.* New York, 1954.

Button, S. J., "The Bearing Capacity of Footings on a Two-Layer Cohesive Subsoil," *Proc. 3rd Intern. Conf. on Soil Mech. and Foundation Engineering*, I, 332. Switzerland, 1953.

Caquot, Albert, and Jean Kerisel, *Tables For the Calculation of Passive Pressure, Active Pressure and Bearing Capacity of Foundations* (translated from French edition by Chief Scientific Advisers' Division, Ministry of Works, London). Paris: Gauthier-Villars, 1948.

Casagrande, Leo, "Electro-osmotic Stabilization of Soils," *Journal of the Boston Society of Civil Engineers*, XIX (1952), 1.

Chellis, Robert D., *Pile Foundations.* New York: McGraw-Hill Book Co., Inc., 1951.

Christie, I. F., "Design and Construction of Vertical Drains to Accelerate the Consolidation of Soils," *Civil Engineering and Public Works.* London, March and April, 1959.

Clevenger, William A., "Experiences with Loess as Foundation Material," *Trans. ASCE*, CXXIII (1958), 151.

Corps of Engineers, U.S. Department of the Army. Washington, D.C. Information furnished by Engineering Division, Military Construction, 1957.

Dawson, Raymond F., *Laboratory Manual in Soil Mechanics.* New York: Pitman Publishing Corp., 1960.

D'appolonia, E., C. E. Miller, Jr., and T. M. Ware, "Sand Compaction by Vibroflotation," *Trans. ASCE,* cxx (1955), 154.

Engineering News-Record, "Sonics Drive a Pile 71 Feet, while Steam Drives Another 3 Inches." November 9, 1961.

Fenneman, Nevin M., *Physiography of Western United States.* New York: McGraw-Hill Book Co., Inc., 1931.

———, *Physiography of Eastern United States.* New York: McGraw-Hill Book Co., Inc., 1938.

Gibbs, H. J., and W. G. Holtz, "Research on Determining the Density of Sands by Spoon Penetration Testing," *Proc. 4th Intern. Conf. on Soil Mech. and Foundation Engineering,* I, 35–39. London, 1957.

Gill, Grayson W., "Waterproofing Buildings Below Grade," *Civil Engineering,* Jan., 1959, p. 3.

Gleser, Sol M., "Lateral Load Tests on Vertical Fixed-Head and Free-Head Piles," *ASTM Preprint* 95*d.* 1953.

Gray, Hamilton, "Contribution of the Analysis of Seepage Effects in Backfills," *Geotechnique,* VIII, 4, 166. London, 1958.

Hansen, J. Brinch, *Earth Pressure Calculation.* Copenhagen: The Danish Technical Press, The Institution of Danish Civil Engineers, 1953.

———, "Foundation of Structures–General Report," *Proc. 4th Intern. Conf. on Soil Mech. and Foundation Engineering,* II, 441–47. London, 1957.

Hetenyi, Miklos, *Beams on Elastic Foundation.* Ann Arbor, Michigan: University of Michigan Press, 1946.

Holtz, W. G., and H. J. Gibbs, "Field Tests to Determine the Behavior of Piles in Loess," *Proc. 3rd Intern. Conf. on Soil Mech. and Foundation Engineering,* II, 51–56. Switzerland, 1953.

Hough, B. K., *Basic Soils Engineering.* New York: Ronald Press Co., 1957.

Hrennikoff, A., "Analysis of Pile Foundations with Batter Piles," *Trans. ASCE,* cxv (1950), 351.

Hudson, Robert Y., "Wave Forces on Breakwaters," *Trans. ASCE,* CXVIII (1953), 653.

Huntington, Whitney C., *Earth Pressures and Retaining Walls.* New York: John Wiley & Sons, Inc., 1957.

Hvorslev, Juul, *Subsurface Exploration and Sampling of Soils for Civil Engineering Purposes.* Vicksburg, Mississippi: Waterways Experiment Station, 1949.

International Conference of Building Officials, *Uniform Building Code*. Los Angeles, California, 1961.

Ireland, H. O., "Pulling Tests on Piles in Sand," *Proc. 4th Intern. Conf. on Soil Mech. and Foundation Engineering*, II, 43–45. London, 1957.

Janbu, N., "Earth Pressures and Bearing Capacity Calculations by Generalized Procedure of Slices," *Proc 4th Intern. Conf. on Soil Mech. and Foundation Engineering*, II, 207–13. London, 1957.

Jumikis, A. R., "Rupture Surfaces in Sand Under Oblique Loads," *J. Soil Mechanics and Foundation Engineering*. ASCE, 1956.

Kirkedam, R., Discussion, *Proceedings Brussels Conference 58 on Earth Pressure Problems*, III, 54. Brussels, 1958.

Krynine, D. P., and W. R. Judd, *Principles of Engineering Geology and Geotechnics*. New York: McGraw-Hill Book Co., Inc., 1957.

Lambe, William, *Soil Testing for Engineers*. New York: John Wiley & Sons, Inc., 1951.

Leonards, Gerald A., "Strength Characteristics of Compacted Clays," *Trans. ASCE*, CXX, 1422. 1955.

Lin, T. Y., and W. J. Talbot Jr., "Pretensioned Concrete Piles, Present Knowledge Summarized," *Civil Engineering*. May, 1961.

Lomizé, G. M., A. V. Netushil, and B. A. Rzhanitzin, "Electro-osmotic Processes in Clayey Soils and Dewatering During Excavations," *Proc. 4th Intern. Conf. on Soil Mech. and Foundation Engineering*, I, 62–67. London, 1957.

MacFarlane, Ivan C., "A Review of the Engineering Characteristics of Peat," *Proc. ASCE*, LXXXV (1959), No. SMI, Part I.

Malter, Henry, "Numerical Solutions for Beams on Elastic Foundations," *Journal Soil Mech. and Foundation Div.*, LXXXIV, No. ST2, Part I. ASCE, March, 1958.

Marcus, H., *Die Theorie Elastischer Gewebe*, 2nd ed. Berlin: Springer–Verlag.

Marsland, Arthur, "Model Experiments to Study the Influence of Seepage on the Stability of a Sheeted Excavation in Sand," *Geotechnique*, III, 6, 223. London, June, 1953.

Mason, Martin A., Surface Water Wave Theories, *Trans. ASCE*, CXVIII (1953), 546.

McLaughlin, W. W., "Soils Classification, Exploration, and Testing," *Public Works*. Ridgewood, N. J., Jan., 1958.

McNamee, J., "Seepage into a Sheeted Excavation," *Geotechnique*, I, 4, 229. London, Dec., 1949.

McNulty, James F., "Thrust Loading on Piles," *Journal Soil Mech. and Foundation Div.*, LXXXII. ASCE, April, 1956.

Meyerhof, G. G., "An Investigation of Bearing Capacity of Shallow Footings on Dry Sand," *Proc. 2nd Intern. Conf. on Soil Mech. and Foundation Engineering*, I, 237–43. Rotterdam, 1948.

———, "The Ultimate Bearing Capacity of Foundations," *Geotechnique*, II, 4, 301. London, 1951.

———, "The Bearing Capacity of Foundations Under Eccentric and Inclined Loads," *Proc. 3rd Intern. Conf. on Soil Mech. and Foundation Engineering*, I, 440–45. Switzerland, 1953.

———, "The Bearing Capacity of Concrete and Rock," *Magazine of Conc. Research*, IV. London, 1953.

———, "The Ultimate Bearing Capacity of Foundations on Slopes," *Proc. 4th Intern. Conf. on Soil Mech. and Foundation Engineering*, I, 384–86. London, 1957.

Newmark, N. M., "Numerical Procedure For Computing Deflections, Moments and Buckling Loads," *Trans. ASCE*, CVIII (1942), 1161.

———, *Influence Charts For Computation of Stresses in Elastic Foundations*, Engineering Exp. Station, Bul. No. 338. Urbana, Ill.: University of Illinios, 1951.

Newmark, N. M., and W. J. Austin, Unpublished class notes on numerical methods. Dept. of Civil Engineering, Graduate School, Univ. of Illinois, Urbana, 1954.

Newell, John N., "Pneumatic Caisson Pier for World's Longest Pipeline Suspension Bridge," *Civil Engineering*. May, 1956.

Osterberg, J. O., "New Piston-type Soil Sampler," *Engineering News-Record*, April 24, 1952.

Peck, R. B., "Earth Pressure Measurements in Open Cuts, Chicago (Ill.) Subway," *Trans. ASCE*, CVIII (1943), 1008.

Peck, R. B., W. E. Hansen, and T. H. Thornburn, *Foundation Engineering*. New York: John Wiley & Sons, Inc., 1953.

Peck, R. B., H. O. Ireland, and C. Y. Teng, "A Study of Retaining Wall Failures," *Proc. 2nd Intern. Conf. on Soil Mech. and Foundation Engineering*, III, 296–99. Rotterdam, 1948; Reprinted in *University of Illinois Bull.*, XLVI, Circular No. 53, March, 1949.

Portland Cement Association, *Soil Primer*. Chicago, 1956.

Polshin, D. E., and R. A. Tokar, "Maximum Allowable Non-Uniform Settlement of Structures," *Proc. 4th Intern. Conf. on Soil Mech. and Foundation Engineering*, I, 402–5. London, 1957.

Popov, E. P., "Successive Approximations for Beams on an Elastic Foundation," *Trans. ASCE*, CXVI (1951), 1083.

Prugh, Byron J., "Tools and Techniques for Dewatering," *Trans. ASCE*, CXXVI (1961), 38–52.

Rijhsinghani, Arjun, "Plates Subjected to Concentrated Loads and Moments," M.S. Thesis, Illinois Institute of Technology, Chicago, 1961.

Roark, R. J., *Formulas for Stress and Strain.* New York: McGraw-Hill Book Co., Inc., 1954.

Rombas, *Steel Sheet Piling Rombas.* Paris: Sidelor, 1957. Supplied to author by Western Services Corp., Linden, N. J.

Rowe, Peter W., "Anchored Sheet-Pile Walls," *Proc. Institution of Civil Engineers.* London, 1952.

———, "Sheet Pile-Walls in Clays," *Proc. Institution of Civil Engineers.* London, 1957.

———, "Measurements on Sheet Pile-Walls Driven into Clay," *Proc. Brussels Conf. 58 on Earth Pressure Problems,* II, 127–330. Brussels: Belgian Group of the Intern. Society of Soil Mech. and Foundation Engineering, 1958.

Salas, J. A. J., and J. M. Serratosa, "Foundations on Swelling Clays," *Proc. 4th Intern. Conf. on Soil Mech. and Foundation Engineering,* I, 424–28. London, 1957.

Salmon, Charles G., Leo Schenker, and Bruce Johnston, "Moment-Rotation Characteristics of Column Anchorages," *Trans. ASCE,* CXXII (1957) 132.

Schmertmann, John H., "The Undisturbed Consolidation Behavior of Clay," *Trans. ASCE,* CXX (1955), 1201.

Seed, H. Bolton, and Lymon C. Reese, "The Action of Soft Clay Along Friction Piles," *Trans. ASCE,* CXXII (1957), 731.

Seely, Fred B., and James O. Smith, *Advanced Mechanics of Materials.* New York: John Wiley & Sons, Inc., 1952.

Skempton, A. W., "The $\phi = 0$ Analysis of Stability and Its Theoretical Basis," *Proc. 2nd Intern. Conf. on Soil Mech. and Foundation Engineering,* I, 72–78. Rotterdam, 1948.

———, "The Bearing Capacity of Clays," Bldg. Research Congr., Div. 1. London, 1951.

———, "Discussion on Settlement of Pile Group in Sand," *Proc. 3rd Intern. Conf. on Soil Mech. and Foundation Engineering,* III, 172. Switzerland, 1953.

Skempton, A. W. and A. W. Bishop, "The Measurement of the Shear Strength of Soils," *Geotechnique* II, 2. London, 1950.

Skempton, A. W., and L. Bjerrum, "A Contribution to the Settlement Analysis of Foundations on Clay," *Geotechnique.* VII, 4, 168. London, 1957.

Sowers, George F., "Modern Procedures for Underground Investigations," *Proc. ASCE,* Paper No. 435. 1954.

Sowers, G. F., A. D. Robb, C. H. Mullis, and A. J. Glenn, "The Residual Lateral Pressures Produced by Compacting Soil," *Proc. 4th Intern. Conf. on Soil Mech. and Foundation Engineering,* II, 243–47. London, 1957.

Swiger, W. F., "Control of Ground Water in Excavations," *Proc. ASCE,* Paper No. 2359. 1960.

Taylor, Donald W., *Fundamentals of Soil Mechanics.* New York: John Wiley & Sons, Inc., 1948.

Teng, Wayne C., "A Study of Contract Pressure Against a Large Raft Foundation," *Geotechnique,* I, 4, 222. London, 1949.

———, "How Air Academy Foundations Were Designed," *Engineering News-Record,* April 18, 1957.

Tennessee Valley Authority, "Design of TVA Projects," Technical Report No. 24, *Civil and Structural Design.* Washington, D.C.: U.S. Government Printing Office, 1952.

Terzaghi, Karl, *Theoretical Soil Mechanics.* New York: John Wiley & Sons, Inc., 1943.

———, "Stress Conditions for the Failure of Saturated Concrete and Rock," *Proc. ASTM,* XLV (1946), 777–801.

———, "Anchored Bulkheads," *Trans. ASCE,* CXIX (1954), 1243.

———, "Evaluation of Coefficient of Subgrade Reaction," *Geotechnique,* V, 4, 297. London, 1955.

———, Discussion, *Proc. 4th Intern. Conf. on Soil Mech. and Foundation Engineering,* III, 154–55. London, 1957.

Terzaghi, Karl, and Ralph B. Peck, *Soil Mechanics in Foundation Engineering,* New York: John Wiley & Sons, Inc., 1948.

Tomlinson, M. J., "The Adhesion of Piles Driven in Clay Soils," *Proc. 4th Intern. Conf. on Soil Mech. and Foundation Engineering,* II, 61–71. London, 1957.

Tschebotarioff, Gregory P., *Soil Mechanics, Foundations and Earth Structures.* New York: McGraw-Hill Book Co., Inc., 1952.

Turnbull, Willard J., and Charles R. Foster, "Stabilization of Materials by Compaction," *Trans. ASCE,* CXXIII (1958), 1.

U.S. Department of Agriculture, *Soil Survey Manual,* U.S. Dept. of Agriculture Handbook No. 18. 1951.

Vargas, Milton, "Some Engineering Properties of Residual Clay Soils Occurring in South Brazil," *Proc. 3rd Intern. Conf. on Soil Mech. and Foundation Engineering,* I, 67–71. Switzerland, 1953.

Wakeman, C. M., E. V. Dockweiler, H. E. Stover, and L. L. Whiteneck, "Use of Concrete in Marine Environments," *ACI Journal,* April, 1958.

White, Lazarus and Edmund A. Prentis, *Cofferdams.* New York: Columbia University Press, 1950.

Werblin, David A., "Installation and Operation of Dewatering Systems," *Proc. ASCE,* LXXXVI (1960), SM1, Part I.

Swiger, W. F., "Control of Ground Water in Excavations," *Proc. ASCE*, Paper No. 2359. 1960.

Taylor, Donald W., *Fundamentals of Soil Mechanics.* New York: John Wiley & Sons, Inc., 1948.

Teng, Wayne C., "A Study of Contract Pressure Against a Large Raft Foundation," *Geotechnique*, I, 4, 222. London, 1949.

———, "How Air Academy Foundations Were Designed," *Engineering News-Record*, April 18, 1957.

Tennessee Valley Authority, "Design of TVA Projects," Technical Report No. 24, *Civil and Structural Design.* Washington, D.C.: U.S. Government Printing Office, 1952.

Terzaghi, Karl, *Theoretical Soil Mechanics.* New York: John Wiley & Sons, Inc., 1943.

———, "Stress Conditions for the Failure of Saturated Concrete and Rock," *Proc. ASTM*, XLV (1946), 777–801.

———, "Anchored Bulkheads," *Trans. ASCE*, CXIX (1954), 1243.

———, "Evaluation of Coefficient of Subgrade Reaction," *Geotechnique*, V, 4, 297. London, 1955.

———, Discussion, *Proc. 4th Intern. Conf. on Soil Mech. and Foundation Engineering*, III, 154–55. London, 1957.

Terzaghi, Karl, and Ralph B. Peck, *Soil Mechanics in Foundation Engineering*, New York: John Wiley & Sons, Inc., 1948.

Tomlinson, M. J., "The Adhesion of Piles Driven in Clay Soils," *Proc. 4th Intern. Conf. on Soil Mech. and Foundation Engineering*, II, 61–71. London, 1957.

Tschebotarioff, Gregory P., *Soil Mechanics, Foundations and Earth Structures.* New York: McGraw-Hill Book Co., Inc., 1952.

Turnbull, Willard J., and Charles R. Foster, "Stabilization of Materials by Compaction," *Trans. ASCE*, CXXIII (1958), 1.

U.S. Department of Agriculture, *Soil Survey Manual*, U.S. Dept. of Agriculture Handbook No. 18. 1951.

Vargas, Milton, "Some Engineering Properties of Residual Clay Soils Occurring in South Brazil," *Proc. 3rd Intern. Conf. on Soil Mech. and Foundation Engineering*, I, 67–71. Switzerland, 1953.

Wakeman, C. M., E. V. Dockweiler, H. E. Stover, and L. L. Whiteneck, "Use of Concrete in Marine Environments," *ACI Journal*, April, 1958.

White, Lazarus and Edmund A. Prentis, *Cofferdams.* New York: Columbia University Press, 1950.

Werblin, David A., "Installation and Operation of Dewatering Systems," *Proc. ASCE*, LXXXVI (1960), SM1, Part I.

AUTHOR INDEX

A

AASHO, 36
Akai, K., 18
American Standards Assn., 53
AREA, 108, 134, 136, 318, 320
ASTM, 36, 241

B

Baracos, A., and M. Bozozuk, 18
Beach Erosion Board, 352
Belcher, D. J., 31
Benoto, Inc., 279
Beyer, K., 184
Bjerrum, L., 205
Bjerrum, L., and O. Eide, 400
Bureau of Plant Industry, 31
Button, S. J., 120, 121

C

Chellis, R. D., 198
Clevenger, W. A., 18

D

Dawson, R. F., 46
Department of Highways, Columbus, Ohio, 204

G

Gibbs, H. J., 214
Gibbs, H. J. and W. G. Holtz, 39
Gill, G. W., 108
Gleser, S. M., 273

H

Hansen, J. B., 59, 85, 356, 364, 421
Hetenyi, M., 159, 272
Holtz, W. G. and H. J. Gibbs, 214
Hough, B. K., 17
Hrennikoff, A., 135, 234, 236
Hudson, R. Y., 94
Huntington, W. C., 330
Hvorslev, J., 33, 44

I

Ireland, H. O., 213

J

Janbu, N., 136
Jumikis, A. R., 57

K

Kirkedam, R., 392, 393
Krynine, D. P., 419

L

Lambe, W., 35
Lin, T. Y., and W. J. Talbot, 198

M

McFarland, I. C., 19
McNulty, J. F., 227
Malter, H., 173
Marcus, H., 181
Marsland, A., 402

Meyerhof, G. G., 12, 57, 58, 134, 136, 137, 138
Molitor, 93, 94

O

Osterberg, J. O., 41

P

Peck, R. B., 10, 17, 61, 218, 219, 295, 316, 394
Polshin, D. E., and R. A. Tokar, 63, 64, 65
Popov, E. P., 168
Portland Cement Association, 301, 331
Public Roads, 225

R

Rijhsinghani, A., 181
Roark, R. J., 131
Rombas, 360
Rowe, P. W., 364, 368, 369, 370

S

Salas, J. A. J., and J. M. Serratosa, 18
Salmon, C. G., L. Schenker and B. Johnston, 141
Schmertmann, J. H., 61
Seed, H. B., and L. C. Reese, 214, 222
Seeley, F. B., and J. O. Smith, 166
Skempton, A. W., 9, 42, 120, 121, 221, 259
Skempton, A. W., and L. Bjerrum, 128, 129
Sowers, et al., 86
State Geologic Survey, 31
Stevenson, 93
Sverdrup and Munk, 95
Swiger, W. F., 97, 104

T

Talbot, 198
Taylor, D. W., 357
Teng, W. C., 190, 313
Tennessee Valley Authority, 248
Terzaghi, K., 57, 61, 74, 82, 89, 90, 117, 118, 186, 234, 260, 356, 357, 369, 377, 399, 421
Terzaghi, K., and R. B. Peck, 12, 26, 37, 63, 118, 120, 174, 218, 295, 392, 393
Timoshenko, 296
Tomlinson, M. J., 214
Tschebotariff, G. P., 26, 399
TVA, 93, 412, 424

U

Uniform Building Code, 243
United States Geological Survey, 31
U.S.S.R. Building Code, 63, 65

V

Vargas, M., 435

W

Wakeman, C. M., 200
Werblin, D. A., 101
White, L., and E. A. Prentis, 277, 395, 398

INDEX

A

Active earth pressure, 73, 74
Adhesion, 75, 76
Air lock, 304
Air shaft, 304
Allowable bearing pressure, 118
Anchor piles, 193
Anchorage, 374
  braced piles, 374
Anchored sheetpiling, 351, 364
  free-end method, 364 ff, 366 f
  Rowe's theory of moment reduction, 369
  tie rod support, 372
  tie rod tension, 373
  wales and tie rods, 371
Angle of internal friction, 8
Angle of wall friction, 76
Anvil, 247
Apparent cohesion, 122
Artesian well, 25
Auger boring, 34
Atterberg limits, 16

B

Backfill compaction, 336
Backfill drainage, 332
Base shear, 55
Batter piles, 193, 227
  analytical method, 232
  determination of lateral movement, 236
Bearing capacity, 57 ff
  bedrocks, 21
  caisson bells, 265

Bearing capacity (*cont.*)
  caissons, 296
  drilled caissons, 258
  spread footings, 117, 120, 122, 123, 137
Bearing failure of retaining walls, 322, 323
Bearing pressure for mat foundations, 174
Bell (*see* Caisson bell)
Benoto machine, 277, 279
Borings, 33, 34
  depth, 43
  number, 42
Boussinesq's equation, 125, 126
Box caissons, 291, 307
  base preparation, 308
  stability during floating, 307
Braced cofferdams, 389
  components, 396
  design, 391
  earth pressure in clays, 394
  earth pressure in sand, 393
  earth pressure in stratified soils, 394
  heaving, 398
  lagging, 390
  lateral pressure, 392
  minimum penetration of sheetpiling, 402
  sheetpiling, 390, 396, 401
  soldier beams, 390, 396
  stability, 397
  stability of bottom of excavation, 400
  struts, 390, 397
  types, 390, 391

Braced cofferdams (*cont.*)
  unbalanced external forces, 397
  uses, 389
  wales, 390, 396

C

Caisson bell, 264
  bearing capacity, 265
Caisson construction, 299 ff
  safety problems, 281
  sand island method, 301
Caisson drilling machine, 276
Caissons, 289 ff
  bearing capacity, 296
  concrete seal, 296
  construction, 299
  cutting edge, 298
  design loads, 294
  design of, 293
  drilled (*see* Drilled caisson)
  sinking, 300
  size and shape, 292
  skin friction, 295
  types, 290
  use, 289
Cantilever sheetpiling, 351
  cohesive soils, 362
  granular soils, 359, 362
Cantilever walls, 358
  design, 326
  forces acting on, 327
Cast-in-place concrete piles, 195, 200
Cathodic protection of steel piles, 209
Cellular cofferdams, 352, 409
  average width, 416
  cell fill, 415, 425
  circular type, 412
  closure, 425
  cloverleaf type, 412
  components, 413
  construction, 423
  design, 415
  diaphragm type, 412
  foundation treatment, 425
  Hansen's theory, 422
  interlock stress, 423
  pile driving, 424
  saturation line, 417
  sliding, 417
  slipping, 418
  stability analysis, 417 ff
  template setting, 423
  types, 409
  uses, 409
  vertical shear, 418 ff
  weep holes, 424
Chicago caisson, 274
Clay, bearing capacity, 120
Coefficient of consolidation, 62
Coefficient of earth pressure, 77, 81
Coefficient of horizontal subgrade reaction, 234
Coefficient of permeability, 10
Coefficient of subgrade reaction, 185, 186
Cofferdams, braced (*see* Braced cofferdams)
Cofferdams, cellular (*see* Cellular cofferdams), 352
Cohesion, 8
Cohesionless soil, 9
Cohesive soils, 9, 13
  compressibility, 16
  creep, 13
  expansion, 17
  free swelling, 18
Combined footings, 151, 153
  conventional method of design, 159, 161
  design, 161, 165, 168, 173
  elastic lines, 165, 166
  method of finite difference, 173
  method of successive approximations, 168
  rigid method of design, 159, 161
Compacted fill, 122
  bearing capacity, 122, 123
Compaction, 11, 67
  percentage of, 68, 69
Compaction equipment, 69
Compaction piles, 193, 210
Compaction test, 49
Components of retaining walls, 325
Composite peils, 195, 202
  joint details, 202
  types, 202
Compressibility of cohesive soils, 16
Compression index, 16, 17, 61, 62
Concrete caissons, 291
  design criteria, 297
Concrete sheetpiling, 348, 349
Consolidation, 5, 60, 61
  secondary, 62
  theory of, 130
Consolidation test, 47
Contact pressure, 130
Contraction joints in retaining walls, 335
Core boring, 34
Corrosion of steel piles, 208
Coulomb's theory, 8, 316
Counterfort walls, 315, 326
  design of, 326 ff
  details of, 328
  forces acting on, 327, 328
  Huntington's design procedure, 329

Cracks, settlement, 66, 67
Crib walls, 330, 332
  members of, 331
Culmann's method, 232
Cushion, 247

D

Dampproofing, 97, 109
  retaining walls, 337
Darcy's law, 10
Dead load, 53
Deadmen, 375, 376
Deep well pumps, 101
Defective piles, 251
Degree of saturation, 3, 4
Density, relative, 11, 12
Depths of borings, 43
Depths of footings, 115, 116, 143
Depth of frost penetration, 116
Deterioration and protection of precast concrete piles, 200
Deterioration of sheetpiling, 354
Dewatering, 97
Diesel hammer, 244, 247, 249
Differential settlement, 62, 63, 64, 65, 66
Direct shear test, 47
Dolphins, 193
Drainage:
  foundation, 97, 105
  retaining walls, 333
Drilled caisson, 255 ff, 274
  bearing capacity, 258
  bell, 264
  belling bucket, 278
  bending moment and eccentricity, 270
  Benoto machine, 277, 279
  cap details, 269
  capacity of caisson shaft, 267
  Chicago caisson, 274
  concrete placement, 278
  construction, 273, 281
  design, 258, 264
  detail, 268
  drill, 275
  Gow caisson, 274
  horizontal force, 272
  inspection of construction, 280
  methods of excavation, 273
  on bedrock, 260
  on cohesive soils, 259
  on sand and gravel, 259
  rock bit, 277
  settlement, 263
  shaft, 266
  skin friction, 261
  stress on lower strata, 262
Drilled caisson (*cont.*)
  types, 256, 257
  underwater excavation, 281
  use, 255
Double-acting hammer, 244, 249
Drive cap, 247
Driving points, steel piles, 205
Driving shoe for pipe piles, 205
Drop hammer, 244
Durability of steel sheetpiling, 352

E

Earth pressure, 54
  acting on sheetpiling walls, 355, 356
  active, 74
  at rest, 73, 78
  charts for retaining walls, 318
  coefficients, 81
  during earthquakes, 92
  Hansen's theory, 85
  braced cofferdams in clays, 394
  braced cofferdams in sand, 393
  braced cofferdams in stratified soils, 394
  lateral strain, 73
  passive, 74
  Rankine theory, 77, 78
  retaining walls, 316
  theories, 76
  theories in retaining wall design, 317
Earthquake zones, 55, 445
  earth pressure during earthquakes, 92
  in U.S. (map), 445
  motion, 55
Eccentrically loaded footings, 130, 131, 132, 133
  bearing capacity, 134
Effective pressure, 25
Ejectors, 102
Elastic foundation, 159
Elastic line method, 165
  design of combined footings, 165
Electroosmosis, 102
End-bearing piles, 193
Engineering News formula, 215
Engineering use of soils, 438, 439
Expansion joints in retaining walls, 336

F

Fender piles, 193
Filter material, 106
Fine-grained soils, 9
Finite difference:
  design of combined footings, 173
Fixed-end method:
  anchored sheetpiling wall, 366 ff

Fixity of spread footings, 141
Floating caissons, 291, 307
Flow net, 87
Follower, 247
Footings:
  common types, 114
Foundation drainage, 97, 105
Foundation pier (*see* Drilled caissons)
Foundation stability, retaining walls, 321
Free-end method, anchored sheetpiling wall, 364 ff
Friction piles, 193, 210
  in clays, 213, 222
  in granular soils, 213
  in loess, 214, 215
  in sand and gravel, 222
  in several soil strata, 215
  in silt, 214
Frictional soil, 9
Franki piles, 201
Frost action, 116

G

Geologic profile, 45
Geological soil classifications, 435
Geophysical maps, 443
Gow caissons, 274
Grain size, 7, 13
  analysis, 46
  distribution, 6, 46
Grains, shapes of, 13
Granular soils, 9, 10 ff
  bearing capacity, 117
Gravity walls:
  design, 325
  stresses, 325
Ground freezing, 102
Ground water, 23 ff
  level, 23, 25
  measurement, 43, 44
  table (*see* Ground water level)
Grout curtain walls, 103

H

H-piles, 204
Hansen's theory:
  cellular cofferdams, 422
  earth pressure, 85
Heaving in braced cofferdams, 398
Helmet, 247
Horizontal force on drilled caissons, 272
Horizontal force on piles, 227
Horizontal subgrade reaction, 234
Huntington's design procedure for counterfort walls, 329
Hydraulic gradient, 10
Hydrolithic waterproofing, 109
Hydrostatic pressure, 76

I

Ice thrust, 90
Igneous rocks, 21
Integral waterproofing, 109
Intergranular pressure, 8
Ironite method, 109

J

Joints in retaining walls, 334, 335

L

Laboratory testing of soils, 35
Lagging, braced sheetpiling, 396
Lateral load on piles, 226
Lateral pressures, 73 ff
  braced cofferdams, 392
  equivalent fluid pressure, 78
  due to line load, 89
  due to point load, 90
  due to strip load, 89
  due to surcharge load, 88
  due to uniform load, 88
  see Page pressure, 86, 87, 88
  wedge theory, 76
Leads, 247
Limestones, 22
Liquid limit, 15, 16, 47
Line load, 89, 90
Live load, 53
Load:
  dead, 53
  live, 53
  snow, 54
  wind, 54
Load-settlement curve of a pile load test, 242
Load test, 123
Loads, calculations of, 56
Loess, bearing capacity, 122

M

Mat foundation, 150, 151
  arrangement, 153
  bearing pressure, 174
  construction, 190
  design, 174, 179, 189
  elastic foundation, 159
  factors affecting soil pressure and settlement, 189
  for circular structures, 184
  method of difference equations, 180

Mat foundation (*cont.*)
 method of finite difference, 179
 types, 154
Maximum density, 68, 69
Membrane waterproofing, 107
Metamorphic rocks, 21
Modulus of foundation, 185
Moisture content, 3
Moisture-density relationship, 49
Moment reduction, sheetpiling walls, 369
Monotube piles, 201
Mooring pull, 95

N

Negative skin friction, 239
Neutral pressure, 25
Newmark's influence chart, 127, 128, 175, 262
Normally consolidated clays, 16
Number of borings, 42

O

Open caissons, 290
Organic content, 19
Organic soils, 19
 bearing capacity, 122
Oslo point, steel pile, 205
Osterberg piston, 41
Overburden, 19
Overcompaction, 86
Overconsolidated clays, 16
Overdriving of timber piles, 198
Overturning of retaining walls, 317

P

Passive earth pressure, 74
Passive resistance, 74
Pedestal piles, capacity of, 213
Pedological soil classifications, 432, 433, 440, 441
Penetration of sheetpiling, braced cofferdams, 402
Penetration resistance, 12
Percentage of recovery, 33
Perched water, 25
Percussion drill, 34
Permeability, 9
Permeability of rock, 21
Physical divisions of U.S. (map), 446
Piezometer, 44
Pile capacity, 211 ff
 allowable stress in piles, 211
 Engineer-News formula, 215, 216
 pedestal piles, 213
 structural strength, 211
 supporting strength of soil, 212
Pile caps, design of, 223, 224, 225
Pile driving:
 cellular cofferdams, 424
 effect of pile driving, 250, 252
 equipment, 244
  Diesel hammer, 244, 247, 249
  double-acting hammer, 244, 249
  drop hammer, 244
  single-acting hammer, 244
  vibratory hammer, 247
Pile driving formula, 215
Pile foundations, 193 ff
 construction, 244
 Culmann's method, 228, 232
 design, 209
 negative skin friction, 239
 Oslo point, 205
 settlement analysis, 220
 stress on lower strata, 218, 219
 types, 210
Pile group, 216, 217, 218
 arrangement, 223
 capacity, 217
 selection, 247, 248
 uplift capacity, 226
Pile load test, 242
Pile spacing, 216, 217
Piles:
 alignment, 251
 defective, 251
 function, 196
 improper driving, 250
 lateral load, 226
 length, 210
 types, 193
 use, 193
 waterjetting, 250
Pipe piles, 204
 splice and driving shoe, 205
Piping in retaining wall foundation, 332
Piston sampler, 41
*p*H values, 208
Plastic limit, 15, 16, 47
Plasticity, 15
Plasticity index, 15, 16
Pneumatic caissons, 290, 303, 306
 air lock, 304
 air shaft, 304
 construction, 304 ff
 working chamber, 303
Point-bearing piles, 193, 210, 212
 in sand and gravel, 220
 on bedrock, 220
 on hard clay, 221
 settlement of, 221
Point load, 90
Pore pressure, 8, 9, 25
Pore water pressure (*see* Pore pressure)

Porosity, 3, 4
Preconsolidated clays, 16
Preloading, compaction, 68
Preservatives, timber piles, 197
Pressure, water, 54
Pressure bulbs, 128
Precast concrete piles, 194, 198
  cut-offs and splices, 199
  design, 198
  deterioration and protection, 200
Pumps:
  deep well, 101
  submersible turbine, 102

Q

Quick-clays, 17

R

Raft foundation (*see* Mat foundation)
Ram, 247
Rankine theory, 77, 78, 79, 316
  sheetpiling walls, 362
Raymond concrete piles, 201
Reinforced concrete caissons, 291
Relative density, 11, 12
Relieving platform, 352
Retaining walls, 311 ff
  backfill compaction, 336
  backfill drainage, 332
  cantilever wall design, 326
  construction joints, 335
  counterfort wall design, 326 ff
  dampproofing and waterproofing, 337
  design, 313
  drainage, 333
  earth pressure charts, 318
  earth pressure computation, 316
  expansion joints, 336
  foundation stability, 321
  gravity wall design, 325
  joints, 335
  overturning, 317, 320
  piping, 332
  proportions, 314 ff
  Rankine theory, 316
  settlement, 334
  sliding, 317, 321
  stability against bearing failure, 322
  stability of, 317
  stability of foundation of two-layer system, 323
  structural components, 325
  structural stability, 317
  Swedish circle method, 323
  tilting, 333, 334
  types, 312, 313
  uses, 311
Rock bit, drilled caissons, 277
Rocks, 19, 20
  bearing capacity, 21, 123
  engineering properties, 19 ff
  permeability, 21
Rollers, 69
Rotary drill, 34

S

Sampler, 33
Sampling of soils, 33, 34
Sand drains, 68
Sand Island method, 301
Sandstones, 22
Secondary consolidation, 62
Sedimentary rocks, 22
Seepage in backfills, 86
Seepage pressure, 86, 87, 88
  sheetpiling walls, 357
Sensitive clays, 17
Sensitivity, 17
Service load for settlement analysis, 65, 66
Settlement, 53
  causes, 60
  differential, 62 ff
Settlement analysis, 65
  pile foundation, 220
Settlement cracks, 66, 67
Settlement of drilled caissons, 263
Settlement of footings, 128, 129, 130
Settlement of mat foundation, 189
Settlement of retaining walls, 333, 334
Shales, 23
Shapes of grains, 13
Shear strength, 8, 14
Sheet piles:
  common types, 347
  concrete sheet piles, 348
  steel sheet piles, 349
  Wakefield sheet piles, 348
  wood sheet piles, 348
Sheetpiling, braced cofferdams, 396, 401
Sheetpiling walls, 347 ff
  anchorage, 377
  construction, 378
  deadmen, 375, 376
  earth pressure, 355, 356
  in cohesive soils, 368
  method for reducing lateral pressure, 373
  Rankine theory, 362
  relieving platform, 352
  types, 351
  unbalanced water pressure and seepage pressure, 357
  uses, 347

Shelby tube, 33, 40, 41, 120
Ship impact, 95
Shrinkage of soils, 17
Shrinkage limit, 18
Silt:
  bearing capacity, 122
Single-acting pile hammer, 244, 247
Skin friction of caisson shaft, 261
Sliding of retaining walls, 317, 321
Slip circle method, 59
Snow load, 54
Soil classification, 431
  engineering use chart for soils, 438, 439
  geological, 435
  Highway Research Board (or AASHO) Classification, 432
  pedological classification, 432, 433, 440, 441
  unified classification, 431
Soil distribution in U.S. (map), 444
Soil exploration, 29, 34
  reconnaissance, 32
Soil exploration program, 49
Soil profile, 45
Soil sampler, 33
Soil survey maps, 433
Soil tests, 33, 34, 46
Soldier beams, braced cofferdams, 390
Splices:
  pipe piles, 205
  precast concrete piles, 194, 198
  steel piles, 205
  timber piles, 197
Split-spoon, 33, 34
Spread footings, 112, 113 ff
  bearing capacity, 117 ff, 137
  construction, 142
  depths, 115, 116, 143
  design, 115
  eccentric loading, 130 ff
  fixity, 141
  inclined load, 135
  immediate settlement, 129
  Jambu's analysis, 137
  minimum depth, 115, 116, 117
  Newmark influence chart, 127, 128
  on slopes, 137, 138
  relative depth, 143
  settlement, 128, 129, 130
  size proportions, 124
  stress on lower strata, 125
  stress under footings, 126
  stresses in pedestaled footings, 140
  structural design, 138, 139
  Terzaghi theory of consolidation, 130
  uplift capacity, 137, 139
Stability of bottom of excavation, 103
Stability of retaining walls, 317, 322, 323
Standard penetration resistance, 37, 120
Standard penetration test, 12, 15, 37, 120
Steel caissons, 292
  design, 297
Steel sheet piles, 349
  Arbed-Belval piles, 350
  deterioration, 354
  durability, 352
  European piles, 350
  Hoesch piles, 350
  Krupp piles, 350
  Larsen piles, 350
Steel piles, 195, 203
  cathodic protection, 209
  common types, 204
  corrosion, 208
  driving points, 205, 206, 207
  splices, 205
Stiff clays, 9
Strap footings, 151 ff
  design, 154, 155
Stresses on lower strata pile foundations, 219
Stresses under footing, 126
Strip load, 89
Struts, braced cofferdam, 397
Sub-pier (*see* Drilled caissons)
Subgrade modulus, 185
Submersible turbine pumps, 102
Sumps, 98
Surcharge load, 88
Swedish circle method, 323
Swelling pressure, 18, 95

T

Tension cracks, 78
Tension piles, 193
Terzaghi bearing capacity factors, 58, 59
Terzaghi bearing capacity theory, 137
Test piles, 240 ff
Test pit, 34
Thermal expansion, 95
Thin-walled tube, 34, 40
Tie rod support, 372
Tie rods, anchored sheetpiling, 371
Tilting of retaining walls, 333, 334
Timber piles, 194, 196
  classes of, 196
  overdriving, 198
  preservative treatment, 197
  splices, 197
Time factor, 62
Top soil, 19
Traction forces, 95
Tremie concrete, 301
Triaxial tests, 121

U

Unbalanced water pressure, sheetpiling walls, 357
Unconfined compression test, 14, 15, 47, 120
Unified Soil Classification System, 431
Uniform load, 88
Unit weight, 4, 5, 46
Uplift capacity of pile group, 226
Uplift of piles, 225
Uplift of spread footings, 137, 139
Uplift piles, 193
Useful width, 134

V

Vane test, 42
Vibroflotation, 68
Void ratio, 3, 4

W

Wakefield sheet pile, 348
Wales:
  anchored sheetpiling, 371
  braced cofferdams, 396
Wall footings, 144
Wall friction, 75, 76
Wash boring, 34
Water content, 3, 4, 5, 47
Water jetting of piles, 250
Water pressure, 54
Waterproofing, 105, 107, 108
  integral, 109
  membrane, 107
  retaining walls, 337
  waterstops, 109, 110
Wave pressure, 93
Wedge theory, 76, 78, 80 ff
Well points, 99, 100
Western piles, 201
Wind load, 54
Winkler foundation, 159
Wood sheet piles, 348